Bio-resources: feeding a sustainable chemical industry

London, UK

19–21 June 2017

FARADAY DISCUSSIONS

Volume 202, 2017

ROYAL SOCIETY OF CHEMISTRY

The Faraday Division of the Royal Society of Chemistry, previously the Faraday Society, was founded in 1903 to promote the study of sciences lying between chemistry, physics and biology.

Faraday Discussions (Print ISSN 1359-6640, Electronic ISSN 1364-5498) is published 10 times a year by the Royal Society of Chemistry, Thomas Graham House, Science Park, Milton Road, Cambridge, UK CB4 0WF.

Volume 202 ISBN 13: 978-1-78262-954-2

2017 annual subscription price: print+electronic £1010 US $1782; electronic only £962, US $1693. Customers in Canada will be subject to a surcharge to cover GST. Customers in the EU subscribing to the electronic version only will be charged VAT.

All orders, with cheques made payable to the Royal Society of Chemistry, should be sent to the Royal Society of Chemistry Order Department, Royal Society of Chemistry, Thomas Graham House, Science Park, Milton Road, Cambridge, CB4 0WF, UK
Tel +44 (0)1223 432398; E-mail **orders@rsc.org**

If you take an institutional subscription to any Royal Society of Chemistry journal you are entitled to free, site-wide web access to that journal. You can arrange access via Internet Protocol (IP) address at **www.rsc.org/ip**

Customers should make payments by cheque in sterling payable on a UK clearing bank or in US dollars payable on a US clearing bank.

Printed in the UK

Faraday Discussions are unique international discussion meetings that focus on rapidly developing areas of chemistry and its interfaces with other scientific disciplines.

Bio-resources: feeding a sustainable chemical industry

Faraday Discussions

www.rsc.org/faraday_d

A General Discussion on Bio-resources: feeding a sustainable chemical industry was held in London, UK on the 19th, 20th and 21st of June 2017.

RSC Publishing is a not-for-profit publisher and a division of the Royal Society of Chemistry. Any surplus made is used to support charitable activities aimed at advancing the chemical sciences. Full details are available from www.rsc.org

CONTENTS

ISSN 1359-6640; ISBN 978-1-78262-954-2

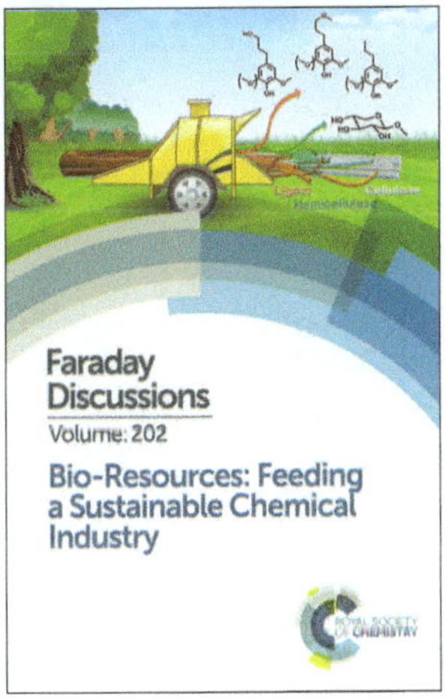

Cover
See Emiel J. M. Hensen *et al.*, *Faraday Discuss.*, 2017, **202**, 141–156.

The cover image describes a catalytic method to extract and upgrade lignin into monoaromatics from lignocellulosic biomass.

Image reproduced by permission of Xiaoming Huang from *Faraday Discuss.*, 2017, **202**, 141–156.

INTRODUCTORY LECTURE

PAPERS AND DISCUSSIONS

SPONSOR

ADDITIONAL INFORMATION

Faraday Discussions

PAPER

Feeding a sustainable chemical industry: do we have the bioproducts cart before the feedstocks horse?

Bruce E. Dale

Received 8th June 2017, Accepted 11th July 2017
DOI: 10.1039/c7fd00173h

A sustainable chemical industry cannot exist at scale without both sustainable feedstocks and feedstock supply chains to provide the raw materials. However, most current research focus is on producing the sustainable chemicals and materials. Little attention is given to how and by whom sustainable feedstocks will be supplied. In effect, we have put the bioproducts cart before the sustainable feedstocks horse. For example, bulky, unstable, non-commodity feedstocks such as crop residues probably cannot supply a large-scale sustainable industry. Likewise, those who manage land to produce feedstocks must benefit significantly from feedstock production, otherwise they will not participate in this industry and it will never grow. However, given real markets that properly reward farmers, demand for sustainable bioproducts and bioenergy can drive the adoption of more sustainable agricultural and forestry practices, providing many societal "win–win" opportunities. Three case studies are presented to show how this "win–win" process might unfold.

How we got here: a very brief history of sustainable bioproducts and bioenergy

The current sustainable bioproducts effort descends from the "farm chemurgic" movement in the United States during the mid-1930s.[1,2] U.S. farms produced vastly more agricultural commodities than Americans could consume, and as a result, agricultural prices were depressed. The four U.S. Department of Agriculture Regional Laboratories were established to develop industrial uses of agricultural materials, increase demand for crop products and raise crop prices to benefit rural economies. Raising crop prices and strengthening rural economies were explicit policy goals. Since then, many bioproducts have been developed but have had to compete against inexpensive, entrenched petroleum-based products, and market penetration has been slow in most cases.

Department of Chemical Engineering and Materials Science, Michigan State University, East Lansing, Michigan 48910, USA. E-mail: bdale@egr.msu.edu

Crop yields have continued to grow, leading to increased U.S. emphasis on exporting its crop commodities, including through various food aid projects. A serious unintended consequence of this food aid is that domestic agriculture in the countries receiving aid has sometimes been undermined.[3] Since agriculture is the most basic economic activity of many countries, whole societies have also been undermined. Driven by corn surpluses, U.S. sugar tariffs enabled the high fructose corn syrup (HFCS) sweetener industry. A serious unintended consequence of the flood of HFCS has been increased obesity and diabetes, both of which are directly linked to over-consumption of HFCS.[4,5]

I raise these issues to point out the fact that there are serious unintended consequences of not finding good industrial uses of our agricultural products. It is neither necessary nor wise to use all agricultural output to produce food and feed.

This Faraday Discussion is devoted to supplying feedstocks for biobased products. A little simple arithmetic is sufficient to reveal an uncomfortable fact about biobased products relative to feedstock supply.

In recent years, the U.S. corn crop has averaged about 14 billion bushels per year, or roughly 356 million tonnes per year. The amount of available starch increases by about 1.0 percent per year as corn yields continue to increase, or about 2.67 million tonnes of "new starch" per year. The Cargill–Dow wet milling plant in Blair, Nebraska, produces about 127 000 tonnes per year of polylactic acid (PLA) from corn starch at about 90% yield. Thus, the total demand of the Blair plant for corn starch to make PLA represents about 5% of one year's increase in the corn crop.

Corn and the other grains can meet just about any conceivable market demand for biobased products made from starch or sugar. There is really no reason to look elsewhere for feedstocks, except for a multitude of sustainability concerns. I will return to this subject later in this paper, but in the meantime, we will focus on agricultural oversupply. As we have seen, the demand for biobased products cannot, by itself, change the problem of agricultural oversupply. Agriculture needs other markets. And by far the largest of these markets is energy.

Using agriculture to produce modern energy has been underway for decades. In the early 1970s, a series of oil supply restrictions began that raised oil prices four fold in just a few years. One result was increased U.S. policy and legislation that encouraged increased blending of gasoline with corn-based ethanol. That effort has continued, with ups and downs, and about 56.8 million cubic meters of ethanol are currently blended with gasoline in the US, roughly 10% of all gasoline used. Most recently, corn ethanol production rose from about 6.1 million cubic meters per year in 2000 to 50.3 million cubic meters per year in 2010.[6] This was truly exponential growth with a doubling time of about 3.0 years, as shown in Fig. 1. Ethanol use now consumes about 35–40% of the U.S. corn crop. There will not be any more rapid doublings of American corn ethanol production, given the competing uses for the crop. A different model is needed to continue to grow ethanol or other biofuels in the U.S.

A similar situation unfolded in Brazil in the 1970s when greatly increased oil prices essentially wiped out the country's hard currency reserves, causing social disruption and a lot of hardship. Brazil turned to its sugarcane industry to produce ethanol to substitute for imported oil. From 1975 to 2007, the Brazilian ethanol program grew from about 0.6 million cubic meters per year to 18 million

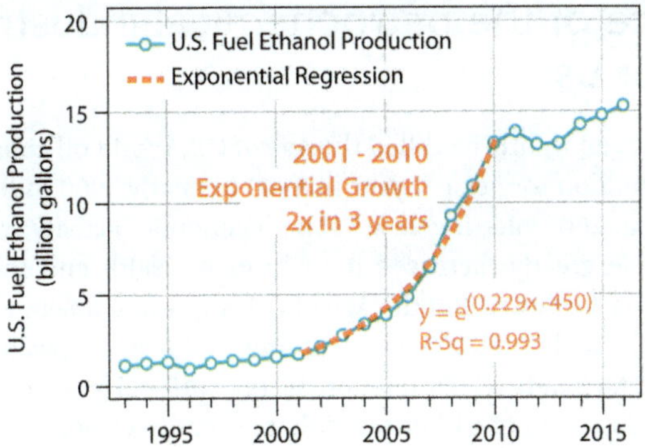

Fig. 1 Recent growth of corn ethanol production in the United States.

cubic meters per year.[7] Today, about 25% of all motor fuel consumed in Brazil is ethanol and all gasoline there is blended with some ethanol.

However, growth in sugarcane ethanol production has slowed greatly. The primary reason is that Brazilian ethanol is a coproduct of cane sugar and the global markets for sugar are saturated. Like the U.S., Brazil needs a different model for ethanol production from cane in order to continue to grow.

Thus, agricultural markets and energy markets, especially petroleum markets, are increasingly intertwined. Let's consider now some lessons from the petroleum industry. (Note: hereafter I will refer to all energy derived from agriculture or forestry as "bioenergy", regardless of whether the fuels are solids, liquid biofuels or gases.)

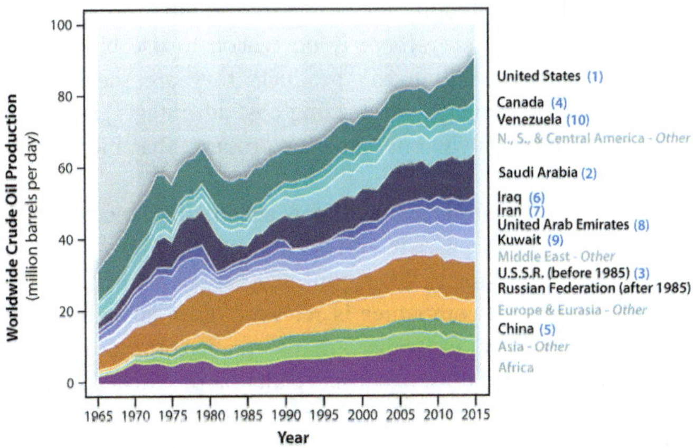

Fig. 2 Historical growth of the petroleum industry worldwide. Image reproduced with permission from ©John Wiley and Sons, 2014 (ref. 8).

Emergence of the petrochemical industry—some lessons for us

The past century can rightly be called the Age of Oil. World oil consumption grew from about 20 million tonnes per year in 1900 to nearly 4000 million tonnes per year in 2005—a 200 fold increase.[8] The economic activity enabled by oil consumption also greatly increased both human wealth and the human population over the last century. But the Age of Oil is winding down. Fig. 2 shows the total world liquid fuels production, including biofuels, deepwater oil, oil from tight shales, and tar sands oil. Production of conventional (cheap) oil has been flat since about 2005 in spite of trillions of dollars spent to explore for new oil fields. We apparently cannot produce cheap oil any faster than we are now producing it. All growth in oil supply has been from higher cost and more environmentally-problematic resources.

The bioproducts industry often seeks to replace petroleum-derived chemicals and materials. It may be instructive therefore to understand how the petrochemical industry emerged. The answer is clear. Petrochemicals grew out of the established petroleum refining industry—not the other way around. The same thing will probably occur for commodity bioproducts also: a robust bioproducts industry will most likely be built on robust bioenergy systems. The large corn wet mills producing multiple products are one example. We do not see commodity bioproducts (such as PLA) made in a dedicated plant. They are produced as coproducts in a large integrated biorefinery.

Why is this so? Important reasons include access to feedstock at marginal cost and very large scale, availability of supporting technical talent under one roof, sharing of expensive utilities including co-generation plants and wastewater treatment plants, access to internal capital for investment, and the marketing prowess and vertical integration of big firms. While I can imagine a specialty bioproduct being produced in a dedicated, smaller plant, I cannot imagine a commodity bioproduct being made anywhere other than in a large-scale biorefinery.

Since the existing markets for materials and chemicals have been largely satisfied, the bioproduct newcomers must adapt themselves to the existing market realities. Unless the bioproducts industry can offer totally new products, with greatly superior properties, there is no reason to use biobased products unless they cost much less than the products they are seeking to replace. Otherwise, drop-in replacements, with strong cost advantages, will be preferred over new products. Fortunately, one superior property that bioproducts might offer is sustainability.

So I argue that increased demand for sustainable bioenergy will pull commodity bioproducts along with it. I do not see the opposite trend occurring, in spite of much hopeful thinking in this regard. The experience of Amyris and other would-be cellulosic biofuel companies is relevant. They encountered the hard realities of the marketplace and especially the reality of scale, and are either defunct or have refocused themselves as specialty bioproducts companies.

If this is true, then to scale up bioproduct commodities we must first scale up bioenergy production and/or make bioproducts along with bioenergy commodities, as corn wet mills are currently doing. We must therefore understand why energy generally, and bioenergy in particular, is so critical to our past, present and future. Why do we need and use so much energy?

Energy, bioenergy and prosperity

All economic activity requires energy. The more energy is used, the more economic activity. As a result, the energy consumption rate (power) is strongly and positively correlated with annual gross domestic product (annual rate of wealth production). The developed world owes its prosperity to massive use of energy, about 85% of which is from non-renewable fossil fuels.[8]

More comprehensive measures of prosperity, such as the United Nations Human Development Index (HDI), show the same relationship. The more power a society consumes, the greater that society's wealth and the greater its citizens' opportunities to develop their human potential. Fortunately, the relationship between power consumption and HDI is not linear: human well-being does not increase continually with increased power consumption. Instead the HDI 'saturates' (achieves a near maximum level) at a power consumption of about 5.0 kilowatts,[9] as shown in Fig. 3. We do not have to consume unlimited power to achieve our human potential, but we must consume a great deal of it.

Since societies become more prosperous as they consume more power, at least up to the saturation power level, they will very likely become less prosperous as they consume less power. Higher oil prices and higher fossil energy prices generally (as all fossil energy sources are depleted) will eventually put most developing and developed countries on the wrong side of the power = prosperity equation. Most countries and most people will become inexorably poorer, less educated, and less healthy as there is less energy to do work and thereby generate wealth.

Is this to be our fate or the fate of our children and grandchildren?

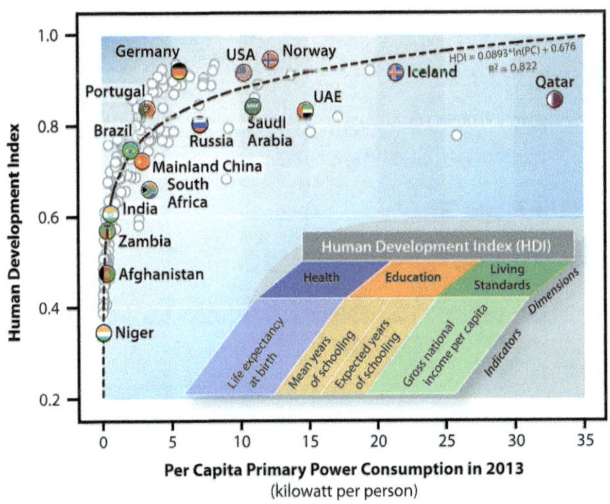

Fig. 3 Human development index *versus* per capita power consumption. Image reproduced with permission from ©John Wiley and Sons, 2014 (ref. 9).

No. Not if we think clearly and act wisely now. The good news is that there is abundant renewable energy potential to provide much more than 'saturation power' for every person on the planet. The sun provides about 100 000 terawatts of renewable power. Just 0.05% of this amount is sufficient to provide saturation power for 10 billion human beings.

Most affluent individuals in the prosperous developed world, if they think about energy at all, focus on greenhouse gas emissions (GHGs) caused by fossil energy use. Unfortunately, these prosperous people do not connect the dots. They do not understand the connection between their energy use and their prosperity. They are living in an artificial world in which the cost of energy bears very little relation to the value of energy in promoting human well-being.

In contrast, the impoverished citizens of the developing world desperately need to use more energy to improve their lives. They understand this. But if the developing world commits itself to fossil energy, it is committing itself to a dead end. Fossil fuels will become too expensive or too polluting (or both) in not many decades. Both the developed and the developing world must find another way.

We require renewable, low carbon energy if we are to enjoy sustainable, long-term prosperity. Bioenergy can provide many of the desired renewable, low carbon energy services. Only high energy density liquid fuels can supply some critical energy services, such as aviation, long-haul trucking, and ocean shipping. As a result, the consensus view of five independent analyses, as summarized in Fig. 4, is that approximately 25% of total energy services (about 138 exajoules of primary energy per year) must be provided by bioenergy[10] to restrain global temperature increases.

The geographically-distributed nature of bioenergy can help ensure that the wealth derived from bioenergy production and use will be more equitably distributed than is currently the case with fossil fuels. Properly produced and used, bioenergy can be sustainable. So, considered from the viewpoints of both the prosperous minority and the impoverished majority, we must have renewable energy, on a very large scale, if we are to enjoy both climate security and

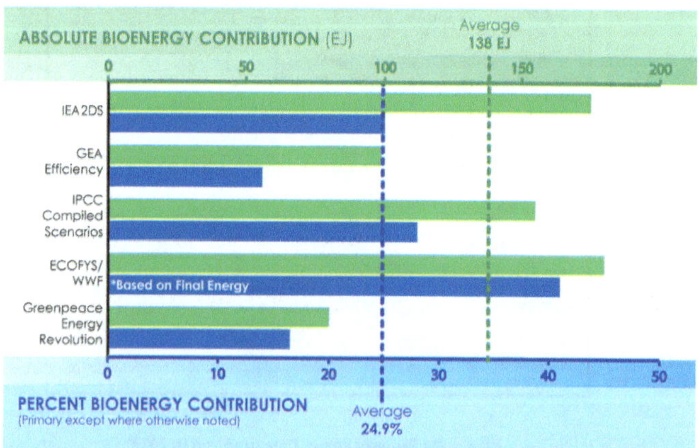

Fig. 4 Bioenergy contribution required to restrain global temperature changes under five different scenarios. Image reproduced with permission from ©American Chemical Society, 2014 (ref. 10).

sustainable prosperity. The renewable energy transition is not an option. It is essential.

The truly sobering reality is that energy transitions take decades and cost trillions of dollars.[11] There is a lot of wishful thinking about the ability of renewable energy technologies to rapidly achieve large scale penetration of the energy markets. Whatever our experience in growth rates of cell phones and other gadgets, the experience from implementing new energy systems tells us something much different. This transition to renewable energy is going to take a long time and it will be very expensive.

So we need demonstrated lower cost technologies with potentially high growth rates. I will provide more information about the absolutely critical importance of high growth rates below.

Now, back to the bioenergy theme. Why must about 25% of all renewable energy be derived from bioenergy? Compared with solar, wind and other renewable energy technologies, biomass seems unique in four ways: (1) it is both the energy collector and the energy storage system (both the solar collector and the battery), (2) it is the only source of high energy density liquid fuels so essential to our wealth and prosperity, (3) it is the only large-scale energy carrier capable of sequestering carbon (taking net carbon from the atmosphere), and (4) it is relatively low cost. Taken as a whole, these features of biomass mean bioenergy will play a key role in future energy systems.

Another feature of bioenergy is less obvious. Because of its energy storage capabilities, bioenergy can provide dispatchable power to balance the intermittent supply of electricity from solar and wind. In addition to being intermittent, solar and wind technologies provide power at very low cost at peak production. They thereby undercut the price of baseload power sources and wreck existing electricity markets, undermining the very industry they need to be able to grow and contribute to a prosperous low-carbon future.[12]

We have abundant evidence that the world does not lack the ability to produce food. "Food *versus* fuel" is apparently logical and emotionally compelling. It is also dead wrong. The problem that we have struggled with for almost 100 years is that there is insufficient demand for agricultural products. Energy services can provide by far the largest demand for agricultural and forestry materials.

People are hungry because they are poor. They are poor because they do not have sufficient access to energy services. Wealth is produced by using energy. Therefore, the best way to attack hunger is to attack poverty by increasing access to energy, including bioenergy. Poor people, as long as they remain poor, will be unable to afford the abundant food which we can produce.

Land and bioenergy/bioproducts: must we really choose between food and fuel?

It is commonly assumed that current agricultural land cannot be used sustainably to increase agricultural output. This is incorrect. Double cropping (also called "sequential cropping") could be applied on much of the U.S. and world cropland and would improve environmental performance (increase soil fertility, improve water holding capacity, reduce erosion, reduce runoff of agricultural chemicals, *etc.*) while substantially increasing total per hectare output without increasing the amount of land under cultivation.

Farmers in much of the world are not double cropping in significant numbers because there is no current market demand for the additional biomass they would generate. Increased production of bioenergy and bioproducts could provide such a demand. Even without significant double cropping, existing croplands are continuing to increase productivity. The trend in the US and the EU over the past couple of decades has been to steadily reduce the amount of land under cultivation while total farm output increases.

It is likewise assumed that food and bioenergy production are necessarily in opposition. This is also wrong. For example, the major source of animal feed protein worldwide is soybean meal. But soybeans are a relatively land-inefficient source of protein. When corn or other grains are fermented to ethanol, a high protein residue called distillers grains is produced and is a valuable animal feed protein. Excess yeast from the biofuel fermentation process is another protein source that can integrate sustainable food and bioenergy systems.

Likewise, a cellulosic biofuels industry can coproduce animal feeds by producing leaf protein concentrates from double crops or by pretreating cellulosic biomass to increase its digestibility by animals and for biofuel production.[13,14] The biomass pretreatment option is discussed in detail below in the case studies. In many places around the world, there are huge tracts of unproductive pasture land. With increased bioenergy demand and better management practices, the productivity of these pasture lands could be greatly increased, more carbon could be fixed and more employment provided in rural areas around the world.

No matter how much data and reasoning are provided to show the contrary, many people are conditioned to think that using land to produce bioenergy is wrong. So here are some questions that might be asked.

Is the only morally acceptable use of arable land to grow the cheapest possible food for human beings? If so, then we should eliminate consumption of beer, wine and other spirits, also chocolate, tea, coffee and tobacco, thereby freeing up that land for increased food production. Tens of millions of acres worldwide are devoted to these crops. Likewise, all domestic pets, companion animals and recreational animals such as horses should be euthanized (and then eaten). Many millions of acres currently provide food for these animals that should be going to human beings instead. All golf courses should close immediately and be converted to farms. All parks of all kinds and all outdoor recreation areas with arable land should close, and start growing food. All family yards and flower gardens, of any size, ought to be converted to growing food.

This is ridiculous, of course. In fact, the world is not short of land, and we are not short of ingenuity and innovations that can coproduce bioenergy, bioproducts, food, and animal feed, while improving environmental outcomes. Two specific innovations discussed below are: (1) coproducing animal feeds and bioenergy, and (2) making more efficient, more sustainable use of existing agricultural land to produce bioenergy.

Reconciling the four big drivers affecting bioenergy and bioproducts: environmental benefits, energy security, food security and rural prosperity

All four of these drivers affecting bioenergy and bioproducts matter, but they vary in relative importance by time and place and product. Put another way, the stakeholders/interested parties are not the same at all times and places.

How can these different drivers be reconciled? I argue that they can be reconciled through more efficient and more sustainable land use. To apply this general principle to specific, real systems, we must first identify and then exploit opportunities to improve how land is used to produce bioenergy and bioproducts.

How might this lofty goal be realized in the real world?

Here are three case studies as examples. In the first, we see how improving the sugar yield by pretreatment of cellulosic biomass increases the value of biomass for both animal feed and bioenergy/bioproduct applications. In the second, the near-term availability of millions of tonnes of underutilized sugarcane leaves and tops may enable new models for bioenergy production from sugarcane in Brazil. Last, an emerging paradigm called Biogasdoneright™ is explored to coproduce bioenergy, bioproducts and food on underutilized croplands.

Case 1: overcoming the recalcitrance and logistics problems of cellulosic biomass to provision commodity markets

Lignocellulosic biomass is the most abundant renewable plant material on earth with a total annual production of about 50 billion metric tonnes, against roughly 4 billion tonnes of petroleum mined each year. About 4 billion tonnes per year of this biomass is crop residues, the direct and widely available by-product of crop production. Cellulose is the major constituent in lignocellulosic biomass (30 to 55%) followed by hemicelluloses, which constitute about 15–35%, lignin (6 to 30%), and a small amount of pectins.[14] Cellulose and hemicellulose are complex

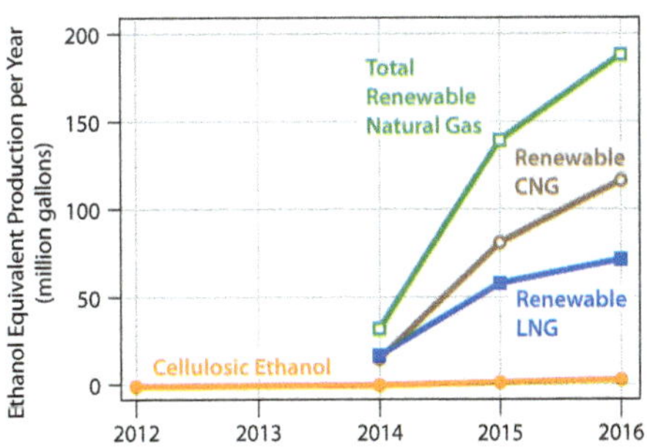

Fig. 5 Recent history of the production of cellulosic ethanol and cellulosic biogas in the United States.

sugar polymers, thus accessing and utilizing these sugars will enable many bioproduct commodities as well as some forms of bioenergy.

In principle, it seems that it should not be difficult to utilize cellulosic biomass to provision a renewable chemicals and fuels industry. After all, biomass is abundant, inexpensive and contains high concentrations of the desired molecules. In practice, there is a key technical problem and also a crucial supply chain problem to solve before cellulosic biofuels can achieve the needed scale.

The primary technical challenge is overcoming the 'recalcitrance' of cellulosic biomass: that is, the cost-effective conversion of biomass into useful sugars. There are literally hundreds of ways of overcoming biomass recalcitrance, but the important question is: at what cost? Basic science can help, but this is primarily an economic problem and must be addressed using system-wide economic tools. The most promising approaches to overcoming recalcitrance (there are comparatively few) need to be thoroughly compared using system-wide metrics, such as cost, yield, concentration and rate.[9,15] Clean, easy to process, inexpensive sugars from cellulosic biomass are also essential to achieve many of the low-cost, sustainable bioproducts we seek.

There is also a supply chain problem. Three years ago I noted that we had built approximately the first 0.1 percent of the cellulosic ethanol capacity we hope ultimately to establish.[16] But that industry is not growing. Fig. 5 shows the growth of the cellulosic biofuels industry in the United States—both cellulosic ethanol and cellulosic biogas. It is instructive that cellulosic biogas has achieved about one hundred times the total energy produced compared to cellulosic ethanol. Why has biogas grown so remarkably over the same time period that cellulosic ethanol has not grown?

Several reasons suggest themselves. The existing U.S. biogas industry is based on wastes such as landfills and manure. Thus, the logistics issues are largely solved for biogas. The raw materials are available on useful scales in well-defined

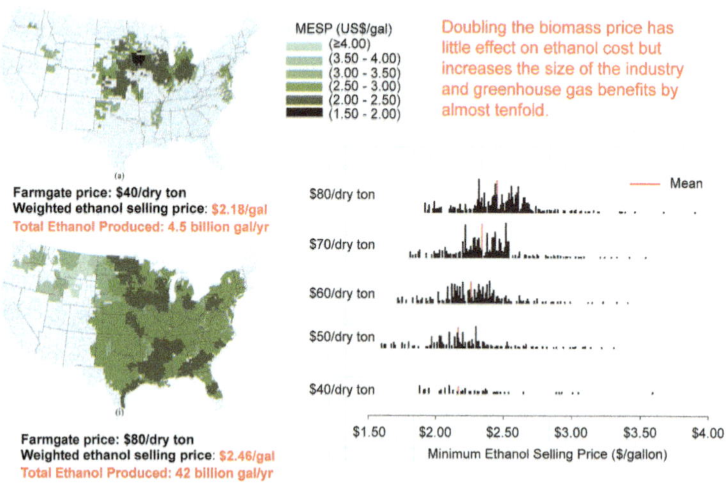

Fig. 6 Doubling the price paid to farmers for cellulosic biomass increases the size of the cellulosic ethanol industry and the associated greenhouse gas benefits by nearly ten fold. Image reproduced with permission from ©John Wiley and Sons, 2014 (ref. 17).

locations. Second and third, the natural gas pipeline infrastructure is extensive and vehicles to use natural gas as a fuel already exist. Fourth and fifth, the technologies to produce biogas and to upgrade raw biogas into pipeline quality renewable natural gas are well-developed, patent-free and relatively low cost.

Broadly speaking, cellulosic ethanol in the U.S. does not enjoy any of these advantages. The logistics issues are not solved. Except for southern yellow pine, one cannot place a future contract for one million tons per year of cellulosic biomass of a defined quality to be delivered to a defined location for a defined price. The pipelines to transport ethanol are not available, largely for ownership reasons rather than technical constraints. Corn ethanol currently occupies most of the available "room" for cellulosic ethanol in the motor fuel pool. And the technologies to produce cellulosic ethanol have not enjoyed the "learning at scale" advantages that biogas, corn ethanol and gasoline have enjoyed.

Another problem is that those who manage the land, the foresters, farmers and others, are not currently organized to supply cellulosic feedstocks. Without their active participation, we will not get the scale we need. Without scale, the costs will not come down as they must. Interestingly, increased farmer participation is so important that if farmers are paid more for their biomass there may be little or no impact on the resulting cellulosic ethanol cost.[17] This is because increased farmer participation leads to shortened supply chains, with reduced transportation costs, and also to larger biorefineries, with better economies of scale. The result is little or no effect on the overall cost to manufacture ethanol. This interesting phenomenon is summarized in Fig. 6.

This surprising result reminds us that not all aspects of renewable bioenergy and bioproducts will be like their petroleum counterparts, particularly in the feedstock supply chains. The reality is that cellulosic biomass does not travel or store well. It is bulky, combustible and degradable. Also, biomass is produced over extensive land areas and must be aggregated at some scale before it can be used. Raw cellulosic biomass has no large, well-established markets to help provide feedstocks for bioenergy production in the way that the large corn market enabled the rapid growth of the corn ethanol industry.

In fact, an "ideal" cellulosic biomass would have properties similar to those of corn. An ideal cellulosic biomass would be dense, stable, storable, shippable, ready for conversion to fuels and other products and would have alternative uses to help create scale. Will such a "corn-like" cellulosic biomass exist any time soon?

Perhaps. Years ago I invented a biomass pretreatment process for bioenergy production called ammonia fiber expansion or AFEX™. AFEX™ treats cellulosic biomass with hot, concentrated ammonia. Minimal degradation of sugars occurs during AFEX™, avoiding the loss of yield and inhibition of fermentation that many pretreatments cause.[18] After the ammonia is removed and recovered, the treated biomass gives much higher biofuel yields and also becomes a much better cattle feed.[14] By mass, the ruminant animal feed market is approximately six times greater than the human food market in the U.S. Therefore, there is a large existing market for animal feeds that can help undergird and support the development of a cellulosic bioenergy/bioproducts industry. Importantly, the ability to coproduce food and fuel using AFEX™ addresses the "food *vs.* fuel" controversy for related bioenergy systems.[19]

Because of its relatively low capital investment costs, it is likely that AFEX™ can be practised in distributed processing facilities, near where the biomass is produced. These facilities, called "depots", could be owned wholly or in part by farmers or other rural interests and thereby provide opportunities for rural economic growth.[20,21] Finally, because of physical and chemical changes in the biomass caused by AFEX™, it is easy and inexpensive to produce stable, dense biomass pellets from AFEX™-treated biomass.[22] These dense pellets can be handled and stored by conventional grain handling equipment, thereby enabling supply chain development for cellulosic bioenergy. AFEX™ is now being scaled up for commercial applications by MBI (Lansing, Michigan, USA).

The AFEX™ process has much to offer the emerging cellulosic bioenergy industry, including the potential to coproduce cellulosic ethanol and purified lignin streams as potential feedstocks for biobased aromatics. Lignin may be the only means of providing such aromatics at scale. If so, it will be critical to generate these aromatic streams in a biorefinery where they are available at scale, relatively reactive and "clean", *i.e.* suitable for further conversion and processing without extensive clean up.

Case 2: cane leaf matter in Brazil—millions of tons of "new" cellulosic biomass with nowhere to go

The sugarcane plant consists of three major parts, the stalk, the juice (mostly sucrose) and the leaves and tops, each roughly having the same mass. The stalk and the leaves/tops are cellulosic materials. Thus, nearly two thirds of the mass of the plant is cellulosic biomass. Previously, the cane field was burned prior to harvest to remove the leaves and tops and facilitate harvesting. Open field

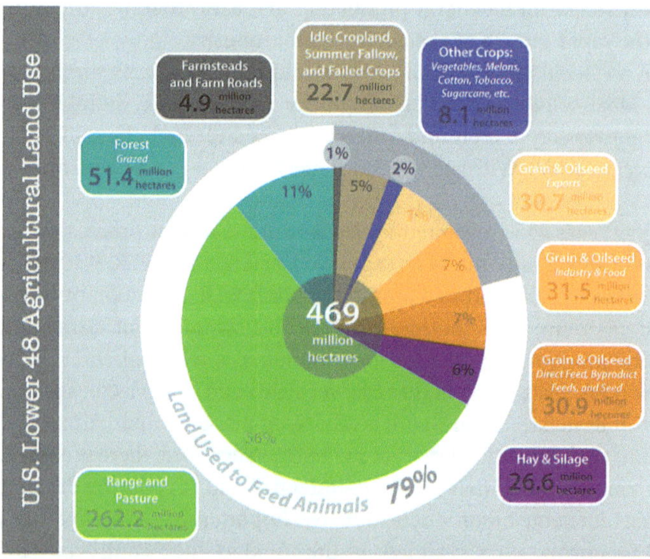

Fig. 7 Land use patterns in the lower 48 United States. Image reproduced with permission from ©John Wiley and Sons, 2014 (ref. 8).

burning has now been outlawed because of air quality issues, making millions of tons per year of "new" cellulosic biomass residue potentially available for other uses, including animal feeds, environmental services and conversion to bioenergy.

The cane leaf and top fraction is becoming available at the same time that Brazil is seeking new models for converting sugarcane to useful products. World sugar markets are clearly oversupplied and the opportunities for further growth of ethanol coproduced with sugar are limited. What other models might be available for further expansion? Two approaches are being explored.

In the first approach, sucrose is fermented to ethanol and/or biobased products, and the cellulosic parts of the plant are converted to second generation biofuels such as cellulosic ethanol and/or burned to produce electricity which is exported to the grid.[23] This is an energy/bioproducts scenario only. In the second approach, parts of the plant are used to produce ruminant animal feeds, while the remainder is converted to energy products. This is a food and fuel scenario, and is of particular interest in this paper.

Livestock production is the world's largest user of land resources, with grazing land and cropland dedicated to the production of animal feeds representing about 80% of all agricultural land use world-wide. The global figure is similar for the U.S., as shown in Fig. 7—about 80% of U.S. land devoted to food production is actually used to produce animal feeds.

In the Brazilian context, Brazil has the second biggest cattle herd in the world, with about 210 million head in 2015, distributed over 167 million hectares (Mha) of pasture. These animal numbers include beef cattle in extensive, intensive and feedlot systems as well as dairy cattle. In general, Brazilian livestock production has low average productivity because pastures are not managed to increase productivity, few inputs are used and newer technologies have not been adopted.

Integration of bioenergy and livestock production can improve land use in Brazil, since more intensive livestock production systems can maintain, or even increase, food production while reducing land use. It is thereby possible to expand cropland for bioenergy production without displacing food crops or livestock production, which otherwise might cause clearing of forests.[13] The Brazilian Center for Bioethanol Technology (CTBE, Campinas, Brazil) is about to begin a large scale modeling and experimental project to optimize various potential integrated systems for sustainable production of food and bioenergy by taking advantage of the synergies between the sugarcane production chains for both first and second generation liquid biofuels, beef cattle and other crops (*e.g.* corn, energy cane and soybean).

Preliminary calculations indicate that this approach is very promising. The cane leaves and tops, perhaps upgraded by pretreatments such as AFEX™, might provide sufficient nutrients for beef cattle to permit one ha of land in sugarcane to provide equivalent nutrients to displace four ha of land currently in pasture. If so, then the country could greatly expand the production of first and second generation biofuels, while actually shrinking the total land required to meet human needs for food and fuel. If successful, this approach might be used elsewhere to increase the efficiency of food and fuel production, while shrinking the human planetary footprint.

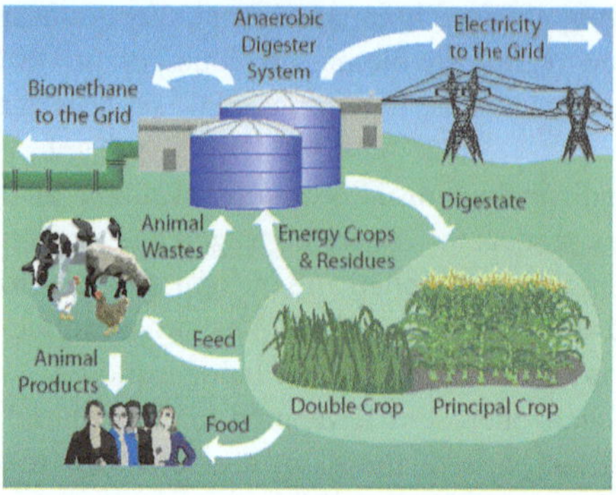

Fig. 8 Biogasdoneright™: system outline. Image reproduced with permission from ©John Wiley and Sons, 2016 (ref. 24).

But if biomass is to be converted at a very large scale to bioenergy, bioproducts and perhaps animal feeds, those who grow the biomass must participate at scale. And that is not happening.

Case 3: Biogasdoneright™ in Italy—an emerging farmer-led effort to produce sustainable bioenergy and bioproducts

If farmer participation in the bioenergy/bioproducts value chain is so important, what will it take to motivate farmers to participate more in this emerging industry? Obviously, greater market access for their products, especially energy products, because of the size of the energy markets, would help greatly. Here is one example.

During the past two years, I have become well-acquainted with an innovative farm-based, farmer-led bioenergy system now emerging in Italy.[24] Energy market access was established in 2007 by a feed-in tariff for renewable electricity provided by the Italian government. As a result, more than 1000 Italian farms organized as

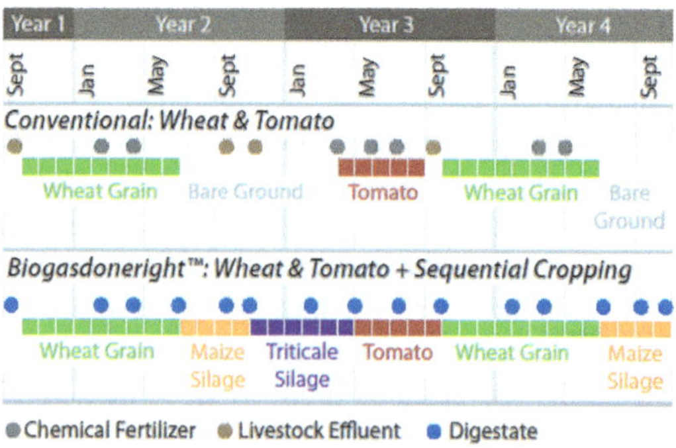

Fig. 9 A sequential cropping system using Biogasdoneright™ principles. Image reproduced with permission from ©John Wiley and Sons, 2017 (ref. 25).

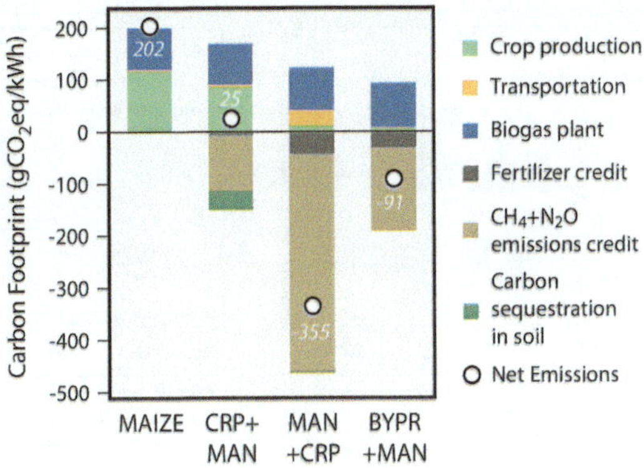

Fig. 10 Three different configurations for Biogasdoneright™ and their sources of greenhouse gas emissions. Image reproduced with permission from ©John Wiley and Sons, 2017 (ref. 25).

the Italian Biogas Consortium (CIB) are producing about 1.4 gigawatts of new renewable electricity using a set of principles called Biogasdoneright™ (BDR), as outlined in Fig. 8.

These systems are based on biogas produced by anaerobic digestion (AD) of animal and agricultural wastes in addition to purpose-grown energy crops. The biogas is combusted on farms and the electricity exported through the electric grid. These energy crops are double crops, also called sequential crops. Sequential crops are additional crops grown after the main food/feed crop is harvested. Food production continues as before; thus there is no food *versus* fuel issue and no indirect land use change (ILUC).

Farmers recycle the soluble and solid nutrients as well as stabilized, recalcitrant carbon from the anaerobic digesters onto their land. This recycle loop significantly reduces purchased fertilizer inputs, while also building soil organic matter by carbon sequestration. Farm profitability, soil fertility and overall environmental performance are improved. The logistics issue of cellulosic bioenergy is solved because all of the processing takes place on the farm. Scale is achieved because the system can be (and has been) widely replicated on many farms. New, stable jobs are created, about 12 000 jobs thus far.

Fig. 9 illustrates how bioenergy crops are grown in those periods during which the land was previously left idle. These bioenergy crops are harvested, ensiled on farm to preserve them, and then fed to the digester along with locally-available wastes and residues. Thus, the previously "wasted" land resource, and waste organic material are now used more productively and with greater environmental benefits than they were prior to bioenergy production, a "win–win" arrangement.[25]

The greenhouse gas benefits of the BDR approach are impressive. Fig. 10 shows three different configurations of BDR, using primarily energy crops as a feedstock, or primarily manure, or primarily agrowaste residues. These three configurations achieve between −335 and 25 grams of total carbon dioxide equivalents per megawatt hour compared to fossil fuel-based electricity consumed in the EU. While no biomethane is currently produced by the CIB, the

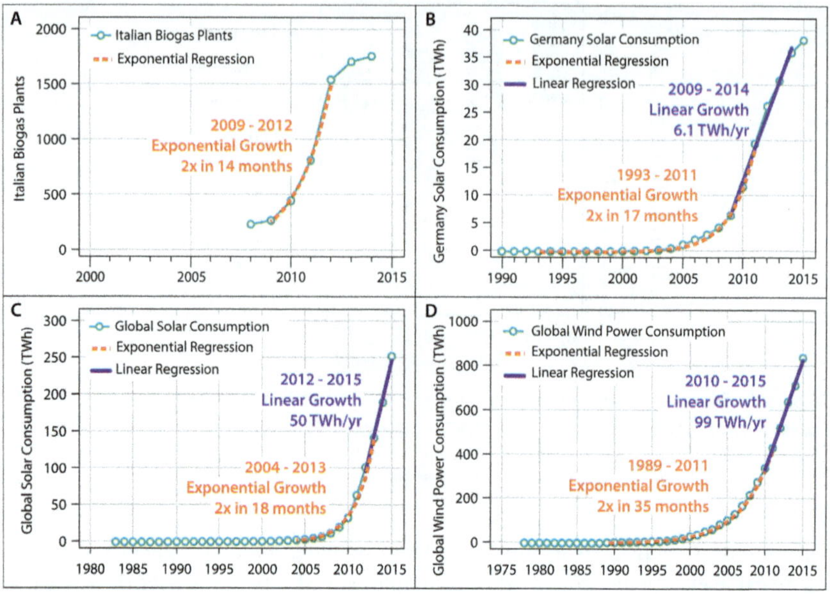

Fig. 11 Comparing the growth rates of renewable energy consumption: global wind and solar plus German solar and Italian biogas.

projected GHG emissions of biomethane (renewable natural gas) are −36 to 10 grams of carbon dioxide per megajoule (MJ) compared to their fossil natural gas counterpart.

Biogasdoneright™ is a scalable, low cost bioenergy system with carbon capture and storage (BECCS). How scalable is it? The growth history of the Italian biogas industry is summarized in Fig. 11 where it is compared with the growth history of the German solar industry and also the world solar and wind energy production. As with the growth of the U.S. corn ethanol industry, this is true exponential growth. In fact, with a demonstrated doubling time of about 14 months, the growth of the BDR system is the fastest demonstrated growth of renewable energy to the gigawatt scale that I am aware of, faster than global wind and solar energy, and even faster than German solar energy consumption.

Why am I so concerned about exponential growth? Because of the huge worldwide need for energy to underpin human prosperity and the need to scale up all renewable energy quickly in the face of declining fossil fuel supplies and increasing environmental damage from their use.

The Italian government is expected soon to announce new policies to maximize the potential of biogas as a transportation and heating fuel. This will involve upgrading the biogas to biomethane, essentially renewable natural gas, and injecting it into the natural gas grid. Natural gas has long been used as a transportation fuel in Italy, so much of the necessary infrastructure for distributing, refueling and driving with biomethane, renewable natural gas, is already in place. Thus, the Italian biogas/biomethane industry may be in for another rapid growth spurt.

What can we learn from the Biogasdoneright™ experience?

First, it is possible to grow bioenergy production very rapidly under favorable circumstances. The factors that seemed to play the biggest role in the success of BDR include: (1) energy market access under terms that allowed the farmers to go to local lenders and get the loans necessary to buy the capital equipment, (2) readily available, proven anaerobic digestion (AD) conversion technology, (3) short supply chains—the farmer controls all aspects of the harvesting, transport and processing of the biomass, and (4) recycle of nutrients and carbon makes the land more fertile, more resistant to flooding and drought, and also reduces the farmer's out-of-pocket costs significantly.

Second, the logistical advantages of on-farm production and processing of the biomass are critical. It simply would not be possible to move the raw, wet biomass feedstock for AD very far without significant cost penalties. Instead of moving bulky materials to a distant biorefinery, an energy-dense product (electricity) is exported through an existing infrastructure and all of the bulky, wet materials are handled on-farm.

The importance of market access cannot be overstated. The farmers involved in the BDR movement could not have received the loans necessary to build out the infrastructure without guaranteed markets for their electricity. If bioenergy is to compete effectively with fossil fuels, there must be market access on terms that permit biofuel producers, including farmers, to compete. As renewable fuel providers gain experience and scale, the costs will come down and the industry will grow, hopefully in time to offset the inevitable decline of fossil fuels in the coming decades.

Finally, the farmers involved in BDR also produce grains, oilseeds and other crops that are good feedstock candidates for bioproducts. Because of the way they are produced, these commodities have unique sustainability advantages. Their GHG emissions in particular will be much lower than conventionally-produced grains and oilseeds, but they will also improve water quality and biodiversity compared with conventional agricultural feedstocks. Bioproducts made from these crop commodities grown in BDR systems are good candidates for "green" labeling and promotion. Likewise the electricity generated *via* BDR systems is "green power" which can and should be used in the production of biobased products.

Feedstocks and platform molecules

I believe the key feedstocks include crop residues, such as corn stover and sugarcane leaf matter, as well as energy crops, especially if the residues and energy crops are grown in sequential cropping systems on existing agricultural land using improved sustainability practices such as those exemplified by Biogasdoneright™. Given the much smaller amounts of feedstock required for bioproducts compared to bioenergy, agricultural commodities such as corn, wheat and soybeans, as well as other crop commodities grown on farms employing BDR principles are sustainable feedstocks. They can and should be used for bioproducts. Lignin represents perhaps the only source of renewable aromatic feedstocks for biobased products, but the conversion of native lignin to finished products is daunting.

As far as platform molecules are concerned, these would certainly include starches, sugars and oils, as long as the feedstocks from which these molecules are derived are sustainably produced. Based on my experience with BDR, I would have to add two more "platform molecules" that I might not have considered previously. These are electricity and biomethane. We need to consider renewable electricity as a platform "molecule" in our bioproducts, both as a process utility and also for the reducing power it can provide to make bioproducts.

Conclusions and an exhortation

My entire career has been devoted to producing sustainable biofuels from plant matter. Because of the strong relationships between biofuels and agriculture, I am necessarily concerned about food production and all the links between energy and agriculture.

We take the modern world for granted. However, the amenities (wealth, education, healthcare, mobility, *etc.*) enjoyed by people in the prosperous nations depend on consuming huge amounts of energy, about 85% of it from fossil resources: oil, coal, and natural gas. The Human Development Index (HDI) is a composite measure of prosperity, and includes metrics for health, education, and living standards. Prosperity is strongly and positively correlated with power consumption. Assuming that HDI is a reasonable surrogate for prosperity, human prosperity rises very rapidly as power consumption increases and then levels off or "saturates" at a power consumption of about 5 kW per capita.

These are sobering numbers. The total world power consumption is currently about 17 terawatts (TW). We will soon have 8 billion people on planet earth. If all 8 billion people were to consume 5 kW, that would total 40 TW, about two and a half times the current power consumption from all sources. However, most of humankind currently consumes much, much less than 5 kW. For most of humankind, their daily reality is poverty caused by a lack of access to energy services.

This should be unacceptable to all of us—a relative few enjoying abundant energy and the resulting prosperity (for a bit longer, at least), while the mass of humankind is mired in poverty because they lack abundant energy.

I believe it is deeply irresponsible to think that fossil fuels can or should provide over twice the power that they currently do. A society built on non-renewable power is simply unsustainable. It cannot continue. It will not continue. A global society built on fossil energy will become increasingly fragile and perilous as fossil energy is depleted. We must have renewable power, many terawatts of it and soon, if we are to enjoy the prosperity that abundant energy provides. Renewable energy is therefore not an option; it is essential if more people, now and in the future, are to achieve their potential as human beings.

I believe that renewable energy, especially bioenergy, will bring along with it many opportunities for sustainable bioproducts. This was the experience with petroleum refining and petrochemicals. For many reasons, I think it is unlikely that commodity bioproducts will emerge without large scale bioenergy. Indeed, current commodity bioproducts are coproduced with bioenergy.

There are many renewable sources of electricity, especially solar and wind power. These can provide heating, cooling, lighting and some mobility services. However, only bioenergy, due to its high energy density, offers a complete renewable replacement for petroleum in all mobility applications, including agriculture. Since we must have renewable energy, and we must have food, it is essential that we find ways to integrate food production with renewable energy production, including bioenergy—not in competition with each other but to their mutual benefit. And we must do so in ways that are environmentally, economically, and socially sustainable.

This is an enormous challenge. It is the primary technical and social challenge around which my research and thinking are oriented. For bioenergy, we must greatly increase the efficiency and sustainability with which we use land. The three case studies outlined here give some idea of the potential for this approach. We can do this. We must do this.

Acknowledgements

Professor Dale gratefully acknowledges support by Michigan State University AgBioResearch and by the USDA National Institute of Food and Agriculture (NIFA).

Notes and references

1 M. R. Finlay, Old Efforts at New Uses: A Brief History of Chemurgy and the American Search for Biobased Materials, *J. Ind. Ecol.*, 2004, 33–46.
2 D. E. Wright, Agricultural Editors Wheeler McMillen and Clifford V. Gregory and the Farm Chemurgic Movement, *Agr. Hist.*, 1995, **69**(2), 272–287.
3 C. W. Dugger, U.S. aid hurts instead of helps, charity says; Group declines funds for African farmers, in *The International Herald Tribune*, Malela, Kenya, 2007.
4 R. E. Kavey, How sweet it is: sugar-sweetened beverage consumption, obesity, and cardiovascular risk in childhood, *J. Am. Diet. Assoc.*, 2010, **110**(10), 1456–1460.
5 F. B. Hu and V. S. Malik, Sugar-sweetened beverages and risk of obesity and type 2 diabetes: epidemiologic evidence, *Physiol. Behav.*, 2010, **100**(1), 47–54.
6 U.S.E.I. Administration, *Table 10.3 Fuel Ethanol Overview*, D.o. Energy, April 2017, p. 155.
7 J. Goldemberg, S. T. Coelho and P. Guardabassi, The sustainability of ethanol production from sugarcane, *Energy Policy*, 2008, **36**(6), 2086–2097.
8 B. E. Dale and R. G. Ong, Design, implementation, and evaluation of sustainable bioenergy production systems, *Biofuels, Bioprod. Biorefin.*, 2014, **8**(4), 487–503.
9 B. E. Dale and R. G. Ong, Energy, wealth, and human development: why and how biomass pretreatment research must improve, *Biotechnol. Prog.*, 2012, **28**(4), 893–898.
10 B. E. Dale, *et al.*, Take a closer look: biofuels can support environmental, economic and social goals, *Environ. Sci. Technol.*, 2014, **48**(13), 7200–7203.
11 G. J. Kramer and M. Haigh, No quick switch to low-carbon energy, *Nature*, 2009, **462**, 568–569.
12 Staff, Clean Energy's Dirty Secret, in *The Economist*, London, England, 2017.
13 A. Egeskog, *et al.*, Integrating bioenergy and food production—A case study of combined ethanol and dairy production in Pontal, Brazil, *Energy Sustainable Dev.*, 2011, **15**(1), 8–16.

14 M. Blümmel, Opportunities from second-generation biofuel technologies for upgrading lignocellulosic biomass for livestock feed, *CAB Rev.*, 2014, **9**(041), 1–8.

15 L. Tao, *et al.*, Process and technoeconomic analysis of leading pretreatment technologies for lignocellulosic ethanol production using switchgrass, *Bioresour. Technol.*, 2011, **102**(24), 11105–11114.

16 B. E. Dale, A New Industry Has Been Launched: The Cellulosic Biofuels Ship (Finally) Sails, *Biofuels, Bioprod. Biorefin.*, 2015, **9**, 1–3.

17 S. Kim and B. E. Dale, All biomass is local: The cost, volume produced, and global warming impact of cellulosic biofuels depend strongly on logistics and local conditions, *Biofuels, Bioprod. Biorefin.*, 2015, **9**(4), 422–434.

18 S. P. Chundawat, B. Bals, T. Campbell, L. Sousa, D. Gao, M. Jin, P. Eranki, R. Garlock, F. Teymouri, V. Balan and B. Dale, Primer on Ammonia Fiber Expansion Pretreatment, in *Aqueous Pretreatment of Plant Biomass for Biological and Chemical Conversion to Fuels and Chemicals*, ed. C. E. Wyman, 2013, pp. 169–200.

19 B. E. Dale, *et al.*, Biofuels done right: land efficient animal feeds enable large environmental and energy benefits, *Environ. Sci. Technol.*, 2010, **44**(22), 8385.

20 B. D. Bals and B. E. Dale, Developing a model for assessing biomass processing technologies within a local biomass processing depot, *Bioresour. Technol.*, 2012, **106**, 161–169.

21 P. L. Eranki, B. D. Bals and B. E. Dale, Advanced Regional Biomass Processing Depots: a key to the logistical challenges of the cellulosic biofuel industry, *Biofuels, Bioprod. Biorefin.*, 2011, **5**(6), 621–630.

22 B. Karki, *et al.*, Physical characteristics of AFEX-pretreated and densified switchgrass, prairie cord grass, and corn stover, *Biomass Bioenergy*, 2015, **78**, 164–174.

23 J. G. D. B. Leite, *et al.*, Sugarcane: a way out of energy poverty, *Biofuels, Bioprod. Biorefin.*, 2016, **10**(4), 393–408.

24 B. E. Dale, *et al.*, Biogasdoneright™: An innovative new system is commercialized in Italy, *Biofuels, Bioprod. Biorefin.*, 2016, **10**(4), 341–345.

25 L. Valli, *et al.*, Greenhouse gas emissions of electricity and biomethane produced using the Biogasdoneright™ system: four case studies from Italy, *Biofuels, Bioprod. Biorefin.*, 2017, DOI: 10.1002/bbb.1789.

Faraday Discussions

PAPER

Unavoidable food supply chain waste: acid-free pectin extraction from mango peel *via* subcritical water†

H. Xia and A. S. Matharu *

Received 26th January 2017, Accepted 7th February 2017
DOI: 10.1039/c7fd00035a

Mango peel is the major by-product of mango processing, and compromises 7–24% of the total mango weight. In this study, pectin was extracted from mango peel waste by using subcritical water extraction (SWE) in the absence of mineral acid. A highest yield of 18.34% was achieved from the Kesar variety and the pectin was characterised using ATR-IR spectroscopy, TGA and ^{13}C solid-state NMR spectroscopy to confirm the structure. The degree of esterification (DE) of the pectin was analysed with both titrimetry and ^{13}C solid-state NMR spectroscopy, and a high DE (>70%) was observed for all three varieties (Keitt, Sindhri and Kesar). This is the first report on acid-free subcritical water extraction of pectin from mango peel, which provides a green route for the valorisation of mango peel waste and contributes to a source of biobased materials and chemicals for a sustainable 21st century.

Introduction

Sustainability is a much-used modern-day buzz word that encompasses economical, social and environmental values, *i.e.* the three pillars of sustainability. Sustainable development is development that meets the needs of the present day without compromising the ability of future generations to meet their own needs. These definitions, developed at the end of the last century, are continually being refined through the outcomes of the UN Sustainable Development Summit (25th September 2015), which adopted the document entitled "Transforming our world: the 2030 Agenda for Sustainable Development" setting out 17 Sustainable Development goals (SDGs) to end poverty, protect the planet, and ensure prosperity for all.

In order to establish biobased materials and chemicals for a sustainable 21st century, we must take heed of SDG 12, which strives to ensure sustainable consumption and production patterns by promoting resource and energy

Green Chemistry Centre of Excellence, Department of Chemistry, University of York, York YO10 5DD, UK. E-mail: avtar.matharu@york.ac.uk

† Electronic supplementary information (ESI) available. See DOI: 10.1039/c7fd00035a

efficiency, promoting sustainable infrastructure, and providing access to basic services, green and decent jobs, and a better quality of life for all. Key targets of SDG 12 include the substantial reduction of waste generation through prevention, reduction, recycling and reuse, and the notion of a systematic approach and cooperation among actors operating in the supply chain, from producer to final consumer, which are relevant to sustainable chemistry for the 21st century.

Valorisation of unavoidable food supply waste represents an interesting Faraday Discussion as food supply waste is a valuable source of biobased chemicals, materials and bioenergy. Food waste is a global problem with in excess of 1.3 billion tonnes of waste produced every year, which is rather ironic as close to 1 billion people suffer from severe chronic malnutrition.[1] Unavoidable food supply chain wastes are those predominantly as a result of processing from farm to fork, which account for approx. 40% losses in some cases. There is potential chemical value proposition associated with this resource beyond bioenergy, for example anaerobic digestion, and most certainly landfill, in the form of chemicals and materials. In this discussion paper we focus on unavoidable food supply chain waste, in particular mango peels.

Mangoes (*Mangifera indica*) are the second most important tropical fruit after bananas. According to the FAO, global mango production in 2014 exceeded 45 M tonnes.[2] Peel is the major by-product of industrial mango processing, which consists of 7–24% of the total mango weight.[3] In recent years, it has been widely reported as a rich source of many valuable compounds, such as polyphenols, carotenoids, enzymes and dietary fibre.[4,5]

Pectin is a family of complex heteropolysaccharides consisting of a few hundred to about 1000 D-galacturonic acid (GalA) units.[6] It is the major component of cell walls, and is distributed in the primary cell wall of all plants.[7] Pectin is widely used as a gelling, thickening and stabilizing agent in the food and cosmetic industries[8] as well as in pharmaceutical industries.[9] It has been reported to be effective for removing lead and mercury from the gastrointestinal tract and respiratory organs, favourably influencing cholesterol levels in blood, and treating iron deficiency anaemia and many other health issues.[10]

Pectin extraction from mango peel has been studied by many researchers with various techniques. Rehman *et al.* used conventional heating (80 °C, 120 min) in the presence of aqueous sulfuric acid (pH 2.5) to afford 21% (by wt) pectin yield.[11] Banerjee *et al.*, using lemon juice instead of conventional mineral acids as the solvent and combining sonication, reported a pectin yield of 26% but with a low degree of esterification.[12] Maran *et al.* reported microwave-assisted extraction (413 W, pH 2.7, 134 s and solid–liquid ratio of 1 : 18 g mL^{-1}) to furnish 28% (by wt) of pectin.[13]

In recent years, subcritical water extraction (SWE) has attracted increasing attention as a promising green extraction technique due to its selectivity and efficiency.[14] Pectin has been reportedly extracted by SWE from apple pomace and citrus peel.[15,16] However, there is no report on the SWE of pectin from waste mango peel using acid-free conditions.

This is the first study that reports the subcritical water extraction (both batch and flow) of pectin from mango peel waste. Prior to extraction, the peel was subjected to heptane and ethanol hot extraction to remove surface waxes and alcohol sugars. The extracted pectin was analysed using ATR-IR spectroscopy, TGA and ^{13}C solid-state NMR spectroscopy to confirm the structure. The degree of esterification of the pectin was also determined by both titrimetry and NMR methodology.

Materials and methods

Materials

Three different mango varieties (Keitt, Sindhri and Kesar) were bought from local supermarkets. The mangoes were peeled using a stainless steel knife and the excess flesh was removed with the blunt edge of the knife. After that, the peel was air dried for about two weeks until a constant weight was observed.

Ethanol/heptane extraction

The dried mango peel (40 g) was soaked in 400 mL of ethanol and heptane, separately, and heated under reflux for 2 hours. Thereafter, the mixture was cooled and filtered, and the extract was concentrated (rotary evaporator), while the mango peel residues were air dried until a constant weight was observed.

Subcritical water extraction (SWE)

The dried mango peel post ethanol/heptane extraction was then subjected to subcritical water extraction with both flow and batch systems. For the flow system, the dried mango peel (6–10 g) was loaded into a pressurised stainless steel vessel and heated at 175 °C, with a flow rate of 3 mL min^{-1} for 15 min. For the batch system, the dried mango peel (5 g) was loaded into a pressurised stainless steel vessel with 200 mL of deionized water. The mixture was heated at different temperatures (75, 100, 125 and 150 °C) for 2 h.

Pectin precipitation

The extract from the SWE was added to twice the volume of ethanol and left overnight to allow precipitation. Thereafter, the resultant mixture was centrifuged (3500 rpm for 16 min) and the ensuing pellet was washed with 20 mL of ethanol before being centrifuged again under the same conditions. The crude pellet (pectin) was then washed with hot ethanol, filtered and dried under a warm air flow until a constant weight was observed.

ATR-IR analysis

ATR-IR analysis was performed using a Bruker VERTEX 70 infrared spectrometer including an ATR probe with a Golden Gate attachment. The scanning number for each sample was 4 and the wavelength range was between 4000 cm^{-1} and 800 cm^{-1}.

Thermogravimetric analysis (TGA)

Thermogravimetric analysis (TGA) was performed on a PL Thermal Science STA 625. About 10.0 mg of the sample was weighed accurately into an aluminium pan and analysed against an empty aluminium reference pan under a flow of nitrogen gas, from 20 °C to 625 °C at a heating rate of 10 °C min^{-1}.

^{13}C solid-state NMR CPMAS

^{13}C solid-state NMR CPMAS was performed on a 400 MHz Bruker Avance III HD spectrometer equipped with a Bruker 4 mm H(F)/X/Y triple-resonance probe and 9.4T Ascend® superconducting magnet. The CP experiments employed a 1 ms linearly-ramped contact pulse, spinning rates of 12 000 ± 2 Hz, optimized recycle delays of 3 seconds, spinal-64 heteronuclear decoupling (at $\nu_{rf} = 85$ kHz) and are a sum of 600–800 co-added transients. The chemical shifts are reported with respect to TMS, and were referenced using adamantane (29.5 ppm) as an external secondary reference.

GC-MS analysis

GC-MS analysis was performed on a Perkin Elmer Clarus 500 Gas Chromatograph and Clarus 560S Mass Spectrometer equipped with a DB5-HT column (30 m, 0.25 m, 0.25 μm) and autosampler. The oven was programmed to maintain a temperature of 50 °C for 4 minutes, followed by heating at 10 °C min^{-1} up to 290 °C, which was then held for 10 minutes. The identified compounds were indicated by comparison with the NIST library of compounds.

HPLC analysis

HPLC analysis was performed to analyse the sugar content in the SWE extracts using a HP1100 instrument, equipped with a 100-sample autosampler, quad-pump and both diode array (DAD) and evaporating light scattering (ELSD) detectors. Standard samples of levoglucosan, rhamnose, xylose, fructose, glucose and sucrose were used to prepare the calibration curve.

Degree of esterification (DE)

The degree of esterification (DE) of the pectin was determined by both titrimetry and NMR methods. The titrimetry was carried out according to Mizote *et al.*'s method. Pectin (2 g) was added to 200 mL of a mixture of water–concentrated hydrochloric acid–2-propanol (90 + 10 + 100) with stirring for 15 min. The mixture was then filtered and washed with 65% 2-propanol until the filtrate was free of chloride. The pectin was then dried in an oven until a constant weight was observed. The dried pectin (0.5 g) was loaded into a conical flask, moistened with a small amount of 65% 2-propanol and then dissolved in 100 mL of distilled water. The solution was titrated to neutrality with 0.1 M NaOH solution and the volume was recorded as V_1. Then, 30.0 mL of 0.1 M NaOH was added into the flask with stirring for 30 min. After that, an equivalent amount of 0.1 M HCl solution was added and the mixture was titrated again with 0.1 M NaOH. The volume of NaOH consumed was recorded as V_2. The degree of esterification is calculated according to eqn (1):

$$\mathrm{DE} = \frac{V_2}{V_1 + V_2} \times 100\% \qquad (1)$$

For the NMR method, the DE was calculated from the ^{13}C solid-state NMR spectra, according to eqn (2):[17]

$$\mathrm{DE} = \frac{I(\text{esterified carboxyl group})}{I(\text{esterified carboxyl group}) + I(\text{non-esterified carboxyl group})} \tag{2}$$

Results and discussion

Ethanol/heptane extraction

The yields of the extracts from the ethanol/heptane extractions are summarised in Table 1. For all three varieties, the yields of the extracts from the ethanol extraction (EE) are much higher than those from the heptane extraction (HE). The Keitt mango peel has the highest EE yield among the three varieties (15.4%). The extract was analysed by GC-MS and the result is shown in Fig. 1 along with the major components listed in Table 2. It can be observed that the ethanol extract of the Keitt mango peel mainly consists of some compounds from the breakdown of cellulose and hemicellulose (**1**, **2** and **4**)[18] and some fatty acids and their esters (**5** and **6**).

Sugar analysis

The sugar content in the ethanol extract was analysed by HPLC and the result is shown in Fig. 2. It can be seen that fructose is the most abundant sugar in all

Table 1 Ethanol/heptane extraction yields for the waste mango peel

	Yield (% by wt)		
Variety	Keitt	Sindhri	Kesar
Ethanol extraction	15.4	6.4	6.0
Heptane extraction	1.4	1.6	1.7

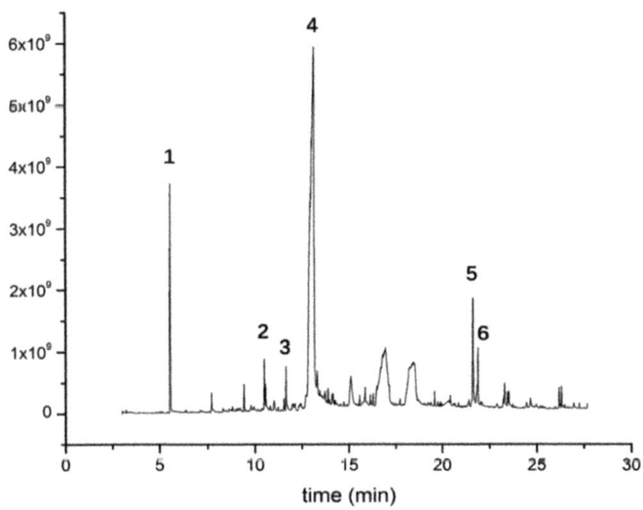

Fig. 1 GC-MS chromatogram of the mango peel ethanol extract (Keitt variety).

Table 2 Major components in the mango peel ethanol extract (Keitt variety)

Peak	Compound
1	Furfural
2	2-Furaldehyde diethyl acetal
3	2,3-Dihydro-3,5-dihydroxy-6-methyl-4*H*-pyran-4-one (DDMP)
4	5-(Hydroxymethyl)furfural
5	Palmitic acid
6	Hexadecanoic acid, ethyl ester

three varieties, accounting for at least 75% of the total amount. In particular, Keitt has much higher contents of both levoglucosan and glucose compared with the other two varieties, but rhamnose was not detected in it. Kesar has the highest sucrose content among the three varieties, followed by Sindhri.

Pectin yield

The pectin yields of the SWEs with both the flow system (all three varieties) and batch system (Sindhri post ethanol extraction) are shown in Fig. 3 and 4, respectively. In the flow system, the highest yield was achieved with the Kesar variety post ethanol extraction (18.34%). The Keitt variety also gained a similar yield of 18.31% with the post heptane extraction. Sindhri has the lowest pectin yield. A conventional acid extraction was also undertaken for comparison (HCl, pH 2.5, 80 °C, s–w ratio = 1 : 40, 2 h). Interestingly, the yield was only 4.88%, which is much lower than that reported in the literature, and lower than the acid-free SWE.

As for the batch system, the highest pectin yield was gained at 100 °C (8.68%), which is slightly higher than that from the flow system. It is noticeable that the yield undergoes a significant decrease between 125 and 150 °C, indicating that the majority of pectin was broken down during the long process of heating at a high temperature.

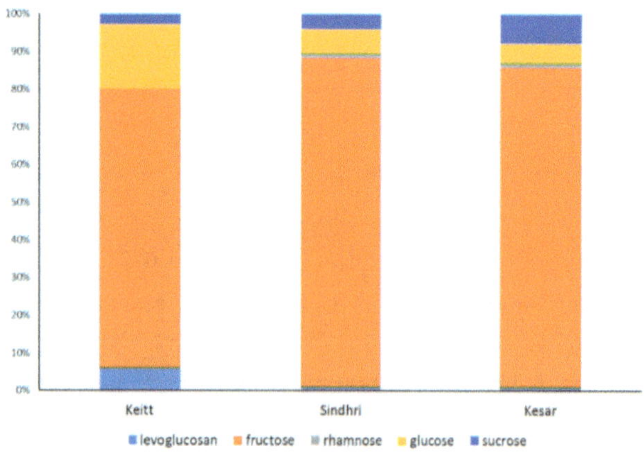

Fig. 2 Sugar content of the mango peel ethanol extracts.

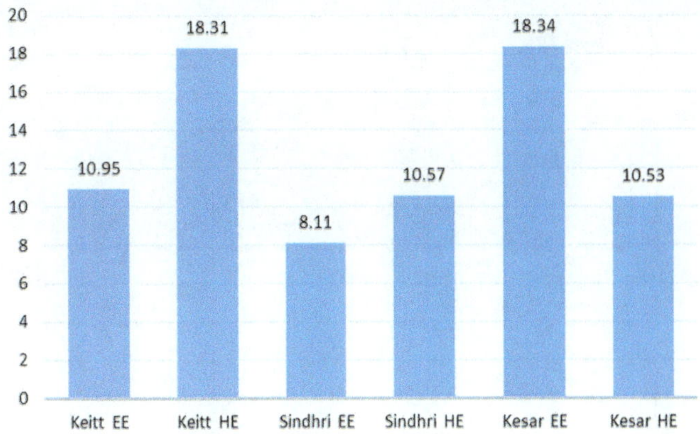

Fig. 3 Mango peel pectin yield (%) from subcritical water extraction (flow system).

ATR-IR analysis

The ATR-IR spectra of the mango peel pectin from the flow system are shown in Fig. 5. The combination of carbonyl absorption bands at 1730 cm^{-1} (methyl ester) and 1600 cm^{-1} (free carboxylic acid) suggests the presence of pectin, and the ratio of their peak areas indicates the degree of esterification for the pectin. Visually, Keitt HE has the highest degree of esterification while Sindhri EE has the lowest. Apart from these two peaks, some other characteristic absorption peaks, such as the broad O–H stretch around 3300 cm^{-1} (hydrogen-bonded hydroxyl groups), C–H stretch at approximately 3000 cm^{-1} (alkyl groups), and intense C–O stretches between 1200 and 1000 cm^{-1}, also confirm the existence of pectin.

The ATR-IR spectra of Sindhri EE pectin from the SWE batch system are shown in Fig. 6. It can be observed that the intensity of the absorption band of the methyl

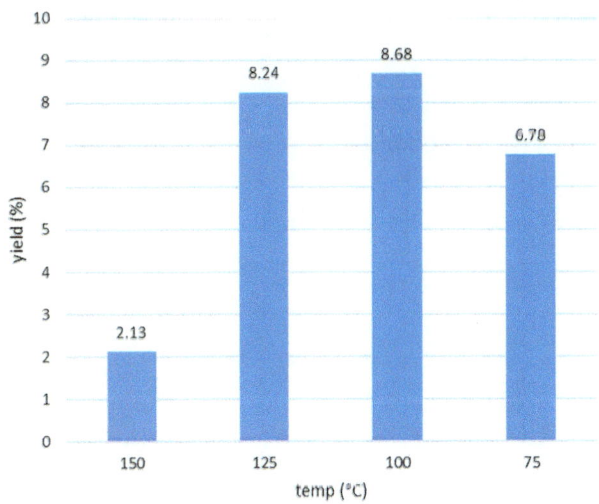

Fig. 4 Mango peel pectin yield (%) from subcritical water extraction (batch system, Sindhri).

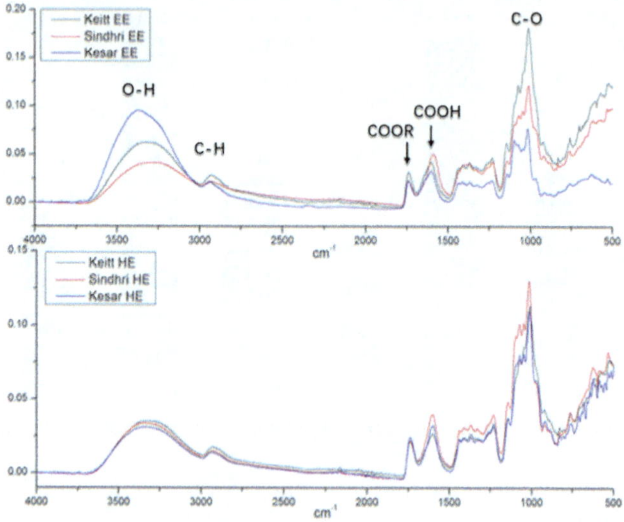

Fig. 5 ATR-IR spectra of the mango peel pectin from the SWE flow system.

ester group (1730 cm^{-1}) gradually increases from 75 to 125 °C. However, when the temperature reaches 150 °C, it dramatically drops to a low intensity, which indicates the low degree of esterification or low pectin content in the 150 °C extract.

Thermogravimetric analysis

The TGA and DTG analysis reveal the thermal properties for the mango peel pectin samples (Fig. 7, flow system). Each sample has three major processes of mass loss. The first, ranging from 60 to 120 °C, is due to the removal of water followed by the second, between 200 and 250 °C, which is attributed to pectin decomposition. The Kesar variety has the largest mass loss during this period

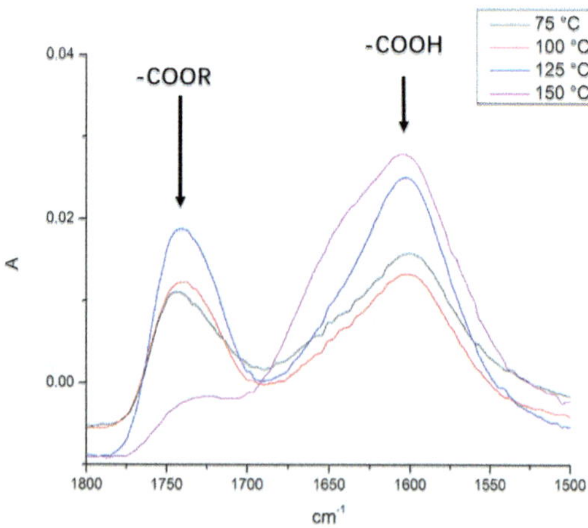

Fig. 6 ATR-IR spectra of the mango peel pectin from the SWE batch system.

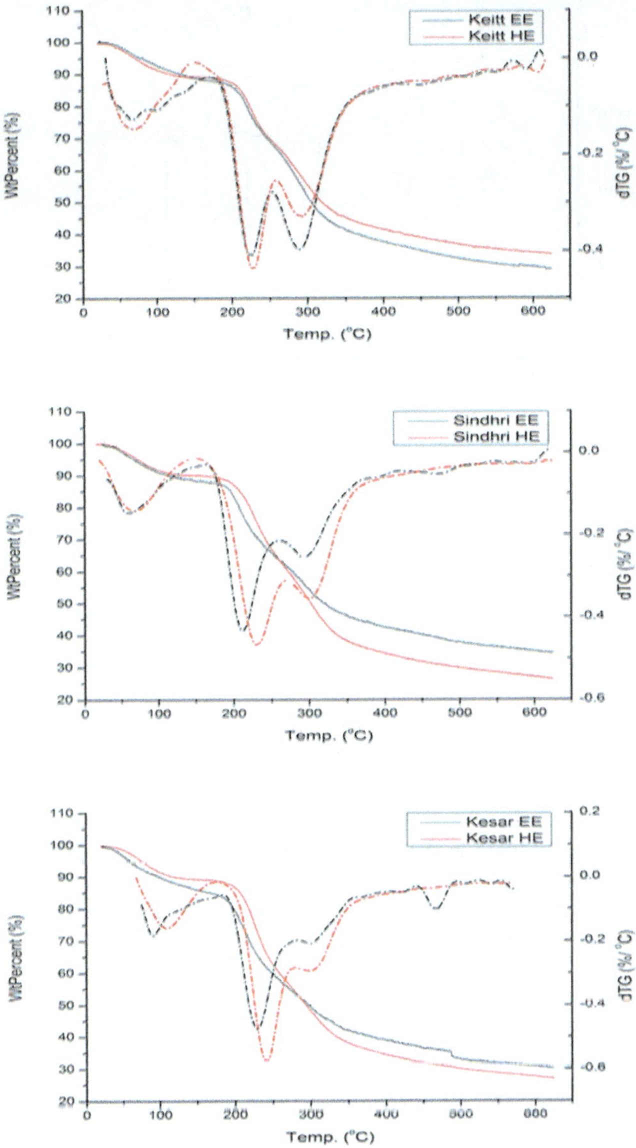

Fig. 7 TGA and DTG traces of the mango peel pectin from the SWE flow system.

(32%), followed by Sindhri (26%) and then Keitt (20%). The third main decomposition process around 300 °C is probably due to some residual cellulose and/or starch in the samples. A qualitative analysis was carried out to confirm the presence of starch in the sample (see ESI†). A blue-black colouration with respect to iodide solution was observed for certain samples.

^{13}C solid-state NMR analysis

The solid-state ^{13}C NMR spectrum of the mango peel pectin is shown in Fig. 8. The strong resonance at 171.08 ppm is attributed to the C-6 carbonyl (ester and

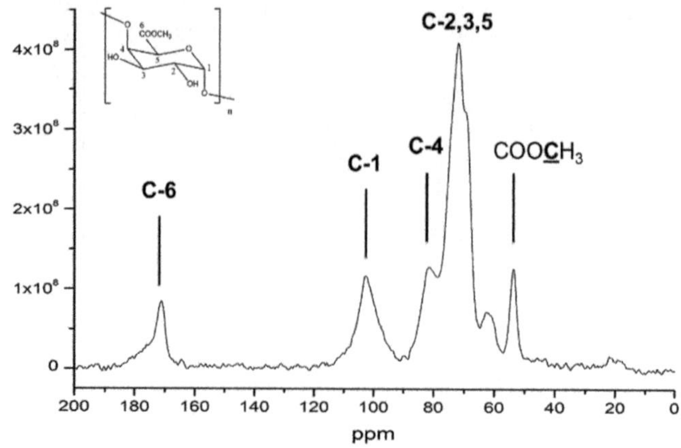

Fig. 8 ^{13}C solid-state NMR spectrum of the mango peel pectin (Keitt EE).

Table 3 Degrees of esterification (DE) of the mango peel pectin

	DE/%	
Variety	Titrimetry	NMR
Keitt	76.80	82.44
Sindhri	70.22	76.43
Kesar	74.35	80.66

carboxylic acid) carbon of pectin, while the resonances at 102.79 and 81.38 ppm can be assigned to the anomeric C-1 carbon and C-4 carbon, respectively. Furthermore, the intense peaks in the region of 60 to 90 ppm are from the carbon atoms of the pyranoid ring (C-2, 3, 5), while the weak resonance at 53.72 ppm is attributed to the methyl carbon of the methyl ester ($COO\underline{C}H_3$).[19]

Degree of esterification

The degrees of esterification of the mango peel pectin are listed in Table 3. All three varieties have high degrees of esterification (DE > 50%). It can be observed that the results from the two different methods have a difference of about 6% but nevertheless are in suitably close agreement.

Conclusions

Waste mango peel, as an exemplar of unavoidable food supply waste, represents an interesting renewable resource that affords pectin with a high degree of esterification. Although not shown here, the resultant pectin can then be re-introduced into the food supply chain as an additive, thus promoting nutrition. In such a way, added beneficial chemical value is gained from unavoidable food supply chain waste, which otherwise may be left to rot or sent for anaerobic digestion. Thus, waste to landfill is minimised, and resource is recovered and reused, promoting a future circular economy, which is commensurate with the UN Sustainable Development Goals.[20]

Thus, we have demonstrated the first subcritical water extraction of pectin from mango peel waste in the absence of any mineral acid. Three different global varieties were studied (Keitt, Sindhri and Kesar) and the highest pectin yield (18.34%) was achieved at a temperature of 175 °C with a flow rate of 3 mL min^{-1} for 15 min from the Kesar variety post ethanol extraction. The degree of esterification of the pectin was determined by both titrimetry and NMR methods and all three varieties have a high DE of over 70%. Meanwhile, some other valuable compounds were also explored *via* ethanol extraction prior to the subcritical water extraction. Fructose, glucose and sucrose are the most abundant sugars in the mango peel waste. Furfural and HMF, which can be used as potential platform molecules, were also obtained as major components in the ethanol extracts from the mango peel.

Acknowledgements

H. X. would like to thank the Department of Chemistry for a Wild Scholarship to pursue Ph.D. studies.

Notes and references

1 T. E. Quested, A. D. Parry, S. Easteal and R. Swannell, *Nutr. Bull.*, 2011, **36**, 460–467.

2 Faostat Mango, http://www.fao.org/faostat/en/#data/QC, accessed on 20/01/2017.

3 N. Berardini, M. Knödler, A. Schieber and R. Carle, *Innovative Food Sci. Emerging Technol.*, 2005, **6**, 442–452.

4 C. M. Ajila, K. A. Naidu, S. G. Bhat and U. J. S. P. Rao, *Food Chem.*, 2007, **105**, 982–988.

5 M. H. A. Jahurul, I. S. M. Zaidul, K. Ghafoor, F. Y. Al-Juhaimi, K. L. Nyam, N. A. N. Norulaini, F. Sahena and A. K. Mohd Omar, *Food Chem.*, 2015, **183**, 173–180.

6 P. Sriamornsak, *Silpakorn Univ. Sci. Technol. J.*, 2003, **3**, 207–228.

7 B. L. Ridley, M. A. O'Neill and D. Mohnen, *Pectins: structure, biosynthesis, and oligogalacturonide-related signaling*, 2001, vol. 57.

8 D. Mohnen, *Curr. Opin. Plant Biol.*, 2008, **11**, 266–277.

9 J. H. Guo, G. W. Skinner, W. W. Harcum and P. E. Barnum, *Pharm. Sci. Technol. Today*, 1998, **1**, 254–261.

10 B. R. Thakur, R. K. Singh, A. K. Handa and M. A. Rao, *Crit. Rev. Food Sci. Nutr.*, 1997, **37**, 47–73.

11 Z. U. Rehman, A. M. Salariya, F. Habib and W. H. Shah, *J. Chem. Soc. Pak.*, 2004, **26**, 73–76.

12 J. Banerjee, R. Vijayaraghavan, A. Arora, D. R. MacFarlane and A. F. Patti, *ACS Sustainable Chem. Eng.*, 2016, **4**, 5915–5920.

13 J. P. Maran, K. Swathi, P. Jeevitha, J. Jayalakshmi and G. Ashvini, *Carbohydr. Polym.*, 2015, **123**, 67–71.

14 A. G. Carr, R. Mammucari and N. R. Foster, *Chem. Eng. J.*, 2011, **172**, 1–17.

15 X. Wang, Q. Chen and X. Lü, *Food Hydrocolloids*, 2014, **38**, 129–137.

16 H. Ueno, M. Tanaka, M. Hosino, M. Sasaki and M. Goto, *Sep. Purif. Technol.*, 2008, **62**, 513–516.

17 A. S. Matharu, J. A. Houghton, C. Lucas-Torres and A. Moreno, *Green Chem.*, 2016, **337**, 695–699.

18 X. Tong, Z. Liu, J. Hu and S. Liao, *Appl. Catal., A*, 2016, **510**, 196–203.

19 A. Sinitsya, J. Copiková and H. Pavliková, *J. Carbohydr. Chem.*, 1998, **17**, 279–292.

20 http://www.un.org/sustainabledevelopment/sustainable-development-goals/, accessed on 20/01/2017.

Faraday Discussions

PAPER

Blend configuration in functional polymeric materials with a high lignin content

Yun-Yan Wang,† Yi-ru Chen and Simo Sarkanen*

Received 28th February 2017, Accepted 6th March 2017
DOI: 10.1039/c7fd00083a

Lignins upgrade the lignocellulosic cell-wall domains in all vascular plants; they embody 20–30% of terrestrial organic carbon. For 50 years, mistaken assumptions about the configuration of lignin have hindered the development of useful polymeric materials with a lignin content above 40 wt%. Now, polymeric materials composed only of methylated softwood lignin derivatives can exhibit better tensile behavior than polystyrene. Marked improvements may be achieved with small quantities (5–10 wt%) of miscible blend components as simple as poly(ethylene glycol). These observations contradict commonly held views about crosslinking or hyper-branching in lignin chains. The hydrodynamic compactness of the macromolecular lignin species arises from powerful noncovalent interactions between the lignin substructures. Individual lignin components undergo association to form macromolecular complexes that are preserved in plastics with a very high lignin content. Material continuity results from interpenetration between the peripheral components in adjoining lignin complexes. Through interactions with the peripheral domains, miscible blend components modulate the strength and ductility of these utterly original lignin-based plastics.

Introduction

A viable solution to the pressing need for lignin valorization[1] may be imminent. Formulations have now been developed for plastics with an 85–100 wt% content of methylated native softwood lignin.[2] Cheap blend components can yield polymeric materials with tensile properties better than those of many in current use. The new lignin-based plastics are simple nano-biomaterials. Associated lignin complexes are the predominant macromolecular species in the solid domains; they are composed of individual components held together by noncovalent forces between interior cofacial aromatic rings.[2] The peripheral chain segments exhibit

Department of Bioproducts and Biosystems Engineering, University of Minnesota, Saint Paul, Minnesota, 55108-6130, USA. E-mail: sarka001@umn.edu

† Current address: Department of Forestry, Wildlife and Fisheries, University of Tennessee, Knoxville, Tennessee, 37996-4542, USA.

a higher incidence of edge-on orientations between aromatic rings, which enable continuity to be established through interlocking between neighboring complexes.

Biorefineries must develop ways of obtaining significant value from co-product lignins in order to achieve economic viability for commodity organic chemical production from plant lignocellulose.[1] Promising conditions can be found for the depolymerization of lignin derivatives into low-molecular-weight chemical precursors.[3] However, an auspicious outcome may be thwarted by the recalcitrance of the polymeric lignin and the complexity of the product mixtures resulting from extensive cleavage.[4] On the other hand, the likelihood of success in transforming lignin derivatives into (inexpensive) biodegradable plastics should be advanced by cognizance of the unique macromolecular characteristics of the starting materials.[5] Regrettably, during the past 60 years a confusing assortment of hypotheses have been proposed to describe the configurations of lignins or readily-produced lignin derivatives. These models include (unbranched) random coils,[6–8] three-dimensionally branched polymer chains,[9] linear oligomers,[10] associated macromolecular complexes,[11] and cleavage products[12] formed from crosslinked networks of lignin chains.[13]

Indeed, covalent crosslinking in lignins has been the most frequently endorsed hypothesis[12] because the polymeric entities in lignin derivatives possess small hydrodynamic volumes.[13] Moreover, the molecular weight distributions of ball-milled lignins (minimally modified with respect to the native biopolymer) were not altered by opportunities for hydrogen bonding.[14] Thus, the (compact) macromolecular species in such lignin preparations were wrongly judged to be covalently discrete components rather than associated complexes.

For the past 40 years, the presumption that lignins are crosslinked macromolecules has dictated what types of formulations may be sanctioned for developing lignin-containing plastics. The strategy for achieving a high modulus and strength became one of covalent reinforcement of the soft-segment matrices with hard-segment lignin domains.[15] Almost without exception, however, it proved impossible to exceed a 40 wt% incorporation limit for lignin in the ensuing materials.[16] The results obtained with compatible lignin-containing polymer blends were very similar.[16] Yet, in a more recent case, three lignin–polybutadiene polyurethane thermoplastics were produced with a 65–75 wt% lignin content, but their tensile strengths were over 10-fold lower than that of polystyrene.[17]

Nevertheless, it has now been established that plastics embodying a simple lignin-derivative content of around 100 wt% can be produced with better tensile behavior than polystyrene.[2] An early indication that this could be achieved came to light on using a blend of high-molecular-weight ethylated methylated kraft lignin with 20 wt% >10 kDa poly(ethylene glycol); the resulting homogeneous material possessed a 51 MPa tensile strength with 6% elongation at break.[18] From the variation in the glass transition temperature (T_g) with composition for various homogeneous blends, it was concluded that the macromolecular species in such simple lignin-based plastics are associated complexes rather than individual (entangled) lignin components.[18] Polyanionic ligninsulfonates have been more challenging as raw materials for producing lignin-based plastics. Apart from being negatively charged (at least initially), the sulfonate groups are large enough to affect the interactions between lignin chains. Thus, 85 wt% ligninsulfonate-based blends with miscible low-T_g polymers can match the tensile behavior of polyethylene,[19] however, the mechanical properties of polystyrene are more difficult, but not impossible, to attain.[2]

It is reported herein that the situation with a typical ball-milled softwood lignin (M_w 3400) is quite different. After methylation, this simple lignin preparation can be transformed through solution casting (on its own) into a polymeric material that manifests better tensile behavior than polystyrene (46 MPa strength, 2% elongation-at-break).[20] In the presence of 5–15 wt% miscible blend components, the methylated ball-milled lignin-based materials can approach 70 MPa in tensile strength as the elongation-at-break draws near to 10%. The comparison is rendered more noteworthy by the fact that the molecular weight of commercial polystyrene is typically ~100-fold greater than that of ball-milled softwood lignin.

Experimental

Materials

Jack pine (*Pinus banksiana*) sapwood (1.5 cm^3 blocks) was ground in a Wiley mill to obtain a 40-mesh particle size. After 48 h of Soxhlet-exaction with acetone, the wood meal was subjected to cooled vibratory ball-milling in 0.3 L jacketed stainless steel grinding vessels (Siebtechnik GmbH, Germany) under N_2 for 48 h. The resulting wood meal was extracted with dioxane–water (96 : 4 v/v) in the form of a stirred suspension, three times consecutively over 96 h. After centrifuging (3000 rpm, 30 min, Beckman J6B), the solvents were removed from the extracts by rotary evaporation. The remaining lignin (18% yield) was treated with 9 : 1 : 4 : 18 v/v/v/v pyridine/acetic acid/water/chloroform; upon removing the solvent mixture, the purified lignin was dissolved in 2 : 1 v/v dichloroethane/ethanol and precipitated from the solution with ether.[21] Acid catalyzed hydrolysis did not liberate chromatographically detectable quantities of monosaccharides from this final product.

A $M_w = 4800$ ($M_w/M_n = 1.4$) ball-milled lignin fraction was obtained by ultrafiltration of the parent preparation in aqueous 0.10 M NaOH through a 10 000 nominal molecular weight cutoff membrane (Amicon YM10); the operation was prolonged until the volume of the permeate had exceeded 10-fold that of the retentate. Ultrafiltration was continued with (triply) distilled water until the pH had fallen to 7.8, whereupon the retentate was freeze-dried.

Dibromobenzophenone (DBBP), poly(ethylene glycol) (PEG), poly(ethylene glycol) methyl ether, polystyrene and tetrabromobisphenol A (TBBP-A) (Sigma-Aldrich) were used as received without further purification. Similarly, poly(-ethylene oxide-*b*-1,2-butadiene-*b*-ethylene oxide) (EBE) M_n PEO(4300)-*b*-PBd(800)-*b*-PEO(4300) (Polymer Source) and dibromoacetophenone (DBAP) (American Custom Chemicals Corp.) were used as received.

Methylation of ball-milled lignin

Ball-milled lignin was dissolved (20 g L^{-1}) in aqueous dioxane (60%) containing 0.10 M NaOH, whereupon dimethyl sulfate (Sigma-Aldrich) was added (2 mL per g of lignin) under N_2. In response to the decreasing pH, the solution was repeatedly basified with 2 M NaOH in aqueous dioxane (60%) to pH ~ 13.5, until the pH stopped falling. Another volume of dimethyl sulfate (2 mL per g of lignin) was introduced and the pH was adjusted as before. The overall procedure was repeated 4 times. Thereafter, the pH of the solution was adjusted to 9.0 with aqueous 0.5 M H_2SO_4, whereupon the dioxane was exhaustively removed through rotary evaporation at 35 °C. The resulting precipitate was thoroughly washed with distilled water and air-dried.

The partially methylated ball-milled lignin was further methylated with diazomethane in a chloroform solution, which had been generated by heating *N*-methyl-*N*-nitroso-*p*-toluenesulfonamide (Diazald, Sigma-Aldrich) in an aqueous KOH–ethoxyethanol–chloroform mixture at 80 °C.

Solution-casting of lignin-based polymeric materials

Methylated ball-milled lignin portions with or without other blend components (0.6 g total weight) were dissolved in 4.0 mL of dimethyl sulfoxide (DMSO) in 10 × 20 mm Teflon molds. After degassing under reduced pressure at 50 °C, each test piece was produced by solution-casting at a suitable temperature and then filing manually to dimensions of a width of 5 mm, a length of 8 mm between the shoulders, and a thickness of ~1 mm.

Tensile tests

The tensile properties of the methylated lignin-based blend materials were characterized by means of stress–strain curves obtained using an Instron model 5542 instrument equipped with a 500 N static load cell. The test pieces were clamped in place with serrated jaws in preparation for the tests employing a crosshead speed of 0.05 mm min^{-1} (Fig. 1). The tensile strength (σ_{max}) and elongation-at-break ($\Delta\varepsilon_b$) were calculated on the basis of the initial sample dimensions.

Differential scanning calorimetry (DSC)

DSC measurements were carried out with a TA Q2000 differential scanning calorimeter (TA Instruments) equipped with Refrigerated Cooling System 90. About 8 mg of each sample was encapsulated in an aluminum pan with a lid put into place after flushing with N_2 in a glove box overnight. Prior to determining the individual values of T_g (Fig. 2D) for the methylated parent ball-milled lignin alone or in blends with poly(ethylene glycol) (PEG), the sample was pre-scanned from 40 to 200 °C with a heating rate of 10 °C min^{-1} to remove any trace of moisture and furnish a uniform thermal history. The second scan was performed from −90 to 200 °C with a cooling rate of 5 °C min^{-1} and a heating rate of 10 °C min^{-1}. The glass transition temperature was taken to be the half-height point of the glass transition region in the DSC thermogram.

In successive thermal cycles, the uncast parent and higher-molecular-weight methylated ball-milled lignins were scanned from 40 to 180 °C and from 80 to 235 °C (Fig. 2A and B), respectively, while the uncast unmethylated parent ball-milled lignin samples were scanned between 80 and 225 °C (Fig. 2C). The heating rate was 10 °C min^{-1} and the samples were kept in an isothermal state for 2 or 5 minutes (as indicated in Fig. 2A–C and 3) on each occasion when the temperature had reached the upper end of the range.

Beyond the effect of the heat-capacity increase, the enthalpy change that characterizes passage of the BML- and MBML-based polymeric materials through the glass-transition region was calculated from the corresponding area under the thermogram relative to the tangent extending from the heat-flow–temperature curve-segment situated below the T_g region. Using TA Instruments software, the onset and end of the T_g (Fig. 2A–C) were determined from the tangents to the thermogram respectively below and above the T_g region, at their intersection points with the tangent passing through the midpoint of the transition. The tangent to each thermogram above T_g was taken to be the one with zero slope. A value of 6 for the constant of proportionality furnishes enthalpy changes in units of J g^{-1} when the temperature rises at a rate of 10 °C min^{-1} (Fig. 3).

X-ray powder diffraction

X-ray diffraction scans were obtained with a Bruker AXS D5005 diffractometer operating in a reflection mode using Cu Kα radiation ($\lambda = 1.542$ Å) and employing a diffracted beam monochromator. The powdered samples were compressed onto a zero background holder. As the sample rotates at a constant angular velocity, the detector (scintillation counter) rotates at double that angular velocity around the sample so that the diffraction angle is always equal to 2θ, where θ is the angle of the incident beam. The scans were obtained within a 5–36.02° range of 2θ with a step size of 0.06° and dwell time of 6 s.

The diffraction patterns from the amorphous MBML-based polymeric materials were satisfactorily fitted to the sums of two Lorentzian functions in the form given by:[22]

$$I(x) = I(0)/(1 + x^2/\text{hwhm}^2),\ x = 2\theta - 2\theta_k, \quad (1)$$

where $I(x)$ is the scattered intensity at an angular distance x from the Bragg angle $2\theta_k$ at the center of the reflection, 2θ is the scattering angle, and hwhm is the half-width at the half-maximum of the peak. The best fits of the two Lorentzian peaks were determined with respect to an algebraic baseline identified with SigmaPlot 10.0 to be below that for the experimental data; the contributions of the two peaks to the overall diffraction pattern were then taken to be proportional to the Lorentzian-peak areas relative to the experimental baseline.

Atomic force microscopy (AFM)

Selected lignin-based plastics and polystyrene pellets were sectioned and trimmed with a razor blade, and then mounted in an AFM specimen holder fitted with a double-D clamp (Mager Scientific Inc.) for ultramicrotomy. The surface of each specimen was smoothed by ultramicrotome-cutting using a 45° glass blade with a step size of 1 μm on a Leica EM UC6 apparatus to produce successive layers. The mirror-like surface was burnished further using a step size of 300 nm several times, and the resulting new surface was subjected to AFM scanning immediately. The lignin-based plastics had been produced by solution-casting while the polystyrene pellets (Sigma-Aldrich) were used as received.

The AFM experiments were performed under air using a Bruker Nanoscope V multimode 8 scanning probe microscope employed in a tapping mode for generating a tip-oscillation-amplitude image (produced in response to interactions with the sample surface). The monolithic silicon probe (Arrow™ NCR, NanoWorld AG) employed in this work features a 160 μm-long cantilever (with a 42 N m^{-1} force constant and 285 kHz resonance frequency) holding a tetrahedral tip with a height of 10–15 μm and a typical radius of curvature that is less than 10 nm.

AFM scans were collected using a Nanoscope 8.15 unit (Bruker), while online plane-fitting for the images during scanning was turned off. A Gwyddion software package (http://gwyddion.net) was employed to process and analyze the output AFM images. Lines in the height image were corrected (*i.e.*, eliminated) by matching the height median. Other reported images were not post-processed. The center-to-center distances between the adjacent nodules in the amplitude image were measured manually.

Estimated dimensions of nodular structures in the AFM images

The separation distances between nearest-neighbor peak maxima in the tapping-mode AFM amplitude images were determined in the first instance manually and, for confirmatory purposes, through one-dimensional height–height correlation functions applied along the fast-scanning axis:[23]

$$H_x(\tau_x) = \frac{1}{N(M-m)} \sum_{l=1}^{N} \sum_{n=1}^{M-m} (z_{n+m,l} - z_{n,l})^2 \qquad (2)$$

where $\tau_x = x_1 - x_2$ and is the distance between two points, and $m = \tau_x/\Delta x$, where Δx is the sampling interval. The height–height correlation functions were evaluated for regions that were not marked by cliffs or ravines between features. As might be expected, the separation distances estimated from the one-dimensional height–height correlation functions were uniformly shorter (by 10%) than the corresponding average diameters measured manually in two dimensions (*vide infra*, Table 1).

Molecular weight estimates for nodular structures in the AFM images

Polystyrene pellets with different molecular weight distributions were employed for comparative reference purposes during these AFM studies of the surfaces of the solid lignin-based polymeric materials. Their ultramicrotome-cut surfaces exhibit a uniform nodular topology (see, for example, Fig. 6), and the diameters of the nodules tend to increase with the molecular weight of the polystyrene preparation (Table 1). It was proposed in relation to a random-coil model[24] that the flexible chain segments in amorphous polymers adopt the same conformations that they possess in dilute solutions under θ conditions (when the solute–solute and solvent–solute interactions cancel one another out). The measured nodule diameter (D) is approximately equal to (the extrapolated value of) $2R_g$, twice the radius of gyration of polystyrene in cyclohexane at the θ temperature (Table 1). With this connection, the R_g (Å) may be estimated, using the empirical equation eqn (3),[25,26] from the weight-average molecular weight (M_w) of each polystyrene sample.

$$R_g = 0.25{M_w}^{0.51} \qquad (3)$$

Correspondingly, the nodules observed in the AFM images of the ultramicrotome-cut polystyrene surfaces may embody random coils that pack to create the characteristic density of the material. Analogously, it is reasonable to suppose that the nodular topology characteristics of the freshly cut surfaces of the lignin-based materials (Fig. 5) arise from partially coalesced associated lignin complexes.

Estimates for the average molecular weight of these macromolecular lignin species can be deduced from the respective relationships between R_g and M_w for polystyrene (eqn (3)) and the (unmethylated) lignin itself (data not shown), which together imply a lower limit to the range that tentatively falls between 54 kDa and 68 kDa. Thus, the degree of association of the macromolecular lignin complexes is expected to be ~20 with respect to the molecular weight distribution of the unmethylated ball-milled lignin preparation deduced from its (aqueous 0.10 M NaOH/Sephadex G100) size-exclusion chromatography elution profile.[19]

Results and discussion

The ball-milled lignin (BML) used as the starting material was isolated and purified[21] from Jack pine (*Pinus banksiana*). Size-exclusion chromatography (Sephadex G100/aqueous 0.10 M NaOH) of the BML revealed a weight-average molecular weight (M_w) of 3400 (ref. 19) and a polydispersity index (M_w/M_n) of 2.5. The quantitative ^{13}C-NMR spectrum[19] of the preparation was essentially identical to that of spruce milled wood lignin;[27] the ratio of the δ_c 55.6 ppm C-3 methoxyl peak area to the spectral area in the aromatic region (~108–156 ppm) indicated that the BML consisted of guaiacyl units almost exclusively. Consecutive treatment with dimethyl sulfate and diazomethane brought about (*inter alia*) a 1.31-fold increase in the proportion of aromatic methoxyl groups in the derivative as a result of C-4 hydroxyl methylation.[19] This is consistent with the estimated frequency of phenolic hydroxyl groups in spruce milled wood lignin.[27] A higher molecular weight ball-milled lignin fraction (M_w 4800, M_w/M_n 1.4) was methylated in the same way as the parent BML.

Tensile and thermal properties

Solution-casting of the methylated ball-milled lignin (MBML) in dimethyl sulfoxide (DMSO) at 140 °C furnished polymeric materials that, with ~50 MPa tensile strengths and 7–8% elongations at break (Fig. 1A), could readily surpass polystyrene in terms of mechanical behavior. The tensile strengths of the 100% MBML-based plastics were reduced significantly by the loss of (plasticizing) methylated lignin oligomers during casting (Fig. 1A). A similar result was obtained after removing lower-molecular-weight lignin components by ultrafiltration from the parent BML prior to methylation and casting (Fig. 1C).

A number of non-lignin components, when introduced at low levels to create homogeneous blends, are able to improve the tensile strength or ductility of these methylated lignin-based plastics. For example, 10 wt% levels of the flame retardant 3,3′,5,5′-tetrabromobisphenol A (TBBP-A) engender a ~15% increase in the tensile strength (Fig. 1B). A decisive effect from electron-withdrawal by the sizable Br-substituents in TBBP-A had been anticipated, but compounds like 3′,5′-dibromoacetophenone (DBAP) and 4,4′-dibromobenzophenone (DBBP) did not act in a similar capacity (Fig. 1B).

A blend of MBML with 15 wt% of a triblock copolymer EBE, namely poly(ethylene oxide-*b*-1,2-butadiene-*b*-ethylene oxide), resulted in a tensile strength approaching 70 MPa with a 7% elongation at break (Fig. 1C). Increasing the EBE

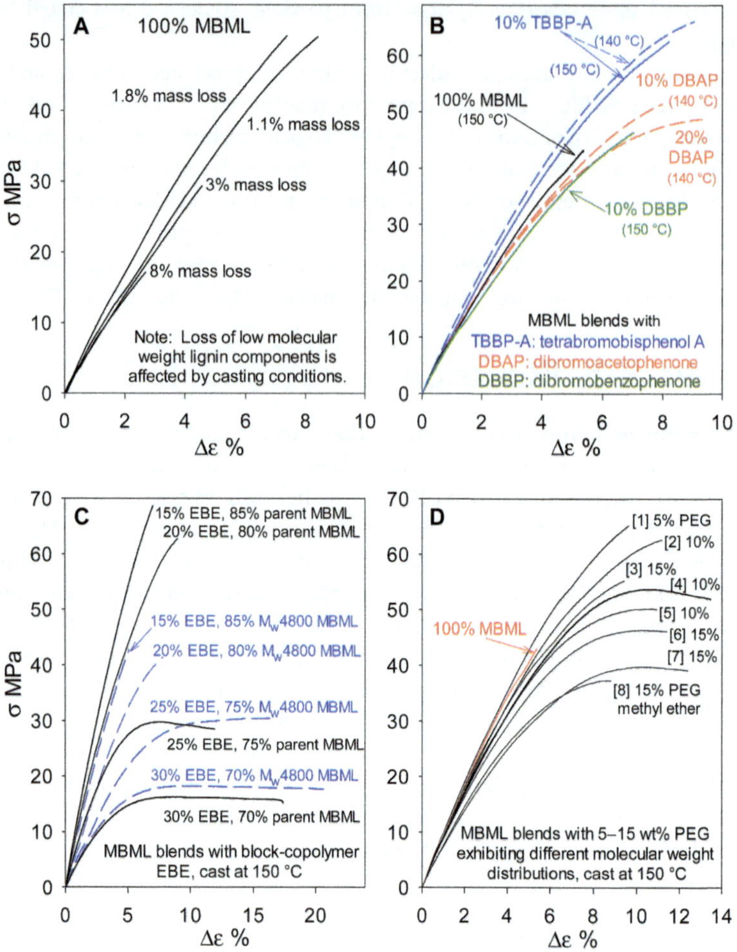

Fig. 1 Tensile behavior of the methylated ball-milled lignin (MBML)-based polymeric materials. (A) 100% MBML-based plastics prepared by solution casting at 140 °C. (B) Blends of 80–90 wt% MBML with bromo-substituted aromatic compounds cast at 140 or 150 °C. (C) Progressive plasticization of parent and higher-molecular-weight MBML-based materials with triblock poly(ethylene oxide-*b*-1,2-butadiene-*b*-ethylene oxide) (EBE). (D) Plasticization effects of 5–15 wt% of poly(ethylene glycol)s (PEGs) with different molecular weight distributions on the MBML-based blends: 100% MBML; [1] 5% PEG, M_n 400 Da; [2] 5% PEG, M_n 400 Da and 5% PEG, M_n 10 kDa; [3] 15% PEG, M_n 4.6 kDa; [4] 5% PEG, M_n 400 Da and 5% PEG M_n 4.6 kDa; [5] 10% PEG, M_n 400 Da; [6] 15% PEG, M_n 2 kDa; [7] 15% PEG, M_n 10 kDa; [8] 15% poly(ethylene glycol) methyl ether, M_n 5 kDa.

level to 20, 25 and 30 wt% brought about progressive decreases in the tensile strength with concomitant extensions of the elongation at break for the material. Interestingly, blends of a higher-molecular-weight (M_w 4800, M_w/M_n 1.4) polydisperse MBML fraction with 15 and 20 wt% EBE exhibited substantially worse tensile behavior than the corresponding materials created with the parent MBML (Fig. 1C). This contrast presumably arises from plasticizing effects engendered by the low-molecular-weight methylated lignin components. On the other hand, the ductility of the 25 and 30 wt% EBE blends with the higher-molecular-weight MBML fraction was significantly better than that obtained when the parent MBML was used. Such comparisons may prompt critical insight into how formulations for lignin-based plastics are best optimized.

The impact of having small proportions of poly(ethylene glycol) (PEG) in blends with the parent MBML is striking and varied. Both the quantity and molecular weight distribution of the PEG affect the tensile behavior in significant ways (Fig. 1D). Three observations stand out. First, just 5 wt% of $M_n = 400$ PEG results in a material with a 65 MPa tensile strength and an elongation at break above 9%. Second, 5 wt% of $M_n = 400$ PEG and an equal weight of $M_n = 4600$ PEG together produce a blend with a tensile strength of 53 MPa and >13% elongation at break. Finally, in these plastics, hydrogen-bonding involving the PEG hydroxyl groups, although they are few in number, plays a beneficial role. Thus, the material containing 15 wt% of $M_n = 5000$ PEG methyl ether exhibits the worst tensile behavior among the PEG-containing blends examined (Fig. 1D).

Interpreting the results of changes in the composition of lignin-based plastics requires the significance of the diagnostic parameters employed to be clearly understood. This is particularly important for evaluating differential scanning calorimetry (DSC) thermograms of MBML preparations and their blends (Fig. 2). The thermograms of the uncast parent MBML undergo systematic displacement toward higher temperatures during successive thermal cycles (Fig. 2A); the apparent glass-transition temperature (T_g) persistently rises during the process. A similar change in the glass-transition temperature occurs with the uncast higher-molecular-weight MBML (Fig. 2B), although the range over which the T_g increases is situated almost 40 °C higher than that observed with the parent MBML.

As the temperature passes through the glass transition temperature region of a polymer, there is an increase in heat capacity caused by the onset of long-range coordinated chain-segmental motions in the rubbery plateau region. A rise in the T_g resulting from thermal cycling of a lignin derivative past the glass-transition region may occur for one of two reasons that are distinguishable by way of the accompanying enthalpy changes. An enthalpy-change reduction taking place during traversal across the transition region could signify a growing covalent-crosslink density.[28,29] On the other hand, greater stability arising from more favorable arrangements of the interacting lignin components[2] in the glassy state would have less of a tendency to diminish the enthalpy change that accompanies the transition to the rubbery plateau. The magnitude of this enthalpy change was calculated from the area between the glass-transition onset and end under the thermogram relative to the tangent extending from the heat-flow–temperature curve-segment that approaches the T_g region from below.

During DSC thermal cycling, the enthalpy change characteristics of the glass transition region in the 10 °C min^{-1} thermograms (Fig. 3) show little dependence on the rising T_g of the ball-milled lignin (BML) (Fig. 2C). Nor are they appreciably affected by methylating the preparation to produce MBML. Indeed, there is only a slight difference between the parent MBML and a higher molecular weight fraction prepared from it (Fig. 3). Such observations indicate that little or no covalent crosslinking occurs during DSC thermal cycling; it is likely that this interpretation may also be extended to lignin-based plastics produced by solution casting from DMSO (Fig. 1A).

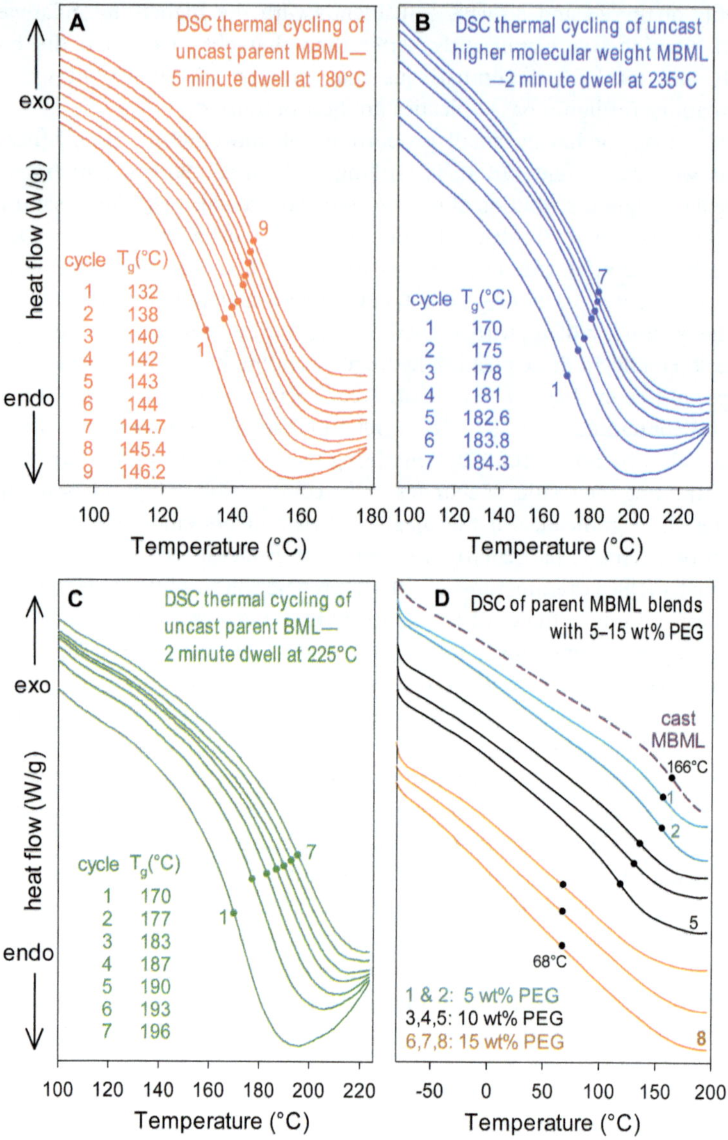

Fig. 2 DSC thermograms characterizing T_g variations of methylated and unmethylated ball-milled lignin-based polymeric materials. Successive thermal cycles of (A) uncast parent methylated ball-milled lignin (MBML), (B) uncast higher-molecular-weight M_w 4800 MBML and (C) uncast unmethylated ball-milled lignin (BML). (D) Thermograms of cast parent MBML alone and in blends containing 5–15 wt% poly(ethylene glycol)s (PEGs) with different molecular weight distributions: [1] 5% PEG, M_n 400 Da ($T_g = 157$ °C); [2] 5% PEG, M_n 1 kDa ($T_g = 156$ °C); [3] 5% PEG, M_n 400 Da and 5% PEG, M_n 1 kDa ($T_g = 136$ °C); [4] 5% PEG, M_n 400 Da and 5% PEG, M_n 4.6 kDa ($T_g = 131$ °C); [5] 10% PEG, M_n 400 Da ($T_g = 119$ °C); [6] 15% PEG, M_n 2 kDa ($T_g = 68$ °C); [7] 15% PEG, M_n 4.6 kDa ($T_g = 67$ °C); [8] 15% PEG, M_n 10 kDa ($T_g = 68$ °C).

The foregoing findings depend on effectual dwell times at the upper-temperature limit of each cycle. It is under these circumstances that the results are clearly at odds with the formation of crosslinks between the lignin

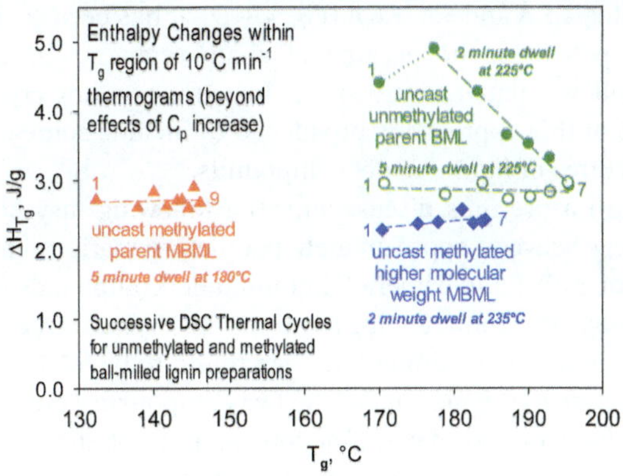

Fig. 3 Enthalpy changes (beyond the requirement for a heat-capacity (C_p) increase) accompanying passage through the T_g region of the 10 °C min^{-1} thermograms that reflect DSC thermal cycling of the uncast methylated parent ball-milled lignin (MBML), uncast higher molecular weight MBML and uncast unmethylated parent ball-milled lignin (BML).

components during successive laps of the program. For example, when held for 2 minutes instead of 5 at 225 °C, the unmethylated BML (in which hydrogen bonding may strengthen the intermolecular forces between lignin components) can experience enthalpy changes that approach numbers that are 1.7-fold larger initially before falling to constant values that match the thermal cycles incorporating 5 minute dwell times (Fig. 3).

Thus, the rise in T_g during thermal cycling reflects an increase in the degree of association between the lignin components which is facilitated by enhanced thermal motions. The corresponding associative interactions will also be affected significantly through the casting of mixtures that form plastics with a very high MBML content (Fig. 2D). Thus, homogeneous blends with small quantities of low-T_g polymers such as PEG (*cf.* Fig. 1D) exhibit dramatic reductions in the T_g. Actually, the transition region becomes so broad for MBML-based plastics containing 15 wt% PEG that its position becomes uncertain and a value for the T_g in the thermogram becomes difficult to assign (Fig. 2D). Despite the vague DSC transitions for MBML blends with 15 wt% PEG, the interior domains of the methylated lignin complexes (observed using X-ray powder diffraction patterns) maintain their integrity quite convincingly (see Fig. 4). Consequently, it is apparent that the T_g values of such MBML-based plastics are governed by the thermal motions of the peripheral chain segments in these compact assemblies of individual lignin components. Indeed, the X-ray powder diffraction pattern of a cast softwood (*Douglas fir*) ligninsulfonate preparation (M_w = 9700, M_w/M_n = 5.0) was found to reflect domains consisting only of cofacially disposed aromatic rings, and no T_g was discernible from the corresponding DSC thermogram.[19]

Macromolecular species in the lignin-based materials

The X-ray powder diffraction patterns of the MBML-based materials blended with 0–15 wt% PEG are composed of two broad Lorentzian peaks reflecting contrasting distributions of the separation distances between the aromatic rings (primarily) in the lignin components (Fig. 4). These peaks have maxima at equivalent Bragg spacings (d) of 4.0–4.1 Å and 5.3–5.6 Å (Fig. 4B–D). It has been suggested that, for semicrystalline polymers, the packing of chain segments in the amorphous domains may follow a pattern intrinsic to the corresponding crystal structure.[30] Here, the scope of this approach is broadened to include some inferences from the crystal structures of lignin-model compounds.

In accord with a previous discussion,[18] the following may be recalled. The average d-spacing between approximately parallel aromatic rings in the crystal structures of dimeric[31,32] and trimeric[33] lignin-model compounds is ~4.0 Å, while that between edge-on aromatic rings within the crystal structure of a (non-phenolic) β-O-4 lignin-model dimer has been found to be ~5.3 Å.[34] These values are remarkably close to the maxima of the two component Lorentzian peaks (4.1 and 5.5 Å) in the X-ray powder diffraction patterns of the cast MBML-based plastics (Fig. 4B–D). Thus, the two Lorentzian peaks may be assigned, in part, to random distributions[35] of interior cofacial aromatic-ring arrangements within the associated complexes (centered at $2\theta \approx 21.8°$, $d \approx 4.1$ Å) and corresponding edge-on orientations of equivalent structures within the peripheral components

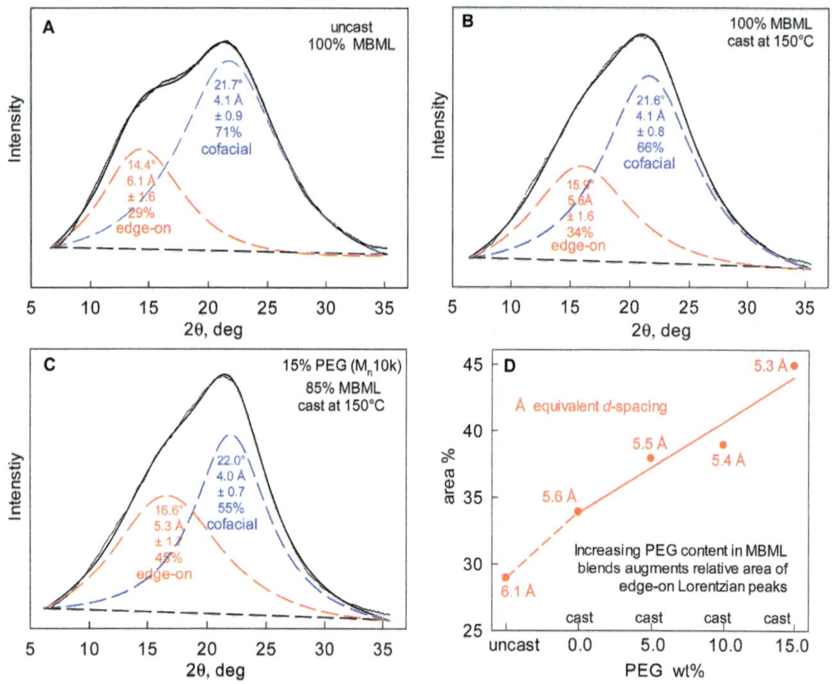

Fig. 4 X-ray powder diffraction patterns of methylated ball-milled lignin (MBML)-based materials. X-ray powder diffraction patterns of (A) uncast 100% MBML, (B) cast 100% MBML and (C) a MBML blend with 15 wt% poly(ethylene glycol) (PEG). The X-ray diffraction patterns of the amorphous polymer blends were analyzed using two Lorentzian functions $I(x) = I(0)/(1 + x^2/\mathrm{hwhm}^2)$ in which $x = 2\theta - 2\theta_K$, where $I(x)$ is the scattered intensity at x from the Bragg angle $2\theta_K$ for the peak, 2θ is the scattering angle, and hwhm is the half-width at the half-maximum of the peak. (D) Relative areas of the lower-θ Lorentzian component peaks in the X-ray powder diffraction patterns of methylated ball-milled lignin-based material blends with an increasing PEG content (each point depicts the average area and equivalent d-spacing for blends with the same amount of PEG).

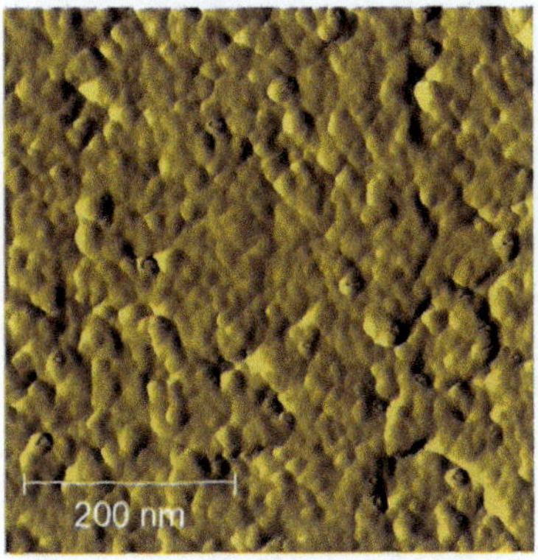

Fig. 5 Packing of associated complexes in the lignin-based polymeric material. Tapping-mode AFM amplitude image of the ultramicrotome-cut surface of methylated ball-milled lignin cast alone from a DMSO solution at 150 °C. The distribution of manually measured distances between the peak maxima (13 ± 3 nm) is consistent with the corresponding parameters determined from the height–height correlation function for the image (Table 1).

(centered at $2\theta \approx 16.2°$, $d \approx 5.5$ Å). It should be kept in mind that, when blended with MBML, amorphous aliphatic PEG contributes very little to the X-ray powder diffraction pattern of the polymeric material; it exhibits a broad peak with a maximum around $2\theta \approx 23°$.

The Lorentzian peak encompassing edge-on aromatic rings undergoes a marked shift (8% between Fig. 4A and B) to a smaller *d*-spacing (from 6.1 to 5.6 Å) simply as a result of casting the MBML powder to form a cohesive material at 150 °C; its area concomitantly increases significantly (by 17% in Fig. 4B relative to 4A), presumably because continuity is established between adjoining complexes in the 100% MBML-based polymeric material. The area of the edge-on Lorentzian peak then increases by another 38% as a result of blending with PEG at levels reaching 15 wt% (Fig. 4C); the effect is accompanied by a further (5%) decrease in

Table 1 Comparison of the manually measured distances (D) between peak maxima in adjoining nodules with the corresponding parameters (τ_x) determined from one-dimensional height–height correlation functions for AFM-images[a] of the MBML-based material and polystyrene samples

Sample	Measured D^b, nm	$\tau_x{}^b$, nm	M_w
MBML (Fig. 5)	12.8 ± 3.3	11.5 ± 3.1	68 kDa[c]
PS M_n 190 kDa	23.3 ± 6.1	21.1 ± 1.6	170 kDa[d]
PS M_n 280 kDa	25.3 ± 6.6	22.5 ± 3.8	200 kDa[d]
PS M_n 350 kDa (Fig. 6)	35.0 ± 10.4	31.4 ± 6.1	380 kDa[d]

[a] Regions unmarked by cliffs or ravines between features on the freshly ultramicrotome-cut surfaces of the methylated ball-milled lignin (MBML)-based material and polystyrene (PS) samples. [b] The ranges of dimensions should not be construed as standard errors. [c] Estimated from the empirical relationship between R_g and M_w determined through light-scattering measurements of ultrafiltered paucidisperse ball-milled lignin fractions in solution. [d] Estimated from eqn (3).

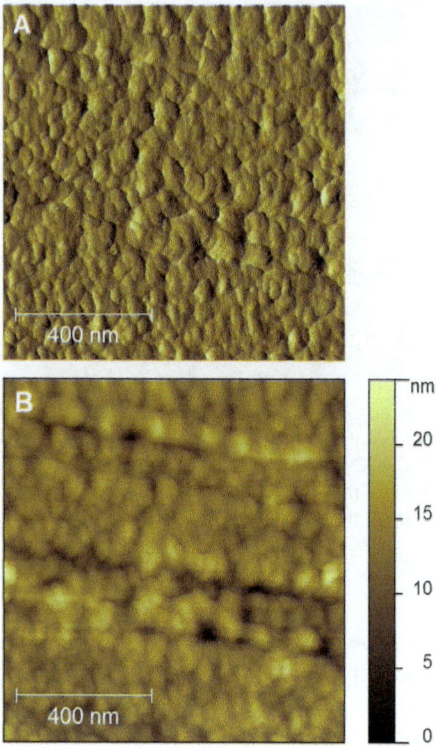

Fig. 6 Tapping-mode AFM (A) amplitude and (B) height images (1 μm × 1 μm) of the ultramicrotome-cut surface of a 350 kDa M_w polystyrene pellet. Ultramicrotomy was performed at −60 °C. The nodule diameter (35.0 ± 10.4 nm) represents the average of 775 center-to-center distances between adjacent nodules in image (A).

the *d*-spacing at the peak maximum. On the other hand, there is little change in the *d*-spacing for the center of the Lorentzian peak representing the interior cofacial aromatic rings. However, a 5% narrowing of the width of this peak is observed over the entirety of the physicochemical changes that take place as a corresponding (22% overall) reduction in peak area occurs.

These findings suggest that continuity between neighboring associated MBML complexes requires a greater proportion of edge-on arrangements of the aromatic rings among the peripheral lignin components (Fig. 4D). The distributions of the interacting aromatic-ring configurations in the cast MBML-based plastics result in a greater overall stability for the polymeric-material blends relative to the uncast MBML. The small but significant (13%) reduction in the edge-on *d*-spacing (from 6.1 Å in uncast MBML to 5.3 Å in the blend with 15 wt% PEG) reflects a progressive increase in stabilization energy for the peripheral components of the associated complexes. This fits with the virtual disappearance of the DSC thermal transition through which T_g should be identifiable in the MBML blend containing 15 wt% PEG (Fig. 2D).

Atomic force microscopy (AFM) was employed in the tapping mode to probe the surface of the parent MBML material that was solution cast at 150 °C. The surface of the material was created through ultramicrotome-cutting with a 45°

glass blade. The path of the blade during microtomy was expected to pass through regions with the lowest densities of covalent and strong noncovalent bonds,[36] *i.e.* whenever possible, between the associated macromolecular MBML complexes in the polymeric material. The distances between adjacent local maxima of the nodular structures on the surface (Fig. 5) were determined to be consistent with a value of 13 $\pm$ 3 nm for their average diameter.

The topology in the AFM image represents a combination of the geometry of the MBML surface features and the smoother trace made by the oscillating tip with an $\sim$8 nm radius. Thus, the shapes of the nodules are affected by tip convolution,[37] but a magnitude of >13 nm for the average separation distance between these features (equivalent to their average diameter) will not be erroneous to a substantial extent. Consequently, the AFM image in Fig. 5 depicts a confirmatory representation of the manner in which the associated complexes in the cast MBML material are packed. It provides support for the interpretation presented here of the X-ray powder diffraction patterns in Fig. 4. The estimate provided for the average diameter of these macromolecular lignin complexes is reasonably sound. The corresponding AFM analysis of a 350 kDa polystyrene sample resulted in a molecular-weight estimate that differs from the expected value by less than 10% (Table 1 and Fig. 6). Thus, the mechanical behavior of plastics with the highest attainable lignin content depends pre-eminently on the interactions between peripheral components circumscribing the surface domains of adjoining lignin complexes.

Conclusions

Plastics with a very high lignin-derivative content differ from typical polymeric materials in that they are composed of well-defined associated complexes rather than individual polymer molecules. The more stable inner domains of the lignin complexes embody cofacially offset configurations of interacting aromatic rings,[38] while there are higher frequencies of edge-on arrangements[38] in the peripheral regions. Material continuity between adjoining complexes is established through these peripheral domains, which also interact preferentially with miscible blend components. Thus, blend components strongly influence the mechanical properties of the lignin-based polymeric materials that were formed, as continuity emerges at the molecular level through sifting in adjacent peripheral domains. The capacity of these peripheral domains to expand has been revealed through the X-ray powder diffraction patterns of the lignin-based blend formulations. The most reliable manifestation of this is reflected in the reduced area of the Lorentzian peak emanating from the cofacial aromatic rings that maintain inner-domain integrity in the associated lignin complexes.

Acknowledgements

This work was funded by Agriculture and Food Research Initiative Grant no. 2011-67009-20062 from the USDA National Institute of Food and Agriculture, and a Subaward (115808 G002979) from the "Northwest Advanced Renewables Alliance" led by Washington State University and supported by the Agriculture and Food Research Initiative Competitive Grant no. 2011-68005-30416 from the USDA National Institute of Food and Agriculture. X-ray powder diffraction studies and atomic force microscopy were carried out in the Characterization Facility at the University of Minnesota, which receives partial support from NSF through the MRSEC program.

References

1 A. J. Ragauskas, G. T. Beckham, M. J. Biddy, R. Chandra, F. Chen, M. F. Davis, B. H. Davison, R. A. Dixon, P. Gilna, M. Keller, P. Langan, A. K. Naskar, J. N. Saddler, T. J. Tschaplinski, G. A. Tuskan and C. E. Wyman, *Science*, 2014, **344**(6185), 1246843, DOI: 10.1126/science.1246843.
2 S. Sarkanen, Y.-r. Chen and Y.-Y. Wang, *ACS Sustainable Chem. Eng.*, 2016, **4**, 5223–5229.
3 L. Shuai, M. Talebi Amiri, Y. M. Questell-Santiago, F. Héroguel, Y. Li, H. Kim, R. Meilan, C. Chapple, J. Ralph and J. S. Luterbacher, *Science*, 2016, **354**(6310), 329–333.
4 J. G. Linger, D. R. Vardon, M. T. Guarnieri, E. M. Karp, G. B. Hunsinger, M. A. Franden, C. W. Johnson, G. Chupka, T. J. Strathmann, P. T. Pienkos and G. T. Beckham, *Proc. Natl. Acad. Sci. U. S. A.*, 2014, **111**, 12013–12018.
5 S. I. Falkehag, *Appl. Polym. Symp.*, 1975, **28**, 247–257.
6 J. Moacanin, V. F. Felicetta, W. Haller and J. L. McCarthy, *J. Am. Chem. Soc.*, 1955, **77**, 3470–3475.
7 S. Sarkanen, D. C. Teller, E. Abramowski and J. L. McCarthy, *Macromolecules*, 1982, **15**, 1098–1104.
8 J. Mlynár and S. Sarkanen, *ACS Symp. Ser.*, 1996, **635**, 379–400.
9 O. H.-H. Hsu and W. G. Glasser, *Wood Sci. Technol.*, 1976, **9**, 97–103.
10 C. Crestini, F. Melone, M. Sette and R. Saladino, *Biomacromolecules*, 2011, **12**, 3928–3935.
11 Y. Li and S. Sarkanen, in *Biodegradable Polymers and Plastics*, ed. E. Chiellini and R. Solaro, Kluwer Academic/Plenum, New York, 2003, pp. 121–139.
12 J. F. Yan, F. Pla, R. Kondo, M. Dolk and J. L. McCarthy, *Macromolecules*, 1984, **17**, 2137–2142.
13 A. Rezanowich and D. A. I. Goring, *J. Colloid Sci.*, 1960, **15**, 452–471.
14 K. H. Ekman and J. J. Lindberg, *Pap. Puu*, 1966, **4**, 241–244.
15 V. P. Saraf, W. G. Glasser, G. L. Wilkes and J. E. McGrath, *J. Appl. Polym. Sci.*, 1985, **30**, 2207–2224.
16 Y.-r. Chen and S. Sarkanen, *Cellul. Chem. Technol.*, 2006, **40**(3–4), 149–163 and references therein.
17 T. Saito, J. H. Perkins, D. C. Jackson, N. E. Trammel, M. A. Hunt and A. K. Naskar, *RSC Adv.*, 2013, **3**, 21832–21840.
18 Y. Li and S. Sarkanen, *Macromolecules*, 2005, **38**, 2296–2306.
19 Y.-Y. Wang, Y.-r. Chen and S. Sarkanen, *Green Chem.*, 2015, **17**, 5069–5078.
20 *Tensile Testing*, ed. J. R. Davis, ASM International, Materials Park, OH, 2nd edn, 2004, pp. 137–154.
21 K. Lundquist, in *Methods in Lignin Chemistry*, ed. S. Y. Lin and C. W. Dence, Springer-Verlag, New York, NY, 1992, pp. 65–70.
22 N. S. Murthy and H. Minor, *Polymer*, 1990, **31**, 996–1002.
23 P. Klapetek, *Quantitative Data Processing in Scanning Probe Microscopy*, William Andrew Elsevier, Burlington, MA, 2013, p. 93.
24 P. J. Flory, *J. Macromol. Sci., Phys.*, 1976, **B12**, 1–11.

25 P. C. Hiemenz and T. P. Lodge, *Polymer Chemistry*, CRC Press, Boca Raton, FL, 2nd edn, 2007, p. 233.
26 Y. Miyaki, Y. Einaga and H. Fujita, *Macromolecules*, 1978, **11**, 1180–1186.
27 E. A. Capanema, M. Y. Balakshin and J. F. Kadla, *J. Agric. Food Chem.*, 2004, **52**, 1850–1860.
28 L. Shan, C. G. Robertson, K. N. E. Verghese, E. Burts, J. S. Riffle, T. C. Ward and K. L. Reifsnider, *J. Appl. Polym. Sci.*, 2001, **80**, 917–927.
29 M. Rico, J. López, R. Bouza and R. Bellas, *J. Therm. Anal. Calorim.*, 2011, **105**, 599–606.
30 N. S. Murthy, H. Minor, C. Bednarczyk and S. Krimm, *Macromolecules*, 1993, **26**, 1712–1721.
31 R. Stomberg and K. Lundquist, *Acta Chem. Scand., Ser. B*, 1987, **41**, 304–309.
32 K. Lundquist and R. Stomberg, *Holzforschung*, 1988, **42**, 375–384.
33 J.-P. Roblin, H. Duran, E. Duran, L. Gorrichon and B. Donnadieu, *Chem.–Eur. J.*, 2000, **6**, 1229–1235.
34 R. Stomberg and K. Lundquist, *Acta Chem. Scand., Ser. A*, 1986, **40**, 705–710.
35 F. H. Chung and R. W. Scott, *J. Appl. Crystallogr.*, 1973, **6**, 225–230.
36 P. J. Aspbury and W. C. Wake, *Br. Polym. J.*, 1979, **11**, 17–27.
37 D. Haba, J. Kaufmann, A. J. Brunner, K. Resch and C. Teichert, *Polymer*, 2014, **55**, 4032–4040.
38 Y.-r. Chen and S. Sarkanen, *Phytochemistry*, 2010, **71**, 453–462.

Faraday Discussions

PAPER

New bio-based monomers: tuneable polyester properties using branched diols from biomass†

Sacha Pérocheau Arnaud,[a] Linglin Wu,[b] Maria-Angelica Wong Chang,[b] James W. Comerford,[a] Thomas J. Farmer, *[a] Maximilian Schmid,[a] Fei Chang,[b] Zheng Li[b] and Mark Mascal *[b]

Received 6th February 2017, Accepted 7th March 2017
DOI: 10.1039/c7fd00057j

A family of monomers, including 2,5-hexandiol, 2,7-octandiol, 2,5-furandicarboxylic acid (FDCA), terephthalic acid (TA), and branched-chain adipic and pimelic acid derivatives, all find a common derivation in the biomass-derived platform molecule 5-(chloromethyl) furfural (CMF). The diol monomers, previously little known to polymer chemistry, have been combined with FDCA and TA derivatives to produce a range of novel polyesters. It is shown that the use of secondary diols leads to polymers with higher glass transition temperatures (T_g) than those prepared from their primary diol equivalents. Two methods of polymerisation were investigated, the first employing activation of the aromatic diacids *via* the corresponding diacid chlorides and the second using a transesterification procedure. Longer chain diols were found to be more reactive than the shorter chain alternatives, generally giving rise to higher molecular weight polymers, an effect shown to be most pronounced when using the transesterification route. Finally, novel diesters with high degrees of branching in their hydrocarbon chains are introduced as potential monomers for possible low surface energy materials applications.

Introduction

Current industrial processing methods enable mega-scale production of diverse classes of polymeric materials with a wide range of properties. The plastics industry in the UK alone had a turnover of £23.5 billion in 2015, producing 1.7 MT

[a]*Green Chemistry Centre of Excellence, Department of Chemistry, University of York, Heslington, York, YO10 5DD, UK. E-mail: thomas.farmer@york.ac.uk*

[b]*Department of Chemistry, University of California Davis, 1 Shields Avenue, Davis, CA 95616, USA. E-mail: mjmascal@ucdavis.edu*

† Electronic supplementary information (ESI) available: Additional supporting figures and tables: details of the characterisation of the polymers (NMR spectra, TGA traces, DSC traces); description of the nomenclature used for naming the polymers; characterisation of the new monomers. Raw experimental data available on request at DOI: 10.15124/20fccc77-5902-42cc-adb9-8e58cf27bac0. See DOI: 10.1039/c7fd00057J

of plastics and employing some 170 000 staff distributed across 5200 manufacturers.[1] However, the production of industrial polymers maintains a disquieting reliance on non-renewable fossil resources, raising concerns regarding the long-term sustainability of this key sector of the chemical industry. Markets in this area have tended to be somewhat conservative, focusing their development around the use of conventional and readily available monomers, such as light olefins, acrylates, styrene, butadiene, ethylene glycol and terephthalic acid, which together account for the great majority of the plastics market. Many of these monomers and their corresponding polymers were first prepared and characterised in the infancy of the petrochemical revolution in the early 20th century.

The last decade however has witnessed the introduction of a wave of new monomers derived from biomass, the unconventional structures of which broaden the range of accessible polymer properties while also supplying a renewable alternative to the use of petroleum as an industrial feedstock.[2] There are already a number of well-established bio-based plastics on the market, such as polylactic acid and polybutylene succinate, but several other systems have been described, and the diversity of chemical functionality in bio-based platform molecules leaves many more to be explored. In 2015, Becer and Isikgor published a comprehensive review of current lignocellulose-derived materials which captured the depth of the functional space available for polymer chemists to explore within this platform.[3] Lignocellulose is heavily oxygenated compared to petroleum-based feeds, and this allows oxygen to be imported from biomass-derived platform molecules through to the monomer and carried finally into the polymer.[4] Herein, we report in detail on the synthesis of novel polyesters from monomers having a common precursor in the form of biomass-derived 5-(chloromethyl)furfural (CMF) **1**. We further introduce two new bio-based diesters that showcase the versatility of the CMF derivative levulinic acid **5** as a source of novel monomers.[5]

CMF **1** is an up and coming platform molecule that is functionally analogous to the well-known fructose derivative 5-(hydroxymethyl)furfural (HMF) but, unlike HMF, can be produced in high yield directly from raw biomass.[6] We have previously demonstrated that CMF can be converted into 2,5-furandicarboxylic acid (FDCA) esters **4** *via* the diacid chloride **3** (Scheme 1).[7] FDCA is currently being commercialized as a renewable substitute for terephthalic acid (TA) **12**, which is produced globally from petroleum on an enormous scale, and which itself can be derived renewably from CMF by direct reduction to 2,5-dimethylfuran **8**,[8] followed by cycloaddition with ethylene to *para*-xylene **11** and ultimate oxidation to **12**.[9,10] Derivatives of FDCA and TA serve as the diacids used in this study.

We have also recently described how the CMF derivative levulinic acid **5** can be coupled electrochemically to give 2,7-octanedione **6**, the hydrogenation of which produces the monomer 2,7-octanediol **7**.[11] 2,5-Hexanediol **10** is likewise derived from its corresponding dione **9**, which is the hydration product of 2,5-dimethylfuran **8**, itself a reduction product of CMF.[12] While secondary diol **7** is virtually unknown to polymer chemistry,[13] diol **10** has seen limited use in materials applications, often in amorphous polymers of low molecular weight.[7a,14] These α-branched diols, along with commercial 2,3-butanediol **14** (ref. 15) and dianhydro-D-glucitol (isosorbide) **15**,[16] are paired with the above mentioned diacids to produce the novel, renewable polyesters which are the focus of this work.

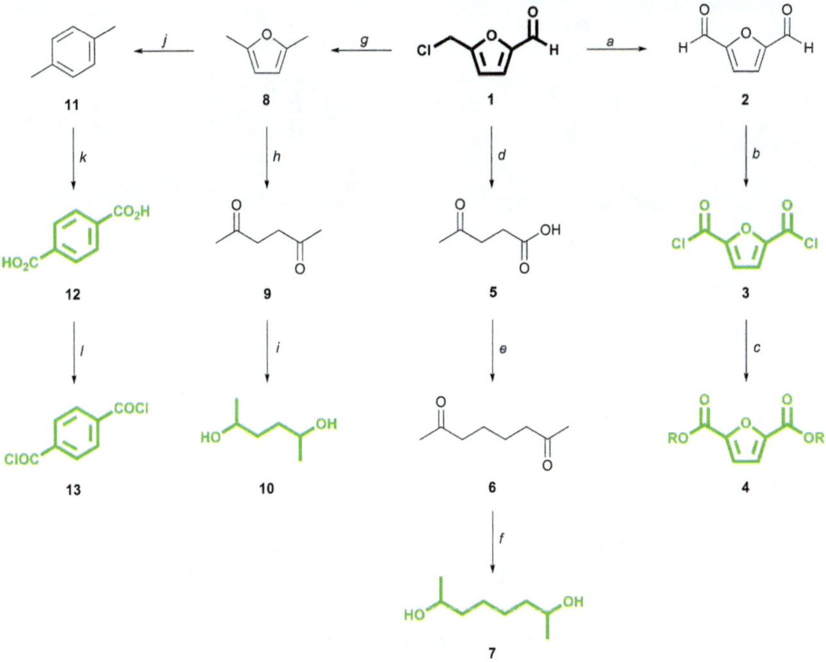

Scheme 1 Conversion of 5-(chloromethyl)furfural (1) to monomers (in green). (a) DMSO, 150 °C, 18 hours, 81% yield;[7b] (b) *t*-BuOCl, 24 hours;[7a] (c) EtOH, 50 °C, 6 hours; 76% yield over 2 steps from **2**;[7a] (d) H_2O, 190 °C, 20 min, 91% yield;[6b] (e) e^-, KOH/MeOH, Pt–Pt, undivided cell, 65% yield;[11] (f) 12 atm H_2, Pd/C, aq. KOH, 80 °C, 94% yield;[11] (g) BuOH, cat. H^+, then H_2, Pd/C, 80% yield;[8] (h) H_2O, 250 °C, 30 min, quantitative yield;[12b] (i) H_2O, H_2, 3 MPa, Pt/C, 120 °C, 2 hours, 80% yield;[12c] (j) ethylene (62 bar), P-BEA or Zr-BEA, 250 °C, 24 h, 97% or 89% yield, respectively;[9] (k) O_2, cat. >90% yield;[10a] (l) $SOCl_2$, DMF, quantitative.[10b]

Results and discussion

Polyesters *via* diacid chlorides

Secondary diols possess intrinsically lower reactivity than α,ω-diols, and the high boiling points of diols like 7 hamper the common strategy of using them in excess to drive the polymerization reaction forward. Both of these issues can, however, be circumvented by the use of reactive diacid chlorides (Scheme 2).[16b]

The method involves the use of a 1 : 1 stoichiometric ratio of diol to diacid chloride in a suitable solvent and with an excess of pyridine to neutralise the HCl condensate. Such conditions have, for example, been successfully employed for

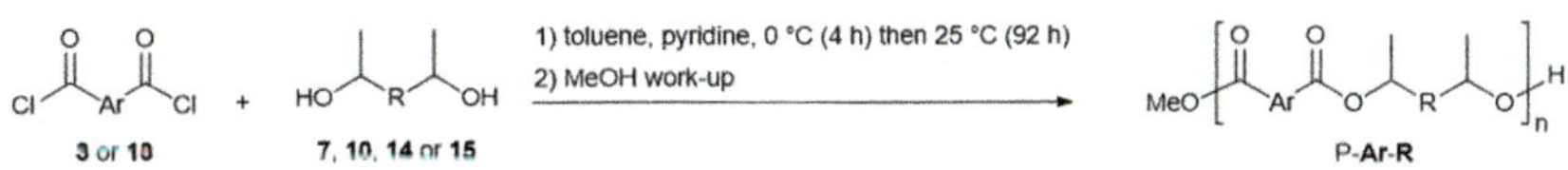

Scheme 2 General solution-phase polymerisation procedure *via* diacid chlorides and diols. **3**: Ar = furan; **13**: Ar = benzene; **7**: R = $(CH_2)_4$; **10**: R = $(CH_2)_2$; **14**: diol = 2,3-butanediol; **15**: diol = isosorbide.

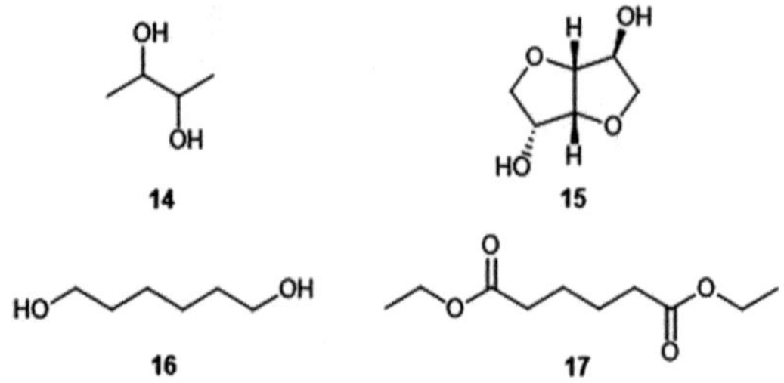

14 15 16 17

Fig. 1 Additional monomers studied.

the preparation of polyesters containing the secondary bicyclic diol isosorbide **15** (Fig. 1).[17] The results of the polymerization reactions between FDCC **3** or terephthaloyl chloride **13** and CMF-derived diols **7** and **10** or the alternative bio-based diols **14** (Fig. 1) and **15** are shown in Table 1.

All diacid chloride polymerisations gave reasonable yields of 50–90%, although a long reaction time of 92 hours was required. When the duration was reduced in one case to 24 hours for the reaction between **3** and **7**, the isolated yield of polymer decreased from 61% to 27%, with a concomitant reduction in the polymer chain length. A likely cause for the non-quantitative yields observed was the loss of polymer during methanol washes/trituration. These were necessary to remove the pyridinium chloride by-product (PyCl) from the samples, and attempts to reduce the number of washes resulted in residual PyCl, evident in the TGA (degradation around 140–200 °C, see Fig. S1†).

Table 1 Solution-phase polymerisation of diacid chlorides with diols^a

Polymer	Ar	R^b	% yield^c	$M_n{}^d$ (g mol^{-1})	$M_w{}^d$ (g mol^{-1})	$Đ^d$	$TD_{10}{}^e$ (°C)	$TD_{50}{}^e$ (°C)	T_g^f (°C)
P-13-14	Benzene	—	84%	3100	4100	1.3	339	365	101 (127)^h
P-13-10	Benzene	$(CH_2)_2$	75%	4600	6900	1.5	326	343	65
P-13-7	Benzene	$(CH_2)_4$	63%	8500	12 000	1.4	312	331	63
P-13-15	Benzene	Iso	79%	2100	2700	1.3	391	420	169 (155–197)^i
P-3-14	Furan	—	56%	8700	13 000	1.5	325	344	84 (87)^j
P-3-10	Furan	$(CH_2)_2$	69%	8300	11 000	1.3	285	300	56
P-3-7	Furan	$(CH_2)_4$	61%	6100	8900	1.5	291	304	40
P-3-7^g	Furan	$(CH_2)_4$	29%	4500	5400	1.2	293	302	31
P-3-15	Furan	Iso	60%	Sample insoluble in GPC solvent			371	401	159 (173)^k

^{a} 2.6 mmol of diacid chloride (**3** or **13**), 2.6 mmol of diol (**7**, **10**, **14** or **15**), 1.7 mL of pyridine, 3 mL of toluene, 6 hours at 0 °C then 86 hours at 25 °C, 2 × 15 mL MeOH wash. ^{b} R = — corresponds to 2,3-butanediol (**14**), R = Iso corresponds to isosorbide (**15**). ^{c} Mass of isolated polymer after methanol wash as a % of anticipated total mass assuming 100% conversion. ^{d} Determined by GPC. ^{e} Determined by TGA, see ESI Fig. S5.† ^{f} Determined by modulated DSC, see ESI Fig. S6.† ^{g} Shortened reaction time (24 hours). ^{h} Literature value of 127 °C for T_g of poly(2,3-butylene terephthalate).[20b] ^{i} Literature value over the range of 155 °C to 197 °C for T_g of poly(isosorbide terephthalate).[19] ^{j} Literature value of 87 °C for T_g of poly(2,3-butylene furanoate).[21] ^{k} Literature value of 173 °C for T_g of poly(isosorbide furanoate).[17b]

As the number of CH_2 groups in the diols increases, there is typically a corresponding increase in M_n and M_w, indicating that the longest diol **7** (R = $(CH_2)_4$) is the most reactive and least hindered by steric effects. This highlights the potential value of this bio-based monomer as it allows for comparatively high chain lengths for a secondary diol, markedly out-performing C_4 diol **14.** Low dispersity ($Đ$) values were observed in the GPC analyses of all these novel polymers, indicating little evidence of issues such as chain branching (which can arise from aldehyde impurities leading to branched acetals) or gelation-induced bi-modal mass distributions.

Modulated differential scanning calorimetry (DSC) was used to investigate the thermal transitions of the polyesters. For both the terephthalate and furan series there is a noticeable drop in T_g between the polyesters of the C_4 diol **14** and the C_6 diol **10** (Fig. 2). However, from **10** to the C_8 diol **7** there was little change, although this effect is skewed somewhat by differences in degrees of polymerisation, which increase the T_g. The effect of higher polymer chain length increasing the T_g can be clearly seen when comparing the standard 92 hour reaction in **P-3-7** with its 24 hour counterpart, where the latter has a lower M_n and M_w and as a result a T_g of 31 °C *versus* 40 °C for the former. The link between increasing polymer chain length and a subsequent increase in T_g is well-established and associated with larger chains resulting in greater restrictions on the long-range coordinated molecular motions associated with progression through the glass transition.[18]

Looking to the literature, Boyd *et al.* compared adipate polyesters of linear 1,6-hexanediol **16** (Fig. 1) with its branched isomer **10** and showed that the former produced highly crystalline polymers (~60%) while polyesters of the latter were entirely amorphous.[14b] It is common for polymers to contain both crystalline and amorphous regions, but this can be in widely varying distributions between the two.

Crystalline regions result in distinct melting points, while amorphous regions result, instead, in a secondary phase transition (T_g) between a hard, glassy state and a rubbery state. Boyd attributed the entirely amorphous nature of polyesters of **10** to its stereo-irregularity. As all diols in this work (with the exception of isosorbide **15**) were used as mixtures of diastereomers, we too observed predominately amorphous behaviour in our polymers, with DSC traces showing

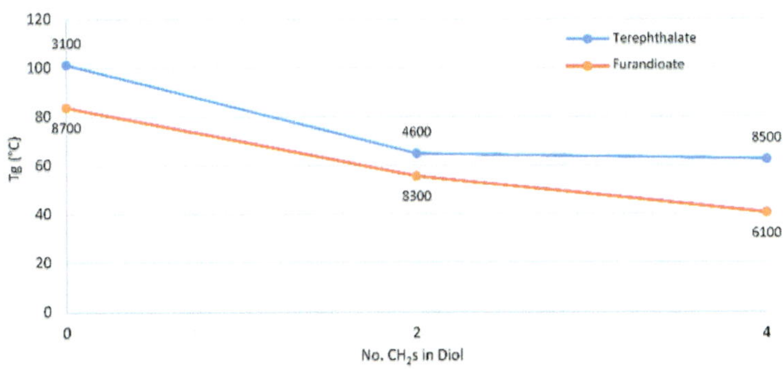

Fig. 2 Effect of increasing secondary diol length on the glass transition temperature for polyesters of **3** and **13**. Labels of each data point are the M_n of the sample as determined by GPC.

distinct T_gs but little evidence of melts. The study by Boyd also indicates how a simple regioisomeric variation in a monomer can drastically alter the T_g of the resultant polymer. Table S1 (ESI†) shows previously reported T_g data for a range of other polyesters produced using both primary and secondary diols. Evident from this comparison is the strong influence secondary diols have on raising the T_g values. Terephthalate and furandioate polyesters of **14**, for example, display T_gs of 84–127 °C, while the equivalent isomeric polyesters made from 1,4-butanediol have T_gs of 26–41 °C, with the terephthalates marginally higher than the furandioates. As the diols progress from C_4 to C_6 to C_8, the T_g values steadily decrease, but in all cases the secondary diols give higher T_gs than their primary diol isomers (Fig. 3). The decrease in T_g in the primary diol series is larger between butanediol and hexanediol than between hexanediol and octanediol for both the terephthalate and furandioate polyesters. This is consistent with our observations in the secondary diol series (Fig. 2), and would suggest that the high T_g associated with polyesters of **14** is both a function of the number of intervening CH_2 groups (A in Fig. 3) and the presence of the CH_3 branches (B in Fig. 3). Thus, the use of linear diols results in increased crystallinity but a reduced T_g for their amorphous regions, while stereoirregular secondary diols show reduced crystallinity but an increased T_g for the amorphous regions of their polyesters.

As anticipated from previously reported data, polyesters of **3** and **13** with isosorbide **15** possessed high T_g values (169 and 160 °C, respectively). The rigidity of the bicyclic structure of this diol is widely observed to increase glass transition temperatures and improve thermochemical properties as the molecular motions associated with the T_g are hindered.[16]

The measured T_g of 169 °C for poly(isosorbide terephthalate) (P-**13**-**15**) was within the range of literature values (155–197 °C).[20] The high T_g value obtained by Ballauf (197 °C) employed a solution-phase polymerisation method similar to our own, but using tetrachloroethane instead of toluene as the solvent. We attribute our lower T_g value to a low polymer chain length for this particular polyester (M_n of 2100 g mol^{-1}). Our value for the T_g of poly(isosorbide furandioate) (P-**3**-**15**) was also somewhat lower than a prior literature value (159 *vs.* 173 °C). This again is likely due to a lower chain length, although we were unable to analyse this particular sample by GPC as it was only sparingly soluble in the eluent (THF).[16b]

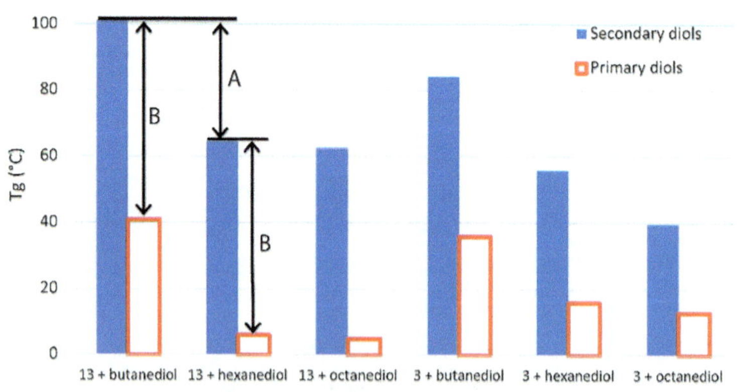

Fig. 3 Comparison of glass transition temperatures for polyesters of **3** and **13** with primary and secondary diol isomers. Data for literature values can be found in Table S1 (ESI†).[19]

Thermogravimetric analysis (TGA) under an inert N_2 atmosphere also indicates a general trend of decreasing thermal stability as the length of the secondary diols increases (Table 1). All TGA traces for these polyesters were monomodal, although very minor (<5%) additional mass losses were occasionally observed between 100–180 °C. These additional losses may be attributed to small amounts of residual toluene (solvent), pyridine or end-groups, the two former were also detected in the ^{1}H-NMR spectra (ESI, Fig. S4†). As was the case for the T_g trend, this decrease in stability was significantly more pronounced between **14** and **10** than **10** and **7**. However, all of the polymer samples had a TD_{10} (a measure of thermal stability indicating the temperature of 10% mass loss) of >280 °C, indicating that all would remain suitable for common applications such as coatings and packaging. The polyesters of isosorbide **15** were found to have the highest thermal stabilities (TD_{10} > 370 °C for P-**13**-**15** and P-**3**-**15**), again demonstrating how this monomer can be used to enhance the properties of plastics. As such, we subsequently investigated the impact of co-polymerisation of **15** with diols **7**, **10**, and **14** to determine whether the thermal stability and T_g values could be tailored for this series of bio-based polyesters.

Isosorbide co-polymers

Isosorbide co-polyesters of the above polymers were prepared using a 2 : 1 : 1 ratio of aromatic diacid chloride : isosorbide : aliphatic diol (Scheme 3). ^{1}H-NMR spectroscopy was used to confirm the successful incorporation of **15** (% iso, see Fig. S2†), with all polyesters showing a % iso in the range of 47–56% (Table 2). The furandioate co-polyesters exhibited marginally higher % iso values than the equivalent terephthalates, although all were acceptably close to the 50% target. GPC analysis showed all *ca.* 50 : 50 co-polymers to have similar chain lengths, with all M_ns in the 3300–4900 g mol^{-1} range (Table 2). This range sits within the values observed for the polyesters in Table 1 without isosorbide, while P-**13**-**5** (% iso = 100%) gave a low M_n of 2100 g mol^{-1}. Thus, we prepared a co-polymer of **3** starting with a 25 : 75 ratio of **15** to **7**, resulting in a copolymer with a % iso of 27% and an improved M_n of 6800 g mol^{-1}, consistent with expectations. For the terephthalate co-polyesters, a trend again was observed that the longer diols gave rise to higher chain length polymers. This mirrors the observation for the simple polyesters of **13**, although the differences in mass in the co polymer series were less prominent (4000, 4500, 4900 g mol^{-1}). The addition of **15** into the co-polyester structure resulted in increased thermal stability (TD_{10} and TD_{50}) for all the polymers studied. This can be seen by comparing the values in Tables 1 and 2 where, for example, P-**3**-**14** has a TD_{10} of 325 °C *versus* 336 °C for the corresponding isosorbide copolymer CP-**3**-**15**-**14**. Of additional note is that the co-

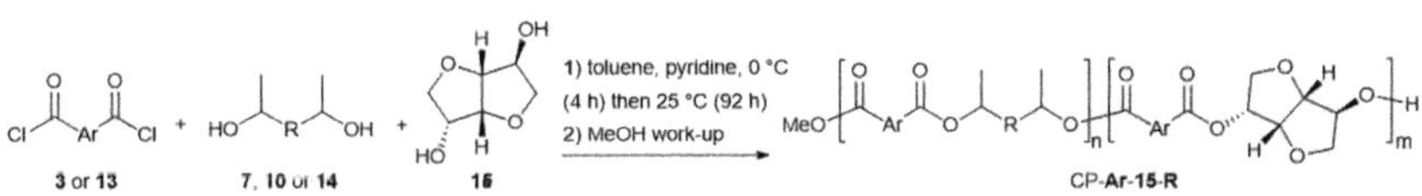

Scheme 3 Co-polymerisation of isosorbide **15** and secondary diols with diacid chlorides **13** or **3**. **3**: Ar = furan; **13**: Ar = benzene; **7**: R = $(CH_2)_4$; **10**: R = $(CH_2)_2$; **14**: diol = 2,3-butanediol; **15**: diol = isosorbide.

Table 2 Co-polymers of isosorbide **15** and secondary diols with **13** or **3**[a]

Polymer	Ar	R[b]	% yield[c]	% iso[d]	M_n[e] (g mol^{-1})	M_w[e] (g mol^{-1})	$Đ$[e]	TD_{10}[f] (°C)	TD_{50}[f] (°C)	T_g[g] (°C)
CP-13-15-14	Benzene	—	86%	53%	4000	6600	1.6	350	399	138
CP-13-15-10	Benzene	$(CH_2)_2$	86%	47%	4500	7300	1.6	324	393	66
CP-13-15-7	Benzene	$(CH_2)_4$	72%	48%	4900	8200	1.7	323	393	43
CP-3-15-14	Furan	—	84%	54%	4400	7800	1.8	336	382	137
CP-3-15-10	Furan	$(CH_2)_2$	76%	56%	3300	4400	1.4	303	382	116
CP-3-15-7	Furan	$(CH_2)_4$	90%	55%	4400	6700	1.5	302	381	79
CP-3-15-7[h]	Furan	$(CH_2)_4$	92%	27%	6800	9700	1.4	294	308	63

[a] 2.6 mmol of diacid chloride (**3** or **13**), 1.3 mmol of isosorbide (**15**), 1.3 mmol of co-diol (**7**, **10** or **14**), 1.7 mL of pyridine, 3 mL of toluene, 6 hours at 0 °C then 86 hours at 25 °C, 2 × 15 mL MeOH wash. [b] When R = — this corresponds to 2,3-butanediol (**14**). [c] Mass of isolated polymer after methanol wash as a % of anticipated total mass assuming 100% conversion. [d] As determined by the ^{1}H NMR spectrum of the collected polymer, see ESI Fig. S4.† [e] Determined by GPC. [f] Determined by TGA, see ESI Fig. S5.† [g] Determined by modulated DSC, see ESI Fig. S6.† [h] 25 : 75 **15** : **7** load ratio (molar).

polymers of **15** typically give two-staged degradations in TGA traces, while the equivalent polyesters from Table 1 were single-staged (Fig. 4 for polymers of **3** and Fig. S3† for polymers of **13**). This suggests differing temperatures of degradation for the different regions of the copolymer. The TGA trace of the *ca.* 25 : 75 **15** : **7** polyester **CP-3-15-7** clearly shows that it is the isosorbide region that possesses the higher stability, as the % mass loss for the first degradation (moiety in the polymer resulting from **7**) increases as the fraction of isosorbide monomer is reduced.

Finally, incorporation of **15** into the polyester also resulted in high values of T_g, with the notable exception of the terephthalate copolymers **CP-13-15-7** and **CP-13-15-10**, which were lower than expected. While the simple terephthalate polyesters

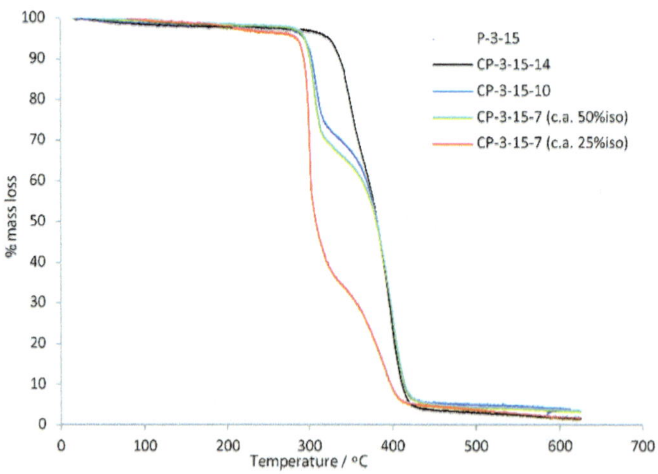

Fig. 4 Comparison of the TGA traces for co-polymers of isosorbide **15** and secondary diols (**14**, **10** and **7**) with **3**.

had shown higher T_gs than the equivalent furandioates using secondary diols (Fig. 3), for the isosorbide co-polymers the opposite was true, as can be seen in Fig. 5. The interpretation of this phenomenon is challenging, but may involve the combination of isosorbide and furandioate leading to more efficient chain packing, or the restriction of interchain mobility by noncovalent interactions. When the fraction of **15** in **CP-3-15-7** reduced to 25% iso the T_g drops, although the value of 63 °C is still about 20 °C higher than the polyester without **15** (**P-3-7**). What remains evident from the experimental data is that **15** can effectively modify the properties of these polyesters, and that the secondary diols introduced in this study continue to demonstrate favourable effects on their resultant polymeric materials.

Polyesters *via* diester transesterification

The above solution-phase diacid chloride polymerisation suffers from drawbacks that could limit its suitability for large-scale synthesis. Activation of the aromatic diacids to the corresponding acid chlorides (**3** and **13**) and the subsequent use in reactions generates salt waste, which reduces efficiency, negatively influences polymer yield, and offsets some of the benefits gained by using bio-based starting materials. This method also required the use of a solvent, and while earlier applications of the protocol employed toxic tetrachloroethane,[17] we were able to use toluene. To further reduce the environmental impact and potential toxicity issues, we investigated whether our branched diols could be used in more industrially relevant transesterifications with 2,5-furandicarboxylic acid diethyl ester (FDEE, **4a**, Scheme 4).

Titanium isopropoxide was chosen as the catalyst as this was found to perform adequately without the toxicity associated with common alternatives such as tin(II) bis(2-ethylhexanoate). Similar to the furandioate polyesters produced using **3**, the molecular weights of the polyesters synthesised *via* **4a** transesterification were found to increase across the series significantly (entries 1–3, Table 3). The

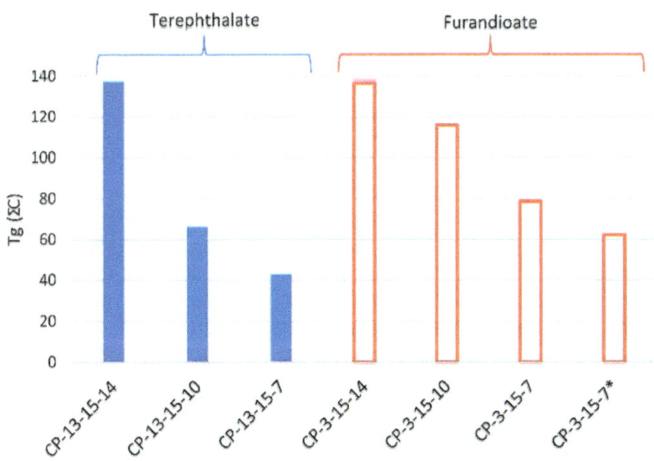

Fig. 5 Effect on the T_g of co-polymers of isosorbide **15** and co-diols (**7**, **10** or **14**) with **3** (furandioate) and **13** (terephthalate). *25 : 75 **15** : **7** molar load ratio.

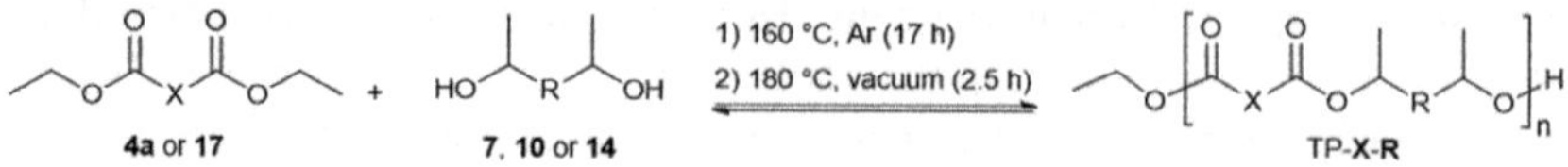

4a or 17 + 7, 10 or 14 → TP-X-R

Scheme 4 General polymerisation *via* transesterification of diols and diesters. **4a**: X = furan; **17**: X = $(CH_2)_4$; **7**: R = $(CH_2)_4$; **10**: R = $(CH_2)_2$; **14**: diol = 2,3-butanediol.

branched hexanediol **7** gave the highest M_n and M_w of the series (TP-**4a**-**7**, entry 3), comparable to the molecular weights achieved using the acid chloride route (P-**3**-**7**). Evidently, our initial misgivings regarding the potential drawbacks of the low volatility of diols **7** and **10** were overcautious. In fact, the most volatile diol **14** was found to give the lowest molecular weight polymer, achieving an M_n of only 830 g mol^{-1} (TP-**4a**-**14**), in agreement with previous studies where **14** was found to typically produce amorphous, low molecular mass polyesters.[16a,21] This observation is best illustrated in Fig. 6 where polymer TP-**4a**-**7** gives a broad monomodal distribution and, despite not having a Gaussian type peak shape (suggesting a slightly higher population of lower molecular mass chains within the sample), the overall molecular masses and polydispersity are good. In contrast, TP-**4a**-**14** shows a series of low molecular weight oligomer peaks, corresponding to a stepwise increase in the number of repeating units. In keeping with the trend, TP-**4a**-**10** sits in between the two, with the majority of the material consisting of chain lengths between 5000–10 000 g mol^{-1}, yet including a quantity of lower molecular weight oligomers.

The low degree of polymerisation seen when using **14** may be attributed to a greater steric hindrance of the alcohol groups by methyl groups in both α and β positions. Steric hindrance appears to have a greater impact when using diester monomers in comparison with highly reactive acid chlorides, which may be partially due to the larger ethyl ester group further restricting attack on the carbonyl. The T_g values appear to follow a similar trend to the polyesters synthesised *via* diacid chlorides, where the C_6 (**10**) and C_8 (**7**) diols gave polymers with

Table 3 Diester and diol transesterification polymerisations[a]

Entry	Polymer	X		R	% yield[d]	M_n[e] (g mol^{-1})	M_w[e] (g mol^{-1})	$Đ$[e]	T_g[f] (°C)
1[b]	TP-**4a**-**14**	Furan	—	—	82	830	1200	1.4	40
2[b]	TP-**4a**-**10**	Furan	—	$(CH_2)_2$	90	3300	5500	1.7	32
3[b]	TP-**4a**-**7**	Furan	—	$(CH_2)_4$	97	6400	9900	1.6	34
4	TP-**17**-**14**	—	$(CH_2)_4$	—	65	2300	3700	1.6	−43
5	TP-**17**-**10**	—	$(CH_2)_4$	$(CH_2)_2$	72	8900	15 000	1.7	−36
6	TP-**17**-**7**	—	$(CH_2)_4$	$(CH_2)_4$	77	18 000	38 000	2.1	−44
7[c]	TCP-**4a**-**17**-**10**	Furan	$(CH_2)_4$	$(CH_2)_2$	87	11 000	16 000	1.5	2
8[c]	TCP-**4a**-**17**-**7**	Furan	$(CH_2)_4$	$(CH_2)_4$	73	9900	19 000	2.0	−8

[a] 20 mmol of diester (**4a** or **17**), 25 mmol of diol (**7**, **10** or **14**), 3 mol% catalyst relative to diester, 160 °C for 17 hours followed by removal of unreacted species under vacuum – see experimental for full conditions. [b] 1% wt Irganox radical scavenger added. [c] 10 mmol of each diester with 25 mmol of diol and 3 mol% catalyst relative to **4a**. [d] Isolated. [e] Determined by GPC. [f] Determined by modulated DSC, see ESI Fig. S6.†

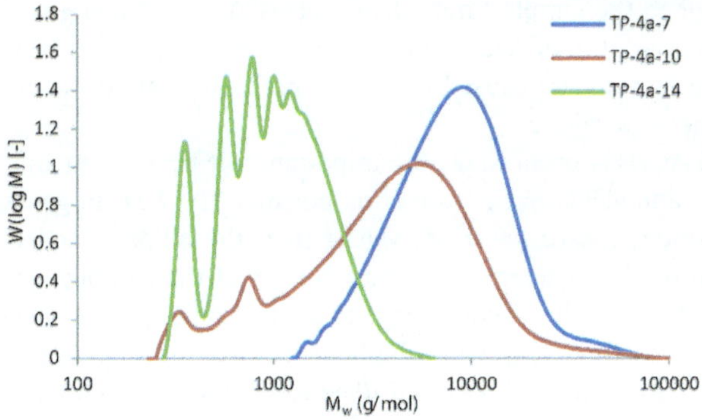

Fig. 6 Comparison of M_w distributions for polymers TP-**4a**-**7**, TP-**4a**-**10** and TP-**4a**-**14**, as determined by GPC.

lower T_gs, presumably due to increased chain flexibility. Again, the T_g data is skewed somewhat due to the lower molecular weights obtained when using **14**, giving a T_g of 40 °C in entry 1 rather than 87 °C as reported in the literature.[22] Promisingly, yields obtained using the transesterification procedure are high, as the washing steps required to remove the salts formed with the acid chloride route are no longer necessary.

Transesterification polymerisations were performed on a larger scale (20 mmol of limiting reagent) compared to the diacid chloride (2.6 mmol), which may have also contributed towards increased yields. With similar molecular weights achieved using either diester **4a** or diacid chloride **3** with diol **7**, transesterification would appear to be the method of choice for the furandioate polyester of diol **7**. However, the lower M_ns observed for TP-**4a**-**10** and TP-**4a**-**14** (3300 and 830 g mol^{-1}, respectively) suggest that the diacid chloride method may give better results when using diols **10** or **14**.

To expand the scope of the transesterification route with diols **7**, **10**, and **14**, a series of aliphatic polyesters were also synthesized using diethyl adipate **17** (entries 4–6, X = $(CH_2)_4$, Table 3). The molecular weights had a similar trend to those previously discussed, with **14** giving significantly lower M_n and M_w values than **10** and **7**. As previously observed, lower degrees of polymerisation had an impact on the T_g of TP-**17**-**14** (−43 °C as opposed to −23 °C reported in the literature[15a]). All the polyesters in the adipate series had higher M_ns compared to their furandioate equivalents, with the octanediol adipate polyester (TP-**17**-**7**) achieving an M_n of >18 000 g mol^{-1}. The T_gs across the adipate polyester series were much lower than the corresponding furandioates, reaching values in the range of −44 to −36 °C, compared to the latter, with T_gs of 32 to 40 °C. This observation was anticipated, as the greater rigidity within the chains offered by the furan moiety would logically lead to higher T_gs. Yet again, the higher degree of polymerisation seen when using **7** resulted in its T_g being only marginally below that of TP-**17**-**10**. With such an impact on the T_g, it is clear that the ability of **7** to produce polyesters of higher chain lengths relative to **10** and **14** highlights a significant benefit in using this bio-based monomer.

In an attempt to incorporate rigidity and yet achieve increased molecular weights, two adipate–furandioate co-polymers were synthesized using diols **7** and **10** (entries 7 and 8, Table 3). The M_ns of polymers TCP-**4a**-**17**-**7** and TCP-**4a**-**17**-**7** were superior to the simple furanodiates, at 9900 and 11 000 g mol^{-1}, respectively. As expected, the introduction of rigidity into the polymer by the incorporation of the furandioate caused the T_gs of both polymers to lie between their homo-polymer equivalents.

Steric effects also appear to play an important role between **14** and **17**, leading to lower M_ns and yields in the formation polymer TP-**17**-**14**. In general, the furandioate monomer gave lower M_n values than the adipate, and this may be attributed to the lower degree of electrophilicity of the carbonyl carbon in the furan. This effect can be circumvented to some extent by employing the above diacid chloride methodology, where higher chain lengths for polyesters of diol **14** were achieved. In contrast, diols **7** and **10** performed well with either of the diester monomers when using the transesterification method, illustrating how high molecular weights and favourable polymer properties can be achieved using secondary diols.

Branched diesters from levulinic acid

Whereas CMF **1** can serve directly as a source of the aromatic diacid derivatives **3**, **4a**, and **13** used in this work, its versatility as a platform for renewable monomers is greatly expanded through its hydration product levulinic acid **5**, which further provided branched diols **7** and **10**, and which we now present as a means to access novel aliphatic diester monomers with branching in the hydrocarbon chain.

We have previously described the reduction of the angelica lactone dimer **19** to provide a mixture of branched C_7–C_{10} hydrocarbons that constitutes cellulosic gasoline.[23] Angelica lactone **18** is accessed from levulinic acid **5** in high yield, as shown in Scheme 5. In place of total deoxygenation of **19**, we have found that gentle hydrogenation in methanol solution in the presence of $La(OTf)_3$ leads cleanly to dimethyl 3-ethyl-4-methylpimelate **20**, a molecule unknown to polymer chemistry. A separate approach that further takes advantage of the remarkable versatility in the reactivity of **5** is the base-induced dimerization of ethyl levulinate **21** to cyclopenadienyl diester **22**. Hydrogenation of **22** gives the annulated adipic ester **23**. Remarkably, compound **22** was first described in 1903 (ref. 24) and was revisited by other workers in 1950, who misassigned the structure.[25] The yield of the condensation reaction was however low (<10%), due mainly to low conversion, and we are currently working to upgrade this method to a standard of preparative usefulness.

Monomers with substantial branching in the hydrocarbon chain are precursors to polymers with weak interfacial forces, resulting in what are referred to as "low surface energy materials" (LSEMs), which possess nominal surface tensions. LSEMs possess a range of useful applications, including as emulsification/suspension agents, lubricants, detergents, antifog coatings, and hydrophobic textiles and fibers.[26] Towards this end, we look forward to preparing polymers of these diacid derivatives, possibly in combination with branched diols **7** and **10**.

Experimental

Materials and methods

Terephthaloyl chloride, 2,3-butanediol, 2,5-hexanediol, isosorbide, diethyl adipate, ethyl levulinate, methanol, anhydrous toluene, anhydrous pyridine,

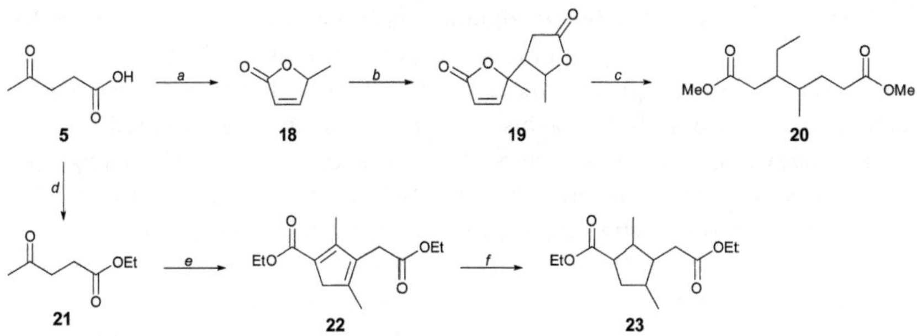

Scheme 5 Synthesis of branched chain diesters **20** and **23** from levulinic acid. (a) K10 montmorillonite, 165 °C, 50 Torr, 92% yield; (b) K_2CO_3, 70 °C, 94% yield; (c) 20 bar H_2, Pd/C, $La(OTf)_3$, MeOH, 220 °C, 60% yield; (d) EtOH, H^+ cat, reflux; (e) NaOEt, CaO, EtOH; 70 h, then H_2SO_4, reflux 24 h; (f) H_2, Pd/C.

titanium isopropoxide, ReadyCal polystyrene standards, tetradecane standard and Irganox radical scavenger were purchased from Sigma-Aldrich and used as received. 10% Pd/C and $La(OTf)_3$ were purchased from Strem Chemicals. 2,5-Furandicarbonyl chloride **3** and 2,7-octanediol **7** were prepared using previously reported methods.[7,11] Diols were dried over molecular sieves prior to use in the polymerisations. Angelica lactone dimer **19** was prepared based on a literature procedure.[23]

Gel permeation chromatography to determine polymer molecular weight was carried out using a set (Polymer Standard Service (PSS) SDV Combination High) of three analytical columns (300 × 8 mm, particle diameter 5 μm) of 1000, 105 and 106 Å pore sizes, plus guard column, supplied by PSS GmbH installed in a PSS SECurity GPC system. Elution was with stabilised tetrahydrofuran at 1 mL min^{-1} with a column temperature of 23 °C and detection by refractive index. 20 μL of a 1 mg mL^{-1} sample in THF was injected for each measurement and eluted for 40 min. Calibration was carried out in the molecular weight range 400–2 × 10^6 Da using ReadyCal polystyrene standards and with a toluene reference.

Modulated differential scanning calorimetry (DSC) experiments were carried out on a TA Instruments Q2000 DSC under a nitrogen atmosphere at a heating rate of 10 °C min^{-1} over a temperature range of 60 to 200 °C and using a sample mass of approximately 10 mg. The T_g values were reported from second heating scans.

Thermogravimetric analysis (TGA) was performed on a PL Thermal Sciences STA 625 thermal analyser. ~10 mg of accurately weighed sample in an aluminium sample cup was placed into the furnace with a N_2 flow of 100 mL min^{-1} and heated from room temperature to 625 °C at a heating rate of 10 °C min^{-1}. From the TGA profiles the temperatures at 10% and 50% mass loss (TD_{10} and TD_{50} respectively) were subsequently determined.

Gas chromatography-mass spectrometry (GC-MS) was performed on an Agilent Technology 6890N equipped with a Varian Factor Four capillary column (VF-5ms, 30 m length, 0.25 mm inner diameter, 0.25 μm film). The injection temperature and the split ratio were 250 °C and 60 : 1, respectively. The oven temperature was held at 60 °C for 2 min, increased to 250 °C (10 °C min^{-1}), and finally increased to

300 °C (20 °C min^{-1}). The initial column pressure was 4.5 psi. The column flow was 0.7 mL min^{-1}. Mass spectrometry was performed using electron impact (EI) ionization. A pure sample of dimethyl 3-ethyl-4-methylpimelate **20** was used to calculate the response factor against the tetradecane internal standard.

For the synthesis of the polyesters, ^{1}H NMR spectra were recorded using a Jeol 400 spectrometer at an operating frequency of 400 MHz. For the synthesis of **20**, **22**, and **23**, ^{1}H NMR spectra were recorded using a VNMRS 600 spectrometer at an operating frequency of 600 MHz. ^{13}C NMR spectra were recorded on the same instrument at an operating frequency of 150 MHz. All NMR data was processed using MestReNova (version 10.0) desktop NMR data processing software.[27]

General polymerization procedures

Polyesters *via* diacid chloride. Diacid chloride **3** or **13** (2.6 mmol) was dissolved in anhydrous toluene (1.5 mL) in a dried 25 mL one-neck round-bottom flask sealed with a rubber stopper. The flask was cooled in a water/ice bath and continually purged with argon. A solution of the selected diol **7**, **10**, **14** or **15** (2.6 mmol) and anhydrous pyridine (1.7 mL) in anhydrous toluene (1.5 mL) was also prepared. The diol mixture was added *via* syringe to the argon-purged diacid chloride solution with slow agitation by a magnetic stirrer bar. Upon mixing, the reaction became cloudy and slightly viscous, with the viscosity increasing over time. After 6 h the flask was removed from the water/ice bath and allowed to reach room temperature and left to stir slowly. After 48 h from the start of the reaction (42 h at room temperature), additional diacid chloride (5 mg) in anhydrous toluene (0.2 mL) was added. This same addition was repeated again after 76 h. After a total reaction time of 92 h (86 h at room temperature), the excess pyridine and toluene were mostly removed *in vacuo* and the solid tacky residue was triturated twice with dry methanol (15 mL). The amount of residual pyridine was reduced by redissolving the product in toluene (5 mL) and evaporation of the solvent *in vacuo*. The solid product was dried under high vacuum (<1 mbar) overnight prior to analysis.

Isosorbide co-polyesters *via* diacid chloride. Diacid chloride **3** or **13** (2.6 mmol) was dissolved in anhydrous toluene (1.5 mL) in a dried 25 mL one-neck round-bottom flask sealed with a rubber septum. The flask was cooled in a water/ice bath and continually purged with argon. A solution of the selected diol **7**, **10** or **14** (1.3 mmol), isosorbide **15** (1.3 mmol) and anhydrous pyridine (1.7 mL) in anhydrous toluene (1.5 mL) was also prepared. The diol mixture was added *via* syringe to the argon-purged diacid chloride solution with slow agitation by a magnetic stirrer bar. Upon mixing, the reaction became cloudy and slightly viscous, with the viscosity increasing over time. After 6 h the flask was removed from the water/ice bath and allowed to reach room temperature and left to stir slowly. After 48 h from the start of reaction (42 h at room temperature), additional diacid chloride (5 mg) in anhydrous toluene (0.2 mL) was added. This same addition was repeated again after 76 h. After a total reaction time of 92 h (86 h at room temperature), the excess pyridine and toluene were mostly removed *in vacuo* and the solid tacky residue was triturated twice with dry methanol (15 mL). The amount of residual pyridine was reduced by redissolving the product in toluene (5 mL) and evaporation of the solvent *in vacuo*. The solid product was dried under high vacuum (<1 mbar) overnight prior to analysis.

Polyesters *via* transesterification. A mixture of diester **4a** or **17** (20 mmol), diol **7**, **10**, or **14** (25 mmol) and titanium isopropoxide (3 mol% relative to the diester) was introduced into a dry, wide-neck round bottom flask with a magnetic stirrer bar mounted with a Dean–Stark condenser. The system was evacuated and backfilled with argon four times and then heated at 160 °C with a stirring rate of 300 rpm. After 17 h the temperature was steadily increased to 210 °C and the stirring rate reduced along with a gradual application of vacuum over 2.5 h until the pressure reached <0.5 mbar. Upon completion of the reaction, the system was backfilled with argon and the mixture was left to cool with stirring. The resulting polymer was analysed as collected. For entries 7 and 8 in Table 3, the adipate ester was added separately after 2 h initial reaction between **4a** and diol.

Conclusions

A range of bio-based monomers, both diacid derivatives and diols, can be accessed from the cellulose-derived platform molecule CMF. A systemic study into the effects of hydrocarbon chain branching in diol monomers yields some novel insights. In terms of reactivity, for polymers prepared using terephthaloyl chloride **13** or by transesterification of diethyl 2,5-furandicarboxylate **4a**, an increase in the number of intervening CH_2 groups between the branch points leads to higher molecular mass polymers. On the other hand, the combination of 2,5-furandicarbonylchloride **3** with branched diols gave products of similar masses irrespective of diol chain length. In all cases, dispersity values of the polymers were low and thermal stabilities were more than adequate for common materials applications. 2,7-Octanediol **7**, in particular, has shown strong potential for tailoring the properties of its resultant polyesters in terms of increasing the glass transition temperature relative to its isomeric non-branched equivalent. Diol **7** was also found to be best suited for use in more industrially relevant transesterification polymerisations, giving markedly higher degrees of polymerisation relative to the other diols studied. As such, **7** represents an excellent candidate for a new commercial diol, allowing for the tailoring of glass transition temperatures whilst also overcoming the low chain length of polyesters typically observed when using secondary diols. This study has demonstrated how a range of diols and diacids accessible from a single renewable platform molecule (CMF **1**) can be combined to produce novel polyesters with tuneable thermal properties for bio-based materials markets.

Access statement

All data used in the preparation of this manuscript for the sections funded by the EPSRC grant EP/L017393/1 and the BBSRC grant BB/N023595/1 is contained within this document, the ESI,† or available on request from DOI: 10.15124/20fccc77-5902-42cc-adb9-8e58cf27bac0.

Acknowledgements

MM acknowledges support from NSF-CBET grant 1335646. TJF and JWC would like to thank the UK Engineering and Physical Sciences Research Council (EPSRC, grant EP/L017393/1) and the Biotechnology and Biological Sciences Research

Council (BBSRC, grant BB/N023595/1) for funding their involvement in this research.

References

1 British Plastics Federation Annual Review, 2015, http://www.bpf.co.uk/about_the_bpf/BPF_Annual_Review.aspx, accessed 14th December 2016.

2 (*a*) I. Delidovich, P. J. C. Hausoul, L. Deng, R. Pfützenreuter, M. Rose and R. Palkovits, *Chem. Rev.*, 2016, **116**, 1540–1599; (*b*) A. Gandini and T. M. Lacerda, *Prog. Polym. Sci.*, 2015, **48**, 1–39; (*c*) Y. Zhang and E. Y.-X. Chen, *Top. Curr. Chem.*, 2014, **353**, 185–227.

3 F. H. Isikgor and C. R. Becer, *Polym. Chem.*, 2015, **6**, 4497–4559.

4 T. J. Farmer, M. Mascal, J. H. Clark, and F. E. I. Deswarte, *Platform Molecules in Introduction to Chemicals from Biomass*, John Wiley and Sons, Chichester, UK, 2nd edn, 2015, ch. 4, pp. 89–156.

5 (*a*) J. J. Bozell, L. Moens, D. C. Elliott, Y. Wang, G. G. Neuenscwander, S. W. Fitzpatrick, R. J. Bilski and J. L. Jarnefeld, *Resour., Conserv. Recycl.*, 2000, **28**, 227–239; (*b*) D. W. Rackemann and W. O. S. Doherty, *Biofuels, Bioprod. Biorefin.*, 2011, **5**, 198–214; (*c*) F. D. Pileidis and M. M. Titirici, *ChemSusChem*, 2016, **9**, 562–582.

6 (*a*) M. Mascal, *ChemSusChem*, 2015, **8**, 3391–3395; (*b*) M. Mascal and E. B. Nikitin, *Green Chem.*, 2010, **12**, 370–373; (*c*) M. Mascal and E. B. Nikitin, *Angew. Chem., Int. Ed.*, 2008, **47**, 7924–7926.

7 (*a*) S. Dutta, L. Wu and M. Mascal, *Green Chem.*, 2015, **17**, 3737–3739; (*b*) C. Laugel, B. Estrine, J. Le Bras, N. Hoffmann, S. Marinkovic and J. Muzart, *ChemCatChem*, 2014, **6**, 1195–1198.

8 S. Dutta and M. Mascal, *ChemSusChem*, 2014, **7**, 3028–3030.

9 (*a*) J. C. Hong, L. Ren, V. Vattipalli, Y.-H. Yeh, N. Gould, B. Xu, R. J. Gorte, R. Lobo, P. J. Dauenhauer, M. Tsapatsis and W. Fan, *ChemCatChem*, 2017, **9**, 398–402; (*b*) C.-C. Chang, H. J. Cho, J. Yu, R. J. Gorte, J. Gulbinski, P. Dauenhauer and W. Fan, *Green Chem.*, 2016, **18**, 1368–1376.

10 (*a*) P. Raghavendrachar and S. Ramachandran, *Ind. Eng. Chem. Res.*, 1992, **31**, 453–462; (*b*) M. G. Rüther, F. Frehill, J. E. O'Brien, A. I. Minett, W. J. Blau, J. G. Vos and M. Panhuis, *J. Phys. Chem. B*, 2004, **108**, 9665–9668.

11 L. Wu, M. Mascal, T. J. Farmer, S. Pérocheau Arnaud and M.-A. Wong Chang, *ChemSusChem*, 2017, **10**, 166–170.

12 (*a*) R. J. Sullivan, E. Latifi, B. K.-M. Chung, D. V. Soldatov and M. Schlaf, *ACS Catal.*, 2014, **4**, 4116–4128; (*b*) B. Kuhlmann, E. M. Arnett and M. Siskin, *J. Org. Chem.*, 1994, **59**, 3098–3101; (*c*) H. Zhou, J. Song, Q. Meng, Z. He, Z. Jiang, B. Zhou, H. Liua and B. Han, *Green Chem.*, 2016, **18**, 220–225.

13 B. A. C. van As, J. van Buijtenen, T. Mes, A. R. A. Palmans and E. W. Meijer, *Chem.–Eur. J.*, 2007, **13**, 8325–8332.

14 (*a*) J. J. O'Malley and W. J. Stauffer, *J. Polym. Sci., Polym. Chem. Ed.*, 1974, **12**, 865–874; (*b*) P. R. H. Boyd and P. A. Aylwin, *Polymer*, 1984, **25**, 330–339; (*c*) P. J. M. Serrano, E. Thüsst and R. J. Gaymans, *Polymer*, 1997, **38**, 3893–3902; (*d*) K. Bouma, M. Regelink and R. J. Gaymans, *J. Appl. Polym. Sci.*, 2001, **80**, 2676–2682.

15 (*a*) E. Gubbels, L. Jasinska-Walc and C. E. Koning, *J. Polym. Sci., Part A: Polym. Chem.*, 2013, **51**, 890–898; (*b*) E. Gubbels, J. P. Drijfhout, C. Posthuma-van Tent,

L. Jasinska-Walc, B. A. J. Noordover and C. E. Koning, *Prog. Org. Coat.*, 2014, **77**, 277–284; (*c*) X. Hu, X. Shen, M. Huang, C. Liu, Y. Geng, R. Wang, R. Xu, H. Qiao and L. Zhang, *Polymer*, 2016, **84**, 343–354; (*d*) M. Köpke, C. Mihalcea, F. Liew, J. H. Tizard, M. S. Ali, J. J. Conolly, B. Al-Sinawi and S. D. Simpson, *Appl. Environ. Microbiol.*, 2011, **77**, 5467–5475; (*e*) X.-X. Wang, H.-Y. Hu, D.-H. Liu and Y.-Q. Song, *New Biotechnol.*, 2016, **33**, 16–22.

16 (*a*) T. Debuissy, E. Pollet and L. Averous, *Polymer*, 2016, **99**, 204–213; (*b*) F. Fenouillot, A. Rousseau, G. Colomines, R. Saint-Loup and J.-P. Pascault, *Prog. Polym. Sci.*, 2010, **35**, 578–622; (*c*) B. A. J. Noordover, V. G. van Staalduinen, R. Duchateau, C. E. Koning, R. T. M. van Benthem, M. Mak, A. Heise, A. E. Frissen and J. van Haveren, *Biomacromolecules*, 2006, **7**, 3406–3416; (*d*) M. Rose and R. Palkovits, *ChemSusChem*, 2012, **5**, 167–176.

17 (*a*) M. Gomes, A. Gandini, A. J. D. Silvestre and B. Reis, *J. Polym. Sci., Part A: Polym. Chem.*, 2011, **49**, 3759–3768; (*b*) R. Storbeck and M. Ballauff, *Polymer*, 1993, **34**, 5003–5006.

18 J. W. Nicholson, *The Chemistry of Polymers*, The Royal Society of Chemistry, Cambridge, edn 3, 2006.

19 (*a*) M. Pyda, E. Nowak-Pyda, J. Heeg, H. Huth, A. A. Minakov, M. L. Di Lorenzo, C. Schick and B. Wunderlich, *J. Polym. Sci., Part B: Polym. Phys.*, 2006, **44**, 1364–1377; (*b*) T. H. Ng and H. L. Williams, *Makromol. Chem.*, 1981, **182**, 3323–3330; (*c*) Y. Jiang, A. J. J. Woortman, G. O. R. Alberda van Ekenstein and K. Loos, *Polym. Chem.*, 2015, **6**, 5198–5211.

20 (*a*) J. Thiem and H. Lueders, *Polym. Bull.*, 1984, **11**, 365–369; (*b*) R. Storbeck and M. Ballauff, *J. Appl. Polym. Sci.*, 1996, **59**, 1199–1202.

21 (*a*) R. W. Watson, N. H. Grace and J. L. Barnwell, *Can. J. Res.*, 1950, **28b**, 652–659; (*b*) W. J. Jackson and J. J. Watkins, *US pat.*, 4,600,768, Eastman Kodak Company, 1986.

22 S. Thiyagarajan, W. Vogelzang, R. J. I. Knoop, A. E. Frissen, J. van Haveren and D. S. van Es, *Green Chem.*, 2014, **16**, 1957–1966.

23 M. Mascal, S. Dutta and I. Gandarias, *Angew. Chem., Int. Ed.*, 2014, **53**, 1854–1857.

24 P. Duden and R. Freydag, *Ber. Dtsch. Chem. Ges.*, 1903, **36**, 944–952.

25 S. Shimizu, *Nippon Nogei Kagaku Kaishi*, 1950, **23**, 288–294.

26 (*a*) S. Alexander, J. Eastoe, A. M. Lord, F. Guittard and A. R. Barron, *ACS Appl. Mater. Interfaces*, 2016, **8**, 660–666; (*b*) S. Alexander, G. N. Smith, C. James, S. E. Rogers, F. Guittard, M. Sagisaka and J. Eastoe, *Langmuir*, 2014, **30**, 3413–3421.

27 *MestReNova (Mnova), version 10.0*, Mestrelab Research, SL, Santiago de Compostela, Spain.

Faraday Discussions

PAPER

Tunable and selective hydrogenation of furfural to furfuryl alcohol and cyclopentanone over Pt supported on biomass-derived porous heteroatom doped carbon†

Xiuyun Liu,[ac] Bo Zhang,[ab] Benhua Fei,[b] Xiufang Chen,*[a] Junyi Zhang[a] and Xindong Mu *[a]

Received 30th January 2017, Accepted 27th March 2017
DOI: 10.1039/c7fd00041c

The search for and exploitation of efficient catalytic systems for selective conversion of furfural into various high value-added chemicals remains a huge challenge for green synthesis in the chemical industry. Here, novel Pt nanoparticles supported on bamboo shoot-derived porous heteroatom doped carbon materials were designed as highly active catalysts for controlled hydrogenation of furfural in aqueous media. The porous heteroatom doped carbon supported Pt catalysts were endowed with a large surface area with a hierarchical porous structure, a high content of nitrogen and oxygen functionalities, a high dispersion of the Pt nanoparticles, good water dispersibility and reaction stability. Benefiting from these features, the novel Pt catalysts displayed a high activity and controlled tunable selectivity for furfural hydrogenation to produce furfuryl alcohol and cyclopentanone in water. The product selectivity could be easily modulated by controlling the carbonization temperature of the porous heteroatom doped carbon support and the reaction conditions (temperature and H_2 pressure). Under mild conditions (100 °C, 1 MPa H_2), furfuryl alcohol was obtained in water with complete conversion of the furfural and an impressive furfuryl alcohol selectivity of >99% in the presence of Pt/NC-BS-500. A higher reaction temperature, in water, favored rearrangement of the furfural (FFA) with Pt/NC-BS-800 as the catalyst, which resulted in a high cyclopentanone yield of >76% at 150 °C and 3 MPa H_2. The surface properties and pore structure of the heteroatom doped carbon support, adjusted using the carbonization temperature, might determine the interactions between the Pt nanoparticles, carbon support and catalytic reactants in water, which in turn could have

[a]Key Laboratory of Bio-based Materials, Qingdao Institute of Bioenergy and Bioprocess Technology, Chinese Academy of Sciences, Qingdao 266101, China. E-mail: chenxf@qibebt.ac.cn; muxd@qibebt.ac.cn
[b]International Centre for Bamboo and Rattan, Beijing 100102, China
[c]Fujian Provincial Key Laboratory of Ecology-Toxicological Effects & Control for Emerging Contaminants, Putian University, Fujian 351100, China

† Electronic supplementary information (ESI) available. See DOI: 10.1039/c7fd00041c

led to a good selectivity control. The effect of different reaction temperatures and reaction times on the product selectivity was also explored. Combined with exploration of the distribution of the reaction products, a reaction mechanism for furfural reduction has been proposed.

Introduction

Energy crises and environmental pollution are major challenges as the unsustainable fossil resources are diminishing.[1,2] Biomass, the most abundant renewable carbon resource in nature, could be converted into clean fuels and chemicals.[3,4] Due to the high oxygen content of biomass, transformation of biomass into oxygen-containing chemicals *via* biomass-derived platform molecules has been found to be an efficient route for utilizing biomass resources.[5–7] Among the bio-based platform chemicals, furfural (FFA), an important renewable C5 resource, is widely used as a raw material to produce various valuable chemicals[8–10] including furfuryl alcohol (FA),[11] tetrahydrofurfuryl alcohol (THFA),[12] 1,2-pentanediol (1,2-PeD),[13] cyclopentanone (CPO)[14] and 2-methylfuran (MF).[15] However, as both the C=O group and the furan ring are easily attacked during a reduction process (Scheme 1), it is difficult to control the reaction to obtain the desired product with high selectivity.[16] Moreover, FFA is generally produced through an acid-catalyzed dehydration of pentose in aqueous solution,[17–19] and the procedures for separating FFA from water would raise the production cost. Thus, it is highly attractive to control the product selectivity by exploring effective catalytic systems for FFA hydrogenation in water.

FA has been identified as a valuable compound with a wide application within precursors of resins, adhesives, foundry binders, farm chemicals, and some fine chemical products like vitamin C and THFA.[20] FA is formed by selective hydrogenation of the C=O bond in FFA, while leaving the furan ring unaffected. So, a suitable catalyst is essential for the selective hydrogenation reaction. In industry, FA was commonly produced using Cu–Cr catalysts, which were active but caused serious pollution of the environment because of the high toxicity of Cr.[21] Thus, much effort has been made to explore Cr-free catalysts such as Pt, Ru, Pd, Co, Cu and Ni loaded on metal oxide supports.[22–27] But these catalysts are

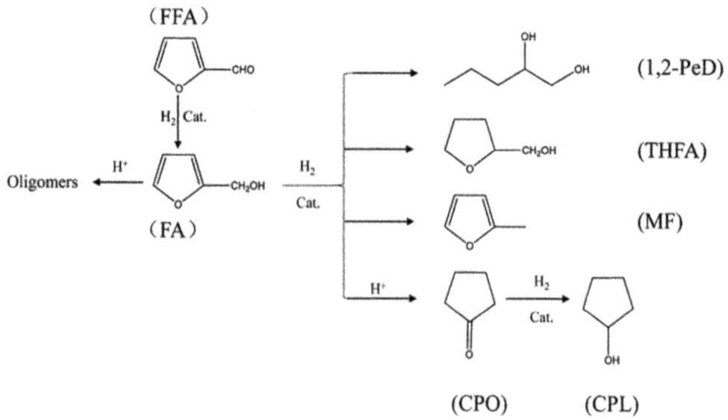

Scheme 1 A schematic pathway for FFA hydrogenation.

usually processed in organic solvents like methanol, 2-propanol, and tetrahydrofuran. In our previous work,[22] we reported a highly selective hydrogenation of FFA to FA over Pt nanoparticles supported on carbon nitride nanosheets in water under mild conditions (100 °C, 1 MPa H_2). An efficient water-tolerant catalyst for selective hydrogenation of FFA to FA was obtained when small-size Pt nanoparticles were loaded on a suitable support with high dispersion. This work provides a green and efficient strategy for the sustainable synthesis of FA.

CPO is also a vital fine chemical intermediate extensively used in the production of perfume, pesticides and medicines.[28] CPO was traditionally synthesized through the oxidation of cyclopentene or the pyrolysis of adipic acid, which were derived from unsustainable fossil fuels, and always required harsh reaction conditions including a high temperature and high pressure.[29] Recently, FFA was found to be an ideal feedstock for producing CPO. Hronec *et al.* reported a one-pot conversion of FFA to CPO *via* hydrogenation and rearrangement in aqueous media using some metal catalysts (*e.g.*, Pt/C and Pd–Cu/C).[14,16,30] Following this pioneering work, other catalysts such as Ni–Cu/SBA-15,[31] Cu–Co,[32] CuNiAl hydrotalcite-type catalysts,[33] Ru/MIL-101,[28] and Au/TiO_2 [29] were further explored for the FFA rearrangement to produce CPO. Although some pioneering work has been done on the synthesis of CPO based on biomass-derived FFA, this area is still in its infancy. Control of the product distribution to yield the desirable product remains a key challenge. Notably, among the various catalysts investigated, supported catalysts constitute the vast majority and the support has played a key role in the catalytic performance. These supported catalysts might suffer from problems such as an inhomogeneous particle size, weak particle–support interactions and poor dispersibility in water, resulting in a low catalytic activity and poor stability in aqueous media.[34] The interactions between the metal and support might modify the electronic structure of the catalytic active metals, and in turn alter the catalytic performance. Thus, the development of an effective and selective catalyst with a suitable support is highly desirable.[35]

Porous carbon materials, which have the advantages of an easy availability, low cost, high surface area and easy-to-design pore structure, are regarded as one of the most appealing types of material for a catalyst support. However, the lack of functional groups and the poor hydrophilicity of pure carbon materials has greatly limited their use in catalysis.[36] Modification of the carbon material through heteroatom (*e.g.* N, B, O and S) incorporation is considered to be the most effective method to overcome these problems.[3] For instance, the incorporation of oxygen or nitrogen atoms into carbon lattice not only modulated the structure, surface acidity/basicity and surface wettability, but also improved the electron conductivity, allowing the materials to be an excellent support for heterogeneous catalysis.[37,38] Owing to the higher electronegativity of oxygen and nitrogen than carbon, oxygen- or nitrogen-doped carbon materials could provide sufficient active sites to anchor and stabilize noble metal nanoparticles with a high dispersion. Up to now, porous nitrogen-doped carbon was commonly fabricated through post-treatment of the porous carbon with a nitrogen source like urea and ammonia, or *in situ* doping by carbonization of nitrogen-rich precursors. These methods have always required tedious and multiple-step processes with the assistance of templates or activation using a considerable amount of chemical reagents like KOH or $ZnCl_2$,[39,40] which would limit mass production. From the standpoint of green and sustainable chemistry, it is therefore highly desirable to utilize porous heteroatom doped carbon materials, as catalytic supports, that are fabricated from renewable sources using a simple, cheap, green and template-free process.[41]

Herein, we report the design of Pt catalysts supported on biomass-derived porous heteroatom doped carbon materials for selective hydrogenation of FFA in water. As a renewable biomass feedstock that can be freely available in nature, bamboo shoots were used as the starting material to prepare porous heteroatom doped carbon materials using a simple and green method without chemical activation or any templates. Apart from the high content of carbon and oxygen, bamboo shoots are rich in nitrogen-containing organic substances including proteins and amino acids, so heteroatoms of nitrogen and oxygen could be doped into the carbon material during the carbonization process without using an exogenous nitrogen source. Highly effective, tunable and selective systems for the transformation of FFA to FA or CPO were obtained by using these novel Pt nanoparticles supported on bamboo shoot-derived heteroatom doped carbons. Various experimental parameters such as the carbonization temperature for the porous heteroatom doped carbons, reaction temperature, hydrogen pressure, and reaction time were studied in detail to tune the product selectivity. Reaction pathways for FFA hydrogenation over the Pt nanoparticles supported on bamboo shoot-derived porous heteroatom doped carbon catalysts have also been proposed. Notably, the novel Pt catalysts exhibit the great advantages of an easy recovery and a high catalytic stability.

Experimental

Materials and reagents

All the reagents used in this work were of analytical grade and used without purification. Fresh bamboo shoots were collected from the Anhui Taiping Test Centre, International Centre for Bamboo and Rattan, Anhui Province, China.

Preparation of the bamboo shoot-derived porous heteroatom doped carbon materials

The porous heteroatom doped carbon materials were prepared in two-steps, including a hydrothermal treatment and a carbonization process. The fresh bamboo shoots were firstly cut into slices and dried at 70 °C for 24 h. The bamboo shoots were then ground into a powder and transferred into a 100 mL Teflon-inner stainless steel autoclave along with 20 mL of deionized water. The autoclave was then heated to 180 °C and maintained at this temperature for 6 h. The resulting brown solid was collected by filtration, washed using distilled water, and dried at 70 °C for 24 h. After that, the hydrochars were calcined at 500 °C or 800 °C for 4 h under a N_2 flow using a heating rate of 5 °C min^{-1}. The porous heteroatom doped carbon materials were obtained and designated as NC-BS-*x*, where *x* represents the calcination temperature.

Preparation of the porous heteroatom doped carbon supported Pt materials

The porous heteroatom doped carbon supported Pt materials with a metal loading of 5.0 wt% were prepared using an ultrasound-assisted reduction method. 0.1 g of the porous heteroatom doped carbon material was dispersed in

10 mL of deionized water under ultrasonic conditions. Then a H_2PtCl_6 aqueous solution (5.0 wt% Pt) was added into the mixture and ultrasonicated for 15 min. After that, 16 mL $NaBH_4$ solution (0.2 wt%) was added slowly into the mixture under ultrasonication. Finally, the product was separated, washed with deionized water and dried at 25 °C for 12 h. The Pt loaded materials were named Pt/NC-BS-*x*. For comparison, Pt supported on commercial activated carbon (Pt/C) was prepared using the same method.

Characterization

The X-ray diffraction (XRD) patterns of all the catalysts were obtained using a Bruker D8 Advance X-ray diffraction diffractometer equipped with CuKα radiation ($\lambda = 1.5147$ Å). The morphology of the catalysts was examined using a H-7600 transmission electron microscope (TEM), a Tecnai G2 F30 high-resolution TEM (HRTEM) and a FEI Tecnai G2 F20 scanning transmission electron microscope (STEM). Nitrogen adsorption–desorption data were obtained using a Micromeritics ASAP 2020 static volumetric sorption analyzer. The specific surface area of the samples was calculated using the Brunauer–Emmett–Teller (BET) method. The micropore volume was calculated using a *t*-plot method. The pore size distribution was determined using non-local density functional theory (DFT). The X-ray photoelectron spectroscopy (XPS) data were obtained using a Thermo ESCALAB250 instrument with a monochromatized Al Kα line source. All the binding energies obtained were calibrated based on the C 1s peak at 284.6 eV. The elemental composition analysis was conducted using a Vario El elemental analyzer. Hydrogen temperature-programmed reduction (H_2-TPR) of the catalysts was conducted using a Micromeritics Autochem 2920 chemisorption analyzer.

Catalytic activity for furfural hydrogenation

Hydrogenation of FFA was conducted in a stainless steel autoclave agitated using a magnetic stirrer. In a typical procedure, 10 mL of a 2 wt% FFA solution and 0.025 g of the catalyst were added into the reactor. The autoclave was then purged with H_2 five times to remove the air. Then the autoclave was pressurized to the desired pressure and heated to the desired temperature under stirring. After completion of the reaction, the products were centrifuged and added to ethanol (volume 1 : 5). Finally, the products were identified using a gas chromatograph coupled with a mass spectrometer (GC-MS, Shimadzu, 2010 plus). Quantification of the reaction products was performed using gas chromatography (Shimadzu, 2010 plus, Fig. S1 and S2, ESI†). The conversions were calculated based on the amount of reactant added into the reactor (ESI†).

Results and discussion

Fabrication and characterization of the NC-BS supported Pt nanoparticles

A simple and green hydrothermal carbonization method was used to prepare porous heteroatom doped carbon materials using bamboo shoots as the starting material without any templates, chemical activation or exogenous nitrogen sources. Oxygen- and nitrogen-containing hydrochars were firstly obtained through hydrothermal treatment of the bamboo shoots with water at 180 °C. Water served as not only a solvent but also a catalyst which promoted the

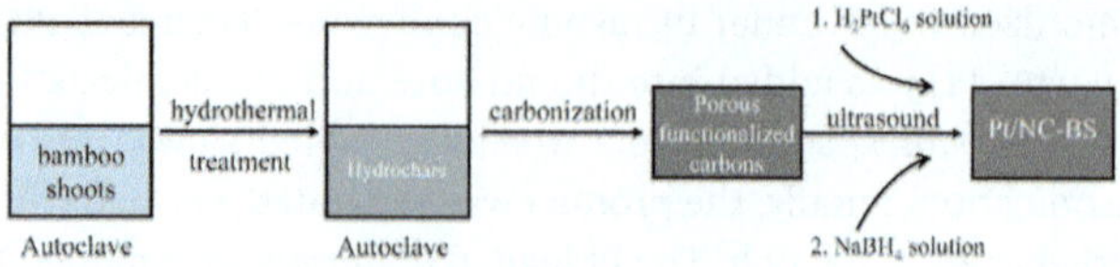

Scheme 2 A schematic illustration of the fabrication of Pt/NC-BS.

hydrolysis and decomposition of the bamboo shoots to small fragments during the hydrothermal carbonization process. Simultaneously, bamboo shoots are mainly composed of cellulose, hemicellulose and lignin, which also contain a certain amount of amino acids and proteins.[42,43] The nitrogen content determined by elemental analysis can reach up to 4.3% based on the dry weight of bamboo shoot. The high content of oxygen- and nitrogen-rich organic substances in the bamboo shoots ensured the incorporation of nitrogen and oxygen into the structure of carbon materials during the carbonization process. Then, the functionalized hydrochars were post-carbonized at 500 °C or 800 °C under an inert atmosphere to increase the graphitization degree and generate more porosity, thus yield porous heteroatom doped carbons. Finally, the Pt nanoparticles were loaded onto the surface of the as-prepared porous heteroatom doped carbons using an ultrasound-assisted method with $NaBH_4$ as the reductant. A schematic illustration of the fabrication of Pt/NC-BS is shown in Scheme 2. The incorporated nitrogen content of the bamboo shoot-derived carbons was examined using elemental analysis. As expected (Table 1), NC-BS-500 had a high nitrogen content of 6.8%. With increase of the carbonization temperature up to 800 °C, the nitrogen content of the carbon material still remained as high as 3.0% for the NC-BS-800 sample. The functional groups on the surface of NC-BS-500 and NC-BS-800 were further analysed using FTIR spectra (Fig. 1). The NC-BS samples had strong characteristic bands for O–H vibrations at 3436 cm^{-1} and N–H stretching at 3260 cm^{-1}. The peaks at 1107 cm^{-1}, 1414 cm^{-1} and 1618 cm^{-1} indicated the existence of C–O/C–N vibrations, C=O/C=N stretching and O–H in-plane bending vibrations.[40,44,45] The results demonstrated that rich oxygen- and nitrogen-containing functional groups were present on the surface of the bamboo shoot-derived carbon materials. Moreover, the functional groups decreased with the rise in the carbonization temperature, as more oxygen and nitrogen species were removed through reaction with oxygenated functional groups or the formation of gaseous products (*e.g.* CO, CO_2, N_2 and NH_3) during the post-carbonization process at the higher temperature.

Table 1 Physical and chemical properties of the NC-BS samples

Sample	$S_{BET}{}^a$ ($m^2\ g^{-1}$)	$V_{total}{}^b$ ($cm^3\ g^{-1}$)	$V_{micro}{}^b$ ($cm^3\ g^{-1}$)	Content (%)		
				C	N	H
NC-BS-500	287	0.245	0.089	76.2	6.8	2.9
NC-BS-800	670	0.306	0.180	78.8	3.0	1.0

[a] Surface area based on the BET method. [b] Based on the DFT method.

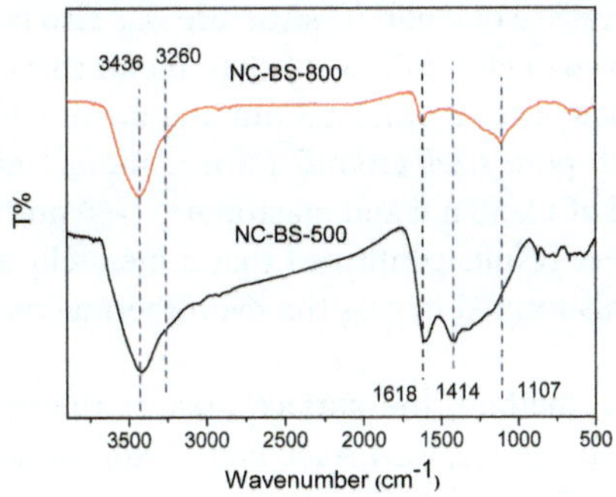

Fig. 1 FT-IR spectra of the NC-BS catalysts.

The textural properties of the NC-BS samples were investigated using N_2 adsorption/desorption analysis. Fig. 2 shows the nitrogen adsorption/desorption results for the NC-BS-500 and NC-BS-800 samples. As can be seen from Fig. 2(a), both samples displayed type IV adsorption isotherms with a H1-type hysteresis loop, clearly indicating the presence of a microporous and mesoporous structure.[39] Besides that, there was a significant amount of N_2 adsorption within the high relative pressure range of 0.9–1.0, implying a large external surface area and the presence of macropores. It is obvious that the adsorption quantity for NC-BS-800 was much higher than that for NC-BS-500, indicating the larger surface area

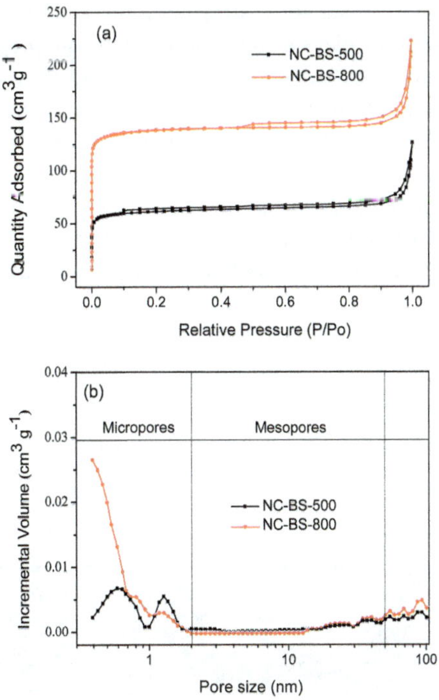

Fig. 2 Nitrogen adsorption/desorption results (a) and the pore size distributions (b) of NC-BS-500 and NC-BS-800.

of NC-BS-800. The presence of a porous structure was also supported by the pore size distributions shown in Fig. 2(b). NC-BS-500 contained a considerable amount of micropores with pore sizes around 1.3 nm and 0.7 nm, while NC-BS-800 had more micropores with pore sizes around 1.3 nm and 0.4 nm. Moreover, mesopores with a pore size of 14–50 nm and macropores (>50 nm) were also present in the two samples. These results confirmed that a hierarchical micro–meso–macroporous structure was formed during the hydrothermal treatment and carbonization process.

Based on the BET method, the surface area of the carbon materials was calculated. As shown in Table 1, NC-BS-500 had a high surface area of 287 $m^2 g^{-1}$ and a pore volume of 0.245 $cm^3 g^{-1}$, indicating that abundant pores with a hierarchical pore structure were successfully achieved through the hydrothermal treatment and carbonization process with bamboo shoots as the raw material without using chemical activation to create pores. With increase of the carbonization temperature to 800 °C, the surface area and pore volume increased up to 670 $m^2 g^{-1}$ and 0.306 $cm^3 g^{-1}$. The porosity of NC-BS-800 consisted of 41 vol% meso-/macro-porosity and 59 vol% microporosity. The results suggest that the carbonization temperature had an effect on the pore structure of the carbons. The greater porosity at a higher temperature might be caused by the decomposition of unstable components to generate gaseous species (*e.g.* CO, CO_2, N_2, and NH_3) during the carbonization process. Thus, the bamboo shoot-derived carbon materials had a porous microstructure with a high surface area, rich in nitrogen and oxygen functionalities, which made them appealing catalytic supports for loading metal particles with a high dispersion. Benefiting from these features, the porous heteroatom doped carbon based catalysts would disperse well in water, which would facilitate the catalytic hydrogenation reaction in water.[46]

An appropriate support plays a significant role in the preparation of controllable sized Pt nanoparticles. Fig. 3 shows the TEM images of the Pt/NC-BS-500 and Pt/NC-BS-800 catalysts and the STEM mapping images of the Pt/NC-BS-800 catalyst. As clearly seen from Fig. 3(a) and (b), the Pt nanoparticles were well dispersed on the both the NC-BS-500 and NC-BS-800 supports with small-sized nanoparticles. The average metal particle size of Pt/NC-BS-500 and Pt/NC-BS-800 was about 4.3 nm and 3.4 nm, respectively. The HRTEM image of the Pt/NC-BS-800 catalyst is presented in Fig. 3(c). The lattice fringes with a spacing of 0.22 nm

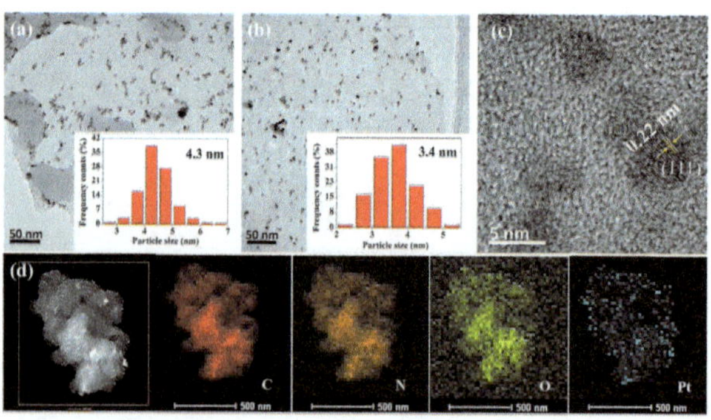

Fig. 3 TEM images of (a) Pt/NC-BS-500 and (b) Pt/NC-BS-800, (c) HRTEM image, (d) HAADF-STEM and corresponding mapping images of Pt/NC-BS-800.

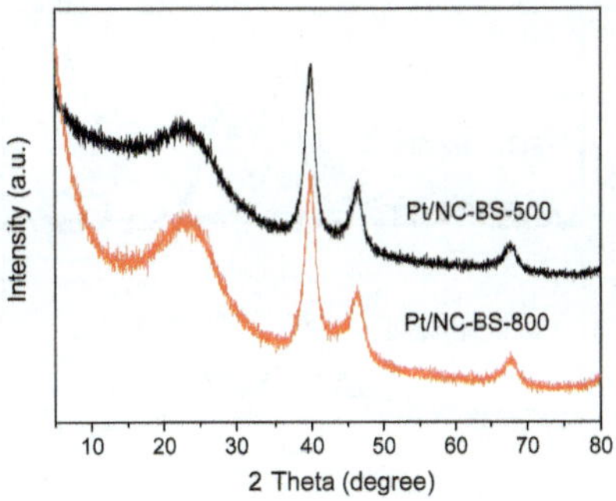

Fig. 4 XRD patterns of the Pt/NC-BS-500 and Pt/NC-BS-800 samples.

were ascribed to the (111) diffraction plane of the Pt nanoparticles. Fig. 3(d) shows the HAADF-STEM image and the corresponding mapping images of the Pt/NC-BS-800. The STEM mapping images supported the existence of Pt on the carbon support (Fig. S3, ESI†), and the N and O atoms were homogenously embedded in the carbon framework of Pt/NC-BS-800.

The presence of amorphous carbon and Pt nanoparticles was also investigated using XRD. XRD patterns of the Pt/NC-BS samples are shown in Fig. 4. A pronounced diffraction peak at 24.0° was observed for both the Pt/NC-BS-500 and Pt/NC-BS-800 samples, which was ascribed to amorphous carbon. The peaks located at 39.7°, 46.2° and 67.4° were assigned to the (111), (200) and (220) planes of metallic Pt,[22] consistent with the results of HRTEM. No other peaks were found, indicating that no impurities were formed during the preparation of the carbon supports and Pt catalysts. Most of the soluble metal ions such as Mg, Na and K in the starting materials were removed when the hydrochars were washed thoroughly with distilled water. The crystallite sizes according to the Scherrer equation using the (111) peak were 5.8 nm and 5.3 nm for Pt/NC-BS-500 and Pt/NC-BS-800, respectively, supporting successful loading of the Pt nanoparticles onto the surface of the bamboo shoot-derived carbon support. The results imply that both the NC-BS-500 and NC-BS-800 supports can act as excellent substrates for anchoring Pt nanoparticles.

To obtain further insight into the structure of the catalyst, XPS was performed to detect the types of N in the Pt/NC-BS. In Fig. 5(a), the nitrogen peaks of Pt/NC-BS could be fitted with three peaks. The peak at 400.9 eV corresponds to quaternary-N, which was the most stable nitrogen species in the N-doped carbons. This supported that nitrogen atoms were successfully incorporated into the carbon lattice. The other two peaks that occurred at 400.3 eV and 398.5 eV were assigned to pyrrolic-N and hexagonal pyridinic-N, respectively, reflecting the nitrogen atoms within the free ligand.[47,48] The surface nitrogen content calculated through XPS measurement was 5.3 at% for Pt/NC-BS-500 and 2.6 at% for Pt/NC-BS-800, which is in good agreement with the results from elemental analysis. Based on the XPS spectra analysis, the proportions of quaternary-N, pyrrolic-N and pyridinic-N were 33.4%, 33.3% and 33.2% for Pt/NC-BS-500 and 31.25%, 37.5% and 31.25% for Pt/NC-BS-800, respectively.

Combined with the FT-IR results, the high content of pyrrolic-N and pyridinic-N species in Pt/NC-BS-500 and Pt/NC-BS-800 supported the presence of rich basic

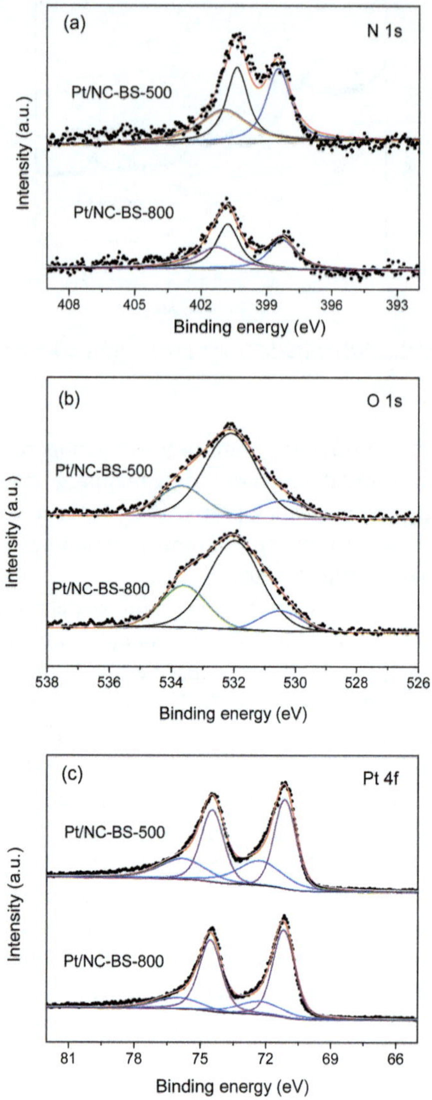

Fig. 5 XPS spectra of (a) N 1s (b) O 1s and (c) Pt 4f for Pt/NC-BS-500 and Pt/NC-BS-800.

functional sites on the surface of the bamboo shoot-derived carbons. It is worth noting that although the nitrogen content decreased as the carbonization temperature was elevated from 500 °C to 800 °C, the relative percentages of the pyrrolic-N and pyridinic-N species remained almost unchanged. Fig. 5(b) shows the O 1s XPS results for the Pt/NC-BS catalysts. The peaks located at around 532.1 eV, 533.6 eV and 530.4 eV can be attributed to hydroxyl and carbonate species, the adsorbed water and the adsorbed oxygen species, respectively.[49] The surface oxygen content was 7.5 at% for Pt/NC-BS-500 and 5.8 at% for Pt/NC-BS-800, indicating the presence of Lewis acidic sites on the carbon materials, and the amount of acidic functional groups also decreased as the carbonization temperature was increased. Thus, the nitrogen and oxygen functionalities on the surface might act as Lewis basic sites and Lewis acidic sites, and were effective in anchoring metal nanoparticles.[44] The surface elements and the chemical states of the Pt species of the Pt/NC-BS were also investigated using XPS. As shown in Fig. 5(c), for Pt/NC-BS-500, the stronger peaks at 71.1 eV and 74.4 eV for Pt 4f7/2 and Pt 4f5/2 belonged to metallic Pt^0. The other doublet at 72.3 eV and 75.8 eV could be assigned to Pt^{2+} states due to surface layer formation of Pt–O or Pt–N bonds.[22] The values were highly consistent with those reported before.[22] The atomic ratio of Pt^0 calculated from the XPS results was 60.2% of the total Pt in the Pt/NC-BS-500 catalyst. On increasing the carbonization temperature of the heteroatom doped carbon support, the atomic ratio of Pt^0 in the Pt/NC-BS-800 increased to 75.0%. The results suggest that the Pt species on the surface of the bamboo shoot-derived carbon materials were mainly in the metallic state and more Pt^0 was formed with the carbon support prepared using a higher carbonization temperature.

To characterize the interaction effects between the Pt and the support for the Pt/NC-BS catalysts, H_2-TPR measurements were conducted. As shown in Fig. 6, the peak at about 100–300 °C in the H_2-TPR profiles of the Pt/NC-BS catalysts was attributed to the H_2 consumed by Pt^{2+} reduction. The peak at 350–700 °C might be assigned to reduction of the surface acidic functional groups of the heteroatom doped carbon. Obviously, the reduction temperature for Pt oxide in Pt/NC-BS-500 was higher than that with Pt/NC-BS-800, indicating stronger interactions between the Pt and carbon support in Pt/NC-BS-500 than in Pt/NC-BS-800.[50,51] In brief, the carbonization temperature affected the surface properties and the structure of the heteroatom doped carbon, and thus changed the interactions between the Pt and carbon support. Note that NC-BS-500 possessed a high surface area (287 $m^2\ g^{-1}$) with a hierarchical pore structure and abundant functional groups on the surface, which could provide sufficient active sites for anchoring small-sized Pt nanoparticles with a high dispersion. For NC-BS-800, although the higher carbonization temperature caused a loss of nitrogen and oxygen species, it also increased the surface area, porosity and graphitization degree of the carbon, and NC-BS-800 has also been proven to be an excellent support for stabilizing Pt nanoparticles of a small size. However, the differences in the surface properties and structure of the support, and in the interactions between Pt and the support might eventually affect the catalytic performance of the Pt catalysts in water.[35]

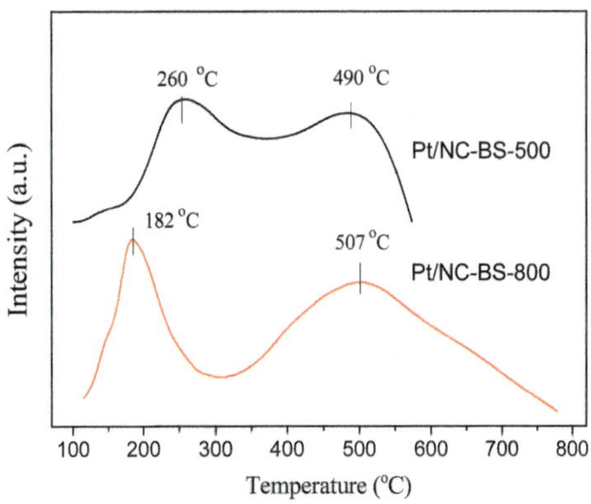

Fig. 6 H_2-TPR results of the Pt/NC-BS catalysts.

Selective hydrogenation of furfural with Pt/NC-BS catalysts in aqueous media

FFA contains both a C=O group and a furan ring, which are unstable during hydrogenation processes. Owing to the inherent reactivity of FFA molecules, selective hydrogenation of FFA to produce the desired products is still tough work.[22] Here, selective conversion of FFA to FA or CPO in water was chosen as a model reaction to study the catalytic performance of the Pt/NC-BS catalysts. To find the optimal conditions for selective C=O reduction of FFA to obtain FA, the catalytic reactions were firstly conducted at 100 °C using 1 MPa H_2 in an aqueous solution in the presence of the Pt/NC-BS catalyst. As expected, the Pt/NC-BS catalysts could disperse well in the reaction solution due to the presence of abundant functional groups on the surface of the support. The results are displayed in Table 2. Pt/NC-BS-500 (entry 4, Table 2) exhibited a high FFA conversion of 91.8% with a high selectivity toward FA of >99%, after 2 h under mild conditions (Scheme 3). No other product was detected in the reaction solution, indicating a highly selective performance toward FA of Pt/NC-BS-500. For comparison, commercial Pt/C exhibited a lower activity (~70% conversion to FA, entry 8, Table 2) than Pt/NC-BS-500 under similar conditions. The results indicate that Pt/NC-BS-500 is a good chemoselective hydrogenation catalyst, which preferred to form FA at 100 °C in water. To further confirm the results, the effect of the reaction time on the FFA conversion and the key products was investigated at 100 °C using 1 MPa H_2 in the presence of Pt/NC-BS-500. As shown in Table 2, at the beginning of the reaction, FFA was converted to FA quickly and >75% conversion of FA was achieved within 1 h. As the reaction continued, the reaction rate decreased gradually and the FFA was completely transformed into FA over 4 h (entry 6, Table 2). Notably, during the process, no other product was detected and the selectivity for FA always remained >99%. The mass balance for FFA and the detected products was >99%. Moreover, no further reaction occurred on prolonging the reaction time to 5 h. These results demonstrate that FA was formed directly through hydrogenation of the C=O bonds in the FFA and that Pt/NC-BS-500 had a strong ability for selective hydrogenation of the C=O bonds while leaving the furan ring unaffected under mild conditions. NC-BS-500 has a hierarchical pore structure with a high surface area and rich surface functional

Table 2 Hydrogenation of FFA over the Pt/NC-BS-500 catalyst with different reaction times[a]

Entry	Time (min)	Conv. (%)	Selectivity (%) FA	Activity (h^{-1})
1	20	27.9	>99	326.1
2	40	61.9	>99	361.8
3	60	75.4	>99	293.8
4	120	91.8	>99	178.9
5	180	95.1	>99	123.4
6	240	>99	>99	97.4
7	300	>99	>99	77.9
8[b]	120	70.0	>99	136.4

[a] Reaction conditions: furfural (0.24 g), 5 wt% Pt/NC-BS-500 catalyst (0.025 g), 10 mL H_2O, 1 MPa H_2, 100 °C. [b] 5 wt% Pt/C catalyst (0.025 g).

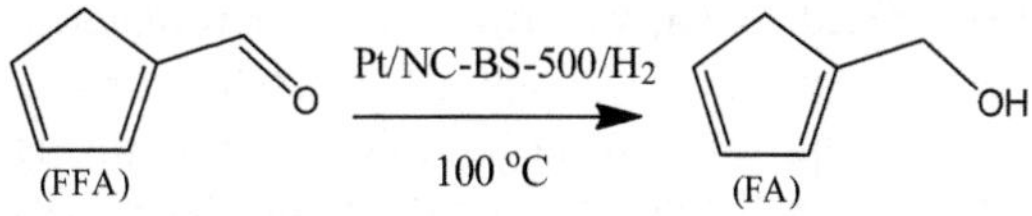

Scheme 3 Selective conversion of FFA to FA in water under mild conditions.

groups, which was favourable for high dispersion and stabilization of the Pt nanoparticles with a small size on the surface. Moreover, the abundant surface functional groups on the support facilitated the Pt catalyst dispersing well in water, which would bring the catalyst into full contact with the substrate during the reaction.[52] These properties led to the high catalytic activity of Pt/NC-BS-500 for FFA hydrogenation to FA in water under mild conditions.

When the Pt/NC-BS-800 catalyst was used to catalyze the hydrogenation of FFA at 100 °C with 1 MPa H_2 (entry 3, Table 3), an inferior catalytic activity was exhibited with a low FFA conversion of 45.2% and FA selectivity of 88.5%. This result suggests that the NC-BS-800 support is not effective for the transformation of FFA to FA under mild conditions. Nevertheless, it is interesting that small amounts of HCP (3.8% selectivity) and CEO (3.6% selectivity) were observed with Pt/NC-BS-800 as the catalyst, which were formed through further rearrangement of FA in water. That the rearrangement reaction even occurred under such mild conditions (100 °C and 1 MPa H_2) implied that Pt/NC-BS-800 had an ability for FFA rearrangement, which is a key step in the transformation of FFA to CPO. These results reflect that the catalyst support has a significant influence on the reaction efficiency and product distribution.

Next, we turned our attention to the more challenging conversion of FFA to CPO using the Pt/NC-BS catalysts. As small amounts of HCP and CEO were formed through hydrogenation and rearrangement of FFA with the Pt/NC-BS-800 catalyst at 100 °C using 1 MPa H_2, the reaction temperature was elevated to promote the reaction. As expected, with the reaction temperature elevated to 150 °C using 1 MPa H_2 (entry 4, Table 3), the FFA was completely converted within 4 h and CPO was found to be the main product with a yield of 53.5%. Certain amounts of 4-hydroxy-2-cyclopentenone (HCP, 14.8%), 2-cyclopentenone (CEO, 15.9%) and CPL (10.8%) were also detected, but no FA was found. These results reflect that the FFA

Table 3 The influence of the reaction conditions on the furfural hydrogenation[a]

Entry	Catalyst	H_2 (MPa)	Conv. (%)	Selectivity (%) FA	HCP	CEO	CPO	CPL	Carbon balance (%)
1	Pt/NC-BS-500	1	99	29.7	17.0	13.6	31.6	3.6	95.5
2	Pt/NC-BS-500	3	>99	13.9	10.3	20.9	49.6	6.9	101.6
3[b]	Pt/NC-BS-800	1	45.2	88.5	3.8	3.6	0	0	95.9
4	Pt/NC-BS-800	1	>99	0	14.8	15.9	53.5	10.8	95.0
5	Pt/NC-BS-800	3	>99	0	6.5	5.4	76.0	8.6	96.5
6[c]	Pt/C	3	98.2	36.8	2.2	4.1	39.3	4.5	86.9

[a] Reaction conditions: furfural (0.24 g), 5 wt% Pt/NC-BS catalyst (0.025 g), 10 mL H_2O, 150 °C, 4 h. [b] Reacted at 100 °C. [c] 5 wt% Pt/C catalyst (0.025 g).

was completely converted to FA, which was further transformed fully to five-membered carbon ring products through a Piancatelli rearrangement of FA at 150 °C in water with the Pt/NC-BS-800 catalyst. When the H_2 pressure was increased from 1 MPa to 3 MPa, the yield of the desired CPO was further enhanced from 53.5% to 76.0% within 4 h (entry 5, Table 3) and the yields of the byproducts (HCP, CEO and CPL) were decreased to 6.5%, 5.4% and 8.6%, respectively (Scheme 4). These results suggest that Pt/NC-BS-800 is highly active for FA rearrangement, supporting that it is a good rearrangement catalyst, preferential for producing CPO. For comparison, Pt/NC-BS-500 afforded a lower CPO yield of 31.6% with 100% conversion of FFA at 150 °C using 1 MPa H_2 after 4 h (entry 1, Table 3), accompanied by the formation of FA (29.7%), CEO (13.6%), HCP (17.0%) and CPL (3.6%). When the hydrogen pressure was increased to 3 MPa (entry 2, Table 3), the yield of the desired CPO was improved to 49.6%, while the FA yield remained as high as 13.9%. The results suggest that FA could not be fully converted to HCP through a rearrangement reaction with the Pt/NC-BS-500 catalyst within 4 h. Therefore, it can be concluded that Pt/NC-BS-500 is effective for selective hydrogenation of the C=O bonds in the FFA to form FA, but ineffective for further rearrangement of FA to HCP, which has a high activation energy, and thus this results in a lower activity for CPO. Based on the analysed results mentioned above, by comparing NC-BS-500 and NC-BS-800, the variation in the nitrogen and oxygen content, adjusted through the carbonization temperature, led to changes in the support including the functional groups, the surface active sites and the acidity/basicity. As reported before,[29] both strong acidic and basic conditions will inhibit the transformation of FFA to the desired CPO in water, while a mild acidity of the catalyst would favor the reaction. In our system, Pt/NC-BS-800 possessed both weak Lewis acidic sites and mild Lewis basic sites, which could be effective in promoting the required FFA conversion and suppressing the undesired side reactions. For Pt/NC-BS-500, its stronger acidity and basicity compared to Pt/NC-BS-800 might primarily contribute to the lower activity toward CPO. Thus, tunable hydrogenation of FFA to produce FA or CPO selectively can be achieved by controlling the surface functionalities of the Pt/NC-BS and the reaction conditions.

In order to clarify the reaction pathway of the FFA hydrogenation, some control experiments were also performed. Different substrates formed in the FFA hydrogenation reaction were employed as the reactants, which were reacted at

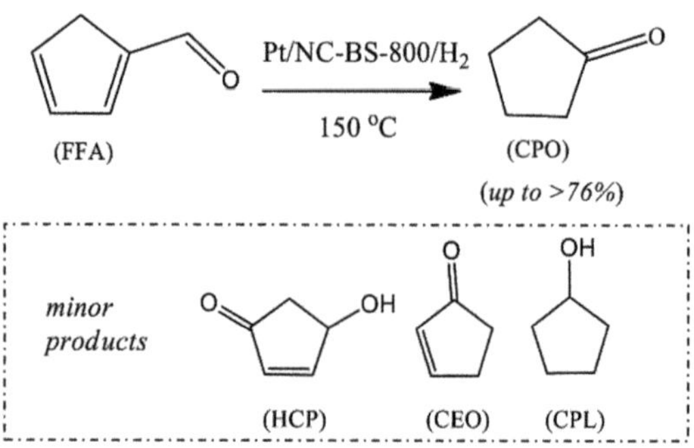

Scheme 4 Selective conversion of FFA to CPO in water.

150 °C using 3 MPa H_2 in water. When FFA was used as the reactant (entry 1, Table 4), FA (21.5% selectivity), HCP (14.2% selectivity), CPO (20.1% selectivity) and CEO (34.3% selectivity) were formed within 0.5 h, along with a small amount of CPL (7.1%). In the case of FA (entry 2, Table 4), the FA also mainly converted into these five-membered carbon ring products through rearrangement reactions, including HCP, CEO, CPO and CPL, along with a higher CPO selectivity of 33.3% after 0.5 h. This meant that FA is the key intermediate in the FFA hydrogenation process. Under suitable reaction conditions, FA could be transformed into HPC through a rearrangement reaction of FA, followed by further hydrogenation and dehydration to generate CPO in water. Additionally, CPL was the only product when CPO was employed (entry 3, Table 4) and no reaction proceeded when CPL was used as the substrate (entry 4, Table 4), indicating that excessive hydrogenation would lead to the formation of CPL.

Based on the reaction results above, a possible reaction pathway for FFA hydrogenation was also proposed (Scheme 5). In the first stage, reduction of FFA would provide the initial intermediate FA through hydrogenation of the C=O with H_2. Secondly, under suitable reaction conditions in water, the initially-formed FA would be converted to HCP through a Piancatelli rearrangement reaction,[16] which was another key intermediate in the conversion of FFA to CPO. During this process, water is essential as the required protons were provided by the auto-dissociation of water. The HCP intermediate would then dehydrate quickly to form CEO, followed by hydrogenation of the unsaturated C=C bonds within the five-membered ring to produce CPO.[31] Due to the presence of C=O bonds in the CPO, the CPO might be further reduced to CPL through an excessive hydrogenation reaction. The product selectivity was primarily determined by the catalyst and reaction conditions. When the reaction was performed under the mild condition of 100 °C, FA was easily produced in water with the Pt/NC-BS-500 catalyst. However, it was difficult to further convert the FA into five-membered carbon rings through the rearrangement reaction at low temperature, which has a high activation energy. A higher temperature along with a suitable catalyst (*e.g.* Pt/NC-BS-800 with mild Lewis acidic sites and Lewis basic sites) would promote the rearrangement reaction of FA in water and favor for the formation of CPO.

In order to verify the reaction pathway for FFA hydrogenation detailed in Scheme 5, we further studied the reaction progress profiles for the hydrogenation of FFA in water using the Pt/NC-BS-800 catalyst. Fig. 7 shows the product

Table 4 Hydrogenation of different substrates[a]

			Selectivity (%)				
Entry	Substrate	Conv. (%)	FA	HCP	CEO	CPO	CPL
1	FFA	66.1	21.5	14.2	34.3	20.1	7.1
2	FA	73.0	—	35.9	16.7	33.3	10.9
3	CPO	85.3	n.d.	n.d.	n.d.	—	99.6
4	CPL	2.1	n.d.	n.d.	n.d.	n.d.	—

[a] Reaction conditions: reactant (2.5 mmol), 5 wt% Pt/NC-BS-800 catalyst (0.025 g), 10 mL H_2O, 3 MPa H_2, 150 °C, 0.5 h.

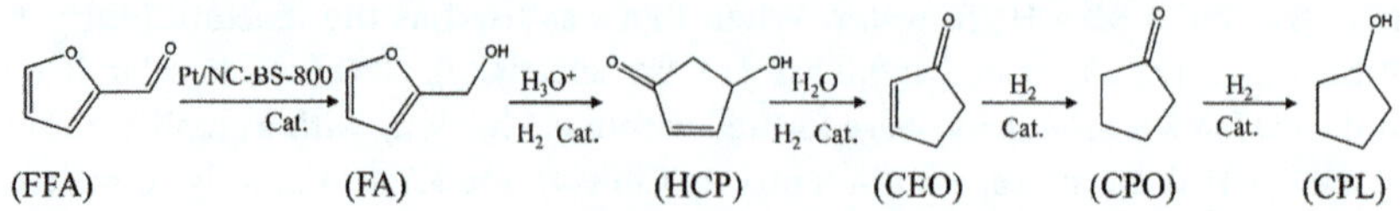

Scheme 5 The proposed reaction pathway for hydrogenation of furfural in water over Pt/NC-BS-800.

distribution with Pt/NC-BS-800 under the conditions of 150 °C and 3 MPa H_2 as a function of time. The product distribution of the FFA hydrogenation is quite complicated, including FA, HCP, CEO, CPO and CPL. At the beginning of the reaction, FFA was quickly converted to FA with about 50% conversion of the FFA within 0.5 h, and the selectivities for FA (21.5%), HCP (14.2%), CEO (34.3%), CPO (20.0%) and CPL (7.6%) were obtained. Only a small amount of FA remained in the product mixture reflecting the high activity for FA rearrangement of Pt/NC-BS-800. As the reaction time was prolonged, the FFA was mostly all transformed within 1 h. The FA concentration in the product mixture decreased rapidly and no FA remained after the reaction had proceeded for 2 h. Meanwhile, the concentration of the desired CPO increased quickly and reached its highest value after reaction for 4 h, correspondingly, the HCP and CEO concentrations decreased gradually during the process. These results indicate that FA was further converted to HCP through a rearrangement reaction, and subsequently dehydration and hydrogenation occurred quickly to form CPO. Notably, CPL was detected in a low concentration when the reaction had proceeded for 4 h, indicating that Pt/NC-BS-800 could prevent excessive hydrogenation of CPO to CPL during the process. On prolonging the reaction time to 6 h, CPO was then converted to CPL *via* hydrogenation. These results were similar to the previous reports by Hronec *et al.*[30]

The reusability of the Pt/NC-BS-500 catalyst for FFA hydrogenation was also investigated and the results are shown in Fig. 8. The catalyst was separated, washed with deionized water and then dried at 40 °C after the reaction. The activity of the Pt/NC-BS-500 catalyst was found to be quite stable under the reaction conditions. No obvious decrease in the FFA conversion was observed

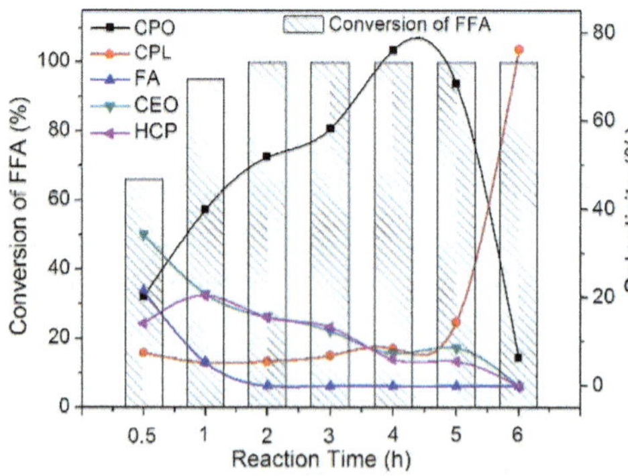

Fig. 7 Product distribution as a function of time with the Pt/NC-BS-800 catalyst. Reaction conditions: furfural (0.24 g), 5 wt% Pt/NC-BS-800 catalyst (0.025 g), 10 mL H_2O, 3 MPa H_2.

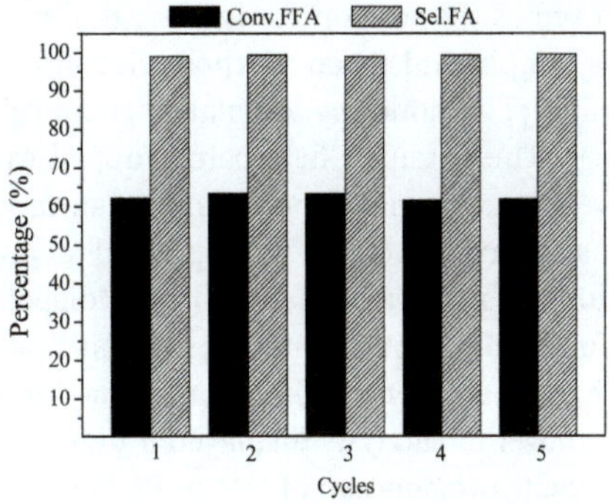

Fig. 8 Recyclability of the Pt/NC-BS-500 catalyst for the reaction of furfural hydrogenation. Reaction conditions: furfural (0.24 g), 5 wt% Pt/C catalyst (0.025 g), 10 mL H_2O, 1 MPa H_2, 100 °C, 40 min.

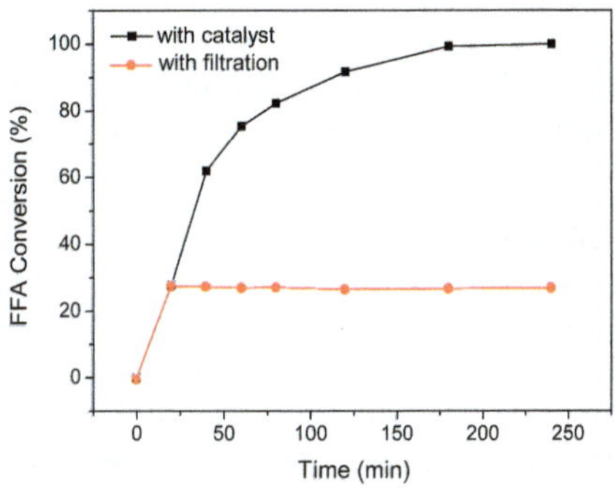

Fig. 9 Activity profile for the conversion of FFA. with the Pt/NC-BS-500 catalyst and with filtration.

after four reactions. At the same time, the selectivity for FA was maintained at >98% during the recycling experiments, indicating that the Pt/NC-BS-500 catalyst was highly stable during FFA hydrogenation in water.

We also designed a parallel experiment to test the robustness of the catalyst. An activity profile for FFA hydrogenation over the Pt/NC-BS-500 catalyst as a function of time is shown in Fig. 9. The conversion of the FFA was found to increase when the reaction time was prolonged and finally reached 100% after reaction for 240 min. However, when the catalysts were removed from the reaction system by filtration at 30 min, the FFA conversion remained unchanged as the reaction time proceeded, indicating that the hydrogenation reaction stopped after the catalyst was removed. ICP analysis results showed that no obvious Pt was detected in the reaction solution after the reaction. These results also demonstrate that the Pt/NC-BS-500 catalyst is stable and robust for FFA hydrogenation in water.

Conclusions

In summary, highly active Pt/NC-BS catalysts were explored for selective transformation of FFA into either FA or CPO in water. A porous heteroatom doped carbon material was used as the support for loading the Pt nanoparticles, which was prepared using a simple and green method with crude biomass (bamboo shoots) as the raw material without any templates, chemical activation or additional nitrogen sources. The obtained heteroatom doped carbon materials had a high BET surface area, large porosity and abundant surface functional groups. The small-sized Pt nanoparticles could be deposited on the biomass-derived porous heteroatom doped carbon material with high dispersion. The introduction of hydrophilic functional groups into the carbon materials endowed the novel Pt catalysts with a good water dispersion and reaction stability. With their excellent features, the novel Pt catalysts displayed a high activity and controlled tunable selectivity for the hydrogenation of FFA to FA or CPO in water, which was modulated by adjusting the carbonization temperature of the support and the reaction conditions. The Pt/NC-BS-500 catalyst was highly efficient for FFA hydrogenation, with a FFA conversion of >99% and FA selectivity of >99% at 100 °C using 1 MPa H_2. The Pt/NC-BS-800 catalyst was favored for transformation of FFA to produce CPO, with a high yield of >76% at 150 °C using 3 MPa H_2. The catalysts were found to be easily recyclable and stable in the liquid-phase system. Our work has provided an effective and green route for the selective conversion of FFA to FA or CPO in water using biomass-derived materials.

Acknowledgements

This work was financially supported by 13th Five-Year the National key R&D projects (2017YFD0600805), the National Natural Science Foundation of China (31370563 and 21433001), the Young Taishan Scholars Program of Shandong Province (tsqn20161052) and the Open Research Fund of Fujian Provincial Key Laboratory of Ecology-Toxicological Effects & Control for Emerging Contaminants (PY16004).

References

1 P. Gallezot, *Chem. Soc. Rev.*, 2012, **41**, 1538–1558.
2 S. Zhu, Y. Xue, J. Guo, Y. Cen, J. Wang and W. Fan, *ACS Catal.*, 2016, **6**, 2035–2042.
3 G. Xu, J. Han, B. Ding, P. Nie, J. Pan, H. Dou, H. Li and X. Zhang, *Green Chem.*, 2015, **17**, 1668–1674.
4 N. Meiri, Y. Dinburg, M. Amoyal, V. Koukouliev, R. V. Nehemya, M. V. Landau and M. Herskowitz, *Faraday Discuss.*, 2015, **183**, 197–215.
5 Y. Zhai, Y. Dou, D. Zhao, P. F. Fulvio, R. T. Mayes and S. Dai, *Adv. Mater.*, 2011, **23**, 4828–4850.
6 A.-H. Lu, G.-P. Hao, Q. Sun, X.-Q. Zhang and W.-C. Li, *Macromol. Chem. Phys.*, 2012, **213**, 1107–1131.
7 B. Hu, K. Wang, L. Wu, S.-H. Yu, M. Antonietti and M.-M. Titirici, *Adv. Mater.*, 2010, **22**, 813–828.
8 C. Aellig and I. Hermans, *ChemSusChem*, 2012, **5**, 1737–1742.
9 J.-P. Lange, E. van der Heide, J. van Buijtenen and R. Price, *ChemSusChem*, 2012, **5**, 150–166.
10 Y. Yang, C. Ochoa-Hernández, V. A. de la Peña O'Shea, P. Pizarro, J. M. Coronado and D. P. Serrano, *Appl. Catal., B*, 2014, **145**, 91–100.

11 S. A. Khromova, M. V. Bykova, O. A. Bulavchenko, D. Y. Ermakov, A. A. Saraev, V. V. Kaichev, R. H. Venderbosch and V. A. Yakovlev, *Top. Catal.*, 2016, **59**, 1413–1423.
12 D. Scholz, C. Aellig and I. Hermans, *ChemSusChem*, 2014, **7**, 268–275.
13 W. Xu, H. Wang, X. Liu, J. Ren, Y. Wang and G. Lu, *Chem. Commun.*, 2011, **47**, 3924–3926.
14 M. Hronec and K. Fulajtarová, *Catal. Commun.*, 2012, **24**, 100–104.
15 V. V. Ordomsky, J. C. Schouten, J. van der Schaaf and T. A. Nijhuis, *Appl. Catal., A*, 2013, **451**, 6–13.
16 M. Hronec, K. Fulajtárová, I. Vávra, T. Soták, E. Dobročka and M. Mičušík, *Appl. Catal., B*, 2016, **181**, 210–219.
17 R. Karinen, K. Vilonen and M. Niemelä, *ChemSusChem*, 2011, **4**, 1002–1016.
18 J. B. Binder, J. J. Blank, A. V. Cefali and R. T. Raines, *ChemSusChem*, 2010, **3**, 1268–1272.
19 K. Yan and A. Chen, *Energy*, 2013, **58**, 357–363.
20 Y. Nakagawa, M. Tamura and K. Tomishige, *ACS Catal.*, 2013, **3**, 2655–2668.
21 A. Corma, S. Iborra and A. Velty, *Chem. Rev.*, 2007, **107**, 2411–2502.
22 X. Chen, L. Zhang, B. Zhang, X. Guo and X. Mu, *Sci. Rep.*, 2016, **6**, 28558.
23 D. C. Elliott and T. R. Hart, *Energy Fuels*, 2009, **23**, 631–637.
24 R. V. Sharma, U. Das, R. Sammynaiken and A. K. Dalai, *Appl. Catal., A*, 2013, **454**, 127–136.
25 K. Yan, C. Jarvis, T. Lafleur, Y. Qiao and X. Xie, *RSC Adv.*, 2013, **3**, 25865–25871.
26 S. Wei, H. Cui, J. Wang, S. Zhuo, W. Yi, L. Wang and Z. Li, *Particuology*, 2011, **9**, 69–74.
27 X. Chen, H. Li, H. Luo and M. Qiao, *Appl. Catal., A*, 2002, **233**, 13–20.
28 R. Fang, H. Liu, R. Luque and Y. Li, *Green Chem.*, 2015, **17**, 4183–4188.
29 G.-S. Zhang, M.-M. Zhu, Q. Zhang, Y.-M. Liu, H.-Y. He and Y. Cao, *Green Chem.*, 2016, **18**, 2155–2164.
30 M. Hronec, K. Fulajtarová and T. Liptaj, *Appl. Catal., A*, 2012, **437–438**, 104–111.
31 Y. Yang, Z. Du, Y. Huang, F. Lu, F. Wang, J. Gao and J. Xu, *Green Chem.*, 2013, **15**, 1932–1940.
32 X.-L. Li, J. Deng, J. Shi, T. Pan, C.-G. Yu, H.-J. Xu and Y. Fu, *Green Chem.*, 2015, **17**, 1038–1046.
33 J. Guo, G. Xu, Z. Han, Y. Zhang, Y. Fu and Q. Guo, *ACS Sustainable Chem. Eng.*, 2014, **2**, 2259–2266.
34 E. Ortel, S. Sokolov, C. Zielke, I. Lauermann, S. Selve, K. Weh, B. Paul, J. Polte and R. Kraehnert, *Chem. Mater.*, 2012, **24**, 3828–3838.
35 J. Kijeński, P. Winiarek, T. Paryjczak, A. Lewicki and A. Mikołajska, *Appl. Catal., A*, 2002, **233**, 171–182.
36 M. Li, F. Xu, H. Li and Y. Wang, *Catal. Sci. Technol.*, 2016, **6**, 3670–3693.
37 Z. Wei, J. Wang, S. Mao, D. Su, H. Jin, Y. Wang, F. Xu, H. Li and Y. Wang, *ACS Catal.*, 2015, **5**, 4783–4789.
38 J. Wang, G. Wang, S. Miao, J. Li and X. Bao, *Faraday Discuss.*, 2014, **176**, 135–151.
39 Q. F. Deng, L. Liu, X. Z. Lin, G. Du, Y. Liu and Z. Y. Yuan, *Chem. Eng. J.*, 2012, **203**, 63–70.
40 X. Fan, L. Zhang, G. Zhang, Z. Shu and J. Shi, *Carbon*, 2013, **61**, 423–430.

41 M. Yang, L. Guo, G. Hu, X. Hu, L. Xu, J. Chen, W. Dai and M. Fan, *Environ. Sci. Technol.*, 2015, **49**, 7063–7070.

42 J. Sun, Z. Q. Ding, Q. Gao, H. Xun, F. Tang and E. D. Xia, *J. Agric. Food Chem.*, 2016, **64**, 2498–2505.

43 W. J. Chang, M. J. Chang, S. T. Chang and T. F. Yeh, *J. Wood Chem. Technol.*, 2013, **33**, 144–155.

44 A. Misra, P. K. Tyagi, M. K. Singh and D. S. Misra, *Diamond Relat. Mater.*, 2006, **15**, 385–388.

45 S. Y. Oh, D. I. Yoo, Y. Shin, H. C. Kim, H. Y. Kim, Y. S. Chung, W. H. Park and J. H. Youk, *Carbohydr. Res.*, 2005, **340**, 2376–2391.

46 M. Li, X. Xu, Y. Gong, Z. Wei, Z. Hou, H. Li and Y. Wang, *Green Chem.*, 2014, **16**, 4371.

47 X. Cui, Y. Li, S. Bachmann, M. Scalone, A. E. Surkus, K. Junge, C. Topf and M. Beller, *J. Am. Chem. Soc.*, 2015, **137**, 10652–10658.

48 R. V. Jagadeesh, A.-E. Surkus, H. Junge, M.-M. Pohl, J. Radnik, J. Rabeah, H. Huan, V. Schünemann, A. Brückner and M. Beller, *Science*, 2013, **342**, 1073–1076.

49 K. Sutthiumporn and S. Kawi, *Hydrogen Energy*, 2011, **36**, 14435–14446.

50 Y. Ji, Z. Zhao, A. Duan, G. Jiang and J. Liu, *J. Phys. Chem. C*, 2009, **113**, 7186–7199.

51 X. Liu, X. Wang, G. Xu, Q. Liu, X. Mu and H. Liu, *J. Mater. Chem. A*, 2015, **3**, 23560–23569.

52 D. Liu, X. Chen, G. Xu, J. Guan, Q. Cao, B. Dong, Y. Qi, C. Li and X. Mu, *Sci. Rep.*, 2016, **6**, 21365.

Faraday Discussions

PAPER

Selective oxidation of lignocellulosic biomass to formic acid and high-grade cellulose using tailor-made polyoxometalate catalysts

Jakob Albert

Received 30th January 2017, Accepted 14th February 2017
DOI: 10.1039/c7fd00047b

The main goal of this project was to identify and optimize tailor-made polyoxometalate catalysts for a fractionated oxidation of lignocellulosic biomass (*i.e.* wood and residues from sugar or paper industries) to produce formic acid (FA) and high-grade cellulose for further processing *e.g.* in white biotechnology to provide bio-ethanol. Homogeneous vanadium precursors like sodium metavanadate and vanadyl sulfate as well as Keggin-type polyoxometalates (POMs) and more exotic structures like Anderson-, Wells-Dawson- and Lindqvist-type POMs were screened for the desired catalytic performance. The most promising behaviour was found using the Lindqvist-type POM $K_5V_3W_3O_{19}$, showing for the first time in the literature a selective oxidation of only hemicellulose and lignin to formic acid, while the cellulose fraction was untrapped. However, this can only be a first step towards the project goal as low product yields were obtained.

Introduction

Biomass is gaining economic significance not only for the production of food and animal feed but also for the production of platform chemicals, energy, and other bio-based products, due to the limited availability of fossil resources. Consequently, economic sectors that have so far relied on fossil resources are now finding themselves part of a shift towards a bio-based economy with biomass as a renewable raw material.[1] However, biomass also includes plants from which food can be produced. Rapid population growth has been accompanied by a growing demand for food and animal feed. This is resulting in increased competition for the limited arable land available between energy-rich biomass and food crops.[2] This conflict between food and energy production from biomass can be solved by using only those raw materials and remnants that are not suitable for the food industry.[3] Biomass consists of many materials of organic origin

Lehrstuhl für Chemische Reaktionstechnik der Universität Erlangen-Nürnberg, Egerlandstrasse 3, D-91058 Erlangen, Germany. E-mail: jakob.albert@fau.de; Fax: +49 9131 85 27421; Tel: +49 9131 85 67417

that contain carbon. Regarding suitable biogenic substrates, lignocellulose seems to be a promising carbon source for the production of liquid fuels and platform chemicals.[4] Lignocellulose forms the main part of plant material and is built up of cellulose (50%), hemicellulose (25–30%) and lignin (15–25%).[5] For further processing of these fractions, selective depolymerization followed by further chemical transformations is required.[6] However, biomass in general has a highly inhomogeneous composition with a low energy density. Therefore, the biorefining of lignocellulosic biomass using catalytic hydrolysis, solvolysis, liquefaction, pyrolysis, gasification, hydrogenolysis and hydrogenation is under intense scrutiny.[7–9] Various concepts have recently been put forward for converting biomass into platform chemicals for use in industrial chemistry or as fuels. One intensively studied approach for converting biomass into fuels involves its conversion into syngas followed by a catalytic Fischer–Tropsch synthesis.[10] Traditional processes for converting biomass into syngas include processes like biomass gasification,[11] reforming in supercritical water[12] or aqueous phase reforming.[13,14] All these processes are characterized by harsh reaction conditions within complex process equipment and by the production of a variety of gaseous, liquid and solid products. Another approach is to develop selective bio- and chemocatalytic process steps to convert lignocellulosic biomass into molecularly well-defined fuel molecules like γ-valerolactone (GVL), 1,4-pentanediol (1,4-PDO) and 2-methyltetrahydrofuran (2-MTHF).[15,16] One of the goals is to preserve nature's synthetic power to the largest possible extent by providing efficient and short defunctionalization sequences. However, to produce these fuel target molecules from largely oxygen-containing biomass feeds, a huge amount of hydrogen is necessary. Only if this hydrogen originates equally from biomass (which is complex and expensive as described above) or is provided from water electrolysis using excess renewable electricity (a requirement that would imply an unsteady fuel production process according to the availability of wind and sun) can this way of biofuel production be regarded as fully sustainable.

The direct technical relevance of these findings is limited as most of the studies deal with pure starting materials,[17] or even use biomass model compounds to study the general nature of the transformations of interest.[18]

Catalytic oxidation of biomass is also a growing field of research. The main efforts have been directed towards overcoming the recalcitrant nature of the feedstock using acid-catalyzed depolymerization followed by oxidative cleavage of the carbon bonds in the biomolecule.[19] Since 2011, our group has been active in exploring an alternative way to convert biomass into valuable chemicals, which differs from the ones mentioned above.[20–25] In this context, we presented first in 2011 a catalytic reaction that converts biomass selectively into formic acid (FA) and CO_2.[20] Later the name "OxFA process" was coined for it.[25] This biomass oxidation process is mildly exothermic and operates under mild temperature conditions, typically below 373 K, using molecular oxygen or synthetic air as environmental benign oxidants. As water is applied as a solvent, biomass feedstocks of different origin, composition and humidity can be applied without drying. As FA is produced as the only product in the liquid phase, tedious separation processes can be avoided. The only side-product at full conversion is pure carbon dioxide, which is formed in the gas phase and can be re-used for the production of new biomass substrates, *i.e.* algae. As the OxFA process converts a very broad range of biogenic raw materials into only two products that separate nicely into the gas and liquid phases, its simplicity and robustness are clear advantages compared to other biomass valorization technologies.

Polyoxometalates (POMs) are organo-metallic complexes of light transition metals like vanadium (V), niobium (Nb), tantalum (Ta), molybdenum (Mo) or tungsten (W) that are in their highest oxidation state and in combination with oxygen. These metal-oxo anion clusters can also contain a multitude of hetero-atoms like phosphorus (P), arsenic (As), silicon (Si) or germanium (Ge). POMs have attracted much attention in recent years because of their fascinating architectures and excellent physico-chemical properties including a strong Brønsted acidity, high proton mobility, fast multi-electron transfer, high solubility in various solvents, and resistance to hydrolytic and oxidative degradations in solution. Moreover, many properties of the polyoxometalate materials can be finely tuned by changing the constituents and counter cations. These features have made polyoxometalates very attractive in acid and oxidation catalysis.[26,27]

A key element of catalytic oxidation of biomass to formic acid is selection of the right catalyst for utilizing molecular oxygen as the oxidant. This relates to the fact that complete combustion of the biomass to produce water and CO_2 is the thermodynamically favored reaction pathway. The selected oxidation catalyst should promote the biomass oxidation to formic acid (partial oxidation). Consequently, the catalyst should have the ability to perform oxidative carbon–carbon bond cleavage and at the same time should prevent further oxidation. This partial oxidation requires an external electron transporter. In previous studies, Keggin-type polyoxometalates of the series $[H_{3+n}PV_nMo_{12-n}O_{40}]$ (also known as HPA-*n*) have been reported to fulfil these requirements.[19,22] For POM catalysed redox processes, it is known from the literature that in most cases the rate determining step is the re-oxidation of the reduced form of HPA-*n*, which should be strongly dependent on the oxygen concentration.[28,29] Moreover, it has been confirmed that for higher degrees of vanadium substitution inside the HPA-*n*, the re-oxidation of the reduced POM becomes faster at lower oxygen pressures. This is due to the higher redox-potential of vanadium compared to molybdenum.[30,31] However, at typical OxFA conditions of 20–30 bar oxygen pressure, POM re-oxidation is faster than oxidation of the substrate and biomass oxidation is consequently the rate determining step.[19,22] Amongst the catalysts of type HPA-*n*, HPA-5 has been reported to have the highest selectivity for formic acid formation.[22,24] It was also observed that for water-insoluble lignocellulosic biomass, HPA-5 can be completely re-oxidized at 10 bar oxygen pressure and the rate determining step is biomass oxidation. This was in contrast to lower vanadium-substituted catalysts from the HPA-*n* series like HPA-2.[25]

The solubility of the substrates in the reaction solvent also plays a major role. For complex water-insoluble biomass materials, such as cellulose or wood, the addition of sulfonic acids as solubility promoters yields significantly higher amounts of formic acid. The best results have been obtained with *p*-toluenesulfonic acid (TSA). Additionally, long-term stability tests proved that the latter is not consumed under the oxidative conditions.[21]

For this contribution, we wanted to expand the OxFA-technology on a higher level by realizing a fractionated conversion of the main components of lignocellulosic biomass, in order to produce biogenic formic acid and high-grade cellulose in one reaction vessel.

Results and discussion

Wood pulping and paper manufacturing processes typically separate the hemicellulose and cellulose fractions from the lignocellulosic feedstock through acid-catalyzed hydrolysis. This is an energy-intensive process with lignin as a by-product and it requires extensive pre-treatment and downstreaming.[32,33] However, from the basic concept of the OxFA process and the previous studies on polyoxometalates as oxidation catalysts, it is known that some homogeneous polyoxometalates firstly catalyze selective decomposition of the amorphous hemicellulose followed by oxidation of the phenolic lignin.[24,25] This offers interesting opportunities for an alternative valorization of lignocellulosic biomass. Consequently, for the present work the aim was to increase the valorization of lignocellulosic biomass in a more sustainable way by simultaneously producing the liquid hydrogen equivalent formic acid along with high-grade cellulose as a feedstock for further processing, such as fermentation to bioethanol.

In this context, appropriate catalyst systems for the stepwise oxidation of model substrates for lignocellulosic biomass have been investigated. As a starting point, monomers of the main components of lignocellulosic biomass with a well-defined molecular structure were used as model substrates. The use of these molecularly well-defined model substrates allows a better understanding of the influence of different catalytically active species on the stepwise degradation of the carbon framework. Initially, the screening experiments were performed with Keggin-type POMs of the HPA-*n* family. Up to now, these polyoxometalate catalysts have shown promising results for the OxFA process with high yields and high selectivity for formic acid.[25] The catalyst screening was further extended to other types of polyoxometalates including Lindqvist-type, Dawson-type and Anderson-type compounds. Furthermore, using catalysts from four different structural categories allows insight into the structural dependency of the catalyst activity and selectivity. To compare the results, two commercial homogenous vanadium-catalysts, namely vanadyl sulfate and sodium metavanadate, were also added to the screening list. The goal of the screening experiments was to find a catalyst that selectively converts hemicellulose and lignin fractions of the lignocellulosic biomass and does not oxidize the cellulose moiety (Fig. 1).

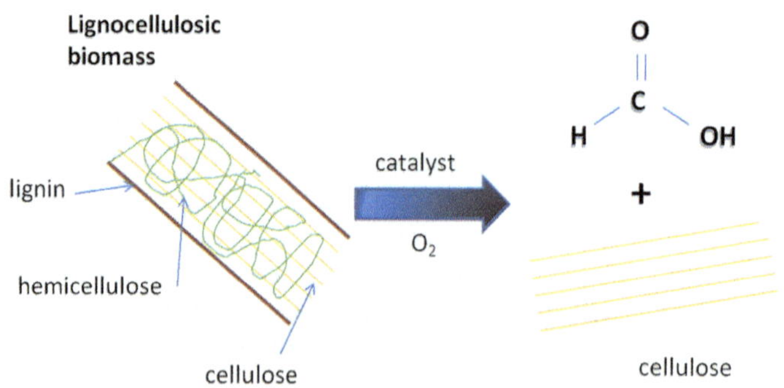

Fig. 1 Illustration representing the goal of the catalyst screening experiments.

Screening of suitable catalysts using model substrates

The screening experiments were carried out using eight model biomass substrates in a batch operated ten-fold reactor system (see Experimental section). As model substrates for cellulose, the water-soluble monomer glucose and the dimer cellobiose were used (Table 1, entries 1 and 2). Xylose, arabinose, mannose and galactose were used as model substrates for hemicellulose (Table 1, entries 3–6). Finally, phenol and 4-hydroxy-3-methylbenzoic acid (4-H3MB) were selected as model substrates for lignin (Table 1, entries 7 and 8). All the screening experiments were performed at 90 °C, with 20 bar oxygen pressure and 1000 rpm stirrer speed, using 0.1 mmol of the catalyst and 1 mmol of the substrate with 10 g of water as the solvent and a reaction time of 8 hours.

In the first set of experiments, various Keggin-type POMs of the series HPA-*n*, the most widely studied class of polyoxometalates for biomass oxidation, were tested. Therefore, vanadium substituted compounds (HPA-*n* with $n = 0$–6) were screened (see Table 1).

The experiments using vanadium substituted heteropolyacids showed consistent behaviour. Non- and mono-substituted Keggin-HPAs did not show any catalytic activity under the applied reaction conditions. This mediocre behaviour was attributed to the slow re-oxidation of HPA-0 and HPA-1 under 20 bar oxygen pressure at 90 °C.[29] The reason for this phenomenon is that the one-electron oxidation of V(IV) to form the $HO_2^{\bullet}$ radical is thermodynamically unfavorable. This behaviour was further confirmed by the dark blue colour of the reaction solution. Therefore, these two compounds are not appropriate as catalysts for the desired catalytic performance. In fact, all the other tested HPA-*n* catalysts with $n =$ 2–6 are stable and re-oxidizable under the applied reaction conditions. All these HPAs showed a clear orange colour without any solid residues after eight hours of reaction time. Hence the presence of a certain number of V ions in the heteropolyanion complexes seems to be essential for a suitable re-oxidizability, as

Table 1 Screening of Keggin-type POMs with various degrees of metal substitution for the oxidation of model substrates

		FA-yield[b] (%)						
		Catalyst						
Entry[a]	Substrate	HPA-0	HPA-1	HPA-2	HPA-3	HPA-4	HPA-5	HPA-6
1	Glucose	0	0	14.6	20.5	25.6	20.0	23.3
2	Cellobiose	0	0	10.0	19.2	20.0	18.3	15.7
3	Xylose	0	0	14.2	19.7	23.8	23.7	22.7
4	Arabinose	0	0	21.8	16.1	23.1	25.2	15.6
5	Mannose	0	0	14.6	18.1	22.7	15.4	22.1
6	Galactose	0	0	20.1	24.7	13.1	23.3	22.2
7	Phenol	0	0	2.8	1.1	2.9	2.6	1.6
8	4-H3MB[c]	0	0	3.3	3.3	2.8	3.3	3.5

[a] Reaction conditions: 1 mmol substrate, 0.1 mmol catalyst, 10 g water as the solvent, 90 °C, 20 bar oxygen pressure, 1000 rpm stirrer speed, 8 hours reaction time. [b] FA-yield was determined by means of ^1H-NMR using benzene as an external standard. [c] 4-Hydroxy-3-methylbenzoic acid.

previously described in literature.[21,29] However, all these catalysts suffer from the fact that they convert all three groups of the model substrates, including glucose and cellobiose. The highest FA-yields of around 25% within 8 h reaction time were achieved using the hemicellulose model substrates arabinose and galactose. Unfortunately, glucose and cellobiose also showed similar conversions. The degree of vanadium substitution plays only a minor role as catalysts HPA-3 to HPA-6 showed nearly the same performance. Moreover, the FA-yields for the two lignin model substrates phenol and 4-H3MB were quite low, below 5%. This is due to the low solubility of aromatic compounds in aqueous solution. Consequently, alternative catalysts had to be identified that convert only the hemicellulose and lignin model substrates.

Therefore, another set of experiments were performed using Anderson-type, Wells-Dawson-type and commercially available vanadium catalysts. Sodium metavanadate ($NaVO_3$) and vanadyl sulfoxide ($VOSO_4$) were used as the latter, and the Anderson-type POM $Na_4Cu_2V_8O_{24}$ and the Dawson-type POM $K_8P_2V_2W_{16}O_{62}$ were applied. Commercially available vanadium precursors have advantages that they are well investigated and cheap sources of vanadium in aqueous solution. Again, 90 °C, 20 bar oxygen pressure and 1000 rpm stirrer speed were applied, using 0.1 mmol of the catalyst and 1 mmol of the substrate with 10 g of water as the solvent and a reaction time of 8 hours. Table 2 shows the results.

The water-soluble vanadium precursors $NaVO_3$ and $VOSO_4$ showed similar results. Firstly, it has to be mentioned that they do not fulfill the requirements for a stepwise oxidation of lignocellulosic biomass. As with the Keggin-type heteropolyacids, they do convert all of the model substrates to formic acid. Nevertheless, the activity was lower than for the HPA-*n* catalysts. This is due to the fact that catalyst re-oxidation can be more easily realized within the Keggin cage than in a free aqueous solution.[22,27] However, the FA-yields from using xylose and arabinose as model substrates are significantly higher than for all other model substrates, while the FA-yields from using the lignin model substrates are nearly

Table 2 Screening of Anderson- and Wells-Dawson-type POMs as well as the water-soluble vanadium salts $NaVO_3$ and $VOSO_4$

		FA-yield[b] (%)			
		Catalyst			
Entry[a]	Substrate	$NaVO_3$	$VOSO_4$	$Na_4Cu_2V_8O_{24}$	$K_8P_2V_2W_{16}O_{62}$
1	Glucose	7.0	7.8	0	4.7
2	Cellobiose	4.5	5.0	0	0.8
3	Xylose	12.2	13.5	0	12.8
4	Arabinose	15.6	13.5	0	26.0
5	Mannose	5.4	4.6	0	11.4
6	Galactose	3.2	6.5	0	13.0
7	Phenol	3.0	3.0	0	3.6
8	4-H3MB[c]	3.8	3.0	0	2.9

[a] Reaction conditions: 1 mmol substrate, 0.1 mmol catalyst, 10 g water as the solvent, 90 °C, 20 bar oxygen pressure, 1000 rpm stirrer speed, 8 hours reaction time. [b] FA-yield was determined by means of ^{1}H-NMR using benzene as an external standard. [c] 4-Hydroxy-3-methylbenzoic acid.

identical. Another interesting observation is that solubility effects were less pronounced, as the conversion rates for glucose and cellobiose were nearly identical.

Regarding the two heteropolyoxometalate species $Na_4Cu_2V_8O_{24}$ (Anderson-type) and $K_8P_2V_2W_{16}O_{62}$ (Wells-Dawson type), a completely different behaviour was observed. While the vanadium-substituted Anderson-type POM did not show any catalytic activity, the Dawson-type POM with two vanadium substitutions showed quite promising results. Again, hemicellulose model substrates provided the highest FA-yields while the lignin model compounds gave only small ones. As a positive outcome, the conversion rates using the cellulose model compounds were significantly decreased, while the FA-yields for all the other model compounds were kept nearly constant. This gave a strong hint that the POM-structure has significant influence on the catalytic performance in biomass oxidation chemistry. However, again the requirements of leaving the cellulose model compounds untapped were not fulfilled.

Finally, a screening of various isopolyoxometalate species of the Lindqvist-type was performed. These screening experiments were performed at 90 °C with 20 bar oxygen pressure and 1000 rpm stirrer speed, using 0.1 mmol of the catalyst and 1 mmol of the substrate with 10 g of water as the solvent and a reaction time 8 hours. The obtained values are summarized in Table 3.

The screening results from using the mono- and non-substituted Lindqvist structures did not show any catalytic activity, as previously described for the Keggin analogues. Increasing the degree of vanadium substitution also increases the oxidation activity. While the two compounds with two vanadium substitutions showed only very little catalytic activity, the three-vanadium substituted structure $K_5V_3W_3O_{19}$ reached the goal of a fractionated biomass conversion. While the cellulose model compounds glucose and cellobiose were left untapped, the model compounds for both hemicellulose and lignin showed conversion. Honestly, this may only be a first step towards the fractionated conversion of lignocellulosic

Table 3 Screening of Lindqvist-type POMs with different degrees of vanadium substitution

		FA-yield[b] (%)			
		Catalyst			
Entry[a]	Substrate	$K_2W_6O_{19}$	$K_3VW_5O_{19}$	$K_4V_2W_4O_{19}$	$K_5V_3W_3O_{19}$
1	Glucose	0	0	0	0
2	Cellobiose	0	0	0	0
3	Xylose	0	0	2.2	13.6
4	Arabinose	0	0	1.8	16.4
5	Mannose	0	0	1.6	16.3
6	Galactose	0	0	0.9	16.3
7	Phenol	0	0	0.3	1.6
8	4-H3MB[c]	0	0	0.2	1.8

[a] Reaction conditions: 1 mmol substrate, 0.1 mmol catalyst, 10 g water as the solvent, 90 °C, 20 bar oxygen pressure, 1000 rpm stirrer speed, 8 hours reaction time. [b] FA-yield was determined by means of ^{1}H-NMR using benzene as an external standard. [c] 4-Hydroxy-3-methylbenzoic acid.

biomass as the conversion rates for the two lignin model substrates were quite low.

Application to the main components of lignocellulose

To round up our study, we probed the so-optimized Lindqvist-type polyoxometalate $K_5V_3W_3O_{19}$ in a transformation involving all the main components of lignocellulosic biomass. This conversion with cellulose, hemicellulose and lignin was performed over a reaction time of 8 hours. As the cellulose feedstock, we used Vitacel L600 powdered cellulose from J. Rettenmaier & Söhne & Co. GmbH with a purity of 98%+. C, H, N, S elemental analysis provided a molecular composition of $C_{1.07}H_{1.88}O_1$. The processed hemicellulose (xylan from birchwood) has a purity of 95%+ and is commercially available from Carl Roth GmbH. Its molecular composition was determined to be $C_{0.99}H_{1.76}O_1$. The lignin was purchased from Sigma Aldrich (alkali, low sulfonate content) and had a purity of 95%+. The molecular composition *via* elemental analysis was $C_{1.70}H_{1.84}O_1$. The results of these experiments are summarized in Table 4.

The experiments using the real lignocellulosic compounds cellulose, hemicellulose and lignin proved the promising performance of the Lindqvist-type POM $K_5V_3W_3O_{19}$. While real hemicellulose and lignin could be depolymerized down to the C_1-product FA, cellulose was not converted within the investigated reaction time. It has to be noted that these investigations can only be the first step for a stepwise conversion of real lignocellulosic biomass like wood or grass. However, a first milestone was reached as a promising subclass of POM catalysts has been identified to fulfil the requirements of the desired task.

Experimental

All the chemicals were obtained commercially and used without further purification. The Keggin-type polyoxometalates were synthesized according to the literature.[22,30] The Lindqvist-type catalysts were prepared based on known literature procedures.[31,34] The Wells-Dawson and Anderson-type catalysts were synthesized using the methods published by Vila-Nadal and Blazevic.[35,36] Vanadyl sulphate was bought from Alpha Aesar and sodium metavanadate was bought from Merck KGaA. The gases were bought from Linde AG. Oxygen (4.5 GA 201) was used as the oxidizing agent. Nitrogen 5.0 was used to provide an inert

Table 4 Transformation of all the main components of lignocellulosic biomass to formic acid and high-grade cellulose using the Lindqvist-type POM $K_5V_3W_3O_{19}$

Entry[a]	Substrate	Catalyst	Molecular composition[b]	FA-yield[c] (%)
1	Cellulose	$K_5V_3W_3O_{19}$	$C_{1.07}H_{1.88}O_1$	0
3	Hemicellulose	$K_5V_3W_3O_{19}$	$C_{0.99}H_{1.76}O_1$	8.3
5	Lignin	$K_5V_3W_3O_{19}$	$C_{1.70}H_{1.84}O_1$	2.4

[a] Reaction conditions: 1 mmol substrate, 0.1 mmol catalyst, 10 g water as the solvent, 90 °C, 20 bar oxygen pressure, 1000 rpm stirrer speed, 8 hours reaction time. [b] Determined *via* C, H, N, S elemental analysis. [c] FA-yield was determined by means of ^{1}H-NMR using benzene as an external standard.

atmosphere. Elemental (C, H, N, S) analyses of the complex biogenic substrates were carried out using a Euro Vector EA 3000. Product analysis of the gases was performed using gas chromatography with a Varian GC 450 equipped with a 2 m × 0.75 mm ID Shin carbon ST column. Thermogravimetric analysis (TGA) was carried out using a SETSYS-1750 CS Evolution from SETARAM Instruments. The ICP-OES measurements were performed using a Perkin Elmer Plasma 400. The NMR spectra were recorded using a JEOL ECX-400 MHz.

Experimental setup for the catalyst screening and performance tests

For the initial catalyst screening with model substrates and the oxidation of the main components cellulose, hemicellulose and lignin, a tenfold screening plant with a batch mode reactor setup was used. It consists of ten 20 ml Hastelloy C-276 autoclaves connected in parallel to a single oxygen supply line. Each reactor was connected to a rupture disk with a burst pressure maximum of 100 bar. All the pipes, valves and fittings were made of stainless steel 1.4571. The autoclaves could be connected to the oxygen line *via* individual couplings attached to tubes that were connected to the same oxygen line. The gaskets for preventing pressure leaks were made of Teflon. The reactors were placed inside a heating plate in order to adjust them to the required temperature. The heating plate was facilitated with a magnetic stirrer. Additionally, external magnetic beads could be used for stirring.

General procedure for the catalytic oxidation reactions

The 20 ml autoclaves were filled with the reactants, catalysts and water (used as a solvent). A magnetic bead was put into each reactor to guarantee effective mixing. A gasket ring was added to each reactor head and the ten closed reactors were kept in the heating block. Metal tubes were connected to the couplings on the closed reactors. Exhaust pipes were attached to the branch on the rupture disk to connect it to the offgas (in case the pressure exceeded the burst pressure of the rupture disks). The adaptor for the pressure logger was connected to the system. The reactors were purged using a 10 bar oxygen pressure to get the residual air out of the reactors. Then they were pre-pressurized to a set oxygen pressure, keeping in mind the increase in pressure with increase of the temperature. The heating and stirring were switched on using the regulators on the heating plate. The heating was regulated by setting the controller to 90 °C. The stirring speed was set to 1000 rpm in order to start the gas entrainment and consequently the experiment.

Conclusions

The aim of this project was to design and optimize tailor-made polyoxometalate catalysts for the stepwise and selective oxidation of lignocellulosic biomass to produce formic acid and high-grade cellulose. This strategy offers the best value for use of neat complex biomass as a substrate without further pretreatment, simply in aqueous media under oxygen pressure. This concept is an extension of previous work on the selective oxidation of biomass to formic acid (OxFA-process).

In this context, we performed a screening of appropriate homogeneous catalysts for the selective depolymerization and oxidation of molecular, well-defined model substrates in order to understand the specific influence of different active species on the stepwise degradation of the carbon-framework. The selection of the catalysts was based on the results from previous studies on selective oxidation of biomass to formic acid. Two commercially available homogeneous vanadium catalysts as well as various polyoxometalates from four structural categories were selected for screening. An important result from these screening experiments was that the presence of at least two vanadium atoms in the framework of the catalyst is important for producing formic acid under mild conditions. Furthermore, the vanadium-substituted Keggin-type polyoxometalates HPA-n ($n \geq 2$) showed the highest FA-yields. Unfortunately, it was observed that these structures catalyze the oxidation of all the model substrates, including those for cellulose. Therefore, they cannot be used for fractionated biomass conversion. Moreover, the water-soluble vanadium precursors $NaVO_3$ and $VOSO_4$ showed the same catalytic performance and are thus not suitable for the desired conversion. Additionally, the Anderson-type POM $Na_4Cu_2V_8O_{24}$ did not show any catalytic activity under the applied reaction conditions, while the Wells-Dawson type POM $K_8P_2V_2W_{16}O_{62}$ showed quite promising results whereby still a little undesired conversion of glucose and cellobiose was observed. The most interesting result of the performed studies was that a fractionated oxidation of biomass was only observed with the Lindqvist-type catalyst $K_5V_3W_3O_{19}$ (selective conversion of the hemicellulose model compounds xylose, arabinose, mannose and galactose as well as the lignin model compounds phenol and 4-H3MB to formic acid). The latter seems to fulfil the requirement of catalyzing oxidation of only the lignin and hemicellulose fractions of lignocellulosic biomass under the applied reaction conditions, and this performance could be confirmed for the three main components cellulose, hemicellulose and lignin. However, the yields of formic acid were quite low. This suggests a need to optimize the reaction parameters and the applied equipment to obtain the desired fractionated conversion.

Acknowledgements

Peter Wasserscheid is gratefully acknowledged for mentoring and support. Financial support from the Max-Buchner-Stiftung (www.dechema.de/mbf.html) and the Exzellenzcluster “Engineering of Advanced Materials” (www.eam.fau.de) is gratefully acknowledged.

Notes and references

1 S. Shafiee and E. Topal, *Energy Policy*, 2009, **37**, 181.
2 G. Huber, S. Iborra and A. Corma, *Chem. Rev.*, 2006, **106**, 4044.
3 D. Alonso, J. Bond and J. Dumesic, *Green Chem.*, 2010, **12**, 1493.
4 R. Palkovits, *Chem. Ing. Tech.*, 2011, **83**, 411.
5 C. E. Wyman, S. R. Decker, M. E. Himmel, J. W. Brady, C. E. Skopec and L. Viikavi, *Polysaccharides: structural diversity and functional versatility*, 2005, vol. 1, p. 1023.
6 J. Chheda, G. Huber and J. Dumesic, *Angew. Chem., Int. Ed.*, 2007, **46**, 7164.
7 K. Nath and D. Das, *Curr. Sci.*, 2003, **85**(3), 265.
8 G. Machado, R. Schaeffer and E. Worrell, *Ecol. Econ.*, 2001, **39**, 409.
9 J. Melero, J. Iglesias and A. Garcia, *Energy Environ. Sci.*, 2012, **5**, 7393.

10 C. M. Balonek, A. H. Lillebo, S. Rane, E. Rytter, L. D. Schmidt and A. Holmen, *Catal. Lett.*, 2010, **138**, 8.
11 R. Toonssen, N. Woudstra and A. H. M. Verkooijen, *Int. J. Hydrogen Energy*, 2008, **33**, 4074.
12 N. Boukis, V. Diem, U. Galla, P. D'Jesús, A. Kruse, H. Müller and E. Dinjus, *Nachr. - Forschungszent. Karlsruhe*, 2005, **37**, 116.
13 R. D. Cortright, R. R. Davda and J. A. Dumesic, *Nature*, 2002, **418**, 964.
14 I. Coronado, M. Stekrova, M. Reinikainen, P. Simell, L. Lefferts and J. A. Lehtonen, *Int. J. Hydrogen Energy*, 2016, **41**(26), 11003.
15 F. Geilen, B. Engendahl, A. Harwardt, W. Marquardt, J. Klankermeyer and W. Leitner, *Angew. Chem.*, 2010, **122**, 5642.
16 K. L. Luska, J. Julis, E. Stavitski, D. N. Zakharov, A. Adams and W. Leitner, *Chem. Sci.*, 2014, **5**, 4895.
17 B. M. Kabyemela, T. Adschiri, R. M. Malaluan and K. Arai, *Ind. Eng. Chem. Res.*, 1999, **38**, 2888.
18 A. Kruse and A. Gawlik, *Ind. Eng. Chem. Res.*, 2002, **42**, 267.
19 A. Sutton, *Nat. Chem.*, 2013, **5**, 428.
20 R. Wölfel, N. Taccardi, A. Bösmann and P. Wasserscheid, *Green Chem.*, 2011, **13**, 2759.
21 J. Albert, R. Wölfel, A. Bösmann and P. Wasserscheid, *Energy Environ. Sci.*, 2012, **5**, 7956.
22 J. Albert, D. Lüders, A. Bösmann, D. M. Guldi and P. Wasserscheid, *Green Chem.*, 2014, **16**, 226.
23 J. Reichert, B. Brunner, A. Jess, P. Wasserscheid and J. Albert, *Energy Environ. Sci.*, 2015, **8**, 2985.
24 J. Albert and P. Wasserscheid, *Green Chem.*, 2015, **17**(12), 5164.
25 J. Albert, *Der Erlanger OxFA-Prozess*, SVH-Südwestdeutscher Verlag für Hochschulschriften, 2016.
26 C. L. Hill and C. M. Prosser-McCartha, *Coord. Chem. Rev.*, 1995, **143**, 407.
27 M. Amman, *J. Mater. Chem.*, 2013, **1**, 6291.
28 A. Shatalov, D. V. Evtuguin and C. Pascoal Neto, *Carbohydr. Polym.*, 2000, **43**(1), 23.
29 I. V. Kozhevnikov, *Chem. Rev.*, 1998, **98**(1), 171.
30 V. F. Odyakov and E. G. Zhizhina, *React. Kinet. Catal. Lett.*, 2008, **95**, 21.
31 J. Andersson, O. Hastings, L. Howarth and J. Pettersson, *Dalton Trans.*, 1996, 2705.
32 T. Heinze and T. Liebert, *Polymer Science: A Comprehensive Reference*, 2012, pp. 83–152.
33 F. H. Isikgor and C. Remzi Becer, *Polym. Chem.*, 2015, **6**, 4497.
34 J. Albert, J. Mehler, J. Tucher, K. Kastner and C. Streb, *ChemistrySelect*, 2016, **1**, 2889.
35 L. Vila-Nadal, S. G. Mitchell, D.-L. Long, A. Rodríguez-Fortea, X. López, J. M. Poblet and L. Cronin, *Dalton Trans.*, 2012, **41**, 2264.
36 A. Blazevic and A. Rompel, *Coord. Chem. Rev.*, 2016, **307**, 42.

Faraday Discussions

PAPER

Plantics-GX: a biodegradable and cost-effective thermoset plastic that is 100% plant-based

Albert H. Alberts[a] and Gadi Rothenberg *[b]

Received 3rd February 2017, Accepted 22nd March 2017
DOI: 10.1039/c7fd00054e

We recount here the story of the discovery and invention of a family of thermoset resins that are fully biodegradable and plant-based. The resin is prepared by polymerising glycerol, the simplest trialcohol, with citric acid, the simplest abundantly available triacid. Mixing these two chemicals at moderate temperatures begins a multi-step esterification that results in a network-like resin foam. The hardness, brittleness, and toughness of this resin can be controlled by changing the alcohol : acid ratio, the temperature, and the presence of additives. Remarkably, this resin is fully biodegradable, breaking down to its molecular components in water. The degradation period can be varied from a few days to a few months, depending on the degree of polymerisation. We have prepared over 800 different formulations of this new material, which is now being produced on the ton scale by Plantics BV in The Netherlands under the trade names Glycix and Plantics-GX. This resin is made from inexpensive and widely available starting materials. It is 100% biobased, non-toxic and biodegradable. Thus, it has the potential to replace petro-based thermoset polymers and foams in a variety of real-life applications.

Introduction

In the early 19th century, the Swedish chemist Berzelius reacted tartaric acid with "sweet oil" (now known as glycerol), obtaining a material that could neither be crystallized nor distilled. About 100 years later, the Belgian-American inventor Baekeland obtained foams from the polymerization of phenol and formaldehyde in aqueous media. The reaction was developed further to make a coherent moldable thermoset (the famous Bakelite[1,2]) and the foam was left behind. Nonetheless, phenol-formaldehyde floral foams became a profitable business after WW II.

[a]Plantics BV, Johannes Vermeerstraat 24, 6814JT Arnhem, The Netherlands; Web: http://www.plantics.com
[b]Van't Hoff Institute for Molecular Sciences, University of Amsterdam, Science Park 904, 1098XH, Amsterdam, The Netherlands. E-mail: g.rothenberg@uva.nl; Web: http://www.hims.uva.nl/hcsc

Foam processing as we know it now, in construction, marine applications, and furniture, came of age when Otto Bayer discovered in 1937 the fast reaction of isocyanates with alcohols at room temperature to give so-called urethanes.[3] By manipulating the ratios of bifunctional isocyanates and polyalcohols, chemists could suddenly make closed-cell and open-cell polyurethane (PUR) foams with controlled resilience and density on an enormous scale. Isocyanates are "spring-loaded" reagents,[4] giving the advantage of fast reaction kinetics. But this comes at a price: isocyanate chemistry involves highly hazardous substances, and the resulting foams are a nightmare to work with.[5] Moreover, their end-of-life is problematic.[6] They are not biodegradable on any human time scale, and their incineration produces suffocating and poisonous smoke.[7] The global polyurethane market demand was nearly 16M tons in 2013 and is expected to reach 22M tons by 2020, growing at a CAGR of 5% from 2014 to 2020.[8] Flexible and rigid foams accounted for over 38% and 26% of the market volume in 2013, respectively. Considering their enormous quantities, PUR and PIR-type foams incur an immense environmental burden.

Other foams can be produced from various petro-based thermoplastics, notably polyvinylchloride (PVC), styreneacrylonitrile (SAN), ethylvinylalcohol (EVA) and the ubiquitous polystyrene (PS) used in packaging. All of these can be used as core materials in sandwich panels in marine applications, aviation and construction. But all of them come with a similar trade-off: they are easy to process and relatively cheap, but they are not biodegradable and in many cases cause environmental problems. Foams are among the greatest environmental pollutants. Most of them are not recycled, and end up in landfills.

Because the environmental problems with these foams are usually incurred at the end of the products' life cycles, they are seen as less critical by many manufacturers and users. This view will have to change if we want to become a sustainable society. The chemical industry can change it by adopting alternative foams that are biobased and biodegradable,[9] but these must also be cheap enough to compete in tomorrow's market.[10] The transition to biobased materials is especially problematic in the case of the thermoset sector, because thermosets' complex structures are not as easy to imitate compared to thermoplastics (this is true also for the biomass starting materials: lignin is much more difficult to process into chemicals than cellulose/hemicellulose[11,12]). In this short paper, we present our (admittedly accidental) invention of such a foam: a thermoset resin made from glycerol and citric acid. Here, with the editor's kind consent, we tell the story as it happened. Although some of the technical details of the invention have been described in patents,[13–17] this is the first time that the full story appears in a scientific journal.

Results and discussion

In March 2010, while at the ORCS meeting in Monterey, we heard several uplifting lectures regarding the wish list of the US Department of Energy for reactions of so-called platform molecules. This gave us the idea for a new kind of bio-kerosene, basically trying to connect some of these molecules together. But about two weeks later, we found that mixing glycerol and citric acid at medium temperatures yielded a solid foam that was impossible to remove from the glass vessel. In fact, it bound so strong to the glass that when it cooled down and contracted, the glass cracked. This polymer swelled and ultimately hydrolysed back to its molecular building blocks in warm acidic water. The experiment was serendipitous. We reacted glycerol with citric acid in a 1 : 1 molar ratio (in hindsight, Berzelius-style) at 120–160 °C for 40–300 min in a glass vessel, obtaining an macroporous brittle thermoset network (see Fig. 1).

These observations got us thinking. In itself it is remarkable that two such simple organic compounds can react to produce a coherent foam without a blowing agent such as pentane or CFCs. We were allured by the prospect of controlling the foaming process to produce a 100% bio-based and biodegradable material. Such a foam could replace the polluting and/or hazardous ones used today in the construction, isolation and packaging sectors. We were further motivated by the abundance of glycerol as a waste stream from biodiesel production, and the worldwide availability of citric acid as a bulk chemical. Surprisingly, a search of the scientific and patent literature revealed that this was a new process (the biodegradation of glycerol/citric acid polymers was studied in drug and fertilizer release).[18] We then filed patent applications on the materials and their applications, and decided to try and develop the process further.

Foam synthesis and properties

First, we had to get away from glass and metal surfaces, because even if you can make the polymer in a glass or stainless steel vessel, you can never get it out. In hindsight, we think that scientists may have synthesised this type of polymer many years ago, before the advent of petroleum-based chemistry. But lab equipment in the 19th century was chiefly glass (and in some cases ceramic and metal). Anyone making this polymer would have thrown it away as useless (sometimes also throwing away the reactor as well). Luckily, we had access to inert space-age materials such as silicone rubber and PTFE. Using sophisticated parallel reactor designs (a silicone rubber muffin baking tray from the corner shop, cost €4.95) we produced our first rigid free-standing machinable foams. The shavings of these could be ground or milled into a fine powder, and mixed in to subsequent runs of the liquid resin, giving essentially quantitative yields. The completion of the esterification reaction is conveniently determined by the weight of water lost (approximately 1/3 of the weight of the initial starting material load). During the foaming process, the liquid resin rises just like bread. It reaches about five times its original height at >98% polymerisation. This gives a gradient in the cell size (large on the bottom, smaller at the top). The top of the fully risen foam is covered

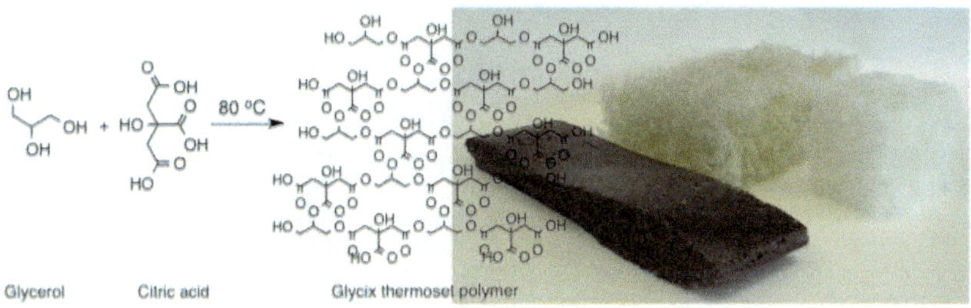

Fig. 1 The chemical reaction between glycerol and citric acid gives a 3D thermoset network polymer.

with a smooth glossy polyester sheet, which turned out to be important for the elastic foams (see below for the details).

Incidentally, adding acid catalysts to the prepolymerized liquid resin does not accelerate the foaming process much, because the rate of evaporation of water vapor is the practical limiting factor. We tried a variety of acids, including HCl, H_2SO_4 (which tans the white foam), H_3PO_4, $HClO_4$ and H_3BO_3 (the latter is incorporated in the final foam, resulting in larger cells).

The prepolymerized glycerol/citric acid resin is an excellent adhesive, most probably due to the numerous –OH and –COOH groups on its surface. It binds strongly to practically any hydrophilic material, including wood, oxidised metal, glass and paper to name just a few examples. Conversely, it does not adhere well to strongly hydrophobic materials such as Teflon, polysilicone or polypropene. This strong binding to surfaces opens the possibility of glueing these surfaces to a machined piece of foam as the core in a sandwich panel construction. For example, aluminum/foam/aluminum sandwiches are easily obtained with heating under light pressure in an oven, using the prepolymerized resin as the adhesive. Using this approach, we constructed 100% biobased sandwich panels with nonwoven hemp, paper/cardboard and plywood (see Fig. 2). This avoids the use of petro-based epoxy or polyester resins. Similarly, three-layer to six-layer plywood can be glued under light pressure, without any phenol-formaldehyde or urea-formaldehyde resins.

Observing the adhesive properties of the resin, we reasoned that it would serve as an excellent matrix for composite materials. Composites made with synthetic fibres often have as their weak point the interface between the fibre and the matrix. For this reason, the glass fibres in so-called fibreglass composites are often coated with a special "sizing", which is often kept as a trade secret.[19,20] But since the surfaces of glass fibres are covered with –OH groups, these can bind exceptionally well to Glycix, with its multiple –OH and –COOH functional groups.[16] Similarly, the polymer also adheres excellently to a variety of biobased fibres and materials, because these also have large numbers of –OH groups on their surfaces.[17] We ran experiments on adding a host of different filler materials, including gelatin, sugar, methylcellulose, wood dust, paper, cotton, regenerated cellulose, glass fibres, metallic fibres, titania, alumina and silica. In total, we

Fig. 2 The foam can be polymerised between layers of wood, paper or metal, giving biobased sandwich materials.

prepared over 800 different composites, using the resin as a matrix and also using its fibres or platelets as the reinforcer. A full description of these composite materials and their applications is out of the scope of this short paper. Detailed technical descriptions are included in the patents.[16,17]

Cellular foams have many applications in various industrial sectors, which depend in turn on the type of cells and on their size. In general, the cell shape and size in these foams and the density depend on the temperature range and processing time (Table 1). Foaming for 48 h at 110–120 °C gives closed-cell foams (0.6 g cm^{-3}), whereas foaming at 120–140 °C for 12–16 h yields robust open-cell reticular foams (see the Experimental section for details). At higher temperatures, the processing time reduces to minutes, giving low-density foams (0.1 g cm^{-3}) with fragilities comparable commercial PUR foams. For comparison, the densities of PUR and PS can reach below 0.05 g cm^{-3}, comparable to Balsa wood. At the other extreme we could suppress the foaming entirely: keeping a thin layer of prepolymerized resin below 100 °C during 5 days of heating on a PTFE tray yielded a colorless transparent PMMA-like sheet, with closed cells that were no longer visible to the naked eye. The material can be machined and polished to various shapes and sizes (Fig. 3).

Table 1 Synthesis and physical parameters of the foams[13]

Entry	Temperature, °C	Time, h	Appearance	Density, mg cm^{-3}
1	90	100	Transparent sheet	800
2	100–120	60	Closed-cell foam	500–700
3	120–140	16	Open-cell foam	300–600
4[a]	180	1	Fragile open-cell foam	100–200

[a] Above 150 °C, the citric acid undergoes decarboxylation to itaconic acid and therefore the network structure changes.

Fig. 3 Photo showing a transparent non-porous resin "wheel" prepared by slow controlled polymerisation followed by machining and polishing, together with a cubic block of one of the original foam pieces prepared by fast polymerisation.

Changing the polymerisation temperature also changes the resin composition. Above 150 °C, citric acid undergoes a slow decarboxylation to itaconic acid (eqn (1)), which has a characteristic odour and brownish color.[21] Indeed, a series of control experiments followed by C, H elemental analysis confirmed this decarboxylation, which can be avoided as long as the process is carried out below 140 °C. Further control experiments starting with itaconic acid instead of citric acid confirmed that no foaming takes place. Instead, the result is a thick yellow disk with a density higher than 1 g cm^{-3} (see the Experimental section for details).[13]

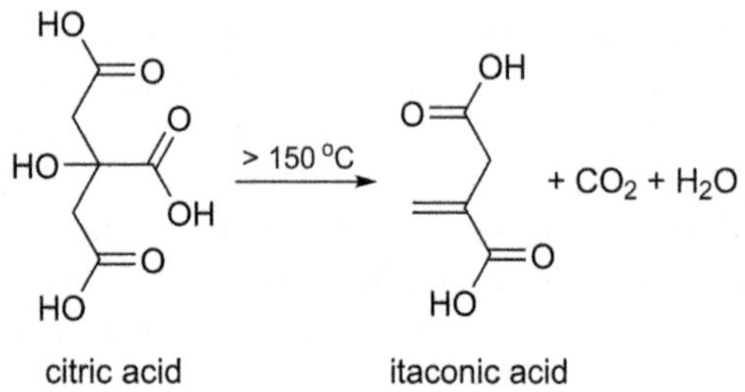

(1)

There is an important difference between this new polyester and most petro-based polyesters, which comes from an inherent difference between biomass and petroleum. Petro-based chemicals are underfunctionalised, and contain practically no oxygen atoms. These are added in the synthesis process (*e.g.* in the making of terephthalic acid from *p*-xylene and ethylene glycol from ethene). Even if these monomers are more hydrophilic than the starting materials, the final polyesters are hydrophobic because all the –OH and –COOH groups have reacted (there are no "surplus" groups). This is done on purpose, and gives highly durable materials. The downside is that these polymers are so durable that they have become persistent pollutants. The hydrophobic chains and groups prevent the hydrolysis of the ester groups that connect the chains.

With our hydrophilic resin, things are different, because the monomers have many more hydroxyl and carboxyl groups. The polymer absorbs water readily, and undergoes a slow hydrolysis. When immersed in water or in humid soil the foams swell up and eventually hydrolyse back to glycerol and citric acid. The hydrolysis rate depends on the density and on the degree of polymerisation. Samples with a density of 600 mg cm^{-3} degrade in about a year, but ones with a density of 100 mg cm^{-3} take only a few days. Similarly, samples that were polymerised to an extent of >95% remained stable in water for months, whereas samples with below 80% polymerisation degraded after a few weeks. At high degrees of polymerisation, the foam keeps its structure and rigidity in air practically indefinitely (the original foam block sample prepared in March 2010, which is shown in Fig. 3 above, is still in pristine condition, after nearly 7 years on the shelf). These observations are consistent with the qualitative description above. At high degrees of polymerisation, both the physical accessibility and the chemical accessibility of the water is impaired. The 3D network is more complete, so there are fewer openings for water molecules to enter. Moreover, incoming water molecules can only form hydrogen bonds with the non-bonding electrons of the ester oxygen atoms. Conversely, at lower degrees of polymerisation there are more –OH and –COOH groups available, and the 3D structure is also more open.

Note that unlike with chain polymers, we believe that this resin cannot reach 100% polymerisation. There is simply not the space to form a perfect crystalline structure. As the network grows, some of the groups will quickly become inaccessible to esterification. Indeed, thermogravimetric analysis confirms that this resin is glassy, with a T_g of *ca.* 70–80 °C. Controlling the reaction kinetics in larger volumes is tricky at best, and can lead to unhappy situations.[22] If the resin forms too quickly, it can cure inside the reaction vessel. Once this happens, there is no way of getting it out. Scaling up the reactions is challenging – in one case, we had to throw away a 50L reactor after the resin polymerised too quickly and formed what was essentially a reactor-sized and -shaped molecule. Eventually, we succeeded in scaling up the process to the ton scale, by careful control of the reaction parameters.

Interestingly, the rigid glycerol/citric acid foams are much stronger compared to PUR foams in the same density range (Fig. 4). This has implications for practical applications in construction and isolation, because one can save on the weight (and cost) of foam core laminates.

Foam formation on metal surfaces

From the start, we observed that the prepolymerized resin adheres strongly to a variety of metal surfaces, particularly aluminum, iron, and copper, and alloys thereof. In hindsight we suggest that this adhesion reflects a chemical binding of surface hydroxyl and carboxyl groups in the resin to hydroxyl and other groups on the metal surface. In several cases we could see clearly that a small portion of the metal dissolved and reacted with the polymer. The most regular and robust foams were obtained from the vitrification of the viscous glycerol/citric acid resin catalyzed by $HCl_{(aq)}$ in aluminum trays (we used disposable baking trays). We attribute this regularity to traces of hydrogen gas evolved by the metal, which act as an

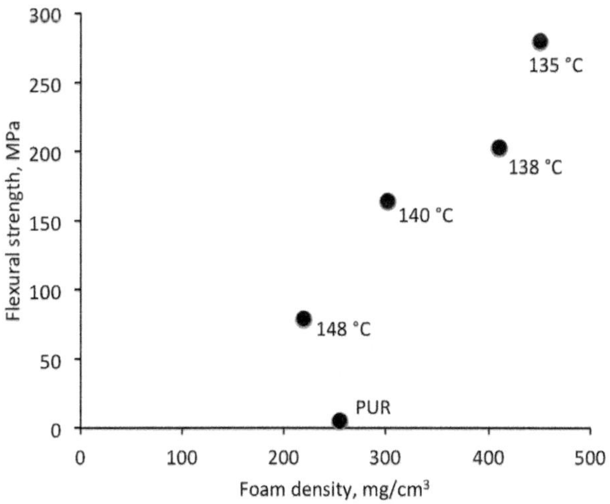

Fig. 4 Three-point test flexural strength as a function of the foam sample density.

assisting blowing agent. Indeed, we observed traces of aluminium salts incorporated in the foam, and contrasted the pale blue-grey color of the samples prepared on aluminum with the white foams prepared on Teflon or silicone rubber surfaces. The foams are free-standing upon peeling off the metal. Alternatively, an aluminum–foam–aluminum sandwich can be easily made by glueing on a top aluminum sheet with prepolymerized resin after the sample is machined to create a flat top surface (see Fig. 5).

Elastic foams from the glycerol/citric acid resin

Inspired by polyurethane diisocyanate/polyalcohol chemistry we decided to vary the molar ratio of glycerol and citric acid. Starting from the original 1 : 1 mole ratio, we synthesised resins with an excess of glycerol of 1.2 : 1, 1.4 : 1, 1.6 : 1 and 1.8 : 1. These materials were synthesised using the same aluminum tray acid-catalyzed procedure as described above for the rigid foams. Our objective was to incorporate an excess of hydroxyl groups in the final foam, as well as to create regions with partial polymerisation. All the samples were rigid when dry, but the 1.8 : 1 glycerol : citric acid foam became elastic after absorbing water vapor under ambient conditions, as well as under controlled humidity conditions in a commercial humidifier. The result was a "memory foam", which deforms under pressure and then slowly regains its original shape. Careful optimisation of the polymerisation showed that the best elastic foams had a density of *ca.* 300 mg cm^{-3}. The densities of the foams prepared with glycerol : citric acid ratios of 1.4 : 1, 1.6 : 1 and 1.8 : 1 were 376, 282, and 317 mg cm^{-3}, respectively. But while the first two samples remained rigid (flexural strength 155 and 191 MPa) the 1.8 : 1 sample was elastic.

The ease of synthesis of these foams is demonstrated by the simplicity of turning them into products. As we mentioned earlier, the pristine foams are covered with a smooth glossy polyester surface when the esterification is driven to completion. When one uses a circular silicone rubber pie tray with the 1.8 : 1 glycerol–citric acid stoichiometry, after humidification a ready-made circular resilient cushion with a spherical closed surface is produced in quantitative yield.

Fig. 5 Coffee table made from a metal frame and polished biobased resin foam. The inset shows how the metal surface reacts with the acid polymer, dissolving into the foam.

Experimental

Materials and instrumentation

Sandwich-type structures, composite sheets and other shapes were pressed using a Rondol Technology heated press (max. 20 tons per square inch, controlled heating up to 200 °C). Humidifying experiments were run using an Intertek G-BS300 humidifier. Thermal conductivity (λ) values were measured following ASTM C-168-15. Three-point flexural strength tests were done following ASTM D-790. Unless stated otherwise, reagent-grade glycerol and citric acid (both monohydrate and dihydrate) were obtained from commercial sources and used as received. Raw waste glycerol from the biodiesel processing of waste oils (a brown smelly viscous mixture containing roughly 80% glycerol) was a gift sample from Petrotec Inc.

Procedure for preparing the glycerol/citric acid foam on the kg scale

Glycerol (0.93 kg; 10.0 mol; reagent-grade) is poured into a 3 L glass beaker equipped with a magnetic stirrer and a hot plate. Then, 2.12 kg (10.0 mol) of citric acid monohydrate is dissolved by adding it in portions while stirring. After completion of the addition the temperature is increased to 130 °C with continued stirring until a colorless homogeneous syrupy liquid is obtained. 50 mL of conc. $HCl_{(aq)}$ is added and the stirring is continued for another 10 min. The liquid is transferred to a 30 × 60 × 5 cm aluminum tray and this tray is placed in a drying oven at 135 °C for 16 h. A small amount of hydrogen (30–40 ppm) will be released over the first 3 h. After cooling the tray down to ambient temperature, the process yields 2.30 kg of hard foam with an average height of 6 cm and a total volume of 7 liters (80.33% void, skeleton density 1.93 cm^3, envelope density 0.38 g cm^{-3}, pore volume 2.11 cm^3 g^{-1}). The yield is $>$98% (the theoretical yield for 100% is 2.33 kg, allowing for a weight loss of 40 mol or 720 grams of water).

Procedure for control experiments with itaconic acid[13]

Glycerol (13.2 gram) and itaconic acid (26 gram) were heated in an open beaker with stirring at 165 °C for 2 h. Then, 100 mg of *p*-toluene–sulfonic acid was added and the viscous liquid was poured in a silicone rubber mould and kept in an oven for 20 h. A yellow thick circular disc was obtained with a density exceeding 1 g cm^{-3}. No foaming took place.

Conclusions

Glycerol and citric acid are easily polymerised under moderate conditions, giving a resin solution which has strong adhesive properties, and ultimately robust macroporous thermoset resin foams. These resins have many advantages. Their synthesis process can be controlled, and they can be made rigid or elastic. They require no foaming agent, and the only byproduct of their synthesis is water. The resulting foams are strong and (depending on the degree of polymerisation) can retain their shapes and properties for years. Yet they also degrade in water in a straightforward hydrolysis, ultimately giving back the original starting materials. Both monomers are bio-based, cheap, and widely available. In its liquid state, the resin is an excellent adhesive matrix for a variety of composites with practically any biobased fibre and many man-made fibres. Considering the above, we foresee bona fide industrial opportunities for these materials in the packaging, construction and agricultural sectors. The resin and its derivatives have been manufactured since 2015 on the ton scale by Plantics BV.[23]

Acknowledgements

We thank Dr M. C. Mittelmeijer-Hazeleger for performing the porosity and skeleton density measurements of the foam samples. This research is part of the Research Priority Area Sustainable Chemistry of the UvA, http://www.suschem.uva.nl.

Notes and references

1 G. Kossmehl, *Kunststoffe*, 2010, **100**, 10–13.
2 P. Reboul, *Spec. Publ. - R. Soc. Chem.*, 1994, **141**, 26–37.
3 O. Bayer, *Angew. Chem.*, 1947, **59**, 257–272.
4 H. C. Kolb, M. G. Finn and K. B. Sharpless, *Angew. Chem., Int. Ed.*, 2001, **40**, 2004–2021.
5 F.-W. Wittbecker and M. Giersig, *Rev. Environ. Contam. Toxicol.*, 2001, **170**, 1–11.
6 L. Ma, Q. Jiang, P. Zhao and C. Zhao, *Mater. Sci. Forum*, 2014, **787**, 176–183.
7 S. T. McKenna and T. R. Hull, *Fire Sci. Rev.*, 2016, **5**, 1–27.
8 *Polyurethane (PU) Market Analysis By Product (Rigid Foam, Flexible Foam, Coatings, Adhesives & Sealants, Elastomers), By End-use (Furniture and Interiors, Construction, Electronics & Appliances, Automotive, Footwear, Packaging) And Segment Forecasts To 2020*, Grand View Research, 2014.
9 M. Patel and R. Narayan, in *Nat. Fibers, Biopolym., Biocompos.*, CRC Press LLC, 2005, pp. 833–853.
10 J. C. Philp, R. J. Ritchie and K. Guy, *Trends Biotechnol.*, 2013, **31**, 65–67.
11 Z. Strassberger, S. Tanase and G. Rothenberg, *RSC Adv.*, 2014, **4**, 25310–25318.
12 R. Beerthuis, G. Rothenberg and N. R. Shiju, *Green Chem.*, 2015, **17**, 1341–1361.
13 A. H. Alberts and G. Rothenberg, *Eur. Pat.*, 2 628 757, Universiteit van Amsterdam, 2013.
14 A. H. Alberts and G. Rothenberg, *US Pat.*, 20 150 005 403, Universiteit van Amsterdam, 2015.
15 A. H. Alberts and G. Rothenberg, *US Pat.*, 20 140 065 392, University of Amsterdam, 2014.
16 A. H. Alberts and G. Rothenberg, *US Pat.*, 20 140 080 973, Universiteit van Amsterdam, 2014.
17 A. H. Alberts and G. Rothenberg, *Eur. Pat.*, 2 697 293, Universiteit van Amsterdam, 2016.
18 D. Pramanick and T. T. Ray, *Polym. Bull.*, 1988, **19**, 365–370.
19 P. Zinck and J.-F. Gerard, *Adv. Mater. Sci. Res.*, 2013, **15**, 251–265.
20 K. Mason, *Sizing Up Fiber Sizings*, 2017, http://www.compositesworld.com.
21 R. R. Umbdenstock and P. F. Bruins, *Ind. Eng. Chem.*, 1945, **37**, 963–967.
22 J.-L. Gustin, *Process Saf. Environ. Prot.*, 2002, **80**, 16–24.
23 http://www.plantics.com.

Faraday Discussions

DISCUSSIONS

Bio-based materials: general discussion

George Huber, Dimitris Argyropoulos, Avtar Matharu, Harry Bitter, Christian Stevens, Servann Herou, Karen Wilson, James Clark, Deepak Pant, Carlos I. Cabrera-Rodríguez, Joseph Samec, Bruce E. Dale, Thomas Farmer, Mark Mascal, Anne Horan, Andrzej Stankiewicz, Florence Gschwend, Xindong Mu, Long Zhou, Xiaoming Huang, Changwei Hu, Terence Cooper, Laura Sparlinek, Vitaliy Budarin, Eero Kontturi, Andrew Hunt, Amalio Garrido, Keith Waldron, Fang Zhang, Anna Zhenova, David Constable, Simo Sarkanen, Magdalena Titirici, Gadi Rothenberg, Jakob Albert and Duncan Macquarrie

DOI: 10.1039/C7FD90047C

Simo Sarkanen opened the discussion of the introductory lecture by Bruce Dale: The supposition that the strength of a national economy is directly related to energy consumption is incomplete. This hypothesis is reminiscent of Marx's suggestion that the value of an article is equivalent to the labor expended in producing it. Actually, the functional demand for the articles produced and services offered is a more realistic basis for evaluating the strength of an economy.

Bruce Dale replied: Perhaps all suppositions are incomplete in one way or another. And if Marx said something true, it remains true nonetheless. Truth does not depend on who speaks it. I am no Marxist, and I am not a capitalist either. I try to observe what is really going on in the world and draw correct conclusions. I observe that the total GDP is directly and linearly related to the total power consumption with a very high (95% plus) correlation coefficient. No other variable so completely and so logically explains GDP as power consumption does. Not one. If someone can point me toward another variable that does a better job of explaining GDP, and show me the supporting data, I will be glad to update my beliefs.

Here is a "thought experiment", as some like to term it, about demand side analysis. I demand a cell phone with all of its wonderful features. But there is no electricity to charge it and run the routers, no fuels to run the machines that will mine, refine, and assemble the hundreds of materials that make up the cell phone, and no fuels and electricity to support the lives of the highly-paid and highly-skilled individuals who will design, build and market the phones and produce the apps. In other words, no energy. How many cell phones will be produced? Not one, in spite of the demand.

So I respectfully but firmly disagree, I think there is demand for many goods and services that remains theoretical until there is energy to meet that demand.

Xindong Mu said: Thank you for your comprehensive and instructive presentation. As you mentioned in your talk, the cost of biomass is crucial for the economic feasibility of biorefineries. What is the optimum proposed price for the cost of biomass feedstocks that could make farmers happy and companies profitable? Should biorefineries be run by farmers? Or what kind of relationship should there be between them?

Bruce Dale answered: I do not think that there is a single universal price for biomass that will satisfy all farmers and biorefiners. The price will probably have to be determined on a regional basis. However, we can be certain that if we pay the farmers more, they will grow more. If they grow more biomass, then transportation distances to the biorefineries will decrease and the size of the biorefineries will increase. Both of these factors (decreased transportation distances and bigger biorefineries with better economies of scale) will tend to decrease the cost of the production of the biofuel. These effects are demonstrated in a paper for the U.S. Midwest.[1] I think most biorefineries are probably too complex to be run by farmers, with the exception of on-farm or local biogas plants. The Italian Biogas Consortium is showing that farmers can readily manage these on-farm digesters to produce biogas and electricity.[2] Also, the capital costs of a competitive lignocellulose to ethanol plant are far beyond the financial resources of even the largest farmer.

So, I think the best way for farmers to participate in a large scale biorefinery is to establish farmer cooperatives to interface with the biorefineries. These cooperatives would aggregate local biomass and then process the biomass in local facilities called "depots" to stabilize and densify the biomass for later sale to the biorefinery. I think this approach would work well for both the farmers and biorefiners. The depot approach using the AFEX pretreatment process has been reported previously.[3] Other technologies might be used instead of AFEX, but this is one example.

1 S. Kim and B. E. Dale, *Biofuels, Bioprod. Bioref.*, 2015, **9**, 422–434.
2 B. E. Dale, F. Sibilla, C. Fabbri, M. Pezzaglia, B. Pecorino, E. Veggia, A. Baronchelli, P. Gattoni and S. Bozzetto, 2016, *Biofuels, Bioprod. Bioref.*, 2016, **10**, 341–345.
3 P. L. Eranki, D. H. Manowitz, B. D. Bals, R. C. Izaurralde, S. Kim and B. E. Dale, *Biofuels, Bioprod. Bioref.*, 2013, **7**, 537–550.

Karen Wilson remarked: With falling costs of crude oil, and tight profit margins, industry often comments that while they would like to use more bio-derived chemicals, they are currently too expensive for their processes. Given there are many potential bio-derived chemicals that can be synthesised, what do you think are the most promising target molecules or classes of molecules that researchers should be focusing their efforts on developing processes for?

Bruce Dale answered: Well, clearly some bio-derived molecules have made it into the chemical industry. I count polylactic acid and propanediol among this small number. These chemicals are made in high yield from corn starch and have some advantages over their petroleum-derived competitors. The various diacids, like succinic acid, are promising. I would currently be looking for substitutes for aromatic compounds also, including PTE.

Keith Waldron asked: Is there an argument to have an interconnected world solar (and maybe wind) energy grid?

Bruce Dale replied: I do not think transmission of electrical power internationally makes much sense. If I understand line losses (electrical resistance heating), we would probably lose too much of the energy to make it worthwhile. And, of course, the capital expense would be enormous, especially for intercontinental transmission lines.

Deepak Pant communicated: Prof. Dale mentioned about improvements in the biorefinery establishment when the price paid to the farmers was doubled from \$40 per ton to \$80 per ton.

A slide showed the remarkable increase in the surface area that was covered by the crop (I assume it was corn). Prof. Dale mentioned that there is no food *vs.* fuel debate but it is more like a food and fuel situation where the increased corn growth does not take away food. However, when the price paid to farmers was doubled, they were incentivized to go for one particular crop, corn in this case. This, in turn, took away the land where probably other food crops were growing earlier and which is now corn dominated. Perhaps this is not such a problem in the US but I assume there will be problems in the rest of the world.

What are your thoughts on the seeming dichotomy between the two statements?

Bruce Dale communicated in reply: Ref. 1 does a much better job of explaining this effect than I was able to do in 30 seconds. A subsequent paper using the same approach estimated the increased employment achieved by increasing the biomass price paid to farmers.[2] What the farmers supplied at higher biomass prices were mostly grasses such as switchgrass and more stover and residues.

1 S. Kim and B. E. Dale, *Biofuels, Bioprod. Bioref.*, 2015, **9**, 422–434.
2 S. Kim and B. E. Dale, *Biofuels, Bioprod. Bioref.*, 2015, **9**, 639–647.

Deepak Pant communicated: Thank you for your answer. Can you direct me towards the source of the figure from your presentation where you made a comparison of *status quo* and different types of biofuel management scenarios?

Bruce Dale communicated in reply: I think the figure you mention is regarding the different biogas production approaches in Italy. If so, the figure can be found in ref. 1.

1 L. Valli, L. Rossi, C. Fabbri, F. Sibilla, P. Gattoni, B. E. Dale, S. Kim, R. G. Ong and S. Bozzetto, *Biofuels, Bioprod. Bioref.*, 2017, DOI: 10.1002/bbb.1789.

George Huber opened the discussion of the paper by Avtar Matharu: What is the purpose of the ethanol and heptane extractions? The residence times in this process seem long. Do you have any ways to reduce the residence times? Another challenge with this technology is recylcing the solvents. Can you explain how you recycle the solvents?

Harry Bitter communicated: Thank you very much for the talk on using mango peels as feedstock. This is a nice example where the chemical and food worlds can meet. From your paper and presentation, I got the impression that you take the heptane/ethanol extraction preceding the sub critical water extraction for granted. Did you consider changing the order of the extraction step? Might there be an advantage to changing the order?

Avtar Matharu answered George Huber and Harry Bitter: The purpose of the ethanol and heptane extractions was to extract any waxes and soluble polyphenolics, oligosugars, *etc.* within the context of a cascade approach to biomass valorisation. These steps are not necessary if we are just chasing pectin. I assume you mean residence times for ethanol and heptane? If so, then yes, these can be optimised, solvents evaporated/distilled and re-used. With respect to sub-critical water, it can recycled within a loop system but once it is treated with ethanol to precipitate pectinaceous matter, distillation must be considered. We have not explored the order, whether we extract waxes/polyphenolics first and then pectin or *vice-versa*.

Changwei Hu asked: Have you extracted in different orders?

Avtar Matharu replied: Yes, we have extracted in different orders using Design of Experiment. The processing temperature is more significant than the concentration (water : biomass ratio).

Andrew Hunt said: In your paper you highlight that up to 15% of the biomass can extracted with ethanol, then superheated water can generate a maximum yield of 18% of pectin. Therefore, what is the remaining 67% residue and what can it be used for?

Avtar Matharu responded: We are converting the remaining depectinated residues into nanocellulose fibrils and crystals. We have just published such work in *Green Chemistry*.[1]

1 E. M. de Melo, J. H. Clark and A. S. Matharu, *Green Chem.*, 2017, **19**, 3408–3417.

Eero Kontturi enquired: What is the amount of pectin extracted with respect to the original amount of pectin in the mango peel? In other words, what is the theoretical yield of pectin in the extraction? Furthermore, what are the reasons behind the theoretical yield? Why is some of the pectin left unextracted after exposure to subcritical water?

Avtar Matharu answered: The amount of pectin extracted with respect to the theoretical amount is a moot point. We often talk about the extractable pectin rather than the total amount as tightly-bound protopectins remain in the material. However, it is suffice to say that based on TGA data of depectinated residues, which show the pectin signature at approximately 208 °C, we are extracting close to 80% of the pectin.

Anna Zhenova asked: Do the different varieties of mangoes grown in each region depend entirely on climate, or would it be possible to convince farmers in one region to grow a different type of mango if that peel is more useful for pectin production? For example, since the Kesar variety has the highest pectin yield overall, could growers around the world switch to that?

Avtar Matharu responded: You raise a good question. Climate plays an important role in the type of geographic cultivar. I am not sure if Kesar can be grown, say, in the wilds of Brazil.

Joseph Samec queried: Have you studied how the chemical structure of pectin is affected by the residence time of the extraction step in supercritical water?

Avtar Matharu replied: Firstly, for clarity, we are using pseudo-sub-critical water not supercritical water. The latter is highly destructive not only to biomass but also to the reaction vessels used.

The residence time will affect the pectin structure based on the assumption that the degree of esterification will be reduced in the case of prolonged residence times (increased hydrolytic activity).

Deepak Pant asked: In your paper, you did a conventional acid extraction for comparison and obtained a yield of only 4.88%, which is much lower than that reported in the literature. So what could be the reason for this low yield in your case? Secondly, what is the average yield that is reported in the literature for such acid based extractions?

Finally, I have a question about the application of this approach. You have shown it works for mangoes, which other fruits could this approach be utilized for?

Avtar Matharu answered: The conventional acid extraction may have been too harsh and cleaved pectinaceous matter, thus reducing the yield. The average literature yields vary quite significantly from paper to paper. A conservative estimate is 12% dry weight basis.

Our approach works for a wide variety of feedstocks.

Christian Stevens said: You analyzed the pectins of different sources and measured, amongst others, the degree of esterification. Did you also check the quality of the pectins by analyzing the gelling properties and characteristics of the gel? Gels can be obtained with calcium ions or with added sugars and can differ in different applications.

Avtar Matharu responded: Yes, we have checked the gelling quality using the standard sucrose (added sugar) method. We are yet to test the gel strength – this is on-going.

Andrzej Stankiewicz asked: Are there any specific equipment design issues related to the technology presented?

Avtar Matharu replied: There are no specific equipment issues as we are working under pseudo-subcritical conditions and not super-critical, which is a major challenge. I do hope that this process can benefit from advances in engineering and process intensification so that we do not always use "off the shelf" but promote innovation.

Mark Mascal said: You reflux the dry peel in ethanol for 2 h and then later the sample is left overnight in ethanol for pectin precipitation. Since pectin is acidic, you could be seeing some transesterification. I wondered whether the small peak at about 63 ppm in the ^{13}C-NMR spectrum could be an ethyl ester CH_2.

Avtar Matharu answered: Yes, the mysterious unidentified peak could well be some transesterification. Thank you for commenting on this.

James Clark opened the discussion of the paper by Simo Sarkanen: What are the opportunities for making small molecules (typically mono-aromatics) from lignin?

Simo Sarkanen replied: Extensive cleavage of lignin derivatives into low molecular weight compounds can certainly be chemically achieved, but the economic viability of such degradative processes is usually confounded by the complexity of the resulting product mixtures. Among the topics presented during this *Faraday Discussions* about bio-resources for the chemical industry, RANEY® Ni-catalyzed reductive cleavage of hardwood lignin (R. Rinaldi *et al.* (DOI: 10.1039/C7FD00069C)) undoubtedly merits further study. Additionally, the biological funneling of lignin-biodegradation products towards specific intermediates in microbial metabolic pathways has been proposed as a likely means of minimizing by-product levels to a substantial extent.[1]

1 J. G. Linger, D. R. Vardon, M. T. Guarnieri, E. M. Karp, G. B. Hunsinger, M. A. Franden, C. W. Johnson, G. Chupka, T. J. Strathmann, P. T. Pienkos and G. T. Beckham, *Proc. Natl. Acad. Sci. U. S. A.*, 2014, **111**, 12013–12018.

Thomas Farmer said: Certain properties for materials can often be opposed to one another. For example, long-term stability for use in building materials may be opposed to biodegradability over a reasonable timescale. Prior literature would suggest very varied rates of biodegradability for lignin. Therefore, do you know about the biodegradability of your materials, including after your treatment with ball-milling and methylation, and what impact this has on the long-term stability for building materials?

Simo Sarkanen answered: Some aerobic bacteria can grow on lignin as a sole carbon source,[1] but white-rot fungi are the most frequently cited biological agents of lignin biodegradation. To degrade lignins, these microorganisms require some carbohydrates, the absence of which will inhibit lignin biodegradation. Whether methylated or not, the lignin materials employed in plastics formulations with a high lignin content are biodegradable before they are incorporated into the new materials. However, we have not yet studied the biodegradability of these lignin-based plastics in their final form.

1 E. Odier, G. Janin and B. Monties, *Appl. Environ. Microbiol.*, 1981, **41**, 337–341.

Vitaliy Budarin commented: Your paper demonstrates a huge potential for making a new polymer based on a lignin–poly(ethylene glycol) blend. This result has also provided some extra information about the lignin structure, showing that the lignin polymer is significantly flexible. This is controversial to the commonly held views about crosslinking or hyper-branching in lignin chains. Therefore, do you have any evidence that the structure of the lignin used in this experiment represents a unique lignin structure within a plant?

Simo Sarkanen replied: In terms of the ^{13}C NMR analyses, pine milled wood lignin (isolated in 22% yield) is similar (although not identical) to the corresponding cellulolytic enzyme lignin (isolated in 23% and 86% yields), as shown in Table 3 of an 11-year-old article by J. F. Kadla *et al.*[1] However, the carbohydrate content of the cellulolytic enzyme lignin is about 4-fold higher than that of the milled wood lignin (4%). Thus, milled wood lignin isolated in yields greater than 20% seems to represent softwood lignin as a whole reasonably well, although the molecular weight after milling is undoubtedly lower than in the native state.

1 J. F. Kadla *et al.*, *Holzforschung*, 2006, **60**, 389–397.

Servann Herou asked: Since lignin has been used for the production of carbon fibres, what do you expect would be the influence of the blending agent on the mechanical strength of lignin based carbon fibres after pyrolysis?

Simo Sarkanen answered: Lignin-based blend formulations that, on casting, produce stronger polymeric materials would be expected to capture improvements in the mechanical properties of carbon fibers formed through pyrolysis. It is worth noting that uniform electrospun fibers can be generated from kraft lignin solutions in DMF without small quantities of poly(ethylene oxide) as a spinning aid.[1]

1 I. Dallmeyer, F. Ko and J. F. Kadla, *Ind. Eng. Chem. Res.*, 2014, **53**, 2697–2705.

Magdalena Titirici asked: What is the potential for supercapacitors? You are using extremely toxic chemicals, do we gain in terms of sustainability? Is it really biodegradable?

Simo Sarkanen responded: Industrial kraft lignin-derived carbon nanofibers have shown promise with regards to supercapacitor electrode applications,[1] but this preliminary evaluation needs to be confirmed through further studies. Otherwise, solution casting of blends to produce plastics with a high lignin content has been employed only to identify successful formulations for functional polymeric materials. Of course, solution casting will be replaced by injection-molding in industrial processes that produce materials for consumer markets.

1 J. I. Dallmeyer, *Ph.D. Thesis*, University of British Columbia, 2013.

Dimitris Argyropoulos enquired: Can you please provide a historical link and a description for the reasons you moved to the very expensive and cumbersome product of methylated ball-milled lignin (that you use now) from your early efforts with kraft and lignosulfonates that date back 15–20 years?

Simo Sarkanen replied: There are two reasons why methylated ball-milled lignin was adopted as a starting material for producing functional polymeric materials with a very high lignin content. First, except in molecular weight, ball-milled lignins are structurally similar to a sizable portion of the native lignins that preserve cell-wall integrity in trees and crops. Thus, as materials, their mechanical properties are more likely to support their utility than those of the derivatives formed through high-temperature exposure to alkaline solutions or aqueous bisulfite under acidic conditions. Second, hydrogen bonding generally increases the overall strength of the powerful intermolecular noncovalent forces between lignin components,[1] and so underivatized lignins, as materials, are more difficult to deform (and are therefore more brittle) than their methylated counterparts. The clarity of the trends in the X-ray powder diffraction patterns displayed by the methylated ball-milled lignin-based blends with poly(ethylene glycol) quickly led to a comprehensive model for the configuration of the macromolecular lignin species in these new materials. Of course, such insights are directly pertinent to the development of functional industrial lignin-based polymeric material blends.

1 Y. Chen and S. Sarkanen, *Phytochemistry*, 2010, **71**, 453–462.

Simo Sarkanen opened the discussion of the paper by Mark Mascal: Do the overall yields of diols and dicarboxylic acids from CMF affect the economic viability of producing these compounds as monomers for synthesizing polyesters? To avoid confusion, the term "branching" should be restricted to its customary meaning with regards to the (linear or branched) configurations of polymer chains. It should not be used to describe the carbon skeletons of individual monomer units.

Mark Mascal answered: The yields at each step in the synthesis of the monomers are good, so the overall process has the potential at least to find traction in renewable markets. While the number of unit operations is a consideration, we are aware that a multistep process to make bio-terephthalic acid *via* CMF is now being commercialized. We are mindful that the use of the term "branching" to describe the nature of the hydrocarbon chain in the monomers has the potential to be confused with the concept of polymer branching. We will look for ways to avoid this matter in future publications.

Karen Wilson asked: How is 5-chloro-methylfurfural (CMF) produced? Given that many early reports of 5-hydroxy-methylfurfural (HMF) production from fructose employed the use of HCl or Lewis acid chlorides in the presence of ionic liquids, is it possible that CMF was also produced in these studies? What is the hydrocarbon solvent used for the extraction of CMF? Solvent extraction is used during HMF isolation, can you selectively partition CMF *versus* and HMF formed in the reaction?

Mark Mascal responded: The CMF process I described in this talk does, in fact, involve HMF as a fleeting intermediate. Earlier reports of the preparation of HMF using HCl or other acidic systems that include chloride generally did not have a sufficient concentration of HCl to generate CMF. One hydrocarbon that I know has been successfully used to extract CMF is cyclohexane. In the CMF process, the product is partitioned effectively into the organic solvent. No HMF is isolated either from the organic or aqueous layer, but some humic material is observed, which is likely due to unconverted HMF.

Magdalena Titirici enquired: How do you detect CMF? How do you quantify the CMF? Do you quantify against other elements in there?

Mark Mascal replied: We always quantify CMF by the isolated yield, confirmed by NMR spectroscopy. Any minor products, such as, for example, levulinic acid, are isolated and quantified separately.

Andrew Hunt said: In your paper, you have CP-3-15-7*, which only appears in one figure. Please can you explain what this is?

Mark Mascal answered: The asterisk denotes a 25 : 75 molar load ratio of compounds **15** : **7**. The figure caption in the paper has been edited to include this information.

Avtar Matharu commented: I see very interesting applications for low surface energy materials. Can you comment on the refractive index of your materials and when blended do you observe phase separation?

Mark Mascal responded: At this point, we have only made the monomers and our collaborators are working toward obtaining the first samples of polymer, so we do not yet have refractive index or phase separation data. These will, of course, be useful data to obtain.

David Constable said: Please consider other metrics as well as yield. Can you comment on the nature of the catalyst?

Mark Mascal replied: As Professor Dale strongly pointed out, there is no more important metric than yield in biomass conversion. Other metrics for the CMF process include low energy intensity. The HCl has multiple recycle paths and the solvent is used in a loop. High yield means minimal waste. The lignin is valorized. Inorganics from biomass and excess water produced by dehydrating the sugars are the only significant waste streams. High carbon efficiency, high atom economy, high E factor. The hydrogenation catalyst was Pd/C. Earth-abundant metal catalysts generally show inferior performance.

Magdalena Titirici asked: Do you think biomass will give significant gains in terms of sustainability?

Mark Mascal answered: Yes. I think this is the whole basis of the biorefinery movement, although we have to be careful to ensure that the carbon footprint of the technologies (energy costs of biomass transportation, processing, *etc.*) is lower than what would be the case from deriving the petrochemical equivalents.

Harry Bitter remarked: During the discussion, the issue came up of the cost of a catalyst. As you rightly stated, in general, the price of a catalyst is not so important since it is not consumed. However, I think that is only partially true. The price is related to the availability and the availability of some elements is endangered. This holds for the noble metals and also, for example, for Co. When an application is found and a scarce element is used, an issue might arise with its supply. How do you feel about that? Will there be a need for replacing endangered elements in catalysts?

Mark Mascal responded: There are a number of misconceptions on this subject. The key metric of importance for industrial catalyst use is performance, not price. Earth-abundant catalysts may be less expensive and in plentiful supply, but if they are inferior in terms of activity and less robust than catalysts prepared from less common elements, then this is a poor value proposition, since high productivity is essential and time offline in a process is very costly. Since expensive metals like Pd, Pt, and Ir are routinely recycled, there is no need to panic that they will become unavailable any time soon. However, when we find opportunities to avoid using such metals in cases where performance is not sacrificed, we should take advantage of that. This means that research in the area of cheap catalysts is indeed valuable, but it should not be taken to the point that we discriminate against research done with noble metals.

Florence Gschwend said: Sustainable sourcing of the feedstock is important to achieve environmental benefits from using biomass. Have you looked at any waste feedstocks?

Mark Mascal replied: We have used various forms of waste, including straw, municipal wastewater solids, corn stover, and sawmill waste. All give CMF yields consistent with their glucan content.

Christian Stevens enquired: What is known about the inherent toxicity of CMF? Although it is only an intermediate in the valorisation chain, it might be important to be aware of toxicity problems and try to work around it if necessary.

Fang Zhang asked: Have you considered the toxicity of CMF due to the existence of the Cl atom?

Mark Mascal answered Christian Stevens and Fang Zhang: CMF is known to be toxic. It hydrolyzes *in vivo* to HMF, which, like other Maillard reaction compounds, has secondary metabolites that show various forms of toxicity. CMF also has primary toxicity due to the release of HCl. However, like many toxic chemicals, it is an intermediate that, when handled responsibly in a multistep chemical process, should not be a significant concern.

Changwei Hu queried: How about the lifetime of the Pt catalyst?

Mark Mascal responded: The only place we use Pt ourselves in this work is as an electrode material for the Kolbe coupling of levulinic acid to 2,7-octanedione, thus not as a catalyst *per se*. In this role, its lifetime is not limited. In fact, we can use graphite instead of Pt as the anode.

Bruce Dale opened a general discussion of the papers by Avtar Matharu, Simo Sarkanen and Mark Mascal: After we have had a chance to chat, I think we found ourselves in agreement. There are a number of ways of measuring yield, and some of them are not very valuable. But for many reasons, the mass yield (kg of salable products per kg of primary feedstock(s)) is critical for both life cycle and economic thinking and should be featured in all research reports for biofuels and bioproducts.

David Constable responded: The point I was trying to make is that yield is a poor measure of "greenness" or of the sustainability of the chemistry. It considers only the reactants and is based on the limiting reactant. It does not matter as to the renewability or sustainability of any of the reactants, the toxicity, the physical hazards, *etc.* It assumes that chemists have no responsibility to conduct their science in a manner that can be sustained. Assessments of chemistry should be multivariate, easy to understand and perform, and include elements of hazard and risk.

Amalio Garrido said: From an engineering company's perspective, the big issues are the logistics of the feedstock and the scale. Have you considered the feasibility of the processes for future industrial applications?

Keith Waldron replied: Regarding my own paper (DOI: 10.1039/C7FD00044H), the economies of scale in biorefining of biomass are critically important due to the low energy density of the feedstock. Likewise, the residues left at the end would be most beneficially exploited in a relatively local situation.

Dimitris Argyropoulos commented: In any efforts of biomass utilization, we need to be cognizant of the fact that species, seasonal, geographic and process variations (as they relate to pulp and paper waste streams) need to be taken into account. As such, we, as a community that develops the science and technology of biomass transformation, need to be aware of these issues and ensure that a good understanding and the foundations for plasticity and flexibility are built into our processes.

Avtar Matharu responded: Yes, biomass variability dependent on geographical location, season, soil, stress, *etc.* is a key factor for the isolation of small molecules that may or may not be present as the biomass grows. We have found that for bipolymers and oligomers global (geographical) variation is less of a concern. We have experimental data from citrus pectin sourced from several global locations, which show no significant variation in pectin yield.

The "plasticity" and flexibility is less of a concern for pectin extraction and in an ideal scenario we envisage multiple, mixed feedstocks, *i.e.* Keitt, Sindhri and Kesar being combined to afford pectin.

Andrew Hunt opened the discussion of the paper by Xindong Mu: Your materials are very highly microporous, is there any method to tune the synthesis to get meso- or macroporous systems?

Xindong Mu replied: The bamboo shoot-derived carbons are prepared by hydrothermal treatment and a subsequent carbonization process in our work, which produces a mainly microporous structure with some meso- and macropores. However, a high meso/macropore proportion of carbons is difficult to synthesize by this method without an activating agent. The chemical activation method using activating agents (*e.g.* $ZnCl_2$, $KHCO_3$) can be used to produce carbons with meso- or macroporous structures.

Deepak Pant said: You used 5 wt% platinum in your system. How did you reach this particular number? Did you try other ratios, like 1% or 10%?

Xindong Mu answered: We have investigated the impact of the Pt content in Pt/NC-BS-500 on the hydrogenation of furfural in water at 100 °C. As the Pt content increases from 1 wt% to 10 wt%, the conversion of furfural after 1 h increases from 21% to 95% with a high furfuryl alcohol selectivity of >99%. The corresponding activity of 1% Pt/NC-BS-500, 5% Pt/NC-BS-500 and 10% Pt/NC-BS-500 is 409, 294 and 185 h^{-1}, respectively. As 5% Pt/NC-BS-500 has both a high furfural conversion in a short time (75.4% in 1 h) and a relatively high specific activity, it was chosen for our system.

Deepak Pant commented: You mentioned surface area as being one of the critical parameters for the performance of your system. Have you tried other feedstocks (other than bamboo) to check if with another feedstock you get a higher surface area and perhaps better conversion?

Xindong Mu responded: The surface area of carbon supports could affect the catalytic performance of the Pt catalyst. In our system, NC-BS-800 (670 $m^2\ g^{-1}$) has a much higher surface area than that of NC-BS-500 (287 $m^2\ g^{-1}$), but Pt/NC-BS-800 shows a much lower activity for the transformation of furfural to furfuryl alcohol under mild conditions than Pt/NC-BS-500. This indicates that other factors (*e.g.* changes in surface properties) might favorably influence the catalytic performance of the N,O-doped carbon supported Pt catalyst. Coconut husk-derived activated carbons ($\sim$860 $m^2\ g^{-1}$) were also used as supports to load Pt nanoparticles and they displayed much lower conversions of furfural to furfuryl alcohol than Pt/NC-BS-500.

Magdalena Titirici asked: What is the morphology of the bamboo shoots before and after hydrothermal treatment? How much nitrogen is in the bamboo shoots before and after?

Xindong Mu replied: The bamboo shoots are relatively loose tissues with some irregular fused sphere-like particles. After hydrothermal treatment, the hydrochars show a loose and interconnected structure, which is made up of numerous relatively uniformly-distributed fused sphere-like nanoparticles.

The nitrogen content of the bamboo shoots is 4.3 at% based on the dry weight of the bamboo shoot, while the nitrogen content of the hydrochars is 3.2 at%.

Magdalena Titirici queried: Do you get uniform spheres?

Xindong Mu answered: The bamboo shoot-derived carbon support is made up of numerous uniformly-distributed sphere-like nanoparticles with small sizes of about 40–80 nm (Fig. 1).

Xiaoming Huang said: You prepared two N-doped activated carbon supported Pt catalysts by changing the carbonization temperature. You found that the amount of nitrogen in the two samples is different. The one calcined at 500 °C contained more nitrogen than the one calcined at 800 °C. Since you claim that the nitrogen species have the function of stabilizing the Pt nanoparticles, I was wondering, did you observe any difference between these two samples in terms of the catalyst stability?

Xindong Mu replied: Both the Pt/NC-BS-500 and Pt/NC-BS-800 catalysts were stable in the furfural hydrogenation reaction, and can be recycled for five runs without any significant loss of activity.

Long Zhou enquired: Bamboo grows very fast, in your work why did you use bamboo shoots but not older bamboo? If you use older bamboo, you will have more feedstock, which I believe is more sustainable.

Xindong Mu responded: Common carbons without nitrogen functionalities are obtained with mature bamboo as the starting material because there are few nitrogen-containing compounds in mature bamboo. By contrast, porous hetero-atom doped carbon materials are prepared by using bamboo shoots as the raw material without any templates, chemical activation or exogenous nitrogen

Fig. 1 Image of the bamboo shoot-derived carbon support.

sources. The bamboo shoots have some unique characteristics in chemical compositions and physical textures, which significantly differ from the mature bamboo. The lignin, crystalline cellulose and xylan predominantly accumulate in the cell wall in the bamboo shoot, which is unlike the high lignification of the mature bamboo. The low degree of lignification in bamboo shoots along with considerable amounts of acid-soluble lignin mean they have unlignified and loose tissues and less extended crystalline cellulose textures, which are beneficial for the decomposition into small carbon nanoparticles during the hydrothermal process. Moreover, bamboo shoots are intrinsically rich in nitrogen-containing organics, including proteins and amino acids. The nitrogen content can reach up to 4.3% based on the dry weight of the bamboo shoot, which ensured the incorporation of nitrogen and oxygen into the structure of the carbon materials.

Harry Bitter said: You showed that your Pt on nitrogen containing carbon has special activity with respect to furfural hydrogenation. You used a pure furfural feedstock. To keep biobased feedstocks low priced, their purification should be limited. Do you know anything about the resistance of your catalyst with respect to potential poisons in the feedstock, such as salts or sulfur containing compounds?

Xindong Mu answered: Furfural directly obtained from a factory (Shandong Baiqian Chemical Co., Ltd.) was also used as the reactant, and was reacted at 100 °C using 1 MPa H_2 in water in the presence of the Pt/NC-BS-500 catalyst. The results showed that 98.2% conversion of furfural and 99% selectivity for furfuryl alcohol were achieved within 4 h, indicating that the Pt catalyst is robust for furfural hydrogenation in water. Pt/NC-BS-500 shows no apparent deactivation in furfural feedstock provided by the factory.

Deepak Pant opened the discussion of the paper by Jakob Albert: You have mentioned carbon dioxide as a by-product, which could be recycled back to grow algal biomass. Wouldn't it make sense to recycle the CO_2 to formic acid electrocatalytically to have an even higher yield of the main product? How does biomass to formic acid compare to current state-of-the-art for CO_2 to formic acid? Also, at what scale is the process currently?

Jakob Albert replied: Of course electrocatalytic conversion of CO_2 to formic acid is an alternative but this is currently not available on an industrial scale compared to algae growing in fermenters.

Biomass to formic acid is, from an industrial point of view, much more interesting compared to CO_2 hydrogenation as biomass is not just seasonally available (compared to fluctuating electricity for electrochemical CO_2 hydrogenation), very flexible concerning quality and origin of feedstock, and very robust against poisoning (compared to CO_2 hydrogenation). Moreover, the current scale of the OxFA process is 60 kg of formic acid per day.

Vitaliy Budarin said: The technology developed in your paper is reasonably harsh: a high final concentration of formic acid, a strong catalyst, the presence of oxygen and an elevated temperature. Under these conditions, a cellulose structure could be significantly degraded. Have you optimized the process for the production of high quality cellulose? It looks like the Lindqvist-type catalyst is colloidal. If this is true, how do you visualize the interaction between the solid biomass and the solid colloid catalyst?

Jakob Albert responded: I would contradict this statement as we are working under mild conditions (below 100 °C, aqueous solution, O_2 and no H_2O_2 as an oxidant). The final concentration of formic acid is typically 1–5 wt% in our process, I would also not state this as "high". The key feature of the Lindqvist POM is that it is not as strong as the Keggin POM, so cellulose conversion is prevented (in contrast to the Keggin POMs). The process is, so far, not fully optimized for cellulose quality but we can improve that further (maybe in collaboration with partners). The Lindqvist-type POM (in its salt form with a K countercation) as well as the hydrolyzed hemicellulose and lignin fragments are completely soluble under the applied reaction conditions, so it is a homogeneous catalysis and not a slurry reaction.

Simo Sarkanen enquired: Neither phenol nor 4-hydroxy-3-methylbenzoic acid are realistic model substrates for lignin. Why did you not use 4-hydroxy-3-methoxybenzyl alcohol (vanillyl alcohol) instead?

Jakob Albert answered: The pre-selection of suitable model substrates was done according to the literature. You are right, phenol is not a realistic model substrate. However, I think that 4H3MB is a suitable model compound. Within the project, we have now changed to 3,4-dimethoxybenzylalcohol.

Carlos I. Cabrera-Rodríguez queried: Is the reported low yield in your paper due to the low oxygen solubility in water or do you have another explanation for it? Have you calculated the maximum theoretical yield based on oxygen solubility at 20 bar? What is the final formic acid concentration that you can obtain?

Jakob Albert replied: The low yield described in the paper is mostly due to the low availability of oxygen in the liquid phase because of inefficient mixing in our screening apparatus. By using another setup with a gas entrainment stirrer, we could significantly improve the yields of formic acid. The maximum theoretical yield of formic acid is dependent on the carbon framework within the substrate and thus different for each feedstock.

The final formic acid concentration that can be achieved is around 40 wt%. Typically, we operate between 5 and 10 wt% in order to keep an optimum performance of our catalyst with respect to the pH value.

Eero Kontturi asked: What happens to the cellulose during the process? Did you check the molecular weight distribution before and after the oxidation? It is quite unusual that nothing happens to cellulose during such treatments, at least some depolymerization generally occurs. Whether this interferes with the quality of the cellulose in further applications is another matter. Generally, slight depolymerization of cellulose is irrelevant from an applications point of view but the figures would be interesting scientifically, especially if nothing happens to the cellulose.

Jakob Albert answered: To the best of our knowledge, the cellulose shows no significant changes after the treatment in our process. This is due to the insolubility of cellulose in water. As we are performing a homogeneous catalysis in aqueous solution, depolymerization/degradation can only occur if (at least) partial solubilisation of the biomass takes place.

Amalio Garrido enquired: In the final acylation, what is the retention/residence time? Do you have any idea of the kinetics? Have you tried to wash the cellulose to remove the acid?

Jakob Albert answered: I am not sure what you mean by "final acylation". The chemical conversion itself is a three-step process with the following steps:

(1) Hydrolysis of the biomass.

(2) Oxidation of water-soluble biomass-fragments by the POM catalyst to formic acid and CO_2.

(3) Re-oxidation of the POM by molecular O_2.

So far, the whole process has been batch/fed-batch, so no "retention/residence time" exists.

We have indeed studied the kinetics of different substrates using Keggin-type POMs. This will soon be published in *ACS Sustainable Chem. Eng.*[1]

For your last question, yes, you are right, we have washed the cellulose several times with water in order to remove the cellulose completely.

1 J. Reichert and J. Albert, *ACS Sustainable Chem. Eng.*, 2017, **5**, 7383–7392.

Avtar Matharu said: You say high grade cellulose many times, can you tell me what you mean by high grade cellulose? Is it just purified cellulose or are you forming nanocellulose crystals and fibres. Are you changing the degree of crystallinity?

Jakob Albert responded: By high grade cellulose we mean cellulose that has:

(1) Unchanged signals using FT-IR spectroscopy.

(2) Unchanged REM images of fibers before and after treatment (in case of the model substrates).

(3) A degree of crystallinity that is still very high.

(4) A high degree of polymerization.

Simo Sarkanen commented: The yields of formic acid obtained by oxidizing cellulose, hemicellulose and lignin for 8 h in the presence of the Lindqvist-type polyoxometalate $K_5V_3W_3O_{19}$ were 0%, 8.3% and 2.4%, respectively. The contrasting results from cellulose and hemicellulose are surprising. Are there any preliminary indications about what occurs when the reaction time is extended beyond 8 h?

Jakob Albert answered: We have already extended the reaction time and optimized the reaction conditions in order to increase the conversion. Luckily, the same trends are observable. So far, we achieved up to 30% formic acid yield from hemicellulose and 12% yield from lignin.

Terence Cooper opened the discussion of the paper by Gadi Rothenberg: The first sample you showed us had a very large pore size. Can you control the pore size? The later samples had much smaller pore sizes. What is producing the foaming? Is there a gas being evolved to do this?

Gadi Rothenberg replied: Yes, we can control the pore size. At this stage I can not give more details about this, sorry.

Harry Bitter said: Thank you for the inspiring presentation with respect to the making and use of a bio-based material. My question is about patenting your material. How did you do that? When I look on the Internet, I find a number of publications and claims about the making and degradation of polymers made from citric acid and glycerol. For example, ref. 1. Does this not conflict with your patent?

1 D. Pramanick and T. T. Ray, *Polymer Bulletin*, 1988, **19**, 368–370.

Gadi Rothenberg answered: I think the difference is that our patents cover the making of a foamed polymer, which necessitates specific conditions. Further patents of Plantics BV cover the composites of this polymer with man-made fibres and with natural fibres, and the possibilities and applications thereof, as well as the applications and process for making the polymer on a metallic substrate.

The patent numbers are all given in the paper and the patents themselves are available online.

Carlos I. Cabrera-Rodríguez asked: Is this reaction specific for citric acid and glycerol? Or have you tried other acids or alcohols?

Gadi Rothenberg replied: The esterification reaction itself is not specific – practically every carboxylic acid and alcohol combination will do. However, the formation of the three-dimensional polymeric foam is, as far as we can see (after testing many combinations), nearly unique to glycerol and citric acid. Combined with the large-scale availability and low market price of these two monomers, this makes Plantics-GX a unique case.

Thomas Farmer said: On page 8 of your article you discuss how altering the molar ratio of glycerol and citric acid altered the properties of the final material, with higher amounts of glycerol producing memory foam characteristics. Have you investigated the substitution of some proportion of the glycerol with alternative polyols, such as sorbitol and xylitol? If so, does this achieve the same outcome but allow for the polyol + glycerol to citric acid ratio to remain at 1 : 1?

Gadi Rothenberg responded: Yes, we have tested numerous combinations and alcohol : acid ratios. In general, the 1 : 1 ratio allows for the most complete polymerisation, and the higher the polymerisation degree, the higher the rigidity. When you increase the alcohol : acid ratio or when you use another alcohol, for example, you will have a less rigid material because you will prevent the formation of more ester bonds.

Anna Zhenova asked: Could you give some adhesion testing data for your material? Specifically, would it be suitable for something like aerospace applications, or is it limited in terms of temperature/strain? I would also love to hear more about the lessons learned from the table that fell apart.

Gadi Rothenberg answered: At this time we cannot give further adhesion data, apart from saying that it is a very good adhesive. The composite patents that are cited in the paper include some adhesion examples.

A major lesson learned from the table is that biodegradable materials are more suitable to some applications than to others. The same goes for all bio-based materials: we should find the applications where such materials truly give better value compared to petro-based ones.

Laura Sparlinek enquired: How is the durability increased when using the material as an isolation material?

Gadi Rothenberg replied: Typically the durability is increased by spraying it with a water-resistant layer or by placing it between two layers of aluminium. If it does not get wet, it will stay strong forever.

Keith Waldron queried: Is the material toxic? Can you eat it? If so, what is it like?

Gadi Rothenberg responded: It is non-toxic. You can eat it but it tastes a little acidic (probably because it breaks down to citric acid). Considering that it breaks down to glycerol and citric acid, I would not recommend eating large quantities.

Mark Mascal said: You appear to have already achieved a measure of commercial success. What is your forward projection in terms of existing and new markets?

Gadi Rothenberg answered: We have made numerous prototypes and secured some investments, but we have not yet achieved commercial success. The markets we are trying to enter are very large and highly conservative. The "green premium" that a lot of people talk about is still zero, no matter how much it is featured in the press. To succeed, Plantics BV needs to scale up production considerably, the current ton scale is really only samples. The company is now in the process of securing more investments for this growth.

Andrew Hunt remarked: Some of the materials were those produced on aluminium sheets and these resulted in traces of aluminium salts present in the polymer. Can you tell us more about the trace aluminium salts? Are they helping to catalyse the process?

Gadi Rothenberg responded: We are currently investigating this, so we hope to be able to discuss it soon, perhaps in another *Faraday Discussions*.

Dimitris Argyropoulos commented: One needs to emphasize that the actual polymerization statistics that are applied in the described system are those developed by Flory and Stockmayer at the turn of the century. These could be very valuable in ensuring that the actual polymerization/gelation process is well described and followed especially when the ratios of the two monomers are varied when different products are sought.

Gadi Rothenberg replied: That is a highly appropriate and correct comment. We are, in fact, now working in this direction, in collaboration with Dr C. P. Lowe.

Anne Horan asked: Could the polymer have medical applications, for example scaffolds for regenerative medicine?

Gadi Rothenberg answered: In principle, yes, as long as the acidity of the citric acid would not be a problem.

Magdalena Titirici asked: Do you have any numbers in terms of tensile strain, *etc.*?

Gadi Rothenberg replied: Yes. The numbers that we can report at this stage are included in the paper.

Avtar Matharu enquired: How many plant pots per week are needed? How many are you making currently, per batch, per hour?

Gadi Rothenberg responded: The market for plant pots in the Netherlands is very large, millions every year.

Right now Plantics is making, in collaboration with partners, sample batches of 100 000 pots. These are still "test samples", as growers use 100 000 pots quickly.

Simo Sarkanen opened a general discussion of the papers by Xindong Mu, Jakob Albert and Gadi Rothenberg: In order to produce high-grade cellulose from lignocellulosic biomass, is it justifiable to convert the hemicellulose and lignin selectively to formic acid and CO_2? Considerable energy and biochemical reducing equivalents are necessary for plants to biosynthesize polysaccharides and lignin from CO_2 in the first place (more than 3 ATP and 2 NADH per molecule of CO_2 incorporated).

Joseph Samec addressed Jakob Albert and Gadi Rothenberg: How do your institutions support your efforts in commercialization? Not just funding, but more – do you get cuts in teaching and administrative duties, for example?

Gadi Rothenberg replied: There is no official policy on that but cases are done on a case-by-case basis. In general, there is good support.

Jakob Albert responded: My university offered me an "Akademischer Rat" position during my habilitation. Concerning commercialization, they supported us in founding the spin-off company OxFA – not financially, only in terms of legal support.

Faraday Discussions

PAPER

Selective production of mono-aromatics from lignocellulose over Pd/C catalyst: the influence of acid co-catalysts

Xiaoming Huang,†[a] Xianhong Ouyang,†[a] Bart M. S. Hendriks,†[a] O. M. Morales Gonzalez,[a] Jiadong Zhu,[a] Tamás I. Korányi,[a] Michael D. Boot[b] and Emiel J. M. Hensen *[a]

Received 28th January 2017, Accepted 20th February 2017
DOI: 10.1039/c7fd00039a

The 'lignin-first' approach has recently gained attention as an alternative whole biomass pretreatment technology with improved yield and selectivity of aromatics compared with traditional upgrading processes using technical lignins. Metal triflates are effective co-catalysts that considerably speed up the removal of lignin fragments from the whole biomass. As their cost is too high in a scaled-up process, we explored here the use of HCl, H_2SO_4, H_3PO_4 and CH_3COOH as alternative acid co-catalysts for the tandem reductive fractionation process. HCl and H_2SO_4 were found to show superior catalytic performance over H_3PO_4 and CH_3COOH in model compound studies that simulate lignin–carbohydrate linkages (phenyl glycoside, glyceryl trioleate) and lignin intralinkages (guaiacylglycerol-β-guaiacyl ether). HCl is a promising alternative to the metal triflates as a co-catalyst in the reductive fraction of woody biomass. $Al(OTf)_3$ and HCl, respectively, afforded 46 wt% and 44 wt% lignin monomers from oak wood sawdust in tandem catalytic systems with Pd/C at 180 °C in 2 h. The retention of cellulose in the solid residue was similar.

Introduction

Lignocellulosic biomass is the most abundant source of renewable carbon with significant potential for the production of sustainable chemicals and fuels.[1,2] Together with cellulose and hemi-cellulose, lignin is a main component of land-based biomass. Lignin makes up about 20–35% of the mass of dry biomass.[3] It is typically obtained as a by-product in the paper and pulping industry and 2nd generation biorefineries. Such technical lignin is usually contaminated by sulphur, ash, proteins and carbohydrates residue.[3,4] During biomass separation,

[a]*Schuit Institute of Catalysis, Inorganic Materials Chemistry, Eindhoven University of Technology, P. O. Box 513, 5600 MB Eindhoven, The Netherlands. E-mail: e.j.m.hensen@tue.nl*

[b]*Combustion Technology, Department of Mechanical Engineering, Eindhoven University of Technology, P. O. Box 513, 5600 MB Eindhoven, The Netherlands*

† Authors contributed equally to this work.

the lignin structure is altered, rendering technical lignins much less reactive than the native lignin contained in the biomass.[5–7] Abundant weak ether linkages, such as β-O-4 ones, are replaced by more recalcitrant C–C linkages.[8] The resulting highly recondensed structure of lignin poses significant challenges for its efficient conversion to chemicals and fuels. Such processes require harsh conditions and typically yield a wide range of aromatic compounds in a low overall yield.[9–14]

Recent studies have demonstrated that mild depolymerization strategies can also be effective when reactive lignin feedstock rich in β-O-4 ether linkages is used. For instance, a two-step strategy has been developed for aspen lignin, consisting of selective oxidation of the secondary alcohol of β-O-4 linkages followed by a redox-neutral bond cleavage at 110 °C.[5,15] A similar strategy was also applied to the depolymerization of a reactive dioxasolv birch lignin (an organosolv lignin extracted by 1,4-dioxane) at 80 °C.[16] The use of a hydrosilane reductant and a Lewis acidic ($B(C_6F_5)_3$) catalyst facilitated the conversion of extracted wood lignin into mono-aromatics with a yield of 7–24 wt% at room-temperature.[17] A combination of triflic acid and a heterogeneous Ru or homogeneous Ir catalyst was also effective in depolymerizing dioxasolv lignin in 1,4-dioxane at 140 °C.[18] A similar catalytic system involving the use of a metal triflate and a homogeneous Rh catalyst was employed for the depolymerization of dioxansolv lignins in a 1,4-dioxane/water solvent mixture at 175 °C.[19] Two-step approaches consisting of a mild pretreatment of woody biomass followed by catalytic depolymerization have also been reported.[20,21] A suitable pretreatment method for the preparation of reactive lignin that preserves high ether linkage content is essential to these mild lignin depolymerization approaches.

An alternative approach is to directly convert the lignin in its native form in the lignocellulosic biomass matrix.[22–25] For instance, Kou *et al.* reported that lignin in birch wood can be selectively depolymerized to alkylmethoxyphenols in 46 wt% yield in a 1 : 1 (v/v) dioxane/water solvent mixture over carbon-supported noble metal (Pt, Ru, Pd and Rh) catalysts at 200 °C for 4 h.[26] Li *et al.* reported the direct catalytic conversion of carbohydrate and lignin fragments from the whole woody biomass into diols and alkylmethoxyphenols in water over a carbon-supported $Ni–W_2C$ catalyst at 235 °C for 4 h.[27] Song *et al.* were able to convert about half of the lignin in birch wood into alkylmethoxyphenols in alcohols over a Ni/C catalyst.[22] Ferrini, Rinaldi and others reported a catalytic upstream biorefining process that converts lignin from woody biomass into low molecular weight lignin oil and a (hemi-)cellulose-rich pulp over a RANEY® Ni catalyst in a mixture of H_2O/2-propanol.[23,28] The group of Abu-Omar used a bimetallic Zn/Pd/C catalyst to convert lignocellulosic biomass into two alkylmethoxyphenols and carbohydrate pulp residue in methanol. A yield of 52 wt% of lignin monomers was obtained from birch hardwood after a 12 h reaction at 225 °C.[24,29] Sels and co-workers reported a lignin-first process that allows simultaneous delignification and reductive depolymerization in the presence of Ru/C or Pd/C in methanol under a H_2 atmosphere at elevated temperature, resulting in a carbohydrate pulp and a lignin oil rich in alkylmethoxyphenol monomers in ~50% yield.[6,30] The addition of acids, such as H_3PO_4, improved the overall efficiency of the process due to enhanced delignification.[31] A similar catalytic lignocellulose fractionation process without the addition of external hydrogen or co-catalysts in a 1 : 1 (v/v) ethanol/water solvent mixture was also reported recently.[32] Related to these lignin-first or catalytic upstream biorefining processes, we have also demonstrated a tandem catalytic system that couples delignification and reductive depolymerization of wood lignin in one pot using methanol as a solvent by a combination of Lewis acid metal triflates (*e.g.* $Al(OTf)_3$) and Pd/C.[33,34] An important function of the acidic co-catalyst is the efficient extraction of lignin from the biomass, which is a relatively slow step. Conventional Lewis acids, such as $AlCl_3$ and $ZnCl_2$, were found to be less effective in deconstructing lignocellulose compared with metal triflates.[33,34] Moreover, these conventional Lewis acids are moisture sensitive and tend to decompose in the presence of water generated in etherification and hydrogenolysis. Another drawback is that their recovery and reuse are difficult.

In this work, we explore a range of Brønsted acid co-catalysts, such as HCl, H_2SO_4, H_3PO_4 and CH_3COOH, as alternatives to the benchmark $Al(OTf)_3$ catalyst. The main goal was to identify a suitable and cheaper acid co-catalyst that helps in the delignification of the biomass but leaves the (hemi)cellulose part as much as possible unconverted. The strategy of this work is to evaluate the performance of these acids as co-catalysts in the cleavage of model compounds relevant to carbohydrate–lignin interlinkages and lignin intralinkages. Finally, oak wood sawdust is used to verify the trends and establish the potential of these cheap acid co-catalysts to replace $Al(OTf)_3$.

Results and discussion

The catalytic process

Fig. 1 illustrates the overall concept of the tandem catalytic system for the valorization of lignin from woody biomass. Metal triflates are effective in the extraction of lignin from the lignocellulosic matrix by cleaving phenyl glycoside, phenyl ether and ester lignin–carbohydrate intralinkages.[31,32] These lignin fragments are then disassembled by Pd-catalyzed hydrogenolysis reactions involving molecular hydrogen. Metal triflates strongly promote the reductive fractionation of lignocellulose, thus shortening the time needed to obtain a high lignin monomer yield.[31,32] The metal triflates are also active for the depolymerization of some of the ether bonds in these lignin fragments and they also modify the performance of the Pd/C catalyst used to depolymerize these fragments. In a typical experiment (2000 mg of birch wood sawdust, 200 mg of Pd/C, 15 mg of aluminum triflate

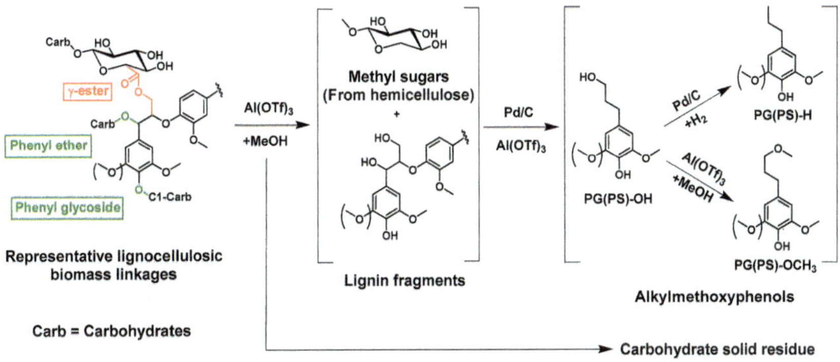

Fig. 1 The cleavage of representative linkages between carbohydrates and lignin (phenyl glycoside, benzyl ether and γ-ester) by a tandem metal triflate and Pd/C catalyst system.

($Al(OTf)_3$), 40 ml of methanol, 180 °C, 2 h, 30 bar H_2), the yield of C9 lignin monomers was 45 wt% (based on 23.6 wt% lignin content of birch wood). The main products are 4-*n*-propylsyringol (PS-H), 4-*n*-propanolsyringol (PS-OH) and 4-*n*-methoxy propyl syringol (PS-OCH_3), and their corresponding guaiacyl counterparts, namely PG-H, PG-OH and PG-OCH_3. The formation of PS-OCH_3 and PG-OCH_3 is caused by the etherification of the propanol side-group with methanol catalyzed by the $Al(TOf)_3$ Lewis acid catalyst. The lignin monomers distribution can be controlled by varying the ratio of Pd/C to $Al(TOf)_3$. A higher Pd/Al ratio results in more PS/G-OH and PS/G-H, and less PS/G-OCH_3 products.[34]

Fig. 2 shows a detailed analysis of the composition of birch wood, the solid residue and the product oil after upgrading by a combination of Pd/C and $Al(OTf)_3$. The solid residue is comprised mainly of cellulose, hemicellulose and a small amount of lignin together with the Pd/C catalyst. About 56 wt% of the hemicellulose and 81 wt% of the lignin were released from the raw birch wood and, respectively, converted into predominantly methyl xylose and mono-aromatics. The cellulose fraction was nearly all unconverted as evident from the unchanged cellulose content.

For the comparison of this optimized catalyst system to common Brønsted acid catalysts, we used another hardwood oak sawdust (125–300 μm) as the starting material. The amount of Pd/C was reduced from 200 mg to 100 mg. The amount of acid was equivalent to the amount of $Al(OTf)_3$ (0.0316 mmol, 15 mg). The effect of the acid concentration was also investigated by increasing the amount of acid to 4, 16 and 32 times the standard concentration. For the model compound studies, the same concentrations of acids were used.

Influence of acid co-catalysts: model compound studies

The selective cleavage of lignin–carbohydrate bonds and lignin–lignin ether bonds is essential for the one-pot reductive fraction process. Typical linkages between lignin and carbohydrates in woody biomass are phenyl glycoside, benzyl ether and γ-ester bonds.[35] With the purpose of understanding the role of acid in the delignification of woody biomass, we selected phenyl glycoside (PG) and glyceryl trioleate (GT) as model compounds, representing phenyl glycosidic and γ-ester type linkages, respectively. The reactions were conducted at 160 °C for 2 h in a small autoclave of 12 ml. When Pd/C was used without an acid co-catalyst, the

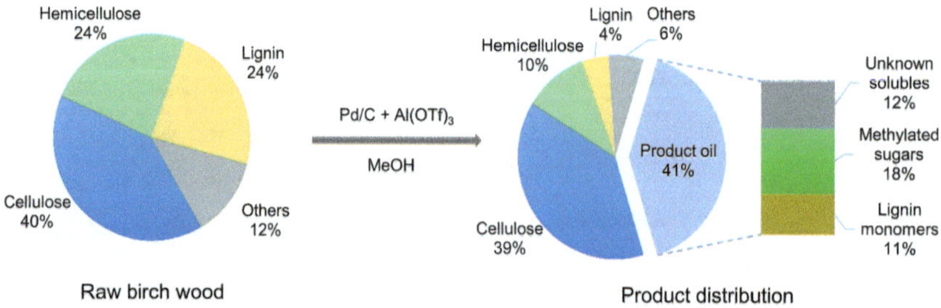

Fig. 2 Composition of birch wood, the carbohydrate residue and product oil mixture after reductive fractionation over a tandem $Al(OTf)_3$ and Pd/C catalyst system. (Reaction conditions: 2000 mg of extracted birch wood sawdust (125–300 μm), 200 mg of 5 wt% Pd/C, 15 mg of $Al(OTf)_3$, 40 ml of methanol, 30 bar H_2, stirring speed: 500 rpm, 2 h.)

conversions of PG and GT were 22% and 5%, respectively (Fig. 3). Methanol is nearly inactive in converting PG.[34] The conversion of PG is mainly due to hydrogenation and hydrogenolysis reactions catalyzed by the Pd/C catalyst (Fig. 3a). Hydrogenation of ester bonds typically requires high reaction temperature and hydrogen pressure.[36] GT can be converted in methanol by trans-esterification catalyzed by the weak acidity of methanol under sub-critical conditions.[16] Trans-esterification can be significantly enhanced by an acid. The use of $Al(OTf)_3$ is very effective in cleaving the phenyl glycosidic bond of PG. The overall product yield of phenol-derivatives is 75% after a 2 h reaction at 160 °C. $Al(OTf)_3$ also shows a high activity towards ester bond cleavage in GT with a 80% yield of methyl octadecanoate, the product of ester bond cleavage.

The use of acids resulted in similar or higher conversions of PG (Fig. 3). Increasing the HCl concentration led to higher conversions and suppressed aromatic ring hydrogenation. Nearly all the PG was converted when H_2SO_4 was the co-catalyst. In this case, all the aromatic products were hydrogenated. H_3PO_4 is very effective in catalyzing the cleavage of the phenol–glycoside bond in PG and it also strongly suppresses aromatic ring hydrogenation. Acetic acid was the least effective acid and phenol was the only product observed from the phenyl part of the reactant.

The conversion of GT was less efficient with HCl and larger amounts were needed to obtain a similar conversion as that obtained for PG. H_2SO_4 was nearly as active as the metal triflate in cleaving the GT model reactant. H_3PO_4 and CH_3COOH were much less active in cleaving the ester bond in GT. Overall, these data show that cleaving the phenyl glycosidic bond is more facile than cleaving the ester bond. The substantial difference (~70%) between PG and GT conversion observed for H_3PO_4 cannot be explained by the small activity difference (~20%) observed when only the Pd/C catalyst was used. Accordingly, we conclude that strong acids, such as HCl and H_2SO_4, are required to catalyze trans-esterification, while a weak acid (H_3PO_4) is sufficient to catalyze trans-etherification of the phenyl glycosidic bond. The use of CH_3COOH as a co-catalyst is not efficient. It appears that acetic acid suppresses the cleavage of the bonds compared to the reference case. A tentative explanation could be that esterification between

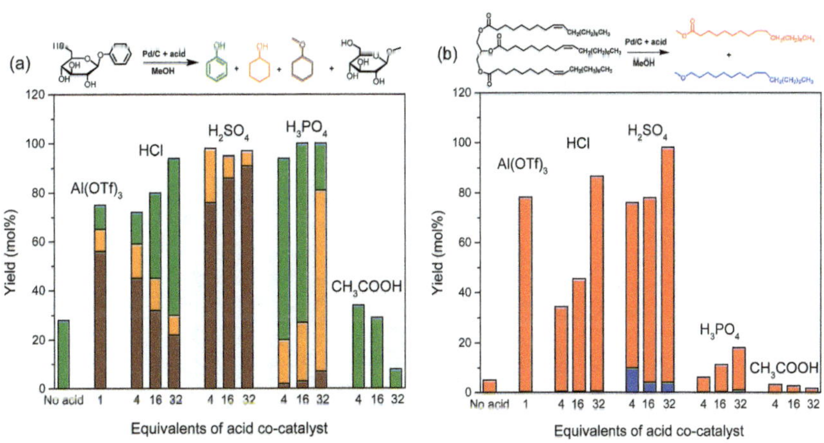

Fig. 3 Reactions of (a) phenyl glycoside (PG) and (b) glyceryl trioleate (GT) in methanol over Pd/C and/or different acid co-catalysts at 160 °C for 2 h in 30 bar H_2.

CH_3COOH and methanol decreases the acidity. In line with this, we observed the formation of methyl acetate by GC-FID. Based on these results, we conclude that HCl and H_2SO_4 acids are potential co-catalysts for the lignin-first approach.

With the purpose of gaining insight into the performance of the acid co-catalyst in the depolymerization of lignin fragments, guaiacylglycerol-β-guaiacyl ether (GG) was used as a model substrate. GG is the most suitable model for the most abundant (50–65%) β-O-4 ether linkage in native lignin.[14] For this reason and also because it is readily available, GG is frequently used in mechanistic studies on lignin deconstruction.[5,16,18,37–39] The reactions were performed under the same conditions as the previous model compound reactions (160 °C, 2 h). Without acid, Pd/C is already active in cleaving the β-O-4 ether linkage in GG. The product of the ether bond cleavage, guaiacol, is obtained in 40% yield (Fig. 4). A combination of Pd/C and $Al(OTf)_3$ resulted in a higher conversion (~60%) of the β-O-4 ether bond. Interestingly, at the lowest HCl concentration, the yield (~20%) was lower than in the Pd-only case. Although a higher HCl concentration slightly improved the GG conversion, the overall product yields remain lower than for Pd/C. In contrast, the use of H_2SO_4 is already effective in low concentrations. At a higher concentration, the yield of GG conversion products decreased. Similar trends were found when H_3PO_4 or CH_3COOH were the co-catalyst (Fig. 4). Notably, the combination of 4 or 16 equivalents of CH_3COOH and Pd/C shows superior performance in cleaving the ether bond of GG (~65%).

In line with the results of PG conversion (Fig. 3a), the addition of acid promoted the hydrogenation activity of Pd/C. With $Al(OTf)_3$, the guaiacol selectivity was about 50% – the remainder being methylated and ring-saturated products such as 1,2-dimethoxycyclohexane and a small amount of methoxycyclohexane (Fig. 4). The promoting effect for hydrogenation is stronger with H_2SO_4: nearly all of the guaiacol-derived products are ring saturated. On the other hand, such a promoting effect was not apparent with HCl, H_3PO_4 and CH_3COOH as co-catalysts, different from the results obtained during PG conversion (Fig. 3a). During PG conversion, the addition of HCl or H_3PO_4 yielded more ring-saturated phenol-derivatives (Fig. 3a). The difference is the higher stability of the aromatic

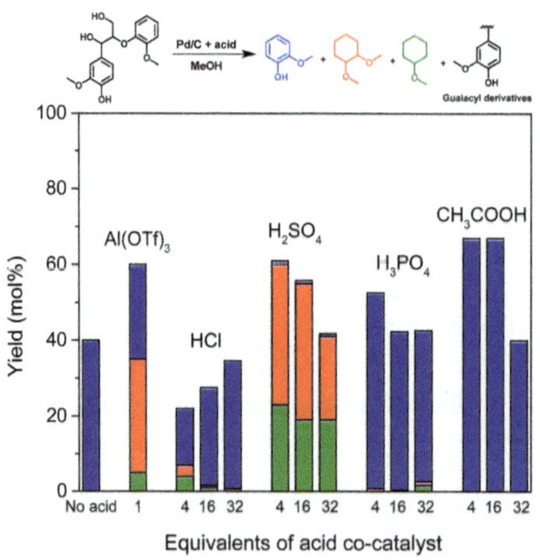

Fig. 4 Reactions of guaiacylglycerol-β-guaiacyl ether (GG) in methanol over Pd/C and/or different acid co-catalysts at 160 °C for 2 h in 30 bar H_2.

ring in guaiacol compared with phenol.[40] This was confirmed by carrying out two separate hydrogenation reactions using guaiacol and phenol as the reactant under identical conditions (Pd/C, 160 °C, 30 bar H_2, 2 h). With guaiacol as the reactant, the use of Pd/C resulted in a conversion to ring-saturated products of 61%, while all the phenol was converted to methoxycyclohexane (Fig. 5a).

A tentative explanation for the promoting effect of acid co-catalysts on the aromatic ring hydrogenation by Pd/C is the removal of hydroxyl groups by their methylation catalyzed by acids. We have shown earlier that hydroxyl groups (*e.g.* as in propanol and sugars) suppress hydrogenation of aromatic rings.[33] This is further supported by two additional experiments in which guaiacol was converted in the presence and absence of an equivalent mass of glucose. Fig. 5b shows the yield of the ring-hydrogenated products as a function of reaction time. The results attest to the lower rate of aromatic ring hydrogenation in the presence of sugars. This result confirms that hydroxyl-rich substrates hinder the hydrogenation activity of Pd/C, presumably by hindering specific adsorption models of the aromatic rings on the Pd surface that would lead to complete ring saturation.[33] In the presence of an acid, the hydroxyl groups are readily converted into methoxy groups in the methanol solvent. This is also supported by the formation of substantial amounts of methylated cyclic products when PG and GG were converted with Pd/C and $Al(OTf)_3$ or H_2SO_4 as co-catalysts (Fig. 3a and 4).

Based on these results, we conclude that combinations of Pd/C and H_2SO_4 or HCl are effective in cleaving two different types of lignin–carbohydrate (phenyl glycosidic and ester) as well as lignin–lignin (β-O-4 ether) linkages. The combination of Pd/C and weak H_3PO_4 acid is only active in cleaving the phenyl glycosidic bond and β-O-4 ether bonds, and not in the cleavage of ester type lignin–carbohydrate linkages. CH_3COOH, the weakest among the tested acids, is not effective in cleaving any of the lignin–carbohydrate linkages.

Oak sawdust upgrading

With these model compound experiment results in hand, we evaluated the performance of the acids as co-catalysts in the upgrading of oak wood. In a typical

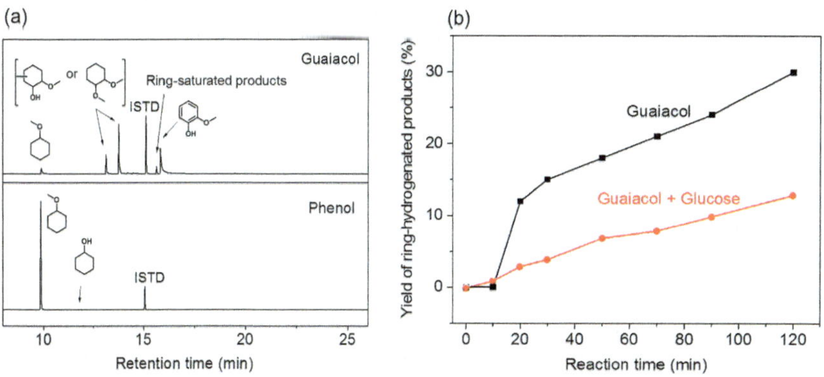

Fig. 5 (a) Gas chromatograms of the reactions in methanol using phenol and guaiacol as model substrates; (b) yield of ring-hydrogenated products obtained from the hydrogenation reaction of guaiacol (black) and guaiacol + glucose (red) in methanol (conditions: 50 mg of substrate, 30 ml of methanol, 10 mg of Pd/C, 160 °C, 2 h, 30 bar H_2).

experiment, 2000 mg of extracted oak wood sawdust, 100 mg of 5 wt% Pd/C and different equivalents of acid (1 equivalent = 0.0316 mmol) were added to a 100 ml autoclave together with 40 ml of methanol (180 °C, 2 h, 30 bar H_2). The acid concentration was the same as in the model compound experiments. An aliquot of the reaction mixture was analyzed by GC-MS and GC-FID to quantify the lignin monomers and methylated sugars. Some of the reaction mixtures were also analyzed by gel permeation chromatography (GPC). Table 1 summarizes the overall results of the tandem-catalyzed reductive fractionation of oak sawdust using different acid co-catalysts under different conditions.

Without a catalyst, the reaction mixture contains only 6 wt% C9 lignin monomers (entry 1). A large amount (1655 mg) of solid biomass residue was obtained (83 wt% based on the mass of the starting biomass). This solid residue was subjected to further analysis to determine the Klason lignin content. The results show that delignification is about 35 wt%, suggesting that methanol is not effective in extracting lignin from woody biomass at 180 °C. The low monomer yield and high solid residue retention obtained in the blank experiment are caused by the low degree of delignification and lignin depolymerization. By adding the Pd/C catalyst, the lignin monomer yield increased to 19 wt%, with PS-H, PS-OH and PG-OH as the main products (Table 1 entry 2, Fig. 6a). The improved yield is mainly due to the enhanced lignin depolymerization. Besides, the extent of delignification was also found to be slightly increased (57 wt%) compared with the blank reaction (entry 1, 35 wt%). This could be due to the fact that Pd/C is also effective in cleaving the phenyl glycosidic linkage. However, the delignification is limited by the poor accessibility of the solid biomass for the solid Pd/C catalyst.

In the benchmark case using 1 equivalent of $Al(TOf)_3$ as the co-catalyst, the yield of lignin monomers increased substantially. A yield of 40 wt% C9 lignin monomer was obtained after a 1 h reaction at 180 °C and the yield increased to 46 wt% after 2 h (entry 3). The delignification degree is 82 wt%. The C9 lignin monomers are mainly PS-H, PS-OH and PS-OCH_3, and small amounts of PG-H, PG-OH and PG-OCH_3 (Fig. 6a). Apart from these mono-aromatics, an amount of 274 mg of methylated C5 sugars was formed. No cellulose-derived C6 sugars nor sugar degradation products were observed. Sugar degradation was only observed when a larger amount of metal triflate was used, consistent with our earlier work.[34] For instance, some aliphatic ketones, such as 2,5-hexanedione, and esters, such as the methyl ester of levulinic acid, were observed when 3 equivalents of $Al(OTf)_3$ were used in combination with 200 mg of Pd/C for the reductive fractionation of birch wood sawdust at 180 °C for 2 h.[34] The incomplete mass balance is due to the presence of lignin dimers and oligomers, which were not identified. Their formation was confirmed by GPC analysis (*vide infra*).

The influence of other acid co-catalysts was evaluated under identical conditions. The use of small amounts of HCl did not improve the lignin monomer yield compared with the Pd/C-only case (entries 4–5). However, larger amounts improved the yields of lignin monomers and also increased the amount of methylated C5 sugars (entries 6–7). With the largest amount of HCl, the lignin monomer yield was 44 wt% (entry 7). The amount of solid residue and the monomeric lignin product distribution (Fig. 5a) are very similar to the results obtained with $Al(TOf)_3$.

Table 1 Influence of different acid co-catalysts on the tandem-catalyzed reductive fractionation of oak sawdust over Pd/C in methanol at 180 °C

		Acid co-catalyst			Solid residue mass (mg)		Lignin monomers yield[e] wt% (mg)				
Entry	Catalyst	Type	Amount (mmol)	pH[a]	Amount[b] (mg)	Delignification[c] (wt%)	0 h[d]	1 h	2 h	Methyl C5 sugars (mg)	Total yield (wt%)
1	—	—	—	6.6	1655	35	1	5	6 (26)	4	84
2	Pd/C	—	—	6.6	1619	57	4	14	19 (81)	4	85
3	Pd/C	$Al(OTf)_3$	0.0316 (1×)	3.4	1175	82	18	40	46 (206)	274	83
4	Pd/C	HCl	0.0316 (1×)	3.0	1546	N/A	4	11	14 (62)	5	81
5	Pd/C		0.126 (4×)	2.2	1422	N/A	6	15	18 (78)	20	76
6	Pd/C		0.506 (16×)	1.7	1211	N/A	12	21	23 (100)	133	72
7	Pd/C		1.011 (32×)	1.4	1148	75	21	41	44 (194)	249	80
8	Pd/C	H_2SO_4	0.0316 (1×)	2.9	1562	N/A	4	9	10 (44)	5	81
9	Pd/C		0.126 (4×)	2.2	1017	76	23	31	35 (153)	257	71
10	Pd/C		0.506 (16×)	1.7	328	N/A	28	35	40 (176)	46	28
11	Pd/C	H_3PO_4	0.0316 (1×)	3.6	1761	N/A	1	6	13 (55)	20	92
12	Pd/C		0.126 (4×)	3.1	1600	N/A	3	9	13 (55)	19	84
13	Pd/C		0.506 (16×)	2.8	1448	N/A	3	13	17 (75)	30	78
14	Pd/C		1.011 (32×)	2.6	1358	69	5	14	26 (117)	31	77
15	Pd/C	CH_3COOH	0.0316 (1×)	5.0	1598	N/A	4	10	15 (65)	5	83
16	Pd/C		0.126 (4×)	4.1	1577	N/A	3	8	10 (41)	6	81
17	Pd/C		0.506 (16×)	3.7	1639	N/A	1	5	8 (35)	7	84
18	Pd/C		1.011 (32×)	3.5	1649	40	1	4	6 (26)	3	84

[a] pH values were measured at room temperature using an electronic pH meter after replacing 20 vol% of the acidic methanol solution with water. [b] Mass of the residue excluding Pd/C catalyst. [c] Delignification was determined by analyzing the Klason lignin content of the solid residue, assuming the Pd/C is homogenously mixed with the biomass residue. [d] Heating time (30 min). [e] Lignin yield is based on the acid-insoluble (Klason) lignin content (22.4 wt%) of the pre-extracted oak sawdust.

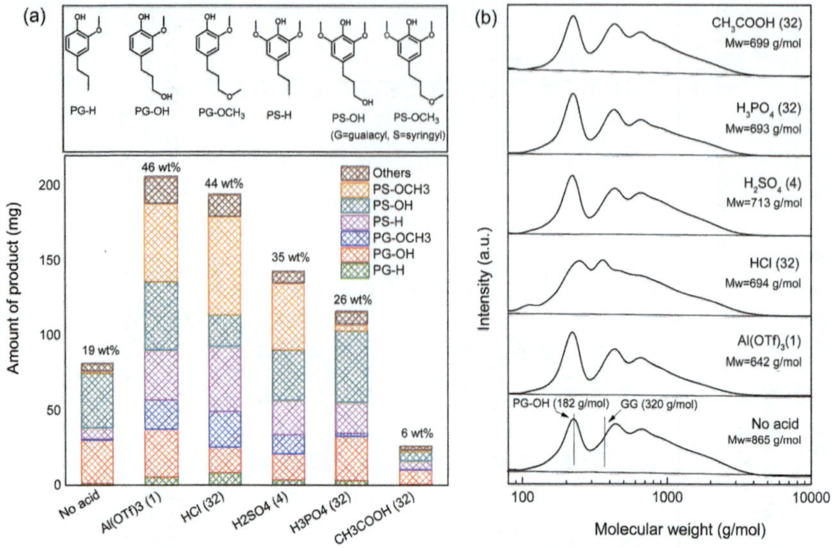

Fig. 6 (a) Lignin monomeric product distributions and (b) GPC chromatograms of the reaction mixtures obtained from the reactions over Pd/C in combination with different acid co-catalysts at 180 °C for 2 h. (Conditions: 2000 mg of pre-extracted oak sawdust, 100 mg of Pd/C, 30 bar H_2, 40 ml of methanol; 1 eqv. = 0.0316 mmol of acid; the numbers on top of the bars in (a) are the lignin monomer yields; GPC results of two reference compounds (GG: guaiacylglycerol-β-guaiacyl ether and PG-OH: 4-*n*-propanolguaiacol) are included.)

At the lowest concentration, H_2SO_4 hardly affected the conversion as with HCl (entry 8). The lignin monomer yield increased drastically to 35 wt% when 4 equivalents of H_2SO_4 were used (entries 9–10). The lignin monomeric product distribution and the amount of residue (1017 mg) are very similar to the best result obtained with HCl (1148 mg, 32 eqv., entry 7) and the Al(TOf)$_3$ reference

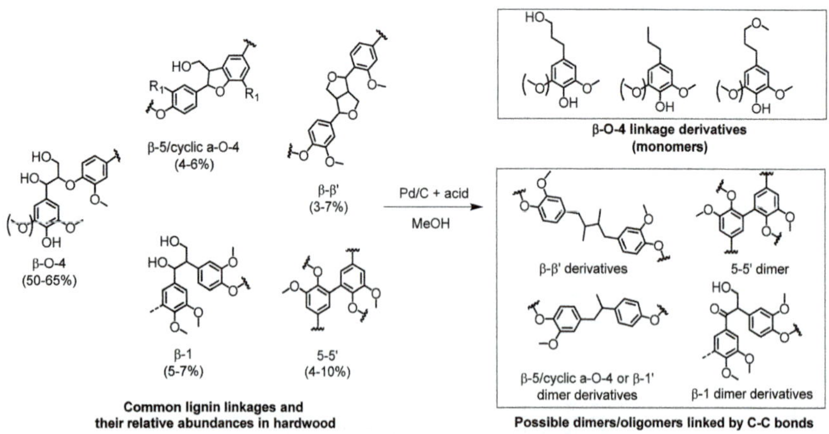

Fig. 7 Formation of lignin monomers and proposed dimers and oligomers in the process of reductive fractionation of lignin from woody biomass. (The proposed structure and relative abundances are based on the literature.[14,22,25])

(1175 mg, entry 3). The delignification degree in this case is 76 wt%. Although increasing the dosage of H_2SO_4 acid resulted in a slightly higher yield of lignin monomers (40 wt%), much less solid residue was left, suggesting that a significant fraction of cellulose was converted. Moreover, darkening of the solution and the very low mass balance (28 wt%) point to humin formation from the released C6 sugars. These findings limit the use of high concentrations of H_2SO_4.

The influence of H_3PO_4 was minor with lignin monomer yields below 20 wt% (entries 11–13). Moreover, it is seen that the solid carbohydrate residue is decreasing with increasing H_3PO_4 content. The lignin monomer yields are even lower than the reference case without H_3PO_4 (19 wt%, entry 2). A maximum yield of 26 wt% lignin monomers was achieved after a 2 h reaction at 180 °C at the highest H_3PO_4 concentration (entry 14). The yield is slightly lower than the one reported in the literature.[31] It was reported that the addition of 2.5 g L^{-1} (equal to 27.8 equivalents) H_3PO_4 in combination with Pd/C resulted in a lignin monomer yield of approximately 40 mol%, compared with the yield obtained from the neutral methanol solvent (26 mol%) at 200 °C using pre-extracted poplar sawdust feedstock. This difference could be caused by the lower reaction temperature (180 °C *vs.* 200 °C), the different feedstock (oak *vs.* poplar sawdust) and the different calculation methods (wt% *vs.* mol%). Nevertheless, it can be said that the addition of an appropriate amount of H_3PO_4 improves the overall efficiency of the reductive fractionation process. Yet, compared with $Al(TOf)_3$ and the other stronger Brønsted acids (HCl and H_2SO_4), the promoting effect of H_3PO_4 is less pronounced. Our model compound experiments show that the combination of H_3PO_4 and Pd/C is active in cleaving the phenyl glycosidic lignin–carbohydrate and β-O-4 ether types of lignin–lignin linkages, but less active in cleaving the ester type lignin–carbohydrate linkages. We surmise that part of the lignin (linked by ester bonds) might not be effectively extracted from the biomass matrix. This is further confirmed by the relatively low degree of delignification (68 wt%), revealed by the compositional analysis of the solid residue obtained at the highest H_3PO_4 concentration. Furthermore, we observed that the lignin monomers are mostly non-etherified products, such as PS-H, PS-OH and PG-OH (Fig. 6a). Also, a small amount of methylated C5 sugars was formed. These results indicate that methylation/etherification is not occurring due to the low acidity of H_3PO_4 (entries 11–14). It is known that sugars and their derivatives (*e.g.* 5-HMF, furfural) can condense with each other and lignin-derived products under acidic conditions into humins.[41,42] Methylation stabilizes sugars.[43,44] In our earlier work, it was also confirmed that methylation of the lignin phenolic intermediates plays a key role in hindering repolymerization of lignin fragments and avoiding char formation.[12] Accordingly, we argue that the low rate of etherification might be another reason for the lower efficiency of H_3PO_4.

With CH_3COOH as the co-catalyst, a lignin monomer yield of 15 wt% was obtained at the lowest concentration (entry 15), which is close to the yield obtained in the case of the Pd/C catalyst. When the concentration was increased, the lignin monomer yield decreased significantly (entries 15–18). The lignin products were mostly non-etherified PS-OH, PS-H and PG-OH (Fig. 5). The amount of solid residue was nearly unchanged, suggesting that the delignification could not be improved by increasing the amount of CH_3COOH. Analysis of the Klason lignin content of the solid residue showed that 40 wt% lignin was released from the woody biomass at the highest CH_3COOH concentration. Based on the decreasing lignin monomer yield and unchanged solid residue yield, we infer that the addition of CH_3COOH negatively affects the lignin conversion, consistent with the model compound results. The lack of delignification and etherification is likely the cause of the low lignin monomer yield.

Based on the lignin monomer yields and the solid carbohydrate residue, lignin oil samples obtained from the following experiments were subjected to GPC analysis: no acid, $Al(OTf)_3$ (1 eqv.), HCl (32 eqv.), H_2SO_4 (4 eqv.), H_3PO_4 (32 eqv.) and CH_3COOH (32 eqv.). Fig. 6b shows the results of the GPC analysis. Surprisingly, all of the selected lignin oils gave very similar results. Three broad peaks were observed related to lignin monomers, dimers and oligomers. For all the acid-catalyzed lignin oil samples, the molecular weights (M_w) were between 642 g mol^{-1} ($Al(OTf)_3$) and 713 g mol^{-1} (H_2SO_4 (4 eqv.)), which are slightly lower than the molecular weight for the Pd/C-only case (865 g mol^{-1}). This confirms that the addition of acid contributes to lignin depolymerization.[34] The dimers and oligomers are likely to be linked by strong C–C bonds. The possible formation of dimers and oligomers is shown in Fig. 7. These C–C linkages either originate from the native structure (*e.g.* β-1, 5–5′) or they derive from the cleavage of some ether linkages (*e.g.* β–β′, β-5/cyclic α-O-4). The cleavage of C–C linkages requires a higher temperature. Accordingly, the fraction of ether linkages (typically 50–65% β-O-4 ether linkages in hardwood) puts an upper limit on the lignin monomers yield.

In brief, the most promising replacement of $Al(OTf)_3$ is HCl, which is able to obtain a 44 wt% yield of lignin monomers from oak wood sawdust in reductive fractionation using Pd/C at 180 °C in 2 h. A drawback of the use of HCl is corrosion of the process equipment, which will add to the overall cost of biomass upgrading. The alkylmethoxyphenols resulting from this approach can be used in the fragrance industry (*e.g.* dihydroeugenol) or upgraded to other valuable base chemicals, such as (alkyl)phenols. One challenge in this lignin-first approach is that the metal catalyst ends up in the solid residue. A solution is to separate the delignification step and hydrogenolysis steps. This approach has been demonstrated in a recent study.[33] The resulting carbohydrate solid residue is an ideal starting material for the paper industry and can also be used in the production of biobased chemicals, such as ethanol, levulinic acid, and γ-valerolactone.

Conclusions

We explored Brønsted acids as alternatives to $Al(OTf)_3$ co-catalysts in the tandem reductive delignification of biomass to aromatics using Pd/C as the hydrogenolysis catalyst. The main role of the acid catalyst is to delignify lignocellulosic biomass, while Pd/C cleaves the ether bonds in the released lignin fragments. The solid residue in this process is a cellulose-rich fraction that also contains some hemicellulose and lignin. We systematically compared HCl, H_2SO_4, H_3PO_4 and CH_3COOH as co-catalysts for the tandem reductive fractionation of woody biomass in methanol using Pd/C at 180 °C for 2 h. Model compound experiments (PG, GT and GG) show that the combinations of Pd/C and strong Brønsted acids (H_2SO_4 and HCl) are suitable to cleave phenyl glycosidic and ester type lignin–carbohydrate linkages as well as lignin–lignin (β-O-4 ether) linkages. The combination of Pd/C and the weaker acid H_3PO_4 was only able to cleave phenyl glycosidic bonds and β-O-4 ether bonds but not ester type lignin–carbohydrate linkages. CH_3COOH is not effective in cleaving the model lignin–carbohydrate linkages. These model compound results explain the better catalytic performance of $Al(OTf)_3$, HCl and H_2SO_4 compared to H_3PO_4 and CH_3COOH for the upgrading of oak wood biomass. The poor performance of the latter two weak acids is due to their low activity in delignification. The most promising replacement of $Al(OTf)_3$ is HCl. Under optimum conditions, a 44 wt% yield of lignin monomers was obtained from oak wood sawdust, similar to the result obtained with the metal triflate.

Experimental section

Chemicals and materials

Oak wood sawdust was obtained from Houtzagerij Mennen (http://www.houtzagerijmennen.nl). Birch wood was obtained from the Energy Research Center of Netherlands (ECN). 5 wt% Pd/C, $Al(OTf)_3$ (≥99%), CH_3COOH (≥99%) and H_3PO_4 (85 wt% in water) were purchased from Sigma Aldrich. HCl (37 wt% in water) was purchased from Merck Emsure. H_2SO_4 (95–97%, 99.99% trace metals basis) was purchased from Merck Emsure. 4-*n*-Propanolguaiacol (PG-OH) was purchased from TCI. 4-*n*-Propylsyringol (PS-H) and 4-*n*-methoxy propyl syringol (PS-OCH_3) were recovered from the reaction mixture; the detailed procedure can be found in our previous work.[33] These phenolic compounds were used to determine the weight response factors relative to *n*-dodecane. Glyceryl trioleate (≥97.0%), phenyl β-D-glucopyranoside (≥97%) and benzyl phenyl ether were purchased from Sigma Aldrich. Guaiacylglycerol-β-guaiacyl ether (GG, ≥97%) was purchased from TCI. Extra-dry absolute methanol was purchased from Biosolve. All the other commercial chemicals were analytical grade and were used without further purification.

Feedstock pretreatment

Oak wood sawdust was obtained in the sawdust form and sieved to obtain particles with sizes between 150 μm and 300 μm. The wood sawdust was subjected to a Soxhlet extraction with water (~4 h) followed by ethanol (~4 h) to remove extractives. After extraction, the feedstock particles were dried at 105 °C overnight for the catalytic reaction.

Catalytic activity measurements

In a typical reaction experiment, the autoclave was charged with a suspension of 2000 mg of feedstock, 100 mg of Pd/C (5 wt% Pd loading), a certain amount of acid (*e.g.* 0.0316 mmol of $Al(OTf)_3$) and 30 μl of *n*-dodecane internal standard in 40 ml of methanol. The reactor was sealed and purged with nitrogen and hydrogen several times sequentially. After leak testing, the pressure was brought to 30 bar with hydrogen and the reaction mixture was heated to the reaction temperature under continuous stirring at 500 rpm within 0.5 h. A sampling valve was installed, which allowed the removal of liquid samples during the reaction. About 0.4 ml solutions were taken out from the autoclave during the reaction and analysed directly by GC-MS. After the reaction, the heating oven was removed and the reactor was cooled to room temperature in an ice-water bath. After releasing the pressure, the autoclave was opened and the mixture was collected and combined with the solution obtained from washing the autoclave with methanol. The collected reaction mixture was filtered by using a filter crucible (porosity 4) under vacuum. The carbohydrate and Pd/C solid residue was left in air to dry for one day to determine the amount of all-solid residue. In some cases, the solid residue was further subjected to a two-stage acid hydrolysis to determine the Klason lignin content following the protocol described earlier.[33]

An autoclave with a volume of 12 ml was used for the reactions with the model compounds. In a typical run, 25 mg (PG, GT and GG) of model compound, 10 mg of Pd/C and 10 μl of *n*-dodecane internal standard were added to 5 ml of methanol, which contained a certain amount of acid (the same concentrations as the corresponding woody biomass reactions described above). After sealing, purging and checking for leaks, the autoclave was heated to 160 °C and maintained at that temperature for 2 h. After the reaction, an aliquot of ~1 ml was taken from the reaction mixture and directly analysed by GC-MS without dilution following filtration with a 0.45 μm syringe filter. In the experiments involving guaiacol and phenol model compounds, a 100 ml Parr autoclave was used as the reactor, which allows sampling during the reaction. In a typical run, 50 mg of reactant (guaiacol, phenol and glucose), 10 mg of Pd/C, and 30 μl of *n*-dodecane internal standard were added to the reactor together with 30 ml of methanol. The reaction was operated at 160 °C for 2 h.

Product analysis and characterization

The liquid phase products were analysed by a Shimadzu 2000 GC-MS system equipped with a RTX-1701 column (60 m × 0.25 mm × 0.25 μm) and a flame ionization detector (FID) together with a mass spectrometer (MS) detector. For the lignin monomers, the product peaks were identified using MS and by authentication with reference compounds. All the quantitative analyses of the liquid phase products were based on GC-FID. Experimentally determined weight response factors (RF) were used for calculation relative to *n*-dodecane, which was the internal standard. For some of the commercially unavailable compounds (side-products obtained in small amounts), the RF of PS-OCH_3 was used for quantification. The yield of lignin monomers was calculated using following equation:

$$\text{Yield of monomers (wt\%)} = (\text{weight of monomers})/(\text{weight of starting feedstock} \times \text{Klason lignin weight percentage in oak wood}) \times 100\%$$

For the model lignin compound reactions (PG, GT and GG), the FID response factors were calculated using the Effective Carbon Number (ECN) method[45] to determine the relative response factors corrected by the molecular weight of the compounds relative to *n*-dodecane, which served as the internal standard.

Gel permeation chromatography (GPC)

GPC analyses were performed using a Shimadzu apparatus equipped with two columns in series (Mixed-C and Mixed-D, polymer laboratories) and a UV-Vis detector at 254 nm. The column was calibrated with polystyrene standards. Analyses were carried out at 25 °C using THF as the eluent at a flow rate of 1 ml min^{-1}. For the lignin residue analysis, the sample was prepared at a concentration of 2 mg ml^{-1}. All the samples were filtered using a 0.45 μm filter membrane prior to injection.

References

1 C. H. Zhou, X. Xia, C. X. Lin, D. S. Tong and J. Beltramini, *Chem. Soc. Rev.*, 2011, **40**, 5588–5617.
2 Y. C. Lin and G. W. Huber, *Energy Environ. Sci.*, 2009, **2**, 68–80.
3 J. Zakzeski, P. C. A. Bruijnincx, A. L. Jongerius and B. M. Weckhuysen, *Chem. Rev.*, 2010, **110**, 3552–3599.
4 A. J. Ragauskas, G. T. Beckham, M. J. Biddy, R. Chandra, F. Chen, M. F. Davis, B. H. Davison, R. A. Dixon, P. Gilna, M. Keller, P. Langan, A. K. Naskar, J. N. Saddler, T. J. Tschaplinski, G. A. Tuskan and C. E. Wyman, *Science*, 2014, **344**, 1246843.
5 A. Rahimi, A. Ulbrich, J. J. Coon and S. S. Stahl, *Nature*, 2014, **515**, 249–252.
6 F. P. Bouxin, A. McVeigh, F. Tran, N. J. Westwood, M. C. Jarvis and S. D. Jackson, *Green Chem.*, 2015, **17**, 1235–1242.
7 J. L. Wen, B. L. Xue, S. L. Sun and R. C. Sun, *J. Chem. Technol. Biotechnol.*, 2013, **88**, 1663–1671.
8 S. Constant, H. L. J. Wienk, A. E. Frissen, P. de Peinder, R. Boelens, D. S. van Es, R. J. H. Grisel, B. M. Weckhuysen, W. J. J. Huijgen, R. J. A. Gosselink and P. C. A. Bruijnincx, *Green Chem.*, 2016, **18**, 2651–2665.
9 R. Rinaldi, R. Jastrzebski, M. T. Clough, J. Ralph, M. Kennema, P. C. Bruijnincx and B. M. Weckhuysen, *Angew. Chem., Int. Ed.*, 2016, **55**, 8164–8215.
10 X. Huang, T. I. Korányi, M. D. Boot and E. J. M. Hensen, *ChemSusChem*, 2014, **7**, 2276–2288.
11 X. Huang, C. Atay, T. I. Korányi, M. D. Boot and E. J. M. Hensen, *ACS Catal.*, 2015, **5**, 7359–7370.
12 X. Huang, T. I. Korányi, M. D. Boot and E. J. M. Hensen, *Green Chem.*, 2015, **17**, 4941–4950.
13 M. P. Pandey and C. S. Kim, *Chem. Eng. Technol.*, 2011, **34**, 29–41.
14 C. Z. Li, X. C. Zhao, A. Q. Wang, G. W. Huber and T. Zhang, *Chem. Rev.*, 2015, **115**, 11559–11624.
15 P. C. A. Bruijnincx and B. M. Weckhuysen, *Nat. Chem.*, 2014, **6**, 1035–1036.
16 C. S. Lancefield, O. S. Ojo, F. Tran and N. J. Westwood, *Angew. Chem., Int. Ed.*, 2015, **54**, 258–262.
17 E. Feghali, G. Carrot, P. Thuery, C. Genre and T. Cantat, *Energy Environ. Sci.*, 2015, **8**, 2734–2743.
18 P. J. Deuss, M. Scott, F. Tran, N. J. Westwood, J. G. de Vries and K. Barta, *J. Am. Chem. Soc.*, 2015, **137**, 7456–7467.
19 R. Jastrzebski, S. Constant, C. S. Lancefield, N. J. Westwood, B. M. Weckhuysen and P. C. A. Bruijnincx, *ChemSusChem*, 2016, **9**, 2074–2079.
20 C. S. Lancefield, I. Panovic, P. J. Deuss, K. Barta and N. J. Westwood, *Green Chem.*, 2017, **19**, 202–214.
21 L. Shuai, M. T. Amiri, Y. M. Questell-Santiago, F. Heroguel, Y. D. Li, H. Kim, R. Meilan, C. Chapple, J. Ralph and J. S. Luterbacher, *Science*, 2016, **354**, 329–333.
22 Q. Song, F. Wang, J. Y. Cai, Y. H. Wang, J. J. Zhang, W. Q. Yu and J. Xu, *Energy Environ. Sci.*, 2013, **6**, 994–1007.

23 P. Ferrini and R. Rinaldi, *Angew. Chem., Int. Ed.*, 2014, **53**, 8634–8639.
24 T. Parsell, S. Yohe, J. Degenstein, T. Jarrell, I. Klein, E. Gencer, B. Hewetson, M. Hurt, J. I. Kim, H. Choudhari, B. Saha, R. Meilan, N. Mosier, F. Ribeiro, W. N. Delgass, C. Chapple, H. I. Kenttamaa, R. Agrawal and M. M. Abu-Omar, *Green Chem.*, 2015, **17**, 1492–1499.
25 S. Van den Bosch, W. Schutyser, R. Vanholme, T. Driessen, S. F. Koelewijn, T. Renders, B. De Meester, W. J. J. Huijgen, W. Dehaen, C. M. Courtin, B. Lagrain, W. Boerjan and B. F. Sels, *Energy Environ. Sci.*, 2015, **8**, 1748–1763.
26 N. Yan, C. Zhao, P. J. Dyson, C. Wang, L. T. Liu and Y. Kou, *ChemSusChem*, 2008, **1**, 626–629.
27 C. Z. Li, M. Y. Zheng, A. Q. Wang and T. Zhang, *Energy Environ. Sci.*, 2012, **5**, 6383–6390.
28 P. Ferrini, C. A. Rezende and R. Rinaldi, *ChemSusChem*, 2016, **9**, 3171–3180.
29 I. Klein, C. Marcum, H. Kenttämaa and M. M. Abu-Omar, *Green Chem.*, 2016, **18**, 2399–2405.
30 S. Van den Bosch, W. Schutyser, S. F. Koelewijn, T. Renders, C. M. Courtin and B. F. Sels, *Chem. Commun.*, 2015, **51**, 13158–13161.
31 T. Renders, W. Schutyser, S. Van den Bosch, S. F. Koelewijn, T. Vangeel, C. M. Courtin and B. F. Sels, *ACS Catal.*, 2016, **6**, 2055–2066.
32 M. V. Galkin, A. T. Smit, E. Subbotina, K. A. Artemenko, J. Bergquist, W. J. J. Huijgen and J. S. M. Samec, *ChemSusChem*, 2016, **9**, 3280–3287.
33 X. Huang, J. Zhu, T. I. Koranyi, M. D. Boot and E. J. Hensen, *ChemSusChem*, 2016, **9**, 3262–3267.
34 X. Huang, O. M. Morales Gonzalez, J. Zhu, T. I. Korányi, M. D. Boot and E. J. M. Hensen, *Green Chem.*, 2017, **19**, 175–187.
35 T. T. You, L. M. Zhang, S. K. Zhou and F. Xu, *Ind. Crops Prod.*, 2015, **71**, 65–74.
36 J. Pritchard, G. A. Filonenko, R. van Putten, E. J. M. Hensen and E. A. Pidko, *Chem. Soc. Rev.*, 2015, **44**, 3808–3833.
37 J. Tsujino, H. Kawamoto and S. Saka, *Wood Sci. Technol.*, 2003, **37**, 299–307.
38 H. L. Wang, L. B. Zhang, T. S. Deng, H. Ruan, X. L. Hou, J. R. Cort and B. Yang, *Green Chem.*, 2016, **18**, 2802–2810.
39 M. Zaheer and R. Kempe, *ACS Catal.*, 2015, **5**, 1675–1684.
40 H. Shafaghat, P. S. Rezaei and W. M. A. W. Daud, *RSC Adv.*, 2015, **5**, 33990–33998.
41 X. M. Zhang, P. Murria, Y. Jiang, W. H. Xiao, H. I. Kenttamaa, M. M. Abu-Omar and N. S. Mosier, *Green Chem.*, 2016, **18**, 5219–5229.
42 P. Sannigrahi, A. J. Ragauskas and S. J. Miller, *BioEnergy Res.*, 2008, **1**, 205–214.
43 G. J. M. Gruter and F. Dautzenberg, EP1834950 A1, 2007.
44 G. J. M. Gruter and F. Dautzenberg, EP1834951 A1, 2007.
45 R. L. Grob and E. F. Barry, *Modern practice of gas chromatography*, Wiley, Hoboken, 2004, pp. 302–303.

Faraday Discussions

PAPER

Challenges in the development of bio-based solvents: a case study on methyl(2,2-dimethyl-1,3-dioxolan-4-yl)methyl carbonate as an alternative aprotic solvent†

Saimeng Jin, Fergal Byrne, Con Robert McElroy, James Sherwood, James H. Clark and Andrew J. Hunt *

Received 30th January 2017, Accepted 17th February 2017
DOI: 10.1039/c7fd00049a

Many traditional solvents have drawbacks including sustainability and toxicity issues. Legislation, such as REACH, is driving the move towards less hazardous chemicals and production processes. Therefore, safer bio-based solvents need to be developed. Herein, a 10 step method has been proposed for the development of new bio-based solvents, which utilises a combination of *in silico* modelling of Hansen solubility parameters (HSPs), experimental Kamlet–Abboud–Taft parameters, a selection of green synthetic routes followed by application testing and toxicity measurements. The challenges that the chemical industry face in the development of new bio-based solvents are highlighted through a case study on methyl(2,2-dimethyl-1,3-dioxolan-4-yl)methyl carbonate (MMC), which can be synthesised from glycerol. Although MMC is an attractive candidate as a replacement solvent, simply being bio-derived is not enough for a molecule to be regarded as green. The methodology of solvent development described here is a broadly applicable protocol that will indicate if a new bio-based solvent is functionally proficient, and will also highlight the importance of early stage Kamlet–Abboud–Taft parameters determination and toxicity testing in the development of a green solvent.

Introduction

Solvents are commonly applied in large volumes in industrial and lab-based procedures as a reaction medium, in addition to their use for extraction, separation and purification.[1,2] In 2012, global consumption of solvent was about 28 million tonnes.[3] Despite their large-scale use, issues relating to their

Green Chemistry Centre of Excellence, Department of Chemistry, The University of York, Heslington, York, YO10 5DD, UK. E-mail: andrew.hunt@york.ac.uk; Tel: +44 (0)1904324456

† Electronic supplementary information (ESI) available. See DOI: 10.1039/c7fd00049a

environmental impact, health, safety and sustainability remain. Many traditional organic solvents are toxic, while some halogenated solvents have been shown to deplete the ozone layer.[4] Furthermore, most conventional organic solvents are non-renewable and therefore at odds with the principles of sustainable development.[5]

Since 2006, European Regulation (EC) No. 1907/2006 "Registration, Evaluation, Authorisation and restriction of Chemicals" (REACH) has been influencing the chemical market within the EU.[6] REACH obliges companies to register and provide comprehensive physical properties, toxicological data and environmental data for every chemical that is manufactured or imported in quantities of one tonne or more per year ("no data, no market"). The REACH authorisation process can place a substance on a list of 'Substances of Very High Concern' (SVHC).[7] This list includes traditional solvents such as 1-methyl-2-pyrrolidone (NMP), *N*,*N*-dimethylformamide (DMF) and *N*,*N*-dimethylacetamide (DMAc). The sale and use of each SVHC will be restricted or effectively prohibited. Restrictions are already in place for many hazardous substances, including the widely used conventional solvents dichloromethane (DCM), chloroform, benzene and toluene.[8] Any products containing inappropriate substances as defined by REACH will be eliminated from the market by the 'Rapid Alert System for Dangerous Non-Food Products' (RAPEX) information scheme.[9] Outside of Europe, other laws, including Schedule I of the "Canadian Environmental Protection Act" in Canada,[10] and the "Code of Federal Regulation Title 40" in the USA,[11] also limit the use of toxic substances. In order to avoid the issues of traditional solvents and abide by relevant legislation, it is important to intelligently develop REACH compliant substitutes to conventional solvents while retaining their desirable properties.

The challenge for developing bio-based solvents

Bio-based solvents have been identified as green candidates to replace petroleum-derived solvents.[12,13] The benefit of bio-based solvents is that they are renewable and potentially do not result in a net increase of carbon dioxide in the atmosphere at the end of their lifetimes. Up to now, a number of bio-based solvents, such as dihydrolevoglucosenone (Cyrene™),[14] *p*-cymene,[15] 2-methyltetrahydrofuran (2-MeTHF),[16] D-limonene,[17] ethyl lactate,[18] and γ-valerolactone (GVL),[19] have been developed. Despite their advantages, bio-based solvents are not necessarily safer, less toxic, or more environmentally benign compared to conventional solvents and so a full analysis must be carried out before they can be classified as green. Some of the main challenges in the development of bio-based solvents are data availability (physical properties and toxicity), performance and cost. Bio-based solvents offer the opportunity to develop renewable and sustainable alternatives to traditional petrochemical-derived solvents. The EU will shortly ratify the European bio-based solvent standard, which will set out the requirements for these solvents in terms of properties, limits, application classes and test methods.[20] It will contain details for assessment and standardise the determination of bio-based content for these molecules. Such standards for solvents will aid to increase the development of this important class of bio-based products.

Herein, a methodology is proposed to develop new bio-based solvents based on our understanding of the challenges involved (Fig. 1). The methodology of solvent testing described here is a broadly applicable protocol that will indicate if

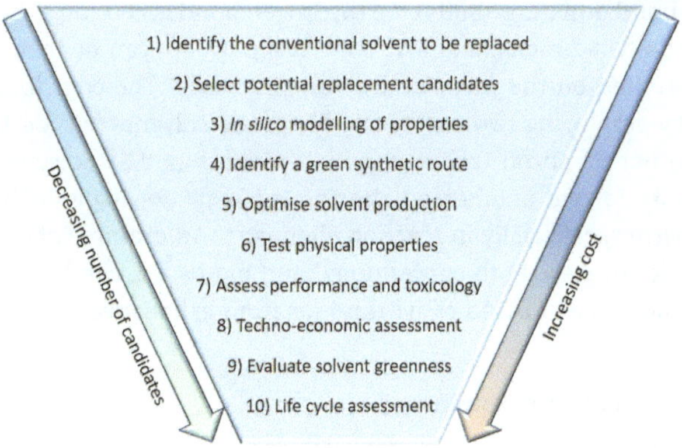

Fig. 1 Flow chart showing the proposed steps for developing new solvents.

a new solvent is functionally proficient (through a combination of *in silico* modelling, property measurements and lab-scale testing), and will also highlight potential health risks of the solvent under investigation. The combination of such a methodology and the use of the European bio-based solvent standard will be a powerful tool for bio-based solvent development. This method is then applied to the development stages of a bio-based solvent, methyl(2,2-dimethyl-1,3-dioxolan-4-yl)methyl carbonate (MMC), in a case study. Recently, MMC was synthesised in two steps from glycerol and was reported to be green due to the renewable nature of the feedstock and the clean synthetic methods used in its synthesis.[21] Herein, the synthesis of MMC was optimised and *in silico* modelling of Hansen solubility parameters relating to dispersion (δ_D), polar (δ_P) and hydrogen bonding (δ_H) interactions predicted that MMC would be an attractive candidate for use as an alternative bio-based aprotic solvent. Kamlet–Abboud–Taft (KAT) polarity scale measurements confirmed that this solvent has properties between dichloromethane, acetone and ethyl acetate. Testing of this solvent demonstrated MMC as a suitable solvent for both the Friedel–Crafts and Diels–Alder reactions.

Step 1: define the polarity and physical properties of the solvent to be replaced

Solvents are selected based on their favourable properties, usually volatility, polarity and flammability. Their negative properties, such as toxicity and environmental hazards, are a consequence of their chemical structure but need not go hand in hand. If the valuable attributes of a solvent can be understood and defined, solvent substitution can be executed more effectively by eliminating the negative properties. The boiling points, melting points, densities and flammability properties of traditional solvents are widely available. However, polarity is often an important property for a solvent and is potentially responsible for improving reaction rates, along with equilibria, solubility, cleaning or extraction efficiency. A solvent polarity map is a convenient tool during the first step of solvent selection. It gives a visual representation of the polarity of the solvent to be replaced and can be used to easily identify other solvents with a similar polarity.

The Kamlet–Abboud–Taft (KAT) solvatochromic parameters are widely used as a tool for understanding solvent polarity. The KAT parameters consist of

α (hydrogen bond donating ability),[22] β (hydrogen bond accepting ability)[23] and π^* (dipolarity).[24] A two-dimensional KAT solvent plot (map) can be then established with β represented on the y-axis and π^* on the x-axis. The contribution of α is recognised by employing two maps, one for protic solvents (with α higher than 0.5), and another for aprotic solvents (with α lower than 0.5). Solvents that are in close proximity to one another on the solvent map are likely to have similar solvent properties, especially in reaction chemistry. An example of a KAT solvent map of aprotic solvents, both conventional and bio-based, can be seen in Fig. 2. The solvent data shown is indicative, and far from exhaustive.[25]

Step 2: identify substitute solvents

In this step, the availability of potential solvents, bio-based or fossil derived, is deduced, also using a solvent polarity map. A comparison of physical properties (should they be known) to the ideal characteristics can be similarly made. However, it is not always the case that an obvious and readily available candidate for solvent substitution will be available. In such an instance, a bespoke synthesis of a new solvent may be required. Although the effort needed is substantial, designing a benign solvent to excel in a particular application is rewarding in the long run. However, the end product must be suitable from a performance, toxicity, environmental and economic perspective. Applying the following steps can help guide this process, but first one must propose molecules that could fulfil

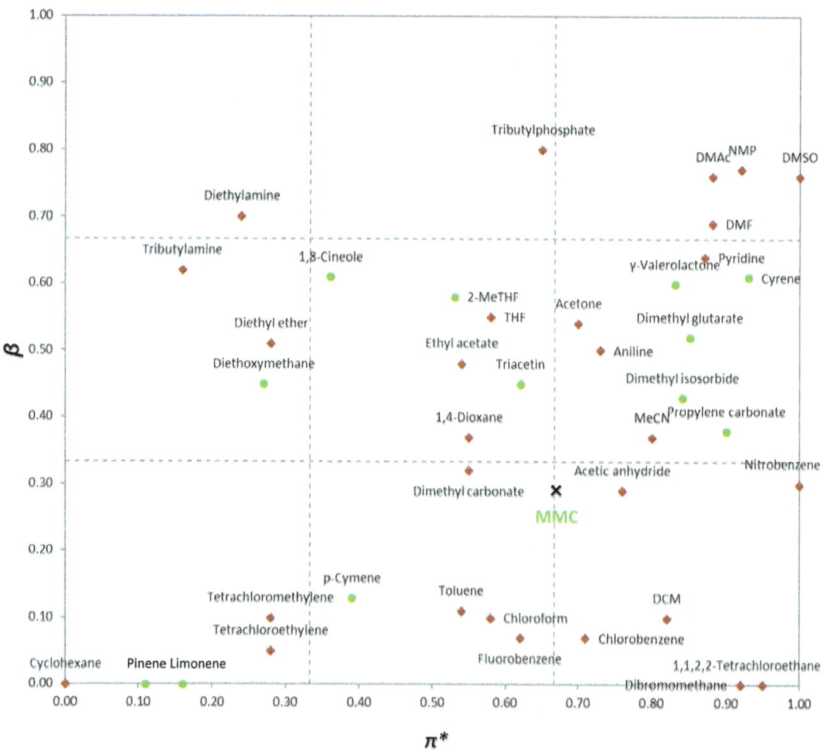

Fig. 2 KAT solvent map showing conventional solvents (red diamonds) and bio-based solvents (green circles).

the requirements that are currently satisfied by the solvent destined for substitution. This could be speculative, but better still is the use of computer programs that generate solvent candidates based on physical property requirements,[26–28] or available transformations of a bio-based platform molecule.[29,30]

Step 3: *in silico* modelling of candidate solvents

It is vital to calculate the properties of potential bio-based solvents before synthesis in order to fast screen through all promising candidates. The Hansen solubility parameters have been employed for over five decades to measure solvation power, and are amongst the most valuable solvent properties that can be accurately predicted.[31,32] The Hansen solubility parameters are three different scales: δ_D (dispersion forces), δ_P (dipole forces) and δ_H (hydrogen bonding forces). They can be used to construct the three-dimensional Hansen space, in effect another type of polarity map. The distance between two solvents, R_a, in the Hansen space is defined in eqn (1):

$$R_a = \sqrt{4(\delta_{D2} - \delta_{D1})^2 + (\delta_{P2} - \delta_{P1})^2 + (\delta_{H2} - \delta_{H1})^2} \tag{1}$$

Generally, if the R_a value between two solvents is low, they are likely to have similar solvency power and dissolve the same types of solute. 'Hansen Solubility Parameters in Practice' (HSPiP, 4th Edition 4.1.04) is a computer program that can be utilised as a powerful tool to predict Hansen solubility parameters.[33] HSPiP can generate the 3D Hansen space and calculate the R_a value between different solvents. As such, it assists users in their comparison of solvent candidates with traditional solvents, and can be applied to postulated molecules as well as 'real' solvents.

Step 4: selection of a synthetic pathway to the candidate solvent

Once a target molecule has been identified, a synthetic route must be devised. To meet the most stringent definition of "bio-based" in relation to a solvent, the raw material must be from biomass, most likely an established platform molecule.[34] A detailed study of the literature will generate numerous potential routes from raw material(s) to product, and the greenest of these must be selected. Applying green chemistry principals in route selection is not facile, requiring a fair and holistic methodology that can be easily applied using the data at hand and in a convenient time frame. This is best achieved using a metrics toolkit, such as that developed for the pharmaceutical industry.[35]

Step 5: optimisation of the solvent synthesis

The devised synthetic pathway to the target compound must then be applied in practice. Literature precedents are more than likely based on shared functionality as opposed to the reactants selected and therefore may not work or may require optimisation. Green chemistry principles must also be applied when changing reaction time, temperature, catalyst, loading, solvent, *etc.* This would most likely be through monitoring using the same metrics from step 4. Sufficient solvent needs to be synthesised at the desired purity to allow for full characterisation and application testing. This could be up to 1 L or even more, although a batch-wise synthesis might be necessary at first.

Step 6: defining the physical properties of the solvent

In order for a solvent to be applied, various physical and solubility properties must be first defined. For a formal list of solvent requirements, the European technical specification for bio-based solvents is helpful (CEN/TS 16766:2015).[20] There are no thresholds to define what physical properties are acceptable, only that the data is presented in a certain way, according to specific test methods. The mandatory solvent characteristics that must be known to adhere to CEN/TS 16766:2015 are composition (for formulations), polarity (Hansen solubility parameters), boiling point, vapour pressure or evaporation rate, colour, density, and viscosity. The one requirement in CEN/TS 16766:2015 that does employ a threshold is the bio-based content of the solvent, which must be at least 25%.[36] Additionally, the biomass feedstock must be classified as sustainable, either by formal certification or an equivalent assessment. Finally, any solvent containing an ether functionality presents a risk of peroxide formation; as such the auto-oxidation potential must be evaluated at an early stage.[37]

Step 7: assessing the solvent application and toxicology

The performance of the candidate solvent must be assessed in applications for which it has been predicted to be useful. When creating replacements for general purpose laboratory solvents, model reactions, such as Friedel–Crafts acylation (see later), Menshutkin *N*-alkylation,[14] Diels–Alder cycloadditions,[38] and cross-coupling reactions,[39] serve well for the demonstration of polar solvents. Esterification reactions and amidations are suitable for the demonstration of weak and non-hydrogen bonding solvents.[15] Multicomponent reactions forming heterocycles can also be used to test the performance of solvents.[40,41]

Equally important to understand is the toxicity and environmental impact of solvents. However, full toxicity testing is very expensive and time consuming. Quantitative structure–activity relationship (QSAR) methods, such as TEST by the EPA, have gained interest in recent years.[42] QSARs are statistical models that use a database of chemicals of a known activity, such as the median lethal dose (LD_{50}) or bioaccumulation factor (BCF), to predict the unknown activities of other molecules. While predictive software, such as TEST, is a valuable tool in assessing toxicity, predictions are not always reliable or have a high margin of error. As such, predictions must be confirmed experimentally, which brings us back to the original problem: cost and time. The Ames test is a simple first test of mutagenicity.[43] Although mutagenicity does not imply carcinogenicity, a strong correlation between the two is well established.[43–46] Two mutated *Salmonella typhimurium* (His^-) strains are employed in the Ames test; they are auxotrophic, which means they are unable to synthesize the histidine required for their growth and so cannot survive in the histidine-free media of the Ames test. Mutagenic compounds can revert these His^- strains back to their prototrophic state (His^+), at which point they can synthesise the histidine required for their growth, enabling them to grow in the histidine-free medium. As bacteria cells are different from mammalian cells, rat liver extracts are often used in combination for a more accurate representation of humans. The reason for this is that the liver is the organ responsible for the breakdown of ingested material in mammals. Some initially non-mutagenic chemicals can metabolize into mutagens during the breakdown process in the liver.[44,47] Including rat liver extracts in the Ames test increases the likelihood of detecting mutagenic metabolites of a test chemical. The Ames test is a relatively cheap test and test kits can be bought with results obtained in 3 days. Therefore, it is a good starting point for toxicity testing of new molecules. If a substance is found to be mutagenic, it may not be worth committing further time and money into its development. A substance that passes the Ames test would be a good candidate to be taken to the next step of toxicology testing.

Step 8: a techno-economic assessment of the solvent

Techno-economic assessments provide a cost–benefit analysis for the potential manufacture of a solvent, utilising various methods.[48] If the solvent candidate is a suitable product (according to the previous steps), then its commercialisation must be achieved through an environmentally and economically sound process for its benefits to be realised. Techno-economic assessments can be difficult and inaccurate based on lab-scale synthetic procedures so it is beneficial to move to a several kilograms' scale for a better understanding. Equipment to carry this out is not widely available in university laboratories and so coordination with industrial partners can play a major role in getting new solvents from the lab to commercialisation.

Step 9: solvent greenness assessed with the CHEM21 solvent selection guide

Chemical Manufacturing Methods for the 21st Century Pharmaceutical Industries (CHEM21) is Europe's largest public–private partnership aiming to develop sustainable manufacturing routes to pharmaceuticals. Along with the metrics toolkit,[35] CHEM21 published a solvent selection guide unifying publicly available solvent selection guides from the pharmaceutical industry.[49] The CHEM21 solvent selection guide is easily applied to rank solvents based on proposed criteria of safety, health and environment in compliance with the Global Harmonized System (GHS) and European regulations. This methodology showed good agreement with classical solvents and was also used to rank novel, less classical solvents using a simple and freely available spreadsheet. This same methodology should be used to evaluate the candidate solvent according to the obtained physical and toxicological data.

Step 10: life cycle assessment (LCA) of the solvent

A life cycle assessment (LCA) is a tool that can be employed to evaluate the environmental impact of a product or process through calculating its emissions. An equivalent tool for the social impact of goods and services, social life cycle assessment (S-LCA), is also possible.[50,51] Before a commitment to manufacturing is made, a pro-active application of LCA is needed to help guide the development of the process. A life cycle assessment can also be applied retrospectively to identify and eliminate areas of concern as they arise.

Results and discussion

In the case study presented, the methodology described above was used to direct the development of a new bio-based solvent. Each step acts as a filtering process, whereby any solvent candidates failing to meet the requirements, whether they be enforced by legislation or imposed by user requirements, can be disregarded to focus resources.

Step 1: identifying halogenated solvents for replacement

Currently there is only one aprotic bio-based solvent (Cygnet 0.0) with a medium to high dipolarity/dipolarizability ($\pi^* = 1.09$) and a low basicity ($\beta = 0.17$).[52] In this area on a conventional aprotic solvent map reside the halogenated solvents dichloromethane (DCM) and chloroform, which are rated as hazardous and highly hazardous, respectively (Fig. 2).[49] In addition, suitable physical properties, such as the boiling point, are desired but not priority in this study. The advantage of a higher boiling point is that less solvent is lost to the atmosphere but with the disadvantage of more difficult removal at the end of a process. Therefore, it is important to develop new bio-based solvents that occupy this area of the map. Any solvent with the polarity of halogenated solvents but without the implicit issues surrounding the presence of a halogen atom would be a highly valuable addition to the current catalogue of bio-based solvents.

Step 2: selecting organic carbonates of glycerol formal and solketal as candidate solvents

Glycerol and its derivatives are well established in the field of bio-based solvent research.[19,53,54] Glycerol is a versatile compound that has many green merits, such as being renewable, non-toxic ($LD_{50} = 12\,600$ mg kg^{-1}), biodegradable and cheap.[55,56] Glycerol is a by-product of biodiesel production through the transesterification of triglycerides and was listed by the National Renewable Energy Laboratory (NREL) as one of the top twelve platform molecules that can be derived from biomass.[57] Approximately 10 kg of crude glycerol can be obtained during the production of 100 kg of biodiesel. Due to the large volumes of glycerol produced, biological or chemical conversion of surplus glycerol to high-value products has received significant attention.[58]

At present, the main research fields of glycerol-derived solvents are alkyl glycerol ethers, glycerol carbonate/esters of glycerol carbonate, glycerol-based ILs, glycerol formal and solketal.[59] The modification of glycerol formal and solketal (solvents in their own right) into new aprotic solvents is an unexploited field. Reacting at the alcohol can produce aprotic molecules, and the extended functionality may well increase the dipolarity without increasing the hydrogen bond basicity (β). The organic carbonates of glycerol formal and solketal were identified as the target molecules in this work (Scheme 1).

Step 3: *in silico* modelling to identify the target solvent

Potential new carbonate solvents, produced from glycerol formal and solketal, and their properties are listed in Table 1. The Hansen solubility parameters of DCM and chloroform were selected as references. The boiling point and R_a value (relative to DCM and chloroform) of each candidate was calculated in HSPiP. After the screening, methyl(2,2-dimethyl-1,3-dioxolan-4-yl)methyl carbonate (MMC) was selected as the target solvent due to its lower R_a value relative to DCM and chloroform. The predicted boiling point of MMC (222 °C) is much higher than that of DCM or chloroform but this is unavoidable given the molecular size and structure. However, the similar solvency power of MMC to these halogenated

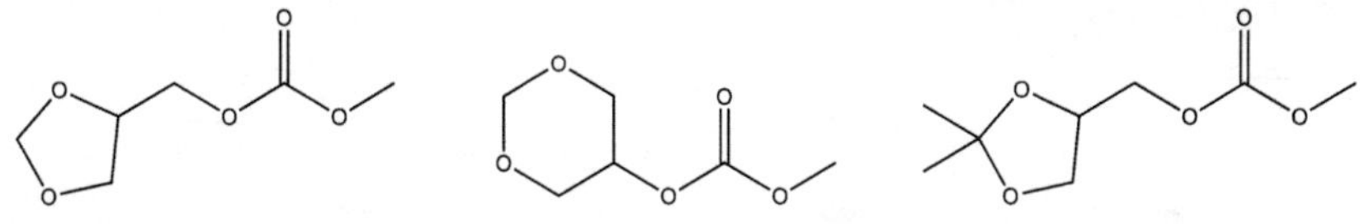

1 2 3

1,3-Dioxolanyl-5-yl methyl carbonate 1,3-Dioxanyl-5-yl methyl carbonate MMC

Scheme 1 Candidate solvents sourced from glycerol to replace halogenated solvents.

solvents could remain interesting. The position of MMC and other nearby conventional solvents in the 3D Hansen space can be seen in the ESI (Fig. S1†).

Step 4: selecting the greenest synthetic pathway to MMC from solketal

As shown in Scheme 2, there are two main methods by which to synthesise MMC from solketal, carboxymethylation *via* methyl chloroformate (MC) (i) or DMC (ii). Metrics analyses applying the CHEM21 toolkit using conditions taken from model reactions found in the literature are displayed in Table 2.[60,61] As is evident, the yield, atom economy and PMI for both routes are very similar. The RME is significantly worse for route (ii) as DMC acts as both a reactant and solvent, but is an acceptable solvent as opposed to acetonitrile, which is problematic. The most significant difference is in the inherent health and safety, with dimethyl carbonate widely accepted as a biodegradable and non-toxic green compound,[62] as opposed to methyl chloroformate, which is highly toxic. Thus, the DMC synthesis is more promising as a green route to MMC.

Step 5: MMC synthesis and optimisation

Although MMC synthesised from solketal *via* DMC chemistry has been previously reported,[21] conditions were applied to preferentially give methylation as opposed to carboxymethylation. Additionally, the products were synthesised as part of a mechanistic study and not considered as solvents. As such, optimisation towards carboxymethylation and determination of further physical properties were required. In this work, MMC was synthesised and the procedure was optimised (see ESI†). The optimised reaction conditions are: reaction time = 20 h, 0.1

Table 1 HSPiP predicted properties of candidate bio-based carbonate solvents synthesised from glycerol derivatives

Potential bio-based solvent	B.P./°C	Hansen δ_D/MPa$^{0.5}$	Hansen δ_P/MPa$^{0.5}$	Hansen δ_H/MPa$^{0.5}$	R_a to DCM	R_a to chloroform
1	206	17.0	10.3	8.1	5.08	7.76
2	211	17.0	10.1	8.1	4.92	7.57
3 (MMC)	222	16.2	7.9	6.0	4.31	5.78
Chloroform	61	17.8	3.1	5.7	—	—
DCM	40	17	7.3	7.1	—	—

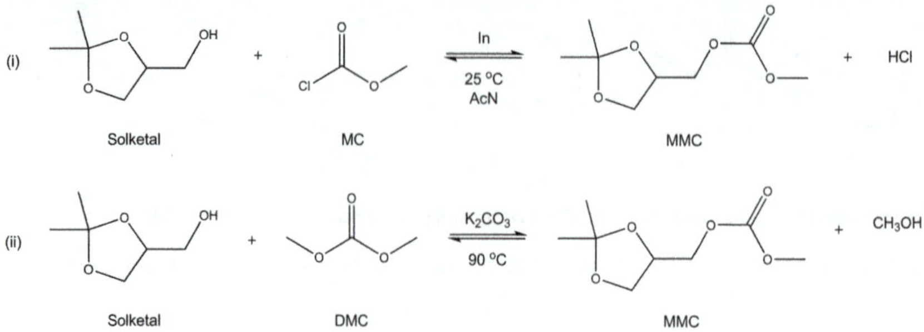

Scheme 2 The two synthetic pathways to MMC from solketal: using methyl chloroformate (i) or dimethyl carbonate (ii).

mol% K_2CO_3 and DMC/solketal molar ratio = 20 : 1. The isolated yield of MMC after distillation was 91% (99% purity by GC).

Step 6: determining MMC's physical properties and tendency towards peroxide formation

Table 3 lists the experimentally observed properties of MMC. The boiling point of MMC was measured to be 232 °C by distillation, only 10 °C higher than the HSPiP estimate. Distillation is extensively used as a product isolation technique in batch processes and therefore solvents with high boiling points, such as MMC, can be problematic unless the product can be crystalized from solution with relative ease. The melting point was determined to be −7 °C by differential scanning calorimetry (DSC), potentially limiting MMC's use in low temperature chemistry. The density of MMC was determined to be 1.14 g cm^{-3} at 298 K, similar to DCM

Table 2 Analyses of routes (i) and (ii) by the CHEM21 metrics toolkit

Pathway	(i)	(ii)
Yield	85%	90%
Rxn. mass efficiency	59%	9%
Atom economy	84%	86%
Solvents	Acetonitrile	Dimethyl carbonate
Health & safety	H330 (methyl chloroformate)	
PMI	15	12
Catalysts used	Indium	Stoichiometric reagent
Reactor	Batch	Batch
Elements	Indium	Potassium
Energy	Room temperature	Reflux
Workup	Quench, distil	Distil

Table 3 The properties of MMC compared to DCM and chloroform

Property	MMC	DCM	Chloroform
MWt	190.2[a]	84.9	119.4
α	0.00	0.13	0.20
β	0.29	0 × 10	0 × 10
π^*	0.67	0.82	0.58
δ_D/MPa$^{0.5}$	16.2[a]	18.2	17.8
δ_P/MPa$^{0.5}$	7.9[a]	6.3	3.1
δ_H/MPa$^{0.5}$	6.0[a]	6.1	5.7
HSP distance[b]	0.0	4.3	5.8
B.P./°C	232[c]	40	61
M.P./°C	−7[d]	−95	−64
ρ/g cm^{-3} 298 K	1.14	1.32	1.48
Viscosity/cP 293 K	3.50	0.44	0.58

[a] Calculated by HSPiP. [b] HSP distance, R_a, to MMC. [c] Measured by distillation. [d] Measured by DSC.

and chloroform. The viscosity of MMC is also comparatively high as compared to other solvents, which may generate issues in processing.

Like many other bio-based solvents, MMC does not contain any halogen atoms or heteroatoms aside from oxygen, thus eliminating the environmental risk to the ozone layer and atmospheric pollution in the form of NO_x and SO_x. Methyl(2,2-dimethyl-1,3-dioxolan-4-yl)methyl carbonate is immiscible with water, enabling applications in liquid–liquid extractions, although more testing is needed in this regard.

As MMC contains an ether functionality, it has the potential to produce explosive peroxide compounds *via* autoxidation by atmospheric O_2. An initial test to investigate the formation of peroxides in MMC was carried out by employing peroxide test strips (Macherey-Nagel, QUANTOFIX® Peroxide-100) to test for any peroxides present in solution. After 224 days of testing without antioxidants or stabilisers, the peroxide concentration in MMC was below the detection limit. This demonstrates that MMC did not readily form peroxides at ambient temperatures over the period of testing, although the routine addition of a stabiliser is recommended.

Its position on the solvent map indicates that MMC has similar solubility properties to dimethyl carbonate, 1,4-dioxane, acetonitrile and acetic anhydride (Fig. 2). These results are evidence that MMC too readily accepts hydrogen-bonds to be considered a replacement for the halogenated solvents DCM and chloroform. Moreover, since MMC is like DMC in terms of polarity, it may not be beneficial to consume DMC to make MMC. However, MMC fulfils the criteria to undergo performance and toxicological testing.

Step 7: assessing the performance of MMC as a solvent and toxicological testing

Friedel–Crafts acylation and Diels–Alder cycloaddition reactions were selected to evaluate the performance of MMC compared to traditional solvents. If, by chance, the performance of MCC exceeded the expectations established by its measured polarity, it would be worth pursuing beyond this stage of solvent development.

These two reactions are commonly performed with halogenated solvents. Analysis of the experimental results allows a comparison of the solvent performance of MMC with a range of traditional solvents. The synthesis of 4-methoxyacetophenone (4-MAP) from anisole and acetic anhydride catalysed by $FeCl_3$ (Scheme 3(iii)) was selected to evaluate the performance of MMC in the Friedel–Crafts reaction. It was found that the reaction conducted in MMC resulted in a yield of 61% 4-MAP (Fig. 3). Although this is lower than when using DCM, it is higher than all other solvents tested.

The performance of MMC was also assessed in the synthesis of 1-(3,4-dimethylcyclohex-3-enyl) ethanone (DE) from 2,3-dimethylbuta-1,3-diene (diene) and 3-butene-2-one, catalysed by anhydrous $YbCl_3$ (Scheme 3(iv)).[63] Dichloromethane, propylene carbonate and acetonitrile all exhibited high yields (>95%), while ethyl acetate, acetone and MMC produced yields of 75–80% (Fig. 3). These results indicate that the solvent properties of MMC are more similar to ketone and ester solvents in the Diels–Alder reaction.

The results of the two experimental case studies show that both reactions are highly dependent on the polarity of the solvent. Specifically, a high π^* is favoured, a trend that is especially true in the case of the Diels–Alder reaction. MMC is competitive in terms of yield, but the superior polarity of DCM makes it the technically more proficient solvent, albeit suspected as a carcinogen. Across both case studies, propylene carbonate, with its strong molecular dipole moment, was also apt as a reaction solvent and worth considering as a solvent for these transformations.

The mutagenicity of MMC was tested using the Ames test. TA98 and TA100 were utilised for the detection of frameshift mutations and base substitution mutations, respectively. Dimethyl sulfoxide (DMSO) was employed as a solvent. A mixture of 2-nitrofluorene (2-NF) and 4-nitroquinoline-*N*-oxide (4-NQO) was utilised as the positive control. This Ames test was conducted without S9 microsomal activation and so metabolites could not be investigated. MMC was not found to be mutagenic using the TA100 strain (ESI, Fig. S3†). However, MMC was found to be mutagenic for the TA98 strain (ESI, Fig. S3†). This indicated that MMC is likely to be a mutagenic solvent and hence, a possible carcinogen. Consequently, although MMC is a bio-based solvent, its potential toxicity means it is unlikely to be

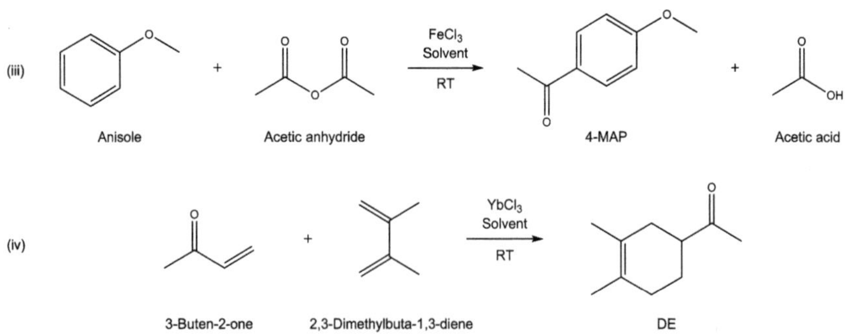

Scheme 3 (iii) Friedel–Crafts reaction between anisole and acetic anhydride to produce 4-MAP and (iv) the Diels–Alder reaction between 3-butene-2-one and 2,3-dimethylbuta-1,3-diene to produce DE.

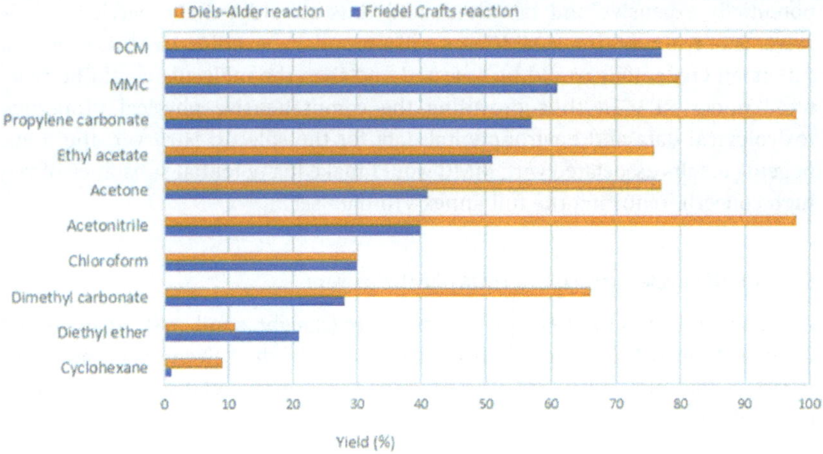

Fig. 3 Results of the test reactions using different solvents. Values shown are reaction yields.

considered for further testing for use as a green solvent. This assay demonstrates the usefulness of the Ames test as a first port of call for toxicity testing and that any new bio-based solvent needs toxicological testing before application. The remaining 3 steps have not been completed for MMC as the solvent failed to pass step 7, however were the techno-economic assessment included, many detailed examples in the literature could have been emulated.

Step 8: a techno-economic assessment of the solvent

Specific examples regarding solvents can be found in the comparison of various small alcohols from biomass,[64] different strategies towards ethanol bio-refineries,[65] and strategies towards production of various fatty esters.[66] The majority of such studies have so far been directed towards fuels where margins are very tight but for a new bulk chemical to be industrially feasible it must meet the triple bottom line, to have an environmental, social and economic advantage.[67] As such, any new bio-based solvent must demonstrate a theoretical economic competitiveness to be further considered. A key factor that is missing from the principles of green chemistry is that any green product or synthesis must be cost effective. It is therefore of vital importance to undertake a techno-economic assessment of the solvent. If at this stage MMC was found to be cost competitive and feasible, further safety, health and environment testing would be required in step 9.

Step 9: solvent greenness assessed with the CHEM21 solvent selection guide

As stated, these are simple assessment criteria to apply using the published methodology. The data collected in step 6 is sufficient to gain an accurate safety score, with the exception of resistivity, which requires specialised equipment to determine as the high impedance requires a high voltage to be applied. With reference to the health and environment ranking, as MMC is relatively novel, it has no Global Harmonized System (GHS) hazard statements nor is it REACH registered, which would result in a default score of 5 (problematic) in both categories. The testing required to generate GHS data and meet REACH criteria is potentially expensive and laborious, and thus should only be carried out on solvents of real promise. If MMC was still a promising candidate at this stage, the full green credentials would be assessed and then the molecule would be registered under REACH, thus providing the comprehensive physical properties, toxicological data and environmental data for the solvent. However, the mutagenic results associated with MMC would make it a potential substance of very high concern, requiring the full annex VIII data set.[68]

Step 10: life cycle assessment (LCA) of the solvent

At the same time as REACH registration is being sought, a full cradle to grave LCA of the solvent will provide a holistic assessment of the solvent, thus providing investors and end users with the confidence to commercialise the process or utilise the solvent in their processes.

Conclusion

In this work, a methodology to focus the development of new bio-based solvents was proposed in order to accelerate the implementation of greener solvents. A case study on the development of a potential bio-based solvent, methyl(2,2-dimethyl-1,3-dioxolan-4-yl)methyl carbonate (MMC), was carried out to exemplify the process. Although MMC is an attractive candidate as a replacement solvent, simply being bio-derived is not enough for a molecule to be regarded as green. This work highlights a systematic method for the development of bio-based solvents, which importantly promotes the use of toxicity testing at an early stage in the development of bio-based molecules. The Hansen solubility parameters and reaction data indicated that MMC could be an attractive bio-based aprotic solvent. The KAT parameters of MMC clarified its polarity and potential reactivity, and were found to be similar to those of dimethyl carbonate. More importantly, MMC was found to be a mutagen in a preliminary Ames test. The methodology of bio-based solvent development described here is a widely applicable approach that highlights the significance of using KAT parameters and toxicology research in the early stages of exploitation of any new bio-based solvent. The combination of such a protocol and the utilisation of the European technical specification (CEN/TS 16766:2015) for bio-based solvents can be an efficient way forward for the rational substitution of hazardous, unsustainable solvents.

Acknowledgements

The authors work like to thank the Biorenewables Development Centre (BDC) for their assistance with the Ames test. The authors also thank Dr Thomas J. Farmer for his beneficial suggestions during this research.

Notes and references

1 C. Reichardt and T. Welton, *Solvents and Solvent Effects in Organic Chemistry*, John Wiley & Sons, 2011.

2 F. M. Kerton and R. Marriott, *Alternative Solvents for Green Chemistry*, Royal Society of Chemistry, 2013.
3 Global Solvents-Opportunities for Greener Solvents, available online at http://www.ihs.com/products/chemical-special-reports-global-solvents.html, accessed September 2016.
4 M. J. Molina and F. S. Rowland, *Nature*, 1974, **249**, 810–812.
5 P. Anastas and N. Eghbali, *Chem. Soc. Rev.*, 2010, **39**, 301–312.
6 European REACH Regulation (EC) No. 1907/2006 of the European Parliament and of the Council on the Registration, Evaluation, Authorisation and Restriction of Chemicals (REACH).
7 Candidate List of substances of very high concern for Authorisation, available online at http://echa.europa.eu/web/guest/candidate-list-table, accessed June 2016.
8 Substances restricted under REACH, available online at http://echa.europa.eu/addressing-chemicals-of-concern/restrictions/substances-restricted-under-reach, accessed June 2016.
9 European Commission Rapid Alert System for dangerous non-food products (RAPEX), available online at http://ec.europa.eu/consumers/consumers_safety/safety_products/rapex/index_en.html, accessed December 2016.
10 Canadian Environmental Protection Act, 1999, available online at http://laws-lois.justice.gc.ca/eng/acts/c-15.31/page-1.html, accessed September 2016.
11 US EPA, Code of Federal Regulations (CFR) CFR Title 40, available online at http://www.archives.gov/federal-register/cfr/subject-title-40.html, accessed September 2016.
12 Y. Gu and F. Jérôme, *Chem. Soc. Rev.*, 2013, **12**, 1127–1138.
13 J. H. Clark, T. J. Farmer, A. J. Hunt and J. Sherwood, *Int. J. Mol. Sci.*, 2015, **16**, 17101–17159.
14 J. Sherwood, M. D. Bruyn, A. Constantinou, L. Moity, C. R. McElroy, T. J. Farmer, T. Duncan, W. Raverty, A. J. Hunt and J. H. Clark, *Chem. Commun.*, 2014, **50**, 9650–9652.
15 J. H. Clark, D. J. Macquarrie and J. Sherwood, *Green Chem.*, 2012, **14**, 90–93.
16 D. F. Aycock, *Org. Process Res. Dev.*, 2007, **11**, 156–159.
17 P. S. Kulkarni, C. Brazinha, C. A. M. Afonso and J. G. Crespo, *Green Chem.*, 2010, **12**, 1990–1994.
18 C. S. M. Pereira, V. M. T. M. Silva and A. E. Rodrigues, *Green Chem.*, 2011, **13**, 2658–2671.
19 Y. Gu and F. Jérôme, *Chem. Soc. Rev.*, 2013, **42**, 9550–9570.
20 European standard CEN/TS 16766, *Bio-based Solvents – Requirements and Test Methods*, European Committee for Standardization (CEN), Brussels, 2015.
21 M. Selva, V. Benedet and M. Fabris, *Green Chem.*, 2012, **14**, 188–200.
22 R. W. Taft and M. J. Kamlet, *J. Am. Chem. Soc.*, 1976, **98**, 2886–2894.
23 M. J. Kamlet and R. W. Taft, *J. Am. Chem. Soc.*, 1976, **98**, 377–383.
24 M. J. Kamlet, J. L. Abboud and R. W. Taft, *J. Am. Chem. Soc.*, 1977, **99**, 6027–6038.
25 P. G. Jessop, D. A. Jessop, D. Fu and L. Phan, *Green Chem.*, 2012, **14**, 1245–1259.
26 R. Gani, C. Jiménez-González and D. J. C. Constable, *Comput. Chem. Eng.*, 2005, **29**, 1661–1676.
27 R. Gani, *Chem. Eng. Res. Des.*, 2004, **82**, 1494–1504.

28 R. Gani, P. A. Gómez, M. Folić, C. Jiménez-González and D. J. C. Constable, *Comput. Chem. Eng.*, 2008, **32**, 2420–2444.
29 L. Moity, V. Molinier, A. Benazzouz, R. Barone, P. Marion and J.-M. Aubry, *Green Chem.*, 2014, **16**, 146–160.
30 L. Moity, V. Molinier, A. Benazzouz, B. Joossen, V. Gerbaud and J.-M. Aubry, *Green Chem.*, 2016, **18**, 3239–3249.
31 C. M. Hansen, PhD thesis, Technical University of Denmark, 1967.
32 C. M. Hansen, *Hansen Solubility Parameters: A User's Handbook*, CRC Press, 2nd edn, 2012.
33 S. Abbott and C. M. Hansen, *Hansen Solubility Parameters in Practice (HSPiP 4.1.04), Hansen-Solubility*, ch. 9, 2008.
34 T. Farmer and M. Mascal, in *Introduction to Chemicals from Biomass*, ed. J. Clark and F. Deswarte, John Wiley & sons, Ltd., Chichester, 2nd edn, 2015, ch. 4.
35 C. R. McElroy, A. Constantinou, L. C. Jones, L. Summerton and J. H. Clark, *Green Chem.*, 2015, **17**, 3111–3121.
36 J. Sherwood, J. H. Clark, T. J. Farmer, L. Herrero-Davila and L. Moity, *Molecules*, 2016, **22**, 48.
37 V. Fábos, G. Koczó, H. Mehdi, L. Boda and I. T. Horváth, *Energy Environ. Sci.*, 2009, **2**, 767.
38 R. Bini, C. Chiappe, V. L. Mestre, C. S. Pomelli and T. Welton, *Org. Biomol. Chem.*, 2008, **6**, 2522–2529.
39 H. L. Parker, J. Sherwood, A. J. Hunt and J. H. Clark, *ACS Sustainable Chem. Eng.*, 2014, **2**, 1739–1742.
40 J. Sherwood, H. L. Parker, K. Moonen, T. J. Farmer and A. J. Hunt, *Green Chem.*, 2016, **18**, 3990–3996.
41 J. H. Clark, D. J. Macquarrie and J. Sherwood, *Chem.–Eur. J.*, 2013, **19**, 5174–5182.
42 Toxicity Estimation Software Tool (TEST), available online at http://www.epa.gov/chemical-research/toxicity-estimation-software-tool-test, accessed January 2017.
43 B. N. Ames, F. D. Lee and W. E. Durston, *Proc. Natl. Acad. Sci. U. S. A.*, 1973, **70**, 782–786.
44 A. J. Griffiths, J. H. Miller, D. T. Suzuki, R. C. Lewontin and W. M. Gelbart, in *An Introduction to Genetic Analysis*, W. H. Freedman, New York, 7th edn, 2000, ch. 16.
45 J. McCann, L. Swirsky Gold, L. Horn, R. McGill, T. E. Graedel and J. Kaldor, *Mutat. Res., Genet. Toxicol.*, 1988, **205**, 183–195.
46 J. McCann, E. Choi, E. Yamasaki and B. N. Ames, *Proc. Natl. Acad. Sci. U. S. A.*, 1975, **72**, 5135–5139.
47 B. N. Ames, W. E. Durston, E. Yamasaki and F. D. Lee, *Proc. Natl. Acad. Sci. U. S. A.*, 1973, **70**, 2281–2285.
48 C. L. Gargalo, A. Carvalho, K. V. Gernaey and G. Sin, *Biochem. Eng. J.*, 2016, **116**, 146–156.
49 D. Prat, A. Wells, J. Hayler, H. Sneddon, C. R. McElroy, S. Abou-Shehada and P. J. Dunn, *Green Chem.*, 2016, **18**, 288–296.
50 A. Jørgensen, A. L. Bocq, L. Nazarkina and M. Hauschild, *Int. J. Life Cycle Assess.*, 2008, **13**, 96.

51 C. Benoît, G. A. Norris, S. Valdivia, A. Ciroth, A. Moberg, U. Bos, S. Prakash, C. Ugaya and T. Beck, *Int. J. Life Cycle Assess.*, 2010, **15**, 156–163.

52 A. Alves Costa Pacheco, J. Sherwood, A. Zhenova, C. R. McElroy, A. J. Hunt, H. L. Parker, T. J. Farmer, A. Constantinou, M. De bruyn, A. C. Whitwood, W. Raverty and J. H. Clark, *ChemSusChem*, 2016, **9**, 3503–3512.

53 A. Wolfson, A. Snezhko, T. Meyouhas and D. Tavor, *Green Chem. Lett. Rev.*, 2012, **5**, 7–12.

54 G. Cravotto, L. Orio, E. C. Gaudino, K. Martina, D. Tavor and A. Wolfson, *ChemSusChem*, 2011, **4**, 1130–1134.

55 R. Ciriminna, C. D. Pina, M. Rossi and M. Pagliaro, *Eur. J. Lipid Sci. Technol.*, 2014, **116**, 1432–1439.

56 Z. Wang, J. Zhuge, H. Fang and B. A. Prior, *Biotechnol. Adv.*, 2001, **19**, 201–223.

57 J. J. Bozell and G. R. Petersen, *Green Chem.*, 2010, **12**, 539–554.

58 Biofuels Barometer 2016, available online at http://www.eurobserv-er.org/pdf/biofuels-barometer-2016-en/, accessed August 2016.

59 A. E. Díaz-Álvarez, J. Francos, B. Lastra-Barreira, P. Crochet and V. Cadierno, *Chem. Commun.*, 2011, **47**, 6208–6227.

60 J.-G. Kim and D. O. Jang, *Tetrahedron Lett.*, 2009, **50**, 2688–2692.

61 P. Tundo, F. Aricò, A. E. Rosamilia, M. Rigo, A. Maranzana and G. Tonachini, *Pure Appl. Chem.*, 2009, **81**, 1971–1979.

62 M. Bilde, T. E. Møgelberg, J. Sehested, O. J. Nielsen, T. J. Wallington, M. D. Hurley, S. M. Japar, M. Dill, V. L. Orkin, T. J. Buckley, R. E. Huie and M. J. Kurylo, *J. Phys. Chem. A*, 1997, **101**, 3514–3525.

63 X. Fang, B. P. Warner and J. G. Watkin, *Synth. Commun.*, 2000, **30**, 2669–2676.

64 L. Tao, E. C. D. Tan, R. McCormick, M. Zhang, A. Aden, X. He and B. T. Zigler, *Biofuels, Bioprod. Biorefin.*, 2014, **8**, 30–48.

65 D. Klein-Marcuschamer, B. A. Simmons and H. W. Blanch, *Biofuels, Bioprod. Biorefin.*, 2011, **5**, 562–569.

66 L. Amer, B. Adhikari and J. Pellegrino, *Bioresour. Technol.*, 2011, **102**, 9350–9359.

67 B. Willard, *The new sustainability advantage: seven business case benefits of a triple bottom line*, New Society Publishers, 2012.

68 ECHA Information requirements, available online at http://echa.europa.eu/regulations/reach/registration/information-requirements, accessed January 2017.

Faraday Discussions

PAPER

Organic waste as a sustainable feedstock for platform chemicals

M. Coma,*[a] E. Martinez-Hernandez, [b] F. Abeln,[c] S. Raikova,[c] J. Donnelly,[b] T. C. Arnot,[bf] M. J. Allen,[d] D. D. Hong[e] and C. J. Chuck[b]

Received 19th February 2017, Accepted 21st February 2017
DOI: 10.1039/c7fd00070g

Biorefineries have been established since the 1980s for biofuel production, and there has been a switch lately from first to second generation feedstocks in order to avoid the food *versus* fuel dilemma. To a lesser extent, many opportunities have been investigated for producing chemicals from biomass using by-products of the present biorefineries, simple waste streams. Current facilities apply intensive pre-treatments to deal with single substrate types such as carbohydrates. However, most organic streams such as municipal solid waste or algal blooms present a high complexity and variable mixture of molecules, which makes specific compound production and separation difficult. Here we focus on flexible anaerobic fermentation and hydrothermal processes that can treat complex biomass as a whole to obtain a range of products within an integrated biorefinery concept.

1. Introduction

Two main challenges for our society are the depletion of fossil resources and increasing waste generation. In order to reduce the dependence on oil but also mitigate climate change in the transport and chemical sectors, alternative production chains are necessary.[1] This involves a shift towards renewable resources, which are not finite and can be easily regenerated. While the energy economy can be based on various alternative raw materials (wind, sun, water, biomass, nuclear fission and fusion), the material economy of substances mainly depends on biomass, in particular plant biomass. Therefore, biorefineries as

[a]*Centre for Sustainable Chemical Technologies (CSCT), University of Bath, Claverton Down, Bath, BA2 7AY, UK. E-mail: m.coma@bath.ac.uk*

[b]*Department of Chemical Engineering, University of Bath, Bath, BA2 7AY, UK*

[c]*Centre for Doctoral Training in Sustainable Chemical Technologies, Department of Chemical Engineering, University of Bath, Claverton Down, Bath, BA2 7AY, UK*

[d]*Plymouth Marine Laboratory, Prospect Place, The Hoe, Plymouth PL1 3DH, UK*

[e]*Algal Biotechnology Department, Institute of Biotechnology, Vietnam Academy of Science and Technology, 18 Hoang Quoc Viet Str., Cau Giay, Hanoi, Vietnam*

[f]*Water Innovation & Research Centre, University of Bath, Claverton Down, Bath, BA2 7AY, UK*

bioresource-converting systems, analogous to petroleum-based refineries, will be the key for access to the bioeconomy: an integrated production of biobased products (food, feed, chemicals and materials) and bioenergy (fuels).[2]

Waste generation is the second major challenge for our society. Nearly 50% of the average composition of global waste, 3 M tons per day, is organic material. This accounts for the main source of greenhouse gases (GHG), and it includes household, food manufacturing and pre-factory wastes, the rest being paper, plastic, glass, metal and others (Fig. 1).[3] These waste streams contain various compounds, most of which have untapped energetic or economic value. Current waste management practices in decreasing order of added value for organic waste include: animal feed, composting, incineration and landfill. Although a number of facilities direct their waste toward land spreading, these facilities only represent between 26% and 46% of the organic waste and most of it is still disposed of in landfills.[4]

1.1. Towards a more sustainable biorefinery

Biofuels such as bioethanol and biodiesel produced from seeds, grains and sugar (the so-called first generation (1G) biofuels) have exponentially increased in use since the 1980s due to their easy applicability within existent engines without modifications, their renewability, their biodegradability, their lower emission generation and the ability to increase the security of the supply and provide

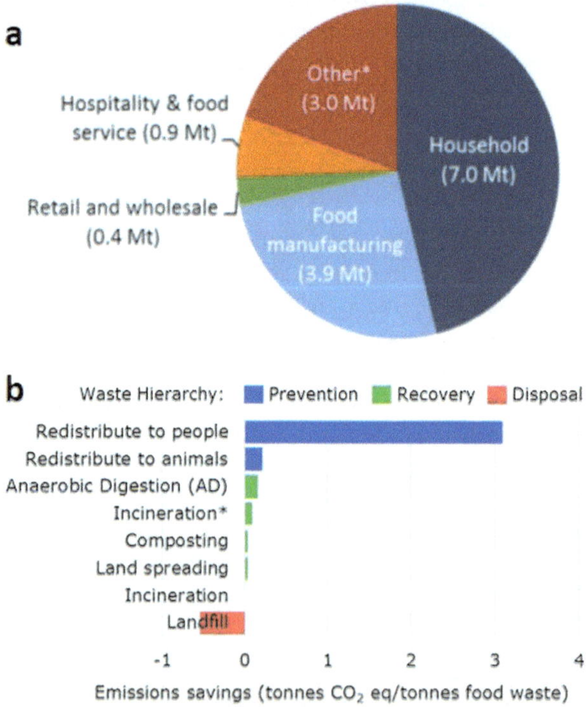

Fig. 1 (a) Sources of organic waste in the UK. *Includes pre-factory waste. (b) Emission savings per tonne of organic waste depending on its management. *Includes energy recovery.[5]

a steady income to farmers.[6] However, energy carriers produced from crops have caused inflation in food prices and led to the food *versus* fuel crisis. Hence, the production of bioenergy from alternative sources such as agricultural and domestic organic wastes, the substrates for second generation (2G) biofuels which are mainly composed of lignocellulosic biomass, has now provided a positive shift towards green energy production.[7] 2G biofuels generated from non-crop feedstocks release the pressure on the food market; however, there is concern over competing land use or required land use changes. Therefore, third generation (3G) biofuels have been derived from past agricultural substrates, waste vegetable oils, microbes or microalgae as a viable alternative energy resource.[6] The biorefinery concept embraces a wide range of technologies to separate biomass resources into their building blocks (carbohydrates, proteins, triglycerides and others) which can be converted not only to biofuels, but also to value added products and chemicals.

Most of the existing biofuels and biochemicals are currently generated in single production chains and not within a biorefinery concept, and thus their exploitation is thereby limited.[1] Modern biotechnology focusses on biofuels. Biogas production uses a resilient ecosystem of diverse microorganisms to co-convert multiple organics into methane. Bioethanol production is only possible with single bio-based substrates and single yeast strains, and its efficiency strongly depends on the bioavailability of carbohydrates. In general, these processes convert only a fraction of organics into biofuel and other outcomes are low value co-products or waste. To overcome this drawback, waste streams should be managed *via* a biorefinery system which integrates technologies flexibly and where all product outcomes are considered. The design and development of multi-purpose biorefineries that generate a variety of products as a consequence of integrated, sequential, non-competitive processes is considered a strategic way to reach this goal.[8]

1.2. Organic waste as a feedstock

Organic waste streams are a sustainable alternative to fossil-based resources as they do not compete directly with food crops. 'Waste' covers any organic material apart from the primary material for which the plants were originally grown (*e.g.* corn stover from maize), but it also applies to any biomass-derived by-product for which supply greatly exceeds demand (*e.g.* glycerol from biodiesel). Nearly all waste streams currently have some value, for instance agricultural waste is used as a soil improver in the fields, but the future looks toward obtaining a higher value from them. Most biorefineries utilise the available feedstocks without upstream concerns, which are grouped into different categories such as lignin, carbohydrates, proteins and triglycerides (from fat and oils). Lignin, from woody biomass, can be used as a fuel but it is also the only large volume renewable that comprises aromatics. Bulk chemicals can be obtained after hydrolysis of the carbohydrate residues to their monomers to obtain, among others, bioethanol, butanol and lactic acid from fermentation, or furfural and 5-hydroxymethyl-furfural produced from acid-catalysed dehydration of pentoses. Proteins residues (*e.g.* distillers grains) are valorised as animal feed, but the ideal scenario would be to isolate the non-essential amino acids as chemical feedstocks.[1,9]

Rapid industrialisation across the developing world has led to a number of adverse effects on the environment. The severe dumping of plastics into water courses has fouled our oceans, rivers, lakes and estuaries; while excessive nutrient run-off from intense agricultural activity has prompted unchecked and persistent micro- and macroalgal (seaweed) blooms, which are potential feedstocks for 3G biorefineries. Harmful algal blooms (HABs) can produce anoxic zones, kill wildlife and produce toxic compounds responsible for death, illness and/or a direct restriction of commercial activities such as fishing and tourism.[10] Whilst they are relatively chemically inert, there is an increasing body of evidence for the detrimental impact of plastics on aquatic wildlife and trophic food webs, in addition to the obvious impact of detritus on the aesthetics of the environment.[11] It is of particular relevance that these vastly different anthropogenic pollutants often become entwined with the complex organic matrix associated with HABs.

Sustainable feedstocks can include energy crops grown on marginal land, agricultural and forestry residues, municipal solid waste, and other novel feedstocks such as algae and other aquatic plants and microbial biomass.[12] However, the variability of the quantity and composition reduces the technological and economic feasibility of potential value conversion processes. The types of biomass sources not falling into the categories mentioned above, but commonly used in biorefineries, are the organic fraction of Municipal Solid Waste (MSW), manure, sewage sludge, wild fruits and crops, proteins and residues from fresh fruit and vegetables or food waste (FW). The physical and chemical characteristics of these wide spectrum biomass resources vary largely and, therefore, they are more suited for systems that can recover the potential of the organics as a whole. Examples might be anaerobic digestion (AD) to produce biogas or hydrothermal systems to produce a crude oil substitute.

1.3. Biologically based conversions for complex biomass

AD is based on a mixed microbial biotechnology (MCB) that originated from the waste treatment field. Compared with pure culture-based industrial biotechnology, MCB does not require sterilisation, has a high adaptive capacity, can use mixed substrates thanks to microbial diversity and it is possible to operate it as a continuous process.[13] The biogas produced during AD is the final product of a long chain of reactions, including hydrolysis of polymers to monomers and oligomers, oxidation of these products during primary fermentation generating volatile fatty acids (VFAs), lactate and ethanol, as well as hydrogen and carbon dioxide, and a final secondary fermentation. All these biochemical reactions are enclosed within the carboxylate platform in which carboxylates are the intermediates or final targets for conversion of biomass to chemicals or biofuel.[14]

Hydrogen generated during acidogenesis (primary fermentation) has the highest energy content per unit weight of any known fuel (143 GJ tonne^{-1}) and is the only fuel that does not contain any carbon. In comparison to the combustion of methane, which is generated by AD, hydrogen combustion is considered a cleaner technology as it does not involve carbon dioxide.[7] However, methane can be generated alongside the hydrogen in low quantities either by typical acetoclastic methanogens (from acetate) or by hydrogenotrophic organisms (from H_2 and CO_2). A mixture of hydrogen and methane is known as biohythane (46–57% H_2, 43–54% CH_4, 0.4% CO_2) and it is a perfect fuel owing to its cleaner nature than methane, high fuel efficiency, improved heat efficiency and ability to make engines easy to ignite with less input energy. The value of these technologies does not rely only on the biogas obtained anaerobically, but also on the other biobased products present in the fermentation broth which may have a high commercial value in the market.[7,15]

The different types of products obtained within the carboxylate platform will depend on the substrate composition and operational conditions, which will control the syntrophy between different organisms and determine the final microbial community.[16,17] In all microbial fermentations where organic carbon is both the electron donor and acceptor for the redox reactions, methane presents the lowest Gibbs energy change and thus a homogeneous end-product will be generated irrespective of the substrate.[18] Thus to switch from biogas to biochemical production in MCB, methane production must be inhibited by working at suboptimal AD conditions. Preliminary studies are required of diverse organic streams as composition variability will modify the synergy between organisms, and thus the biological reactions and final product.

1.4. Hydrothermal based conversions for complex biomass

Algal biomass and plastics each represent different, yet significant, opportunities and problems from a remediation perspective. Algal biomass requires extensive drying to obtain a suitable feedstock for processing, rendering direct combustion or pyrolysis routes, normally used for plastic waste, uneconomic.[19] However, suitable conversion technologies for wet biomass, such as anaerobic digestion (AD) and fermentation, are not able to convert the energetically rich plastics. Hydrothermal liquefaction (HTL) offers an interesting opportunity for simultaneous processing of heterogeneous organic material. In HTL, biomass is processed wet, with solid loadings of 5–20%, at 280–350 °C under pressures of up to 180 bar. HTL results in four product phases, including a bio-crude oil that can be processed into fuels and chemicals similarly to crude oil, an aqueous phase containing nitrate and phosphate based micronutrients, CO_2, and a solid residue containing the inorganic elements and further carbon.[20]

1.5. A multi-product integrated biorefinery for organic waste

Biorefineries must be developed using process design, technology integration, and analysis of the sustainability and economics. We are working towards a biorefinery that will produce different products from various sustainable feedstocks by integrating upstream, processing and downstream stages, taking into account both biological and thermo-chemical technologies and their integration for maximum recovery. Various wild yeast strains have been applied to either produce lipids from depolymerised lignocellulose[21] or microbial palm oil substitutes.[22] Bacterial communities have been investigated for the production of antimicrobials, using the intermediate products of anaerobic digestion and bioethanol obtained from side-streams.[16,23] Fuel precursors as a bio-oil have been obtained from physical treatment of algal biomass.[24] Although recently advances in biobased feedstock processing have been achieved, a holistic perspective is still required to allow efficient technology integration, which is justified by sustainability and economic analysis. To achieve this, previous knowledge of the substrate behaviour in each technology is required. Evaluation of the processes and their integration can be achieved thanks to system modelling. Process system analysis tools enable integration of multi-step processes for maximisation of the energy and resource recovery efficiency, and mitigation of the emissions, waste and cost,[25] achieving an interdisciplinary approach for sustainable feedstock valorisation. In this work, we evaluate two separate biological and thermochemical systems for organic waste streams, the latter is used with plastics as the main impurities in the feedstock. The final aim was to unravel the energy and chemical potential of different substrate compositions to be applied to an integrated biorefinery concept.

2. Results and discussion

2.1. Biological conversion of food and liquid waste to biogas and chemicals

An anaerobic digestion process is applied for complex biomass conversion into methane. Its microbiome containing hydrolisers, fermenters and methanogenic organisms in syntrophy allows the oxidation of variable substrates to a single product. High concentrations of VFAs or a low pH cause inhibition of the methanogenesis. Limiting methane production increases the possibility of recovering chemicals and alternative fuels from these systems. Under conditions of overloading and in the presence of inhibitors (*e.g.* free ammonia, high salts), the methanogenic activity cannot remove hydrogen and volatile organic acids as quickly as they are produced. The result is accumulation of acids, depletion of the buffering capacity and depression of the pH to levels that, sometimes, also inhibit other fermentation processes.[26] Several fermentation processes for carboxylic acids have been pursued at a low pH (*i.e.* below or close to the lowest pK_a) although higher titers were achieved at neutral pH, which in carboxylic acid production is between pH 6–8.[27] At a pH of around 6, short carboxylates can be recovered as bulk chemicals, and further conversion of them is possible to medium-chain VFAs with a higher value[28] or hydrogen can be simultaneously generated as an alternative fuel. Therefore, evaluation of the fermentation products and biomethane potential by adjusting the load and pH within MCB allows determination of the maximum energy and chemical production and composition. Here we investigated the influence of the food to microorganisms ratio during anaerobic conversion of food waste (FW) and determined the acidification potential (AP) and biochemical methane potential (BMP) of various organic substrates.

2.1.1. Influence of inoculum concentration. A substrate concentration as chemical oxygen demand (COD) of 5 g COD L^{-1} of FW was mixed with different concentrations of an AD mixed microbial inoculum to obtain various food to microorganisms (F/M) ratios of ∞, 10, 5, 1 and 0.5, corresponding to a volatile solids (VS) concentration of 0, 0.5, 1, 5 and 10 g VS L^{-1}, respectively. A F/M ratio of 0 was used as an inoculum control with 10 g VS L^{-1} and no substrate. Biogas production was only significant for F/M ratios of 1 and 0.5, steadily increasing at a rate of 112 and 215 mL d^{-1}, respectively, during the seven days of incubation (Fig. 2a). As for the rest of the conditions, the pH decreased below 6 during the first day of fermentation (data not shown). About 100 mL of biogas was produced for F/M ratios of 5 and 10 during the first day, probably H_2 from acidification, and the rates decreased to zero after that day. For F/M $= 0$, biogas (60% CH_4) was produced at a rate of 27 mL d^{-1}, a product of conversion of the soluble substances

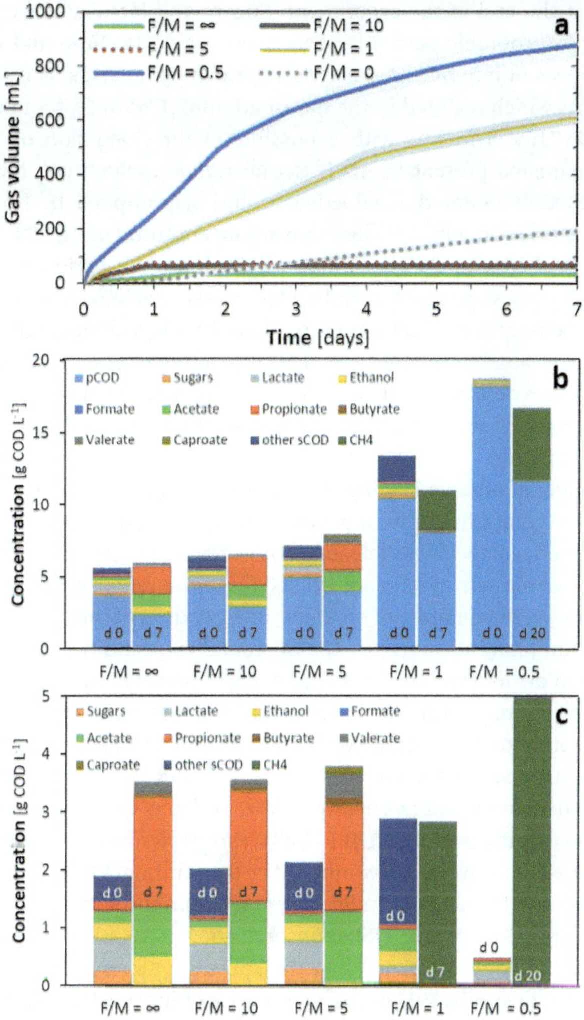

Fig. 2 (a) Gas production during the first few days of anaerobic fermentation; (b) total COD composition at the start and end of each experiment; (c) soluble COD composition at the start and end of each experiment. Tests were carried out at various food to microorganism ratios (F/M, g COD g^{-1} VS). F/M = ∞ refers to only substrate (5 g COD L^{-1}) while F/M = 0 to only inocula (10 g VS L^{-1}).

and auto-digestion of the sludge. Regarding the total COD supplied in the tests, increasing the inoculum concentration not only increased the particulate organics (pCOD, Fig. 2b) but also other soluble organics inherent in the sludge. For the fermentations with low inoculum, the total COD was similar at the start and end of the test, while this balance was not present for F/M ratios of 1 and 0.5 probably due to losses through H_2 and CO_2, which were not accounted during the experiment.

In terms of soluble COD, using the substrate alone or F/M ratios of 10 and 5 solubilised a fraction of the pCOD as can be observed from the increase in soluble substances at the end of the experiment. Sugars and lactic acid were primarily converted into propionic acid (51–54%), acetic acid (24–36%) and ethanol (2–15%) with traces of butyric (2.5–5%) and valeric (3–11%) acids. Not surprisingly, the conditions which resulted in the lowest ethanol (F/M of 5), produced 3.5% of caproic acid. This is in line with a possible chain elongation of VFAs using common organisms present in AD (*e.g. Clostridium*), which can reverse the β-oxidation reaction under the reductive conditions supplied by hydrogen and ethanol as the electron donor.[29] Therefore, a concentration of 1 g VS L^{-1} (F/M of 5) was chosen as the optimal concentration to evaluate the acidification potential (AP) of the substrates, as minimal carbon losses were detected from the total COD while this inoculum concentration even allowed biological upgrade of the short VFAs into more valuable chemicals such as caproic acid. For a F/M ratio of 1, the main loses were detected as being due to biogas production, while optimal methane production was obtained with a F/M of 0.5 in line with values in the literature.[30]

2.1.2. Effect of organic loading. A F/M ratio of 5 was found to be optimal for the AP when comparing the inoculum concentrations. However, microbes inherent in the feedstock when dealing with organic wastes also provide hydrolytic and acidogenic activity when incubated for fermentation purposes (F/M = ∞, Fig. 2). Therefore, the quantity of organics, independent of the F/M ratio provided by the inoculum, may also increase the microbial community and modify the AP outcome when dealing with waste streams. Fig. 3 presents various combinations of organic loading rates and the inoculum, all with F/M ratios equal to or above 1, with the exception of the control inoculum experiment (this time with 1 g VS L^{-1}).

As initial AP tests indicated, F/M ratios of 1 did not lead to fermentation products; at minimum loadings nearly no VFAs, ethanol or biogas were produced, while at higher loadings of 5 g COD L^{-1} all the organics were converted to biogas (Fig. 3a). These results were also confirmed by the minimal VFA production rate or even negative rate from day 3, indicating consumption of the VFAs in methane production (Fig. 3b). From a F/M ratio of 2 and above, all the tests produced similar or higher concentrations than the substrate alone (F/M = ∞, 5(0)). We observed that with similar F/M ratios, an increase of inocula boosted the total VFA production and reduced the presence of propionate by 10–30% in the final composition (*e.g.* 5(0.5) to 10(1) or 5(0.75) to 7.5(1)). Ethanol predominated after the 7-day fermentation for the tests with a maximum organic loading over the inoculum; however, an ethanol peak of 0.7 g COD L^{-1} was observed at day one and depleted afterwards for the tests with 5(0.75), 5(1) and 5(2). Lactic acid was consumed only, from the substrate, except for the test with 10(1) wherein the lactic acid concentration peaked at 0.5 g COD L^{-1} after one day and was depleted during consecutive days. Ethanol and lactate consumption, as electron donors for chain elongation reactions, should have increased the longer chain VFAs. This situation was only correlated with results from 10(1) and 5(1), with the production of caproic acid. Finally, the tests with higher inoculum concentrations but a F/M ratio over 2 provided comparable or higher VFA concentrations than the rest and an improved VFA production rate for the first two days of fermentation as well as during the last day (Fig. 3b). Although higher loading rates were providing an improved VFA concentration, the conversion yield (Y_{VFA}), which stands for the conversion of total COD to VFAs, was reduced below 50% in such cases. A maximum yield of 70% was obtained with 5(1), in line with conversions from the

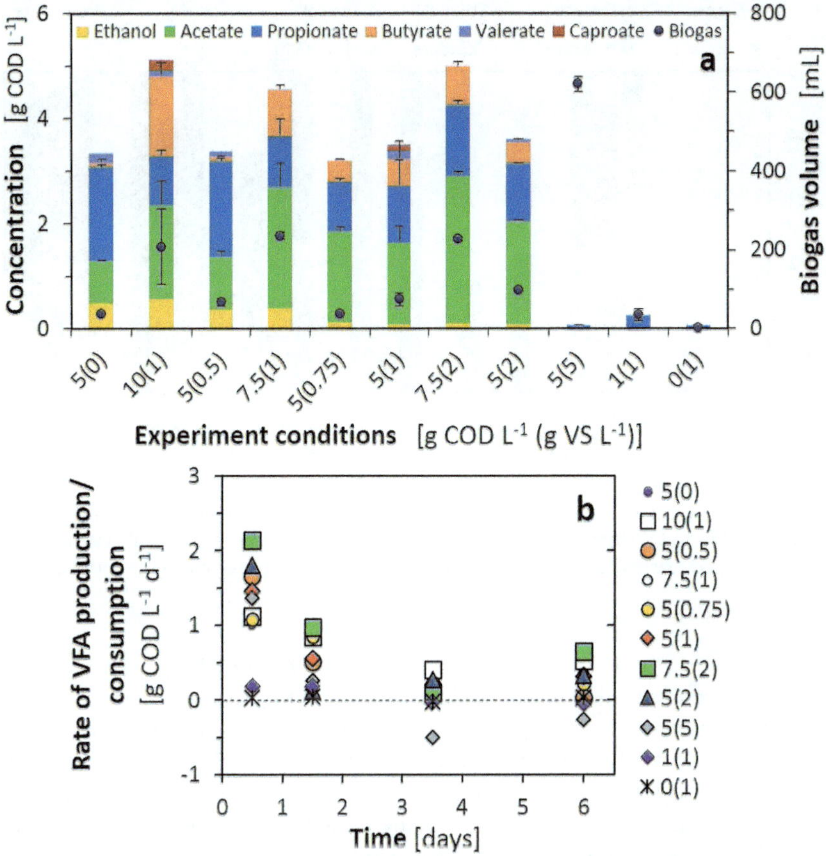

Fig. 3 (a) VFA (left) and biogas (right) production and composition using various 7-day fermentation combinations; (b) rate of VFA production or consumption during the 7-day fermentation tests with various loads and F/M ratios. Experimental conditions indicated are the organics concentration from FW to inoculum [g COD L^{-1} (g VS L^{-1})].

literature obtained for sugar based substrates converted to methane or to VFAs.[17,31] While similar values for the VFA production rate and Y_{VFA} were obtained for 7.5(2) and 5(2), ethanol production, and thus possible further valorisation to long chain fatty acids, was limited. Therefore, optimal test conditions of F/M = 5 with 1 g VS L^{-1} were adopted for further AP tests.

2.1.3. Effect of substrate composition on the VFA and methane production. Optimised AP and BMP tests were carried out for six different liquid wastes (LW1–LW6) to determine the conversion potential to VFAs or methane. Fig. 4 presents the initial composition of the added substrate (5 g COD L^{-1}, except for LW2 which was a diluted substrate) and the end composition for both the AP and BMP tests.

LW2 presented a nearly non-detectable VFA production for the 7-day AP test but converted nearly all its initial COD into methane during the BMP test. This was due to an extremely slow hydrolysis process confirmed by a constant of 0.05 d^{-1} (Table 1). Thus a longer residence time of 30 days allowed substrate degradation to occur. Similar profiles with low amounts of VFA products for the AP and

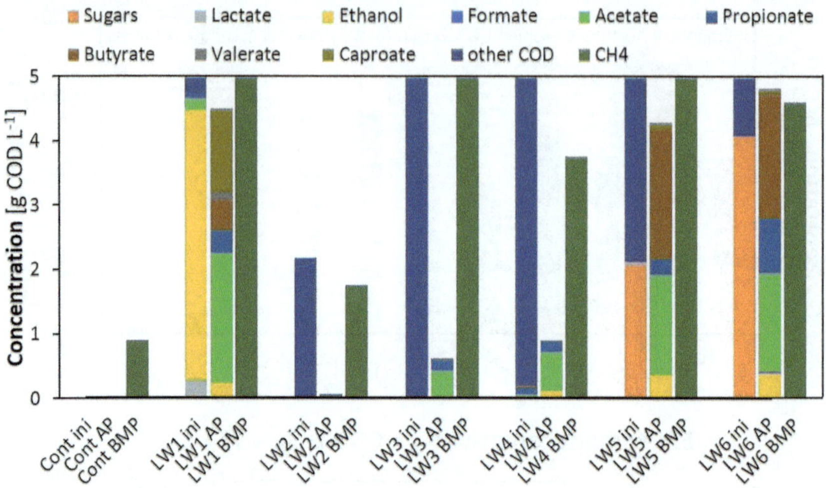

Fig. 4 VFA and methane production, and the obtained compositions, from 7-day acidification potential (AP) tests and a 30 day biochemical methane potential (BMP) test using 6 different liquid wastes. Cont stands for control while ini stands for initial composition.

high yields for the BMP tests were observed for LW3 and 4, again with lower hydrolysis constants that would limit further fermentation steps.

Substrates with a high proportion of ethanol or sugars in their initial composition (LW1, LW5 and LW6) presented the highest VFA production as well as Y_{VFA}, which was over 90% for the substrate mainly composed of ethanol. Ethanol oxidation to acetate did not only improve the increase in VFAs, but provided the required energy to initiate chain elongation reactions[28] and, thus, generate butyrate from acetate and furthermore caproate from butyrate. Nearly no biogas, only 18 mL, was produced from LW2, consisting of 4% CO_2 and the rest gases other than methane, supposedly hydrogen. This would have also enhanced the reductive conditions required for chain elongation. In the case where sugars were present within the initial substrate, the monomers had to undergo acidification to produce the required ethanol and acetate, accompanied with maximal

Table 1 pH variation, efficiency and kinetic parameters of the AP and BMP tests for the different substrates tested

	Acidification potential (AP)				Biochemical methane potential (BMP)			
	ΔpH	DA g COD_{VFA} g^{-1} COD_{fed}	Y_{VFA}%	R^0_{VFA} g COD_{VFA} L^{-1} d^{-1}	ΔpH	SMP mL CH_4 g^{-1} COD_{fed}	Y_{BMP}%	K_h d^{-1}
FW	−2.00	0.59	102	1.45	−0.3	377	108	0.36
LW1	−2.71	0.78	94	1.16	−0.2	407	116	0.64
LW2	0.29	0.00	1	−0.03	0.05	348	100	0.05
LW3	−1.57	0.12	16	0.26	−0.65	462	132	0.24
LW4	−0.61	0.11	37	0.26	−0.4	264	75	0.46
LW5	−3.25	0.74	81	2.07	−0.5	373	107	0.56
LW6	−3.29	0.82	87	1.89	−0.5	320	92	0.57

biogas production other than methane (25% CO_2, 75% H_2), therefore chain elongation was only extended to butyrate during the 7-day fermentation.

Kinetic parameters obtained from both the AP and BMP tests corroborated that VFA production relies on the hydrolytic capacity of the substrate as well as the activity of the microbiome. Lower VFA rates were obtained for those substrates with lower hydrolysis constants (LW2–4). In such cases, the substrates should be pretreated to enhance the initial step of fermentation or only applied for biomethane production with larger retention times. For sugar based substrates, even relatively low hydrolysis constants would allow a high recovery of VFAs together with a potential for biohydrogen recovery. These systems also presented the highest pH decrease during fermentation, thus potential inhibitions should be taken into account. Finally, ethanol based substrates can also be considered for VFA recovery with the added value of production of longer chain VFAs.

2.2. Hydrothermal co-liquefaction (HTL) of algal and plastic wastes

While the hydrothermal co-liquefaction of plastics in water has been examined with lignocellulose,[32] no reports detail the effect of plastics on the HTL of micro- or macroalgae. To assess this application, we studied the microalgal species *Arthrospira platensis* (Spirulina) and the macroalga *Ulva* spp. (Ulva), both of which are common and problematic bloom formers in areas such as Vietnam. All samples were liquefied at 310 °C over 60 minutes. Under these conditions, an oil yield of 34% was obtained for Spirulina, while the liquefaction of Ulva produced only 7% biocrude (Fig. 5a). While this is substantially lower than with Spirulina, it is common for macroalgae to contain a higher polysaccharide and ash content, and correspondingly have lower oil yields. For example, other members of this genus have been reported to provide oil yields of 18–32% under similar conditions.[33,34]

Plastics such as polyethylene (PE) and polypropylene (PP) are more thermally stable than the algal biomass, and under the conditions tested both plastics failed to degrade when processed separately to the biomass. However, on co-liquefaction both PE and PP demonstrated significant degradation. For example, with addition of both PE and PP, the overall oil yield remains approximately stable for Spirulina, despite proportionally less biomass being present in the reaction mixture, where the addition of both plastics increased the bio-crude yield dramatically for Ulva. Presumably, as the biomass begins to degrade, these secondary products impact on the thermal stability of the polymers, which then react and decompose. The hydrocarbon polymer subsequently becomes a hydrogen donor, and this hydrogen can stabilise radicals formed during biomass decomposition and prevent recondensation to solid residues. This effect has been observed from co-liquefaction of lignocellulose with HDPE, for example.[32]

The bio-crude fraction was analysed using elemental analysis (Fig. 6a). On addition of PE or PP to the Spirulina biomass, the C : H ratio decreased compared to the bio-crude from the pure Spirulina. This effect was even more pronounced at higher plastic loadings. A similar trend was also observed for the bio-crude produced from Ulva. The N content was also reduced in the bio-crude fractions with as little as 3.3% being observed. Nitrogen tends to be present in aromatic heterocycles and as such requires extensive hydrotreatment for removal from the

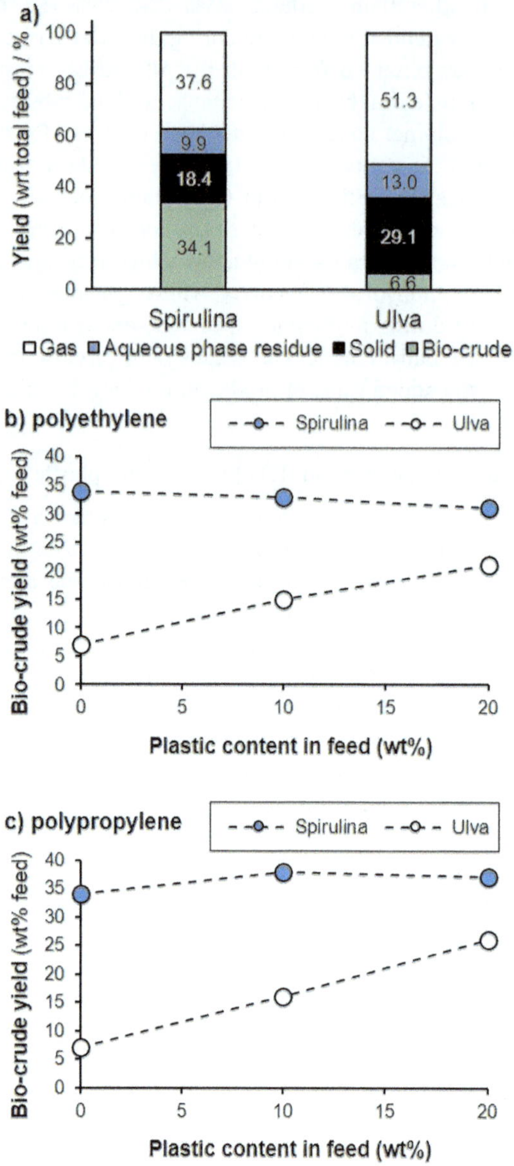

Fig. 5 HTL of Spirulina and *Ulva* spp. (310 °C, 60 min): (a) product mass balance of the biomass; (b) bio-crude yield with increasing PE content; (c) bio-crude yield with increasing PP content.

resulting crude upon chemical upgrading. Bio-oils containing a lower nitrogen content are therefore significantly easier to process.

GC-MS analysis of the bio-crude fractions demonstrated that the majority of the lighter fraction was composed of aromatic compounds, fatty acids and nitrogen containing heterocycles. While this did not change substantially for the co-liquefaction products, an increasing hydrocarbon content (C_{10}–C_{22}) was observed with the addition of PP and PE. For the liquefaction of Ulva, a significant

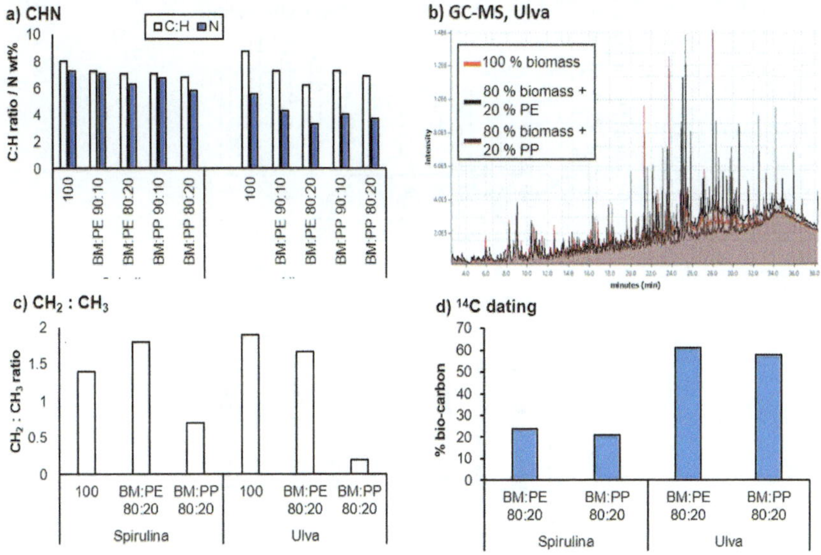

Fig. 6 Analysis of the bio-crude fractions: (a) the carbon : hydrogen ratio and nitrogen content of the bio-crude with a varied plastic content; (b) GC analysis of the bio-crude derived from Ulva; (c) the ratio of CH_2 to CH_3 moieties estimated from the 1H NMR; (d) % fossil carbon in the bio-crude.

proportion of additional saturated hydrocarbons were observed around the C_{20} range for both PP and PE liquefaction (Fig. 6b). This strongly suggests that the plastic is partially fragmenting and partitioning into the bio-crude fraction. ^{1}H NMR analysis of the biocrudes demonstrated that the crude is relatively similar when PE is introduced, with approximately the same CH_2 : CH_3 ratios. However, following addition of PP, a larger proportion of CH_3 groups were present, suggesting either deposition of the PP polymer chain or the production of shorter chain moieties from the biomass. To estimate the total carbon content from the biomass, ^{14}C dating of the bio-crudes was undertaken, and it was demonstrated that between approximately 21–61% of the carbon in the bio-crude came from the biomass source, for the 20% polymer loadings. This equates to approximately 10–15% of the available carbon in the Spirulina biomass depositing into the bio-oil and between 45–55% of the carbon available in Ulva depositing into the bio-oil.

The solid residue fraction was also increased by the addition of plastic, with higher loadings of plastic producing higher yields for Spirulina and Ulva (Fig. 7a), though this was far more pronounced for the macroalgae. The additional solid residues showed an increased carbon and decreased nitrogen content (Fig. 7b and c), both factors suggesting that some of the plastic waste was depositing in this phase, with more plastic available to deposit with the Ulva sample due to lower deposition in the bio-oil.

This exciting preliminary study demonstrates that the co-liquefaction of algae and plastics has significant potential for remediation of complex organic pollutants. Indeed, the additional plastics result in higher conversions to bio-crude, which was generally of a higher quality than the biocrude produced from algal biomass alone.

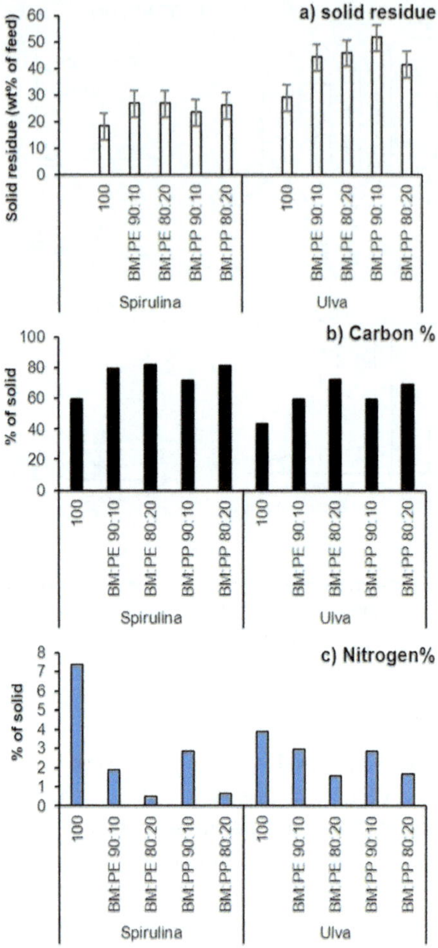

Fig. 7 Analysis of the solid residues from the HTL of Spirulina and Ulva (310 °C, 60 min): (a) solid residue yield with increasing plastic loading; (b) carbon % from the elemental analysis; (c) wt% nitrogen.

2.3. Organic waste biorefinery system integration

A biorefinery is a facility for the sustainable conversion of biomass and waste feedstocks, through the integration of physical, chemical, biochemical and thermochemical processes, into multiple products.[35] The analogy to today's crude oil refineries suggests that adoption of process system engineering principles, such as feedstock fractionation, multiple product portfolios, process flexibility and process integration, should be applied for biorefinery concepts to achieve high efficiency levels. In this section, an integrated biorefinery system is devised based on the process technologies investigated in this work for the co-processing of wastes. The integrated co-processing of different waste streams in a biorefinery fashion has shown potential for improved economics and also as a technological solution towards a circular economy.[36]

Fig. 8 shows the system integration for a waste biorefinery concept combining the processes investigated in this paper to produce platforms for biofuels or

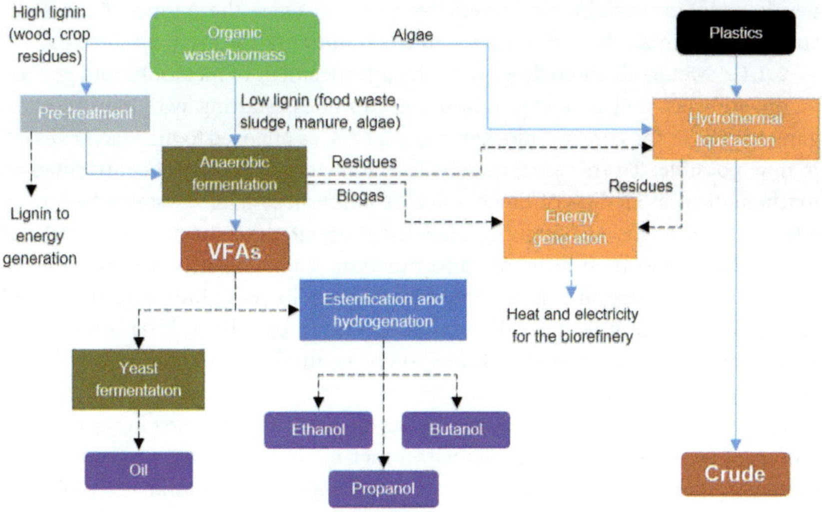

Fig. 8 System integration for a waste biorefinery concept combining biochemical and thermochemical processes to produce platforms for biofuels or chemical production.

chemical production. Such an integrated system provides feedstock flexibility, allowing the ability to use any organic waste or biomass as the input and even co-process plastic wastes. Low lignin waste and de-lignified waste can be processed by anaerobic fermentation to produce short chain volatile fatty acids, as shown in Section 2.1. Although the HTL process was investigated for algae biomass and plastic waste in Section 2.2, this process can also utilize any solid residual streams from the organic waste processing, thus yielding more bio-crude product. Furthermore, solid residues from the HTL process can be used together with biogas from the anaerobic fermentation process for energy generation and to supply heat and electricity for operating the core biorefinery processes. This systematic integration of biochemical and thermochemical processing routes can potentially increase the carbon and energy recovery efficiencies while reducing residual streams and emissions. These efficiency gains are especially important for the sustainability of biorefineries.

Fig. 8 also shows the integration of alternative VFA conversions into other products. Biochemical processing by yeast can produce a palm oil substitute, which can be sold for cosmetics or to the food industry or for biodiesel production. The chemical synthesis route involves esterification and hydrogenation to produce mixed alcohols including ethanol, propanol and butanol, which can be sold as biofuels or as platform chemicals. The chemical synthesis route would require finding a way to supply the hydrogen required. Thus, another integrated route would be needed, possibly *via* anaerobic fermentation, which could be tuned to produce VFAs and hydrogen, or *via* steam reforming of biogas or gasification of solid residues. The selection of the best integration alternatives for a sustainable waste biorefinery would require extensive analysis and optimisation at all levels. Fig. 9 shows a systematic framework for biorefinery process design and integration. The figure also shows the goals at each level and the research areas and tools needed for a truly multidisciplinary approach to accelerate the development of sustainable biorefinery systems. Once the nature of the feedstock(s) is defined, the various steps in the framework involve the following:

2.3.1. **Metabolic modelling.** With the advancement of metabolic and genetic engineering as well as computational capabilities, starting with the microorganism cells as the core of a biochemical process, as opposed to the reactor vessel, is now possible. To this end, metabolic engineering has largely contributed to mathematical modelling of biochemical reaction networks.[37] Systems biology is advancing in terms of obtaining knowledge about the behaviour of microbial communities and their structure and function. Therefore it is now possible to understand the relation between microbial metabolisms, culture conditions and productivity to optimise microbial production in a biorefinery.[38] This will allow synergistic tuning of mixed cultures such as in the anaerobic fermentation process shown in this paper.

2.3.2. **Process system simulation.** VFAs are generally obtained in the reaction effluent in a diluted form together with other by-products, water and unreacted biomass. To obtain a marketable product, an appropriate combination of separation units (for filtration or centrifugation, or flash separators, distillation columns or liquid–liquid extraction columns, among others) is required. Process models are used in simulations to analyse the performance of the process integrated as a whole (reaction and separations) so that it can be optimised. The resulting mass and energy balance values are the basis for the next levels of analysis and optimisation.

2.3.3. **Process integration.** It is important that energy and material inputs are used as efficiently as possible to offset fossil energy needs and greenhouse emissions. This is a standard step in chemical process design and pinch analysis can be employed to target energy recovery and make the biorefinery more energy efficient. Another important aspect is integration of the various processes in the biorefinery through material stream exchanges.[39] For example, some processes can be sources of CO_2 (*e.g.* fermentation) and others can be sinks of CO_2 (*e.g.* algae cultivation) thus balancing the sources and sinks can provide higher efficiency levels.

2.3.4. **Economic analysis.** This is a well-established step towards selecting a process design by evaluating economic performance indicators such as the payback, net present value, minimum selling price and economic margin potential. Traditional life cycle costing as well as a value analysis method for economic margin analysis of biorefinery processes could be used.

2.3.5. **Environmental impact analysis.** An environmental impact analysis is needed to select the process alternatives that have the lowest potential for causing damage to the environment, ecosystems and human health. Traditional tools include carbon footprint and water footprint analyses and the more holistic life cycle assessment (LCA). Trade-offs between the environmental and economic objectives of a biorefinery system may arise during the analyses, thus to support decision making a simultaneous economic value and environmental impact (EVEI) analysis can be used.[39]

The technologies studied in this paper have the potential to enable high productivity and conversion of waste to valuable products in a biorefinery conceptualised through system integration and through employing a holistic multidisciplinary approach for system optimisation. This approach could open new possibilities for biorefining waste to produce biofuels and chemicals in a sustainable manner.

3. Experimental

3.1. Biological anaerobic conversion

3.1.1. Substrates and inoculum. The inoculum from full scale anaerobic digesters (3.14 g TS L^{-1}; 2.19 g VS L^{-1}; 33 g COD L^{-1}) and the food and liquid waste (Table 2) were supplied by GENeco (Wessex Water, UK).

3.1.2. Acidification potential (AP) and biochemical methane potential (BMP) tests. Batch experiments were carried out in 500 mL Schott® bottles immersed in a water bath at 35 °C. The bottles were topped with a rubber stopper containing three ports; one of them was connected to an automatic biogas counter (Bioprocess control AMPTS II), the second port was used for sampling and the third contained the vertical stirrer controlled by the system. The required amounts of substrate and inoculum were added according to the desired proportions (F/M) and the COD and VS concentrations, and topped up with tap water until a working volume of 400 mL was reached. The pH was initially adjusted to neutrality (7–7.5) with 2 M HCl or NaOH. For the AP tests, various F/M ratios and controls without the substrate were tested in triplicate over 7 days, sampling at days 1, 2, 4 and 7 of the experiment. The optimised AP tests consisted of F/M = 5 with 5 g COD L^{-1} of substrate and 1 g VS L^{-1} of inoculum. The degree of acidification was calculated according to Bengtsson *et al.*[40] and the VFA yields according to Scoma *et al.*[17] For the BMP tests, fixed amounts of 5 g COD L^{-1} of the substrate and 10 g VS L^{-1} of the inoculum (F/M = 0.5) were used for each substrate in triplicate, including controls for the inoculum. The tests were carried out over 30 days and sampling was performed at the end of the experiment. Specific Methane Production (SMP) and the kinetic parameters were calculated according to Angelidaki *et al.*[30]

3.2. Hydrothermal liquefaction (HTL)

3.2.1. Substrate. *Ulva lactuca* Linnaeus (Ulva) was collected from Xom Con, Nha Trang, Khanh Hoa province, Vietnam on June 10, 2016. Prior to analysis and conversion, the macroalga was freeze-dried and milled to <1400 μm in diameter. The Spirulina *platensis* strain, classified as *Arthrospira* (Spirulina) *platensis*, was obtained from Hidumi Pharma Green Science Joint – Stock company, Vietnam, and used without subsequent purification.

3.2.2. Reactors. Batch bomb-type reactors were fabricated according to literature precedent using stainless steel Swagelok® tube fittings.[41] The reactor

Table 2 Physico-chemical characteristics of the municipal organic waste

	pH	COD_{TOT} g COD L^{-1}	Sugars g COD L^{-1}	Lactate g COD L^{-1}	Ethanol g COD L^{-1}	VFA g COD L^{-1}	TS%	VS%
FW	4.6	157 ± 11	7.80	11.14	7.65	8.78	10.48 ± 0.03	9.56 ± 0.06
LW1	4.2	45 ± 5	0.03	2.48	35.91	2.16	1.09 ± 0.01	0.86 ± 0.01
LW2	7.5	7 ± 2	0.01	0.00	0.00	0.04	1.82 ± 0.01	0.69 ± 0.01
LW3	12.6	294 ± 1	0.63	0.44	0.00	2.82	12.99 ± 0.23	12.42 ± 0.24
LW4	6.2	87 ± 14	0.00	0.00	0.00	2.82	3.58 ± 0.01	2.91 ± 0.06
LW5	2.7	196 ± 24	78.66	1.26	0.00	0.00	17.47 ± 2.39	15.56 ± 4.81
LW6	2.7	114 ± 1	87.68	0.00	0.00	0.00	8.32 ± 0.07	7.68 ± 0.26

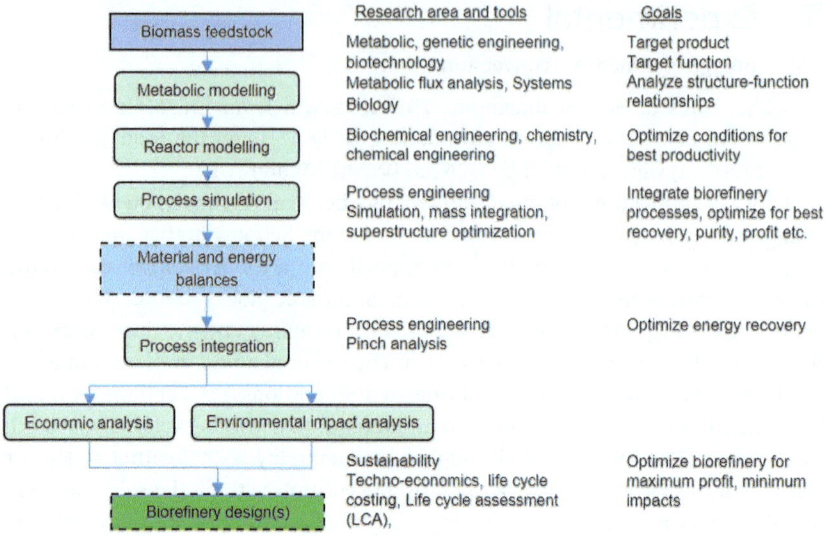

Fig. 9 Systematic framework for process design and integration for an organic waste biorefinery.

body consisted of a length of 1/2″ tubing capped at one end, and connected at the other to a pressure gauge, thermocouple, and needle valve. The total internal volume of the reactors was *ca.* 9 mL.

3.2.3. HTL procedure. An adapted reaction procedure, based on previous studies, was followed.[24] In a typical reaction, the reactor was loaded with 0.5 g of the total solids (made up of biomass and 0–20% plastics) and 5 mL of freshly deionized water. The reactor was pressurised to 30 bar with compressed air and heated within a vertical tubular furnace set to 400 °C until the specified reaction temperature was reached (310 °C, ±10 °C, 60 min), and then removed from the furnace and allowed to cool to room temperature.

After cooling, the pressure was released *via* the needle valve. Following this, the aqueous phase was decanted from the reactor contents and filtered through a filter paper pre-dried overnight at 60 °C. The product yield in the water phase was determined by leaving a 2.5 mL aliquot to dry in a 60 °C oven overnight, and scaling the residue yield to the total aqueous phase mass.

To separate the remaining bio-crude oil and char phase, the reactor was washed repeatedly using chloroform until the solvent ran clear, which was then filtered through the same filter paper used to separate the aqueous phase (after drying for a minimum of 1 h). The filter paper and collected char were washed thoroughly with chloroform to remove all remaining bio-crude. The filtrate was collected and the solvent removed *in vacuo*. The char yield was calculated from the mass of the retentate collected on the filter paper after drying overnight in an oven at 60 °C.

Three repeat HTL runs using Spirulina with no additional plastic were carried out to determine the standard deviation in the mass balance values under the conditions examined.

3.3. Analysis

Total and volatile solid values were determined using gravimetric standard 2540G.[42] The total and soluble COD were analysed using colorimetric kits (LCK014, Hach Lange®). HPLC (1260 Infinity, Agilent) was used to determine the sugars, VFAs and ethanol present with an Aminex® HPX-87H column (Bio-Rad) and a refractive index detector, using 5 mM H_2SO_4 as the mobile phase at 35 °C and 0.6 mL min^{-1}. GC (7890A, Agilent) was used with an HP-PLOT/Q column at 35 °C for 5 min using He as the carrier gas, and FID for CH_4 and TCD for CO_2.

Elemental analysis was carried out externally at London Metropolitan University using a Carlo Erba Flash 2000 Elemental Analyser to determine the CHN content (elemental analyses were carried out at least in duplicate for each sample, and average values are reported).

Analysis of the biocrude was carried out using ^{1}H NMR spectroscopy and GC-MS. ^{1}H NMR spectroscopic measurements were carried out at 298 K using a Bruker AV400 spectrometer, operating at 400 MHz for ^{1}H. Typically, the samples were analyzed in $CDCl_3$ and the spectra were referenced to the residual $CHCl_3$ peak from the solvent (δ 7.26 ppm). GC-MS analysis was carried out using an Agilent 7890A Gas Chromatograph equipped with a CP-Sil capillary column (25 m × 0.250 mm internal diameter) and a He mobile phase (flow rate: 1.2 mL min^{-1}), coupled with an Agilent 5975C MSD. Approximately 50 mg of each sample was dissolved in 100 mL of hexane and 1 µL of each solution was loaded onto the column, pre-heated to 40 °C. This temperature was held for 1 minute, and then the temperature was increased to 250 °C at a rate of 10 °C min^{-1} and then held for 10 minutes.

^{14}C analysis was performed by Beta Analytic Inc. (Florida, USA) according to ISO/IEC 17025:2005.

4. Conclusions

Organic waste valorisation with complex and variable compositions is possible using anaerobic fermentation processes and/or HTL processes. To unravel the chemical and energy potentials during biological conversions, assessment of the substrate using BMP or AP tests is required, and the latter were optimised using a F/M ratio of 5 to allow all the chemical conversions to occur. Feedstocks presenting high hydrolysis kinetics are recommended to be used in fermentation instead of AD for recovering chemicals with added value. This study has also demonstrated that it is feasible to use HTL to convert opportunistic algal biomass as a feedstock, and that not only is there no need to separate out plastic detritus from the organic matrix prior to processing but the plastic itself could improve the economic viability of the process. System integration and a holistic multidisciplinary approach could open new possibilities for biorefining waste to produce biofuels and chemicals in a sustainable manner.

Acknowledgements

MC is currently funded from a personal donation to the University of Bath by Sue and Roger Whorrod. FA is co-funded by H2020-MSCA-CO-FUND-2014, #665992,

MSCA FIRE: Fellows with Industrial Research Enhancement, and FA and SR are funded by EP/L016354/1, EPSRC Centre for Doctoral Training in Sustainable Chemical Technologies. We acknowledge the EPSRC for partially funding this work through an Institutional Global Challenges Research Fund award, and we acknowledge Wessex Water (Geneco) for providing some of the organic wastes and the Hidumi Pharma Green Science Joint – Stock company for proving the Spirulina.

References

1 F. Cherubini, *Energy Convers. Manage.*, 2010, **51**, 1412–1421.
2 B. Kamm and M. Kamm, *Appl. Microbiol. Biotechnol.*, 2004, **64**, 137–145.
3 D. Hoornweg and P. Bhada-Tata, *What a Waste. A global Review of Solid Waste Management*, World Bank, Washington DC, 2012.
4 C. S. K. Lin, L. A. Pfaltzgraff, L. Herrero-Davila, E. B. Mubofu, S. Abderrahim, J. H. Clark, A. A. Koutinas, N. Kopsahelis, K. Stamatelatou, F. Dickson, S. Thankappan, Z. Mohamed, R. Brocklesby and R. Luque, *Energy Environ. Sci.*, 2013, **6**, 426–464.
5 A. Parry, K. James and S. LeRourx, *Strategies to achieve economic and environmental gains by reducing food waste*, WRAP, London, UK, 2015.
6 P. S. Nigam and A. Singh, *Prog. Energy Combust. Sci.*, 2011, **37**, 52–68.
7 B. Sen, J. Aravind, P. Kanmani and C. H. Lay, *Renewable Sustainable Energy Rev.*, 2016, **53**, 547–557.
8 A. Scoma, S. Rebecchi, L. Bertin and F. Fava, *Crit. Rev. Biotechnol.*, 2016, **36**, 175–189.
9 C. O. Tuck, E. Perez, I. T. Horvath, R. A. Sheldon and M. Poliakoff, *Science*, 2012, **337**, 695–699.
10 Y. Kotaki, K. Koike, M. Yoshida, C. Van Thuoc, N. T. M. Huyen, N. C. Hoi, Y. Fukuyo and M. Kodama, *J. Phycol.*, 2000, **36**, 1057–1060.
11 M. Sigler, *Water, Air, Soil Pollut.*, 2014, **225**, 2184.
12 European Parliament Legislative resolution P8_TA PROV(2015)0100 of 28 April 2015 on Fuel quality directive and renewable energy directive.
13 R. Kleerebezem and M. C. van Loosdrecht, *Curr. Opin. Biotechnol.*, 2007, **18**, 207–212.
14 M. T. Agler, B. A. Wrenn, S. H. Zinder and L. T. Angenent, *Trends Biotechnol.*, 2011, **29**, 70–78.
15 S. V. Mohan, G. N. Nikhil, P. Chiranjeevi, C. N. Reddy, M. V. Rohit, A. N. Kumar and O. Sarkar, *Bioresour. Technol.*, 2016, **215**, 2–12.
16 M. Coma, R. Vilchez-Vargas, H. Roume, R. Jauregui, D. H. Pieper and K. Rabaey, *Environ. Sci. Technol.*, 2016, **50**, 6467–6476.
17 A. Scoma, M. Coma, F. M. Kerckhof, N. Boon and K. Rabaey, *Biotechnol. Biofuels*, 2017, **10**, 23.
18 R. Kleerebezem, B. Joosse, R. Rozendal and M. C. M. Van Loosdrecht, *Rev. Environ. Sci. Bio/Technol.*, 2015, **14**, 787–801.
19 M. Kröger and F. Müller-Langer, *Biofuels*, 2012, **3**, 333–349.
20 D. L. Barreiro, W. Prins, F. Ronsse and W. Brilman, *Biomass Bioenergy*, 2013, **53**, 113–127.
21 L. A. Sargeant, M. Mardell, K. M. Saad-Allah, A. H. Hussein, F. Whiffin, F. Santomauro and C. J. Chuck, *Eur. J. Lipid Sci. Technol.*, 2016, **118**, 777–787.

22 F. Whiffin, F. Santomauro and C. J. Chuck, *Biofuels, Bioprod. Biorefin.*, 2016, **10**, 316–334.

23 S. J. Andersen, P. Candry, T. Basadre, W. C. Khor, H. Roume, E. Hernandez-Sanabria, M. Coma and K. Rabaey, *Biotechnol. Biofuels*, 2015, **8**(1), 221.

24 S. Raikova, H. Smith-Baedorf, R. Bransgrove, O. Barlow, F. Santomauro, J. L. Wagner, M. J. Allen, C. G. Bryan, D. Sapsford and C. J. Chuck, *Fuel Process. Technol.*, 2016, **142**, 219–227.

25 J. Sadhukhan, K. S. Ng and E. Martinez-Hernandez, *Bioresour. Technol.*, 2016, **215**, 131–143.

26 I. Siegert and C. Banks, *Process Biochem.*, 2005, **40**, 3412–3418.

27 C. S. Lopez-Garzon and A. J. J. Straathof, *Biotechnol. Adv.*, 2014, **32**, 873–904.

28 L. T. Angenent, H. Richter, W. Buckel, C. M. Spirito, K. J. J. Steinbusch, C. M. Plugge, D. P. B. T. B. Strik, T. I. M. Grootscholten, C. J. N. Buisman and H. V. M. Hamelers, *Environ. Sci. Technol.*, 2016, **50**, 2796–2810.

29 L. T. Angenent, H. Richter, W. Buckel, C. M. Spirito, K. J. Steinbusch, C. M. Plugge, D. P. Strik, T. I. Grootscholten, C. J. Buisman and H. V. Hamelers, *Environ. Sci. Technol.*, 2016, **50**, 2796–2810.

30 I. Angelidaki, M. Alves, D. Bolzonella, L. Borzacconi, J. Campos, A. Guwy, S. Kalyuzhnyi, P. Jenicek and J. Van Lier, *Water Sci. Technol.*, 2009, **59**, 927–934.

31 J. De Vrieze, M. Coma, M. Debeuckelaere, P. Van der Meeren and K. Rabaey, *Water Res.*, 2016, **98**, 293–301.

32 X. Yuan, H. Cao, H. Li, G. Zeng, J. Tong and L. Wang, *Fuel Process. Technol.*, 2009, **90**, 428–434.

33 N. Neveux, A. K. L. Yuen, C. Jazrawi, M. Magnusson, B. S. Haynes, A. F. Masters, A. Montoya, N. A. Paul, T. Maschmeyer and R. de Nys, *Bioresour. Technol.*, 2014, **155**, 334–341.

34 N. Neveux, A. K. L. Yuen, C. Jazrawi, Y. He, M. Magnusson, B. S. Haynes, A. F. Masters, A. Montoya, N. A. Paul, T. Maschmeyer and R. de Nys, *Algal Res.*, 2014, **6**, 22–31.

35 J. Sadhukhan, K. S. Ng and E. Martinez-Hernandez, *Biorefineries and Chemical Processes: Design, Integration and Sustainability Analysis*, John Wiley & Sons, Chichester, UK, 2014.

36 S. Satchatippavarn, E. Martinez-Hernandez, M. Y. L. P. Hang, M. Leach and A. Yang, *Chem. Eng. Res. Des.*, 2016, **107**, 81–90.

37 G. Stephanopoulos, A. A. Aristidou and J. Nielsen, *Metabolic engineering: principles and methodologies*, Academic press, 1998.

38 V. Chubukov, A. Mukhopadhyay, C. J. Petzold, J. D. Keasling and H. G. Martín, *npj Systems Biology and Applications*, 2016, **2**, 16009.

39 E. Martinez-Hernandez, G. Campbell and J. Sadhukhan, *Chem. Eng. Res. Des.*, 2013, **91**, 1418–1426.

40 S. Bengtsson, J. Hallquist, A. Werker and T. Welander, *Biochem. Eng. J.*, 2008, **40**, 492–499.

41 J. Wagner, R. Bransgrove, T. A. Beacham, M. J. Allen, K. Meixner, B. Drosg, V. P. Ting and C. J. Chuck, *Bioresour. Technol.*, 2016, **207**, 166–174.

42 APHA, *Standard Methods for the Examination of Water and Wastewater*, American Public Health Association, Washington DC, USA, 19th edn, 2005.

Faraday Discussions

PAPER

Fractional conversion of microalgae from water blooms†

Yingdong Zhou, Linling Li, Rui Zhang and Changwei Hu *

Received 13th February 2017, Accepted 3rd April 2017
DOI: 10.1039/c7fd00065k

Fractional conversion of natural algae cyanobacteria from Taihu Lake was conducted. The raw Taihu Lake algae (TLA) and pretreated samples were pyrolyzed at 290 °C and 450 °C according to the TGA results. Extraction of lipids or saccharides from the TLA was performed as a pretreatment to obtain lipid extracted algae (LEA) or saccharide extracted algae (SEA). The total yields of bio-oil from fractional pyrolysis were 40.9 wt% from TLA, 42.3 wt% from LEA, and 48.5 wt% from SEA. From TLA, the major components of the bio-oil were fatty acids, amides and hydrocarbons (heptadecane) at 290 °C whereas those at 450 °C were phenols and C_{10}–C_{15} hydrocarbons. Following the lipid extraction, acids, amides and indoles accounted for a large proportion at 290 °C, while the main products obtained at 450 °C were phenols, indoles and pyrroles. It is worth mentioning that the yield of bio-oil from the LEA had increased, and the composition of the bio-oil was simplified. Moreover, the average molecular weight of the bio-oil obtained from LEA had decreased. Interestingly, the extraction of saccharides inhibited pyrolysis of the lipids, so the distribution of the bio-oil from SEA changed only a little. Fractional pyrolysis of pretreated microalgae not only increased the bio-oil yield but also improved the quality of the bio-oil.

1. Introduction

In recent years, a huge number of lakes have suffered from water eutrophication, resulting in an outbreak of water blooms. The thriving algae do harm not only to the coastal ecological environment, but also to the lake water quality and human health.[1,2] Thus urgent environmental management is needed. However, algae could also be regarded as an important resource for biofuels and a feasible alternative to limited fossil fuels, which is attributed to their lipid accumulation potential.[3–5] The natural algae from water blooms have a short growth cycle and can grow on non-arable land or waste water. Furthermore, the cultivation of

Key Laboratory of Green Chemistry and Technology, Ministry of Education, College of Chemistry, Sichuan University, Chengdu, Sichuan 610064, P. R. China. E-mail: changweihu@scu.edu.cn; Fax: +86-28-85411105; Tel: +86-28-85411105

† Electronic supplementary information (ESI) available. See DOI: 10.1039/c7fd00065k

microalgae can be coupled with wastewater bioremediation because algae have the ability to fix nitrogen, phosphorus and heavy metals from waste water.[3,6,7]

The conversion of microalgae to biodiesel *via* conventional transesterification requires algae containing a high fraction of lipids,[8] whereas natural algae from water blooms usually have a low lipid content (2.5–10.7% of total dry weight).[9–11] Consequently, conventional lipid extraction methods are not viable for low-lipid algae and thermochemical conversion is viewed as an option. Pyrolysis is a promising method for converting the whole microalgae into liquid fuels. Noncatalytic pyrolysis, known as direct pyrolysis, degrades microalgae at relatively high temperatures (250–600 °C) in the absence of any catalyst. The pyrolysis temperature has great influence on the distribution of the products and yields. A study on the pyrolysis behavior of blue-green algae blooms was conducted by Hu *et al.*[10] A maximum bio-oil yield of 54.97 wt% was obtained at 500 °C, with an O/C molar ratio of 0.16 obtained under optimum conditions. When pyrolysis of a lipid extracted residue of *Tribonema minus* was conducted in a fixed-bed reactor, the maximum liquid yield was 29.82 wt% at 450 °C, and the liquid product contained a high proportion of oxygen and nitrogen.[12] The bio-oil obtained from direct pyrolysis is complex, and contains hydrocarbons, aromatics and N,O-containing compounds,[13–15] causing a decrease in the HHV of the bio-oil and a poor biofuel quality.

In order to improve the quality and simplify the composition of the bio-oil, fractional conversion has been put forward, which is defined as a selective *in situ* conversion of biopolymers to desired products. Fractional pyrolysis of algae has been applied to convert the three main components, lipids, saccharides and proteins, separately through control of the pyrolysis temperature. In our previous work,[16] the fractional pyrolysis of cyanobacteria and model compounds at a series of temperatures was studied, and the results suggested that although pyrolysis fractionated by temperature was an effective way to enhance the bio-oil quality, further improvements were needed.

Cyanobacteria blooms occur from the eutrophication of Taihu Lake every year, producing cyanotoxins (algal toxins) which harm the surrounding plants, aquatic wildlife and humans with acute and chronic toxicity.[17,18] Collection, and thus removal, of such species greatly benefits the lake ecosystem. Moreover, if this waste could be converted to bio-fuels or useful chemicals, this would provide an economical way of both achieving biofuel production and curbing environmental pollution.

In this study, TLA was pretreated using lipid extraction and saccharide extraction firstly. Then the extracted algal residues were used as feedstocks for fractional pyrolysis at two temperatures, according to the TGA results. This was carried out in order to simplify the composition of the bio-oil. We expected to find information on the interactions between different components in the microalgae.

2. Experimental

2.1. Materials

The microalgae feedstock (Taihu Lake algae, TLA), the cyanobacteria, was collected from Taihu Lake located in Jiangsu Province, China during the summer of 2013. The raw material was washed with freshwater and then dried in sunlight. The dried algal sample was pulverized into a powder with a particle size of less than 80 meshes and dried again in an oven at 80 °C overnight. The algal powder was stored in a sealed glass bottle before use.

2.2. Fractionation of TLA by pretreatment

Pretreatment of the TLA included extraction of lipids and saccharides. Soxhlet extraction[19] and the Folch method[20] were used to extract lipids, while hot water extraction[21] was applied to extract saccharides from TLA.

2.2.1. Lipid extraction. Different solvents and methods were tested, and the method that provided the maximum yield of lipids was optimized and applied to extract the lipids effectively. In a representative run, 10 g of TLA powder was mixed with 100 mL of a solvent mixture (typical volume ratio of chloroform and methanol was 2 : 1) and heated at 70 °C in the Soxhlet extractor for 1 h of reflux. The solid residue was transferred into a centrifugal tube to consecutively extract the lipids using the Folch method 5 times. The extraction was carried out as follows: (a) the solid residue was mixed with the solvent mixture, (b) the mixture was sonicated for 5 min, (c) the mixture was centrifuged for 5 min at 6000 rpm, and (d) the supernatant was mixed with the previously extracted solution. The pooled solution was rotary evaporated to remove the solvent and calculate the lipid yield, and then the removed solvent was recycled for the next extraction experiment. Since the components of the co-solvent are not benign to human beings, the operation was carried out in a closed system and caution must be applied. The solid residue was dried in an oven at 80 °C overnight to obtain lipid extracted algae (LEA). The highest yield of lipids was 10.9 wt%.

2.2.2. Saccharide extraction. For extraction of the saccharides from TLA, 10 g of TLA powder was added to 150 mL of ultrapure water in a round-bottom flask and heated at 90 °C for 6 h of reflux. The mixture was centrifuged and the solid residue was extracted, again using the same method. The pooled extract solution was evaporated at 80 °C and weighed to obtain the saccharide yield. The solid remaining after saccharide extraction was oven dried at 80 °C overnight to obtain saccharide extracted algae (SEA). The highest yield of saccharides was 21.6 wt%.

2.3. Pyrolysis experiment

The apparatus used is similar to that of Li *et al.*[16] The process for the pyrolysis experiment is depicted in Fig. 1. For a typical pyrolytic reaction, 1.5 g of the sample was added into the top pool of the inner quartz tube, which was put into the middle of the outer tube, with a thermocouple put in the flush position at the bottom of the top pool to measure the reaction temperature. The pyrolysis reaction was performed in a tubular electric furnace and the reaction system was flushed with N_2 carrier gas at a flow rate of 40 mL min^{-1} to provide an inert atmosphere. A condenser immersed in an ice trap was applied to collect the condensates. NH_3 from the pyrolytic gas was absorbed using an absorption bottle linked to the condenser. In this study, TLA, LEA and SEA samples were pyrolyzed at 290 °C and 450 °C, according to the TGA results. The samples were firstly heated from room temperature to 290 °C with a heating rate of 10 °C min^{-1} and then kept at 290 °C for 2 h. The solid residue obtained was pyrolyzed at 450 °C according to the procedure above. The bio-oil was recovered by rinsing with acetone, while the gaseous products (except NH_3) were collected using a gas bag.

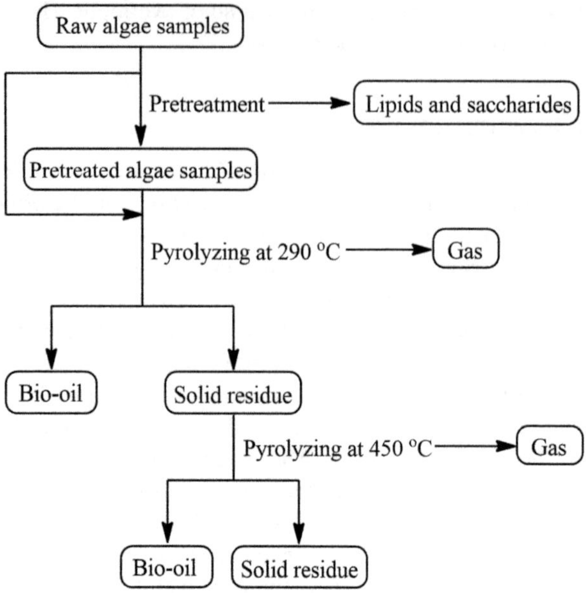

Fig. 1 Scheme of the process for the pyrolysis experiment.

In this study, the weight of the bio-oil was taken as the weight difference between the sum of the outer tube and the condenser before and after the pyrolysis. The weight difference of the inner tube before and after the reaction was recognized as the weight of the solid residue (SR). Besides, the weight of the gaseous products was calculated using the overall mass balance. The yields of the bio-oil, SR and gaseous products were calculated as follows:

$$\text{Yield (bio-oil)} = \frac{W_{\text{bio-oil}}}{W_{\text{algae}}} \times 100\% \tag{1}$$

$$\text{Yield (SR)} = \frac{W_{\text{SR}}}{W_{\text{algae}}} \times 100\% \tag{2}$$

$$\text{Yield (gas)} = 1 - \frac{W_{\text{liquid}} + W_{\text{SR}}}{W_{\text{algae}}} \times 100\% \tag{3}$$

The reported data are average values of at least 3 paralleled runs and the errors for the liquid yields were within 1 wt% (shown in Table S1†).

2.4. Analysis methods

Thermogravimetric analysis (TGA) was used to characterize the TLA, LEA and SEA, and a NETZSCH TG 209F1 instrument was employed. Dried samples were heated from room temperature to 800 °C with a heating rate of 10 °C min^{-1} under a nitrogen atmosphere.

Fourier transform infrared (FT-IR) spectroscopic analysis of the solid samples (TLA, LEA and SEA) and the bio-oil was performed using a Nicolet 6700 infrared spectrometer over a range of 400–4000 cm^{-1}.

The C, H and N elemental content of the solid samples was attained using a FLASH 1112ERIES Element Analyzer.

The composition of the bio-oil was analyzed using an Agilent 5973-6890N gas chromatograph/mass spectrometer with a HP-INNOWAX capillary column (30 m × 0.23 mm × 0.25 μm). Helium was used as the carrier gas with a flow rate of 1.0 mL min^{-1}. The temperature program was to heat from 40 °C to 110 °C at 10 °C min^{-1} followed by temperature ramping to 210 °C at 4 °C min^{-1}, and then it was ramped to 240 °C at 5 °C min^{-1} and held at 240 °C for 7 min. The injector and detector temperatures were both 280 °C. The percentages for the products are based on the peak area from GC-MS.

Gel permeation chromatography (GPC) was also performed to characterize the bio-oil using a HLC-8320 analyzer with two columns of TSK gel super HZM-M and TSK gel super HZ3000 (6.0 × 150 mm).

The composition of the gaseous products (except NH_3) was analyzed using a GC 9710 with a thermal conductivity detector (GC-TCD), and a TDX-01 carbon molecular sieve packed column (2 m × 3 mm I.D.) was employed to separate the gaseous products. Nitrogen with a flow rate of 20 mL min^{-1} was used as the carrier gas. The temperature parameters were as follows: column temperature, 120 °C; inlet temperature, 140 °C; detector temperature, 160 °C.

3. Results and discussion

3.1. Characterization of the samples

3.1.1. TGA of the samples. Fig. 2 shows the thermogravimetry (TG)/differential thermal analysis (DTA) curves for TLA, LEA and SEA. It was observed that the weight losses of TLA, LEA and SEA are similar. The thermal degradation can be mainly divided into 4 stages and the weight losses are listed in Table 1.

The first degradation stage for TLA, LEA and SEA starts at room temperature and continues to 163.8 °C, 178.8 °C and 179.4 °C, and the weight loss was 6.7 wt%, 7.8 wt% and 6.9 wt% based on the weight of the initial TLA, respectively. The DTG

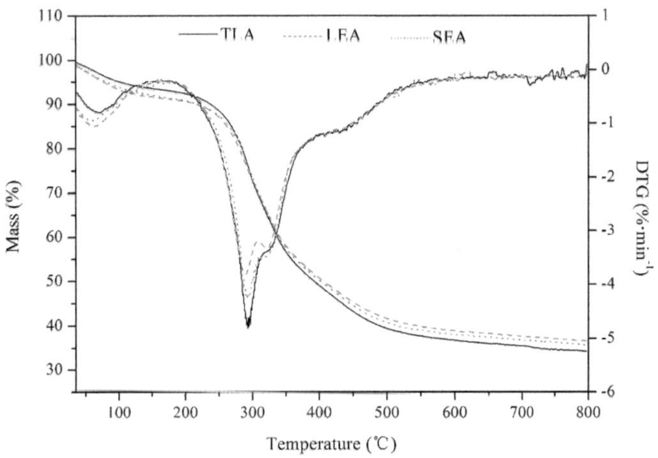

Fig. 2 TG/DTG curves for the raw and pretreated algae samples.

Table 1 The pyrolysis temperature and mass loss of the four stages for the raw and pretreated algae samples (based on the weight of the raw algae sample, %)

Algae samples		TLA	LEA	SEA
First stage	Temp. (°C)	R.T.[a] → 71.3 → 163.8	R.T. → 62.8 → 178.8	R.T. → 62.8 → 179.4
	Weight loss (wt%)	6.7	7.8	6.9
Second stage	Temp. (°C)	163.8 → 292.6 → 308.3	178.8 → 291.3 → 308.3	179.4 → 292.6 → 308.3
	Weight loss (wt%)	23.7	18.8	17
Third stage	Temp. (°C)	308.3 → 550	308.3 → 321.2 → 550	308.3 → 550
	Weight loss (wt%)	32	27.5	26.1
Fourth stage	Temp. (°C)	550 → 800	550 → 800	550 → 800
	Weight loss (wt%)	3.4	3	2.7

[a] Room temperature.

curves show that the temperature that corresponds to the maximum weight loss rate is 71.3 °C, 62.8 °C and 62.8 °C, for TLA, LEA and SEA respectively. This is mainly attributed to water loss in microalgae.[22,23] Furthermore, the weight loss rate of LEA appears to be higher than for TLA and SEA, and this is probably due to the fact that extraction of the lipids makes dehydration of the saccharides easier. In addition, the ending dehydration temperatures of LEA and SEA are higher than for TLA, because the interaction of the components in microalgae changes after extraction of the lipids and saccharides. That is, extraction of the lipids or saccharides stabilizes the other components in the microalgae at this stage. In other words, the interaction between the three components may partially facilitate their pyrolysis within this temperature range.

The second degradation stage for TLA, LEA and SEA ranges from 163.8 °C, 178.8 °C and 179.4 °C to 308.3 °C, and the weight loss was 23.7 wt%, 18.8 wt% and 17 wt%, respectively. During this stage, the weight loss rate rises significantly at first and then decreases gradually. From our previous work,[16] saccharides and lipids degrade mostly within this temperature interval. Moreover, because of the extraction of the lipids and saccharides, the weight losses of LEA and SEA appear slightly lower than that of TLA. It was indicated that extraction of the lipids and saccharides probably facilitates the degradation of protein. TLA, LEA and SEA have a maximum weight loss rate at the same temperature of about 290 °C.

The third degradation stage is between 308.2 °C and 550 °C, and the weight losses of TLA, LEA, and SEA were 32.0 wt%, 27.5 wt%, 26.1 wt%, respectively. Proteins are mainly degraded during this stage.[16,24] The extraction of lipids or saccharides results in a significantly decreased contribution to the overall degradation during this stage. Therefore, the weight losses of LEA and SEA decrease by 4.5 wt% and 5.9 wt% compared to that of TLA. Similarly, LEA has a maximum weight loss rate temperature (321.2 °C). However, different from the last two stages, we cannot see the peaks of TLA and SEA in the DTG curves clearly. This is probably caused by the fact that extraction of the lipids accelerates the degradation of proteins.

The last degradation stage for TLA, LEA and SEA is from 550 °C to 800 °C. The TG curves slightly decline in this stage with a weight loss of around 3.0 wt%. It appears that the carbonaceous substances and mineral components are mainly degraded during this stage.[23,25] From Fig. 2 we can see that the percentage of the residue for TLA, LEA and SEA is 34.3 wt%, 33.1 wt% and 29.4 wt%, respectively, suggesting a higher conversion of the samples after pretreatment.

In accordance with the information above, the fractional pyrolysis of TLA, LEA and SEA was carried out at temperatures of 290 °C and 450 °C.

3.1.2. FT-IR characterization of the samples. The FT-IR spectra of TLA, LEA and SEA are shown in Fig. 3. For the spectrum of TLA, the absorption at 3600–3200 cm^{-1} is ascribed to O–H or N–H stretching vibrations of polysaccharides and proteins.[14] The absorption peaks within 3000–2800 cm^{-1} are ascribed to C–H stretching vibrations of CH_3 and CH_2. The absorption peaks around 1653 cm^{-1} and 1540 cm^{-1} are ascribed to C=O and C–N vibrations possibly due to peptide bonds in proteins. However, absorption peaks of C=O groups from carboxylic acids and esters do not appear in the spectra of the three samples. It is possible that the lipids in the algae do not exist in the form of fatty acids or aliphatic esters. Also, the huge content of proteins is likely to affect the absorption of carbonyl groups of lipids. The absorption peaks at 1410 cm^{-1} and 1450 cm^{-1} are ascribed to bending vibrations of CH_3 and CH_2. The P=O/P–O vibrations at 1240 cm^{-1} indicate the presence of nucleic acids and phospholipids in TLA. In addition, the peaks around 1040 cm^{-1} and 1080 cm^{-1} are ascribed to Si–O vibrations of silicate.[26] The extraction of lipids resulted in weakening of the 3000–2800 cm^{-1} peaks ascribed to C–H stretching vibrations of CH_3 and CH_2. No significant change in the absorption peaks for carbonyl groups appears in the spectrum of LEA. In

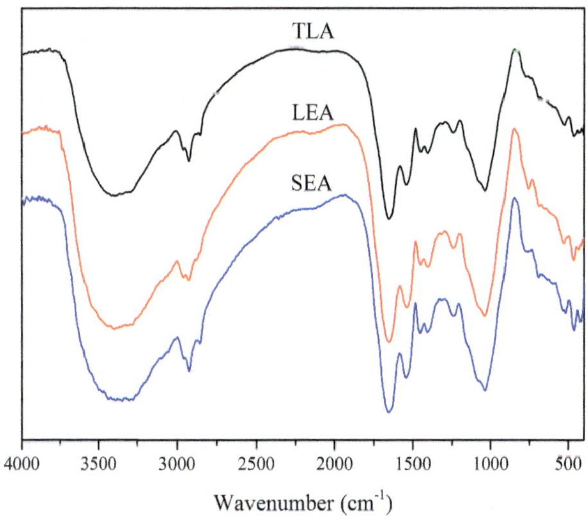

Fig. 3 FT-IR spectra of the raw and pretreated algae samples.

Table 2 Elemental composition of the raw and pretreated samples

Algae samples	C (wt%)	H (wt%)	N (wt%)
TLA	38.9	6.3	6.3
LEA	36.6	5.8	6.3
SEA	37.5	6.3	5.6

addition, there is no obvious change in the absorption peaks of O–H for SEA because of the existence of water and proteins.

3.1.3. Elemental analysis of the samples. Table 2 presents the elemental compositions (C, H and N) of TLA, LEA and SEA. Due to the abundant C and H in lipids, the content of carbon and hydrogen in LEA decreases significantly, while the nitrogen content remains the same as for TLA. However, the nitrogen content of SEA decreases from 6.3 wt% to 5.6 wt%. This is probably caused by the fact that saccharides and a part of the proteins are extracted from the algae at the same time.

3.2. Analysis of the pyrolysis products

3.2.1. Distribution of the pyrolysis products. The yields of the gaseous, liquid and solid products from the pyrolysis of TLA, LEA and SEA at 290 °C and 450 °C are presented in Table 3. The yields were calculated based on the weight of the raw samples, and the yields for the extracted lipids and saccharides were added to the liquid products at 290 °C and the total liquid products. From Table 3, it can be seen that the liquid yields from LEA and SEA at 290 °C (32.1 wt% and 38.3 wt%, respectively) show a noticeable increase compared with that from TLA (27.8 wt%). 8.4 wt% and 8.2 wt% of gaseous products were obtained from TLA and LEA at 290 °C, while only a 6.1 wt% yield of gas was produced from SEA, a significant drop. This is because the pyrolysis of saccharides produces more gaseous products than that of lipids. Also, the solid residue (SR) yields from LEA (59.7 wt%) and SEA (55.6 wt%) are much lower than that from TLA (63.8 wt%) at 290 °C, and the decrease is related to the amount of lipids and saccharides firstly extracted from TLA. Due to the extraction and removal of lipids or saccharides, at 450 °C, the liquid products are from the pyrolysis of proteins and the other remaining

Table 3 The yields of the products from fractional pyrolysis of the raw and pretreated algae samples (based on the weight of the raw algae sample, %)

	290 °C (wt%)			450 °C (wt%)			Total (wt%)		
Algae samples	Gas	Liquid	Solid	Gas	Liquid	Solid	Gas	Liquid	Solid
TLA	8.4	27.8	63.8	7.6	13.1	43.1	16.0	40.9	43.1
LEA	8.2	32.1[a]	59.7	7.8	10.2	41.7	16.0	42.3[b]	41.7
SEA	6.1	38.3[a]	55.6	6.2	10.2	39.2	12.3	48.5[b]	39.2

[a] Because the preparation temperatures of LEA and SEA are far lower than 290 °C, the extraction yield of lipids or saccharides was added into the yield of liquid products obtained at 290 °C. [b] The yield of lipids or saccharides for extraction was added to the total yield.

components, thus a significant decrease in the liquid yield is observed for LEA and SEA (both 10.2 wt%) compared with that of TLA (13.1 wt%). However, the differences in the gas and solid yields at 450 °C are similar to those at 290 °C.

In terms of the total yield, there is a slight increase in the liquid yield from LEA compared with that of TLA, while the gas yield remains the same and the solid yield decreases marginally. In other words, the extraction of the lipids has only a slight effect on the product yields. However, compared with TLA, the liquid yield from SEA shows an obvious increase, while the solid yield drops dramatically. This means that we can obtain more products when considering the extracted saccharides as a liquid product. That is to say, more useful products are obtained from microalgae *via* fractionation.

3.2.2. The composition of the bio-oil. The compounds in the bio-oil were analyzed using GC-MS. Fig. 4 illustrates the composition of the bio-oils obtained from TLA at 290 °C and 450 °C. From Fig. 4, we can see that the main products obtained at 290 °C were water, fatty acids, amides, esters and hydrocarbons. Besides, a small number of phenols, nitriles, indoles, pyrroles and some other compounds were also obtained at 290 °C. At 450 °C, phenols and hydrocarbons were the most abundant compounds in the bio-oil, with a few amides, nitriles, indoles, pyrroles and other compounds. The distributions of the products at the two different temperatures exhibit a large difference, except for the very few nitriles and indoles. Although the amount of hydrocarbons constitutes a large part of the bio-oil at both temperatures, the actual products are not the same. Specifically, the hydrocarbons at 290 °C are mainly heptadecane, while C_{10}–C_{15} hydrocarbons are mainly obtained at 450 °C. Compared with the bio-oil obtained from a direct pyrolysis at 450 °C,[27] the composition is simplified by fractional pyrolysis. According to our previous work,[16] lipids and saccharides are pyrolyzed mainly at 290 °C, whereas proteins are pyrolyzed at 450 °C. Consequently, the two step pyrolysis of the microalgae fractionates the conversion of the different components to some extent, making the distribution of the liquid products more concentrated, and the variety of the product composition declines at either temperature.

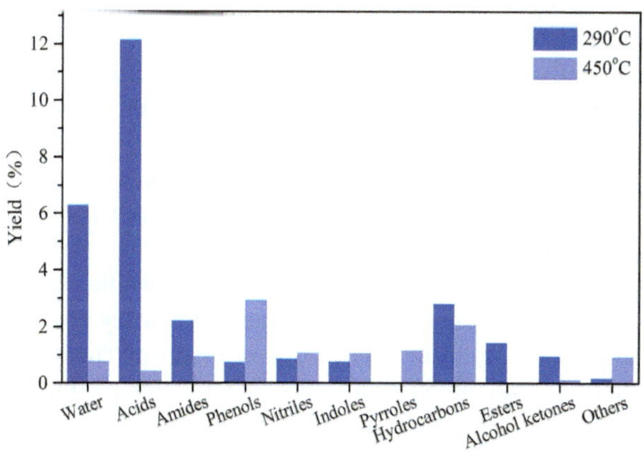

Fig. 4 The component distribution of bio-oil from fractional pyrolysis of TLA (based on the weight of the algae samples, %).

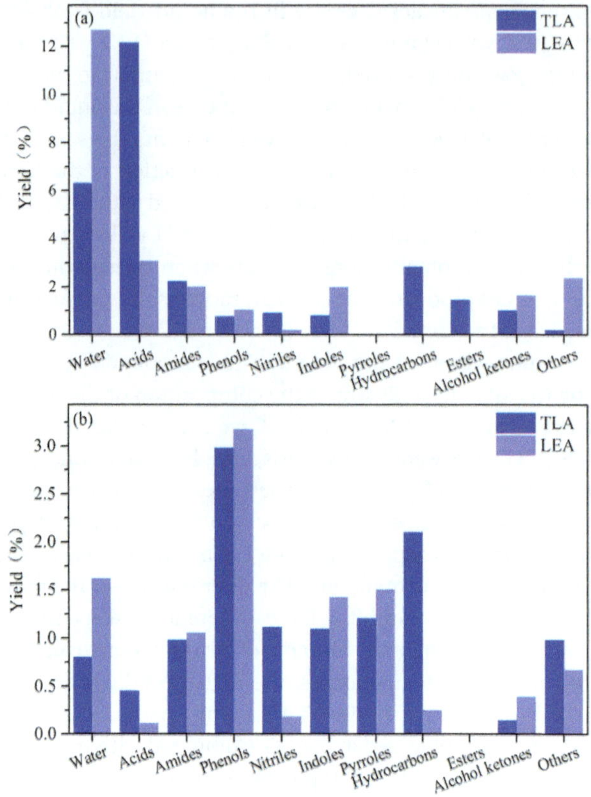

Fig. 5 The component distribution of bio-oil from fractional pyrolysis of TLA and LEA at (a) 290 °C and (b) 450 °C (based on the weight of the algae samples, %).

Fig. 5 shows a comparison of the component distributions for the bio-oils obtained from fractional pyrolysis of LEA and TLA. At 290 °C, the content of water in the bio-oil from LEA has risen significantly compared with TLA, followed by a dramatic decrease in the fatty acids and nitriles, and disappearance of the hydrocarbons and esters, while the amount of indoles from LEA has almost doubled compared with TLA. Besides, the amounts of amides, phenols and carbonyl compounds show slight changes, while the amount of some other N,O-containing compounds has increased. At 450 °C, the water, indoles, pyrroles and alcohol ketones show an obvious increase, whereas the amounts of fatty acids, nitriles and hydrocarbons decline sharply. Also, the amounts of amides and phenols remain nearly unchanged. From the component distribution of the bio-oil obtained from LEA, we can see that the products originating from lipids have decreased significantly because of lipid extraction. Meanwhile, either the method of lipid extraction cannot extract the algal lipids completely or some fatty acids and nitriles are formed from the degradation of proteins. The dramatic drop in nitriles originating from protein indicates that the interactions between the intermediates from the pyrolysis of lipids and proteins are significantly weakened after lipid extraction.[16,26] In addition, the increase in water indicates that extraction of the lipids facilitates the pyrolysis of saccharides, while the rise in the content of indoles and pyrroles suggests that the acceleration of the degradation of proteins is caused by lipid extraction.

A comparison of the distributions for bio-oils obtained from the pyrolysis of SEA and TLA is presented in Fig. 6. From Fig. 6, it can be seen that the amounts of water and fatty acids from SEA decline slightly at 290 °C, whereas the amounts of the other compounds from SEA remain almost the same compared with those from TLA. In other words, the saccharides in TLA do not contribute to the bio-oil obtained at 290 °C from TLA. At 450 °C, the amounts of amides and phenols exhibit an obvious drop, but the content of nitriles rises dramatically while the amounts of the other compounds only fluctuate slightly. The rise in the content of nitriles further confirms that nitriles are the products from interaction between the intermediates of lipids and proteins, and saccharide extraction facilitates this interaction. From the distribution of the bio-oil obtained from SEA, the amount of fatty acids is decreased at first compared to TLA at 290 °C and then is increased at 450 °C, which indicates that saccharide extraction weakly inhibits the degradation of lipids to fatty acids. However, extraction of the saccharides changes the structure or interactions among the proteins themselves, causing a decrease in the phenols and a slight increase in the indoles.

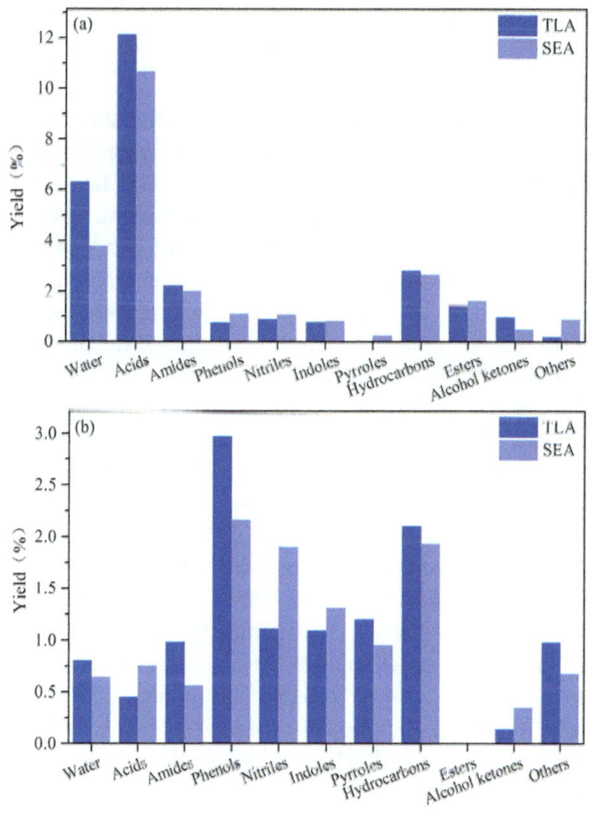

Fig. 6 The component distribution of bio-oil from fractional pyrolysis of TLA and SEA at (a) 290 °C and (b) 450 °C (based on the weight of the algae samples, %).

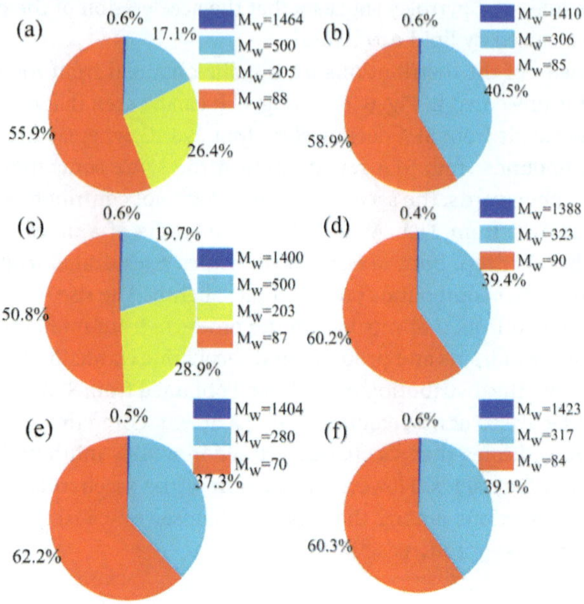

Fig. 7 The molecular weight distributions of bio-oils from fractional pyrolysis of the raw and pretreated samples: (a) TLA, 290 °C, (b) LEA, 290 °C, (c) SEA, 290 °C, (d) TLA, 450 °C, (e) LEA, 450 °C, and (f) SEA, 450 °C.

3.2.3. GPC analysis of the bio-oil. Fig. 7 illustrates the GPC results for the bio-oil samples. From the molecular weight distribution of the bio-oils at 290 °C, it can be seen that the bio-oils from TLA, LEA and SEA all contain 0.6% of oligomers with a M_W over 1000. Meanwhile, the bio-oils from TLA and SEA contain 17.1% and 19.7% of oligomers with a M_W near 500, while this cannot be seen in the bio-oil from LEA. In general, the bio-oils obtained from the three samples at 290 °C have over 70% of compounds with a M_W under 200, that is to say, most products exist in the form of small molecules. The GPC results of the bio-oils obtained at 450 °C have no significant difference. Approximately, 37–39% of the products have a M_W of about 300, whereas 60% of them have a M_W less than 100. The GPC results indicate that lipid extraction has a significant influence on the molecular weight distribution of the bio-oil while changes due to saccharide extraction are not obvious, because lipid extraction facilitates the pyrolysis of saccharides to water and other complex oxygenates instead of oligomers.

3.2.4. FT-IR characterization of the bio-oil. The FT-IR spectra of the bio-oils from TLA, LEA and SEA at 290 °C and 450 °C are presented in Fig. 8. The absorption due to O–H at 3400 cm^{-1} indicates the presence of fatty acids, alcohols and phenols. The absorption at about 3200 cm^{-1} is ascribed to the stretching vibrations of N–H groups from amides. The absorbance peak at 3060 cm^{-1} is ascribed to C–H stretching vibrations of benzene rings. The absorptions at 2960 cm^{-1}, 2870 cm^{-1}, 2920 cm^{-1} and 2850 cm^{-1} are ascribed to stretching vibrations of C–H groups from CH_3 and CH_2. The C=O absorptions at 1710 cm^{-1} and 1665 cm^{-1} are possibly from fatty acids, esters, ketones and amides.[28] The C–N and C=C stretching vibrations within 1560–1510 cm^{-1} indicate the presence of amides

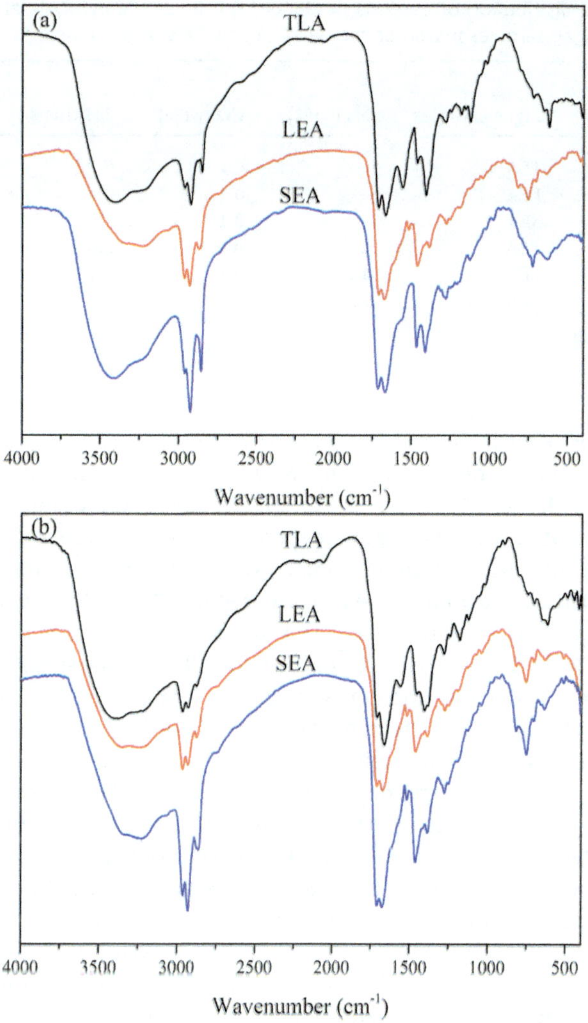

Fig. 8 FT-IR spectra of the bio-oils from fractional pyrolysis of the raw and pretreated algae samples at (a) 290 °C and (b) 450 °C.

and aromatics. The absorption peaks within 1470–1350 cm^{-1} are ascribed to C–H bending vibrations. In addition, the absorbance peaks appearing within the band from 1300–950 cm^{-1} are ascribed to the C–O and O–H vibrations of alcohols, phenols and esters. From the FT-IR spectrum of the bio-oil from TLA at 290 °C, the main compounds in the bio-oil could be fatty acids, alcohols, phenols, hydrocarbons, amides, esters, ketones and aromatics. Comparing the spectra of the bio-oils obtained from LEA and SEA with that from TLA at 290 °C, the peak at 3400 cm^{-1} for the bio-oil from LEA has become weak, suggesting a decrease in fatty acids, which is in accordance with the lipid extraction. Besides, the peak at 1550 cm^{-1} disappears for the bio-oil from LEA, indicating that extraction of the lipids reduces the amides and nitriles in the bio-oil. In particular, the increase in the absorption at 750 cm^{-1} probably suggests a rise in the amount of indoles for

Table 4 The distribution of gaseous products from fractional pyrolysis of the raw and pretreated algae samples (based on the weight of the raw algae sample, %)

Temperature (°C)	Algae samples	H_2 (wt%)	CO (wt%)	CH_4 (wt%)	CO_2 (wt%)
290	TLA	—[a]	0.1	0.2	7.8
	LEA	—	0.1	0.1	7.6
	SEA	—	0.1	0.1	6.5
450	TLA	3.3	0.1	2.6	2.2
	LEA	1.3	0.1	2.9	3.6
	SEA	2.3	0.1	1.6	2.6

[a] Not detected.

the bio-oil of LEA, which is in accordance with the results of GC-MS. However, the spectrum of the bio-oil from SEA at 290 °C shows slight changes, which is in agreement with the GC-MS results. The spectra at 450 °C are similar to those at 290 °C. In addition, the changes in the spectra at 450 °C indicate a decrease in the content of hydrocarbons from LEA and a drop in the amides from SEA.

3.2.5. The composition of the gaseous products. The yields and distributions of the gaseous products from TLA, LEA and SEA at the two temperatures are shown in Table 4. The gaseous products for all the experiments are mainly composed of H_2, CO, CH_4 and CO_2. CO_2 is the major gaseous product at 290 °C, while the yields of H_2 and CH_4 experience a dramatic increase on increasing the temperature to 450 °C. This indicates that decarboxylation occurs at a relatively low temperature, while cracking, dehydrogenation and dehydrocarboxylation occur at high temperatures.[26,29] The amounts of CH_4 and CO_2 obtained from LEA exhibit different levels of increase at 450 °C compared with those from TLA, proving that extraction of the lipids accelerates the degradation of proteins by decarbonylation, decarboxylation and dehydrocarboxylation. In contrast, the yields of CH_4 and H_2 from SEA at 450 °C show a downward trend compared with those from TLA, suggesting that saccharide extraction inhibits pyrolysis of the proteins in microalgae. However, the amount of CO is quite low from all the samples at both temperatures. In addition, the significant drop in the yield of H_2 from LEA at 450 °C is probably due to weakening of the lipid dehydrogenation after extraction of the lipids.

4. Conclusions

In this study, pyrolysis of raw and pretreated microalgae (cyanobacteria from Taihu Lake) was carried out at 290 °C and 450 °C, according to the TGA results, and the total yield of bio-oil was 40–48 wt%. The two step pyrolysis of TLA fractionated the conversion of the different components, making the distribution of the liquid products more concentrated. From TLA, the main compounds in the bio-oil were fatty acids, hydrocarbons (heptadecane) and amides at 290 °C, and phenols and C_{10}–C_{15} hydrocarbons at 450 °C. The composition of the bio-oil obtained from LEA was simplified, with disappearance of the hydrocarbons and esters and a decrease in the fatty acids and nitriles at both temperatures. Moreover, lipid extraction not only reduced the average molecular weight of the bio-oil but also increased the bio-oil yield. Besides, extraction of the lipids accelerated pyrolysis of the saccharides and proteins in microalgae. However, the composition of the bio-oil from SEA showed no obvious change because bio-oil obtained from the pyrolysis of algae mainly originates from lipids and proteins. Additionally, saccharide extraction had a weak inhibiting effect on lipid pyrolysis.

Acknowledgements

This work was supported by the National Basic Research Program (973 program, No. 2013 CB228103). The characterization of the products performed at the Analytical and Testing Center of Sichuan University is greatly acknowledged.

References

1 D. Liu, J. K. Keesing, Q. Xing and P. Shi, *Mar. Pollut. Bull.*, 2009, **58**, 888–895.
2 X. Miao, Q. Wu and C. Yang, *J. Anal. Appl. Pyrolysis*, 2004, **71**, 855–863.
3 S. Sawayama, K. K. Rao and D. O. Hall, *Appl. Microbiol. Biotechnol.*, 2010, **87**, 1291–1301.
4 F. A. Ansari, A. Shriwastav, S. K. Gupta, I. Rawat, A. Guldhe and F. Bux, *Bioresour. Technol.*, 2015, **179**, 559–564.
5 M. Morweiser, O. Kruse, B. Hankamer and C. Posten, *Appl. Microbiol. Biotechnol.*, 2010, **87**, 1291–1301.
6 Y. Chen, Y. Wu, D. Hua, C. Li, M. P. Harold, J. Wang and M. Yang, *RSC Adv.*, 2015, **5**, 18673–18701.
7 R. H. Wijffels and M. J. Barbosa, *Science*, 2010, **329**, 796–799.
8 P. J. B. Williams, *Nature*, 2007, **450**, 478–478.
9 N. Sudasinghe, H. Reddy, N. Csakan, S. Deng, P. Lammers and T. Schaub, *BioEnergy Res.*, 2015, **8**, 1962–1972.
10 Z. Hu, Y. Zheng, F. Yan, B. Xiao and S. Liu, *Energy*, 2013, **52**, 119–125.
11 Q. Guo, M. Wu, K. Wang, L. Zhang and X. Xu, *Ind. Eng. Chem. Res.*, 2015, **54**, 890–899.
12 X. Ji, B. Liu, G. Chen and W. Ma, *J. Anal. Appl. Pyrolysis*, 2015, **116**, 231–236.
13 R. Zhang, L. Li, D. Tong and C. Hu, *Bioresour. Technol.*, 2016, **212**, 311–317.
14 V. Anand, V. Sunjeev and R. Vinu, *J. Anal. Appl. Pyrolysis*, 2016, **118**, 298–307.
15 I. V. Babich, M. van der Hulst, L. Lefferts, J. A. Moulijn, P. O'Connor and K. Seshan, *Biomass Bioenergy*, 2011, **35**, 3199–3207.
16 L.-L. Li, R. Zhang, D.-M. Tong and C.-W. Hu, *Chin. J. Chem. Phys.*, 2015, **28**, 525–532.
17 W. Song, T. Teshiba, K. Rein and K. E. O'Shea, *Environ. Sci. Technol.*, 2005, **39**, 6300–6305.
18 T. P. J. Kull, O. T. Sjövall, M. K. Tammenkoski, P. H. Backlund and J. A. O. Meriluoto, *Environ. Sci. Technol.*, 2006, **40**, 1504–1510.
19 F. Soxhlet, *Dinglers Polytech. J.*, 1879, **232**, 461–465.
20 J. Folch, M. Lees and G. H. S. Stanley, *J. Biol. Chem.*, 1957, **226**, 497–509.
21 X. Y. Yin, F. Wang and Z. G. Cui, *Applied Chemical Industry*, 2011, **40**, 386–387.
22 A. Marcilla, A. Gómez-Siurana, C. Gomis, E. Chápuli, M. C. Catalá and F. J. Valdés, *Thermochim. Acta*, 2009, **484**, 41–47.
23 A. B. Ross, K. Anastasakis, M. Kubacki and J. M. Jones, *J. Anal. Appl. Pyrolysis*, 2009, **85**, 3–10.

24 S.-S. Kim, H. V. Ly, J. Kim, E. Y. Lee and H. C. Woo, *Chem. Eng. J.*, 2015, **263**, 194–199.

25 K. Kebelmann, A. Hornung, U. Karsten and G. Griffiths, *Biomass Bioenergy*, 2013, **49**, 38–48.

26 H. Li, L. Li, R. Zhang, D. Tong and C. Hu, *J. Energy Chem.*, 2014, **23**, 732–741.

27 Y. Zeng, J. Tang, S. Lian, D. Tong and C. Hu, *Biomass Bioenergy*, 2015, **73**, 95–101.

28 D. Zhou, S. Zhang, H. Fu and J. Chen, *Energy Fuels*, 2012, **26**, 2342–2351.

29 P. Duan, X. Bai, Y. Xu, A. Zhang, F. Wang, L. Zhang and J. Miao, *Bioresour. Technol.*, 2013, **136**, 626–634.

Faraday Discussions

PAPER

Developing new platform chemicals: what is required for a new bio-based molecule to become a platform chemical in the bioeconomy?

José-Vitor Bomtempo, * Flavia Chaves Alves and Fabio de Almeida Oroski

Received 30th January 2017, Accepted 13th March 2017
DOI: 10.1039/c7fd00052a

This paper proposes a framework with six dimensions that can be useful for evaluating the potential and the current stage of a bio-based platform chemical. The framework considers the technological and strategic challenges to be fulfilled by a company that intends to lead a platform based on a bio-based chemical. A platform chemical should be an intermediate molecule, with a structure able to generate a number of derivatives, that is produced at a competitive cost, capable of allowing exploitation of the scale and scope economies, and inserted within a complete innovation ecosystem that is able to create value with governance mechanisms that are capable of allowing coordination of the innovation process and facilitation of the value capture by the focal company leading the platform, in our case the producer of the platform molecule. Based on these six dimensions, three potential platform chemicals – succinic acid, butanol and farnesene – are compared and discussed. It is possible to identify important differences concerning the technological dimensions and the strategic dimensions as well. Two of the molecules – farnesene and succinic acid – adhere to most of the conditions required to structure a platform chemical. However, the innovation ecosystem is not complete and the governance mechanisms are still under development, so it is not clear if they will be capable of allowing a favorable position for value capture by the platform leader. Butanol structuring for a platform does not seem promising. The potential of the molecule is apparently not high and the strategic initiatives are in general not focused on innovation ecosystem structuring.

Introduction

Advance of the bio-based industry[1] is driving a great emphasis on the development of platform chemicals, understood as chemical intermediates with the potential to develop new product families. New platform chemicals are an essential opportunity for the bio-based economy. The exploration of

School of Chemistry, Federal University of Rio de Janeiro, Brazil. E-mail: vitor@eq.ufrj.br

opportunities related to these molecules is not simple, since it involves a complex development process, including value chain building and market adoption strategies. The literature on platform chemicals is quite rich in details of the technological and market potential,[2,3] but until now has failed to provide an understanding of the critical strategic dimensions involved in platform chemical development. We suggest an exploration of the literature on technological platforms to understand the structuring process of platform chemicals. According to this vast literature,[4,5] a technological platform must present some key characteristics: a modular technological architecture, varied degrees of openness among the interface components, access to innovation agents, regulation by a governance structure, value creation through scope and scale economies, and amenable to coordination to keep competition under control. How do these dimensions fit to platform chemicals? We propose an analytical framework, based on the technological platform literature but adapted to the particular aspects of platform chemicals,[6] which could be useful for evaluating the potential of platform candidates. This framework considers six dimensions related to the molecule and to the strategic dimensions as well: an intermediate position in the value chain, a flexible chemical structure, a wide range of potential derived products at a competitive cost, a well-developed innovation ecosystem, a governance structure and the ability to create value through scope and scale economies. Three different examples in the biobased industry – biosuccinic acid, farnesene and biobutanol – are explored using the framework. It is possible to evaluate if these molecules have the potential to become a platform chemical and to discuss the challenges concerning the value chain structuring. If a new biobased molecule meets the conditions to become a platform chemical, strategic dimensions have to be deployed. During this process, some dilemmas and challenges which are classic in platform development have to be faced: competition or innovation, value capture or openness, scale production or application focus. We conclude that the building of a new platform chemical is a very challenging process. A capacity for coordination of a multivariate environment is crucial for structuring it. The technological development of a potential molecule is an essential step, but it is only the first one. If the key strategic dimensions are not developed in a timely manner, the platform is unlikely to be built.

This paper is organized as follows. First, we present a brief literature review on technological platforms in order to identify the key dimensions that guide platform building. Based on these dimensions and considering the particular aspects of platform chemicals, we propose an analytical framework that could be applied to evaluate the possibility and the present development stage of a potential platform chemical. We examine three potential platform chemicals – biosuccinic acid, farnesene and biobutanol – to illustrate the platform building process. In the last section, we present our concluding remarks, the implications and some open questions that could be interesting research themes.

Technological platforms

"Platforms are the future – but not for everyone" was quoted in The Economist, May 2016, referring to the wide diffusion of the technological platforms notion over the last 15 years and, as a result, to its overutilization. Many industries, from cars to information technology sectors, have been exploring the idea of a platform as a strategic way to organize and explore new business opportunities. It is important to emphasize that platforms are distinct from conventional products[7] "in that they have the potential to generate positive feedback loops, between the primary product (the platform) and its users as well as with complementary products and services". Positive feedback can create incentives for more users to enter the platform and so to reinforce the ecosystem.

The literature on the subject is now quite vast, from books such as *Platform revolution* and *Matchmakers: the new economics of multisided platforms*,[8,9] which was recently published, to papers in the highest impact economics and management journals. A recent review[4] has identified three different perspectives: market (industrial organization economics), firms (strategic management) and integrative (technology management). Our work uses the technology management viewpoint as a literature reference to present the key dimensions of platforms. In this perspective, platforms are seen as technological architectures which are intended to promote innovation. Nevertheless, Gawer[10] stresses that the interactions between the technological structure and business strategy have to be considered for an integrative perspective.

From this perspective, platforms can be defined[11] as "evolving organizations or meta-organizations that: (1) federate and coordinate constitutive agents who can innovate and compete; (2) create value by generating and harnessing economies of scope in supply or/and in demand; and (3) entail a modular technological architecture composed of a core and a periphery". A typology of technological platforms[11] considers three kinds of platforms: product, supply chain and industry platforms. A supply chain platform is a particular case of internal platform where assemblers and suppliers perform in a quasi-integration way. Thus, we consider in this paper product and industry platforms.

Product or internal platforms[5] can be defined as "a set of assets organized in a common structure from which a company can efficiently develop and produce a stream of derivative products". Industry or external platforms are defined[5] as "products, services, or technologies that act as a foundation upon which external innovators, organized as an innovative business ecosystem, can develop their own complementary products, technologies, or services".

Product and industry platforms can be differentiated[11] by at least five structural characteristics. In product platforms, the constitutive agents are one firm and its constituent sub-units, while in industry platforms, the agents are the platform leaders and the complementors.[12] Complementors are developers and providers of complementary products or services that have to be combined to give a product all the characteristics needed to attain the final users' expectations. Both platforms have a modular architecture organized around a core and a periphery, but the interfaces are closed in product platforms and open in industry platforms. This means that in the first case the interface specifications are defined within the company and they are not known by external agents. In the case of industry platforms, the interfaces are essentially open and are shared with the complementors. Concerning access to the innovative capabilities, product platforms are limited to the company capabilities while with industry platforms, due to the open interfaces, the access is virtually unlimited. Finally, the coordination mechanisms are also diverse. Product platforms are coordinated by an internal authority through the managerial hierarchy. On the other hand, for industry platforms governance mechanisms have to be developed in an innovation ecosystem.

In summary, the above discussion has established a distinction between platforms developed by vertically integrated companies (product or internal platforms) and those developed by companies within innovation ecosystems (industry or external platforms). This last one, to be successful, depends on innovation initiatives from external agents. Nevertheless, notwithstanding the organizational context in which platforms are developed, it is possible to establish, according to Gawer[11] and Gawer and Cusumano,[7] that any platform:

(i) has a modular technological architecture composed of a core and a periphery;

(ii) has technological interfaces, between the core and the periphery, with varying degrees of "openness";

(iii) involves access to innovation agents with varying levels of capabilities;

(iv) involves coordination among constituent agents in order to increase capabilities and control competition;

(v) is regulated by coordination mechanisms which are particular to the platform organizational setting;

(vi) creates value by generating and harnessing economies of scope and scale in supply or/and in demand.

The structuring process of a new platform, based on new bio-based chemicals, for example, challenges companies to deal with an environment where innovation and competition drivers interact. Companies have to manage some key dilemmas in order to conciliate competition and innovation strategies without compromising their economic benefits. They have to consider internal *versus* external platforms and also value capture *versus* the openness degree. There are some technological and strategic challenges to overcome in order to develop the right architecture and interfaces, and to ensure the necessary complements are present to create value as a way to generate demand, develop markets and gain a competitive advantage over the competitors. In the next section, we explore these points for the case of potential platform chemicals. How do these dimensions fit to platform chemicals?

Platform chemicals

The literature on platform chemicals has remarkably increased in the last ten years. For example, there are more than eight thousand citations referring to platform chemicals or platform molecules within Google Scholar, five thousand within the last three years. However, this vast literature barely explores the technological challenges involved in a platform structuring process, not to mention the strategic challenges. At the same time, companies introduce themselves as exploring bio-based platform chemicals and consultation studies[13,14] frequently emphasize the market opportunity for these molecules as potential platforms.

Most of this literature focus on the technological and market potential of the new bio-based molecules,[2,3,15,16] but until now has failed to provide an understanding of the critical strategic dimensions involved in platform chemical development. So, there is no comprehensive definition of a chemical platform. Most papers do not present any definition, taking for granted the notion of a platform chemical. We can take as a reference the very objective definition by Bozell and Petersen:[2] "Compounds that serve as starting materials for the production of derivatives". This definition emphasizes essentially the ability for the production of derivatives, lacking a broad understanding of the challenges involved in platform development.

The terms platform and building blocks have been used interchangeably by most authors. The DOE Report from 2004 [15] is a remarkable reference on the identification of promising bio-based molecules, which presented twelve well-known high-value bio-based chemicals as building blocks, explaining that "Building block chemicals, as considered for this analysis, are molecules with multiple functional groups that possess the potential to be transformed into new families of useful molecules". This is indeed a good definition of platform chemicals as most of the twelve sugar-based chemicals – 1,4-diacids (succinic, fumaric and malic), 2,5-furan dicarboxylic acid, 3-hydroxypropionic acid, aspartic acid, glucaric acid, glutamic acid, itaconic acid, levulinic acid, 3-hydroxybutyrolactone, glycerol, sorbitol, and xylitol/arabinitol – are so called in the current literature.

There is a quite unusual distinction between platform chemicals and building blocks present in the literature. Bozell and Petersen,[2] revisiting the top-10 DOE list from 2004, have proposed a multi-criteria analysis in order to compare and identify the new top bio-based molecules. In Bozell and Petersen's framework, platform chemicals and building blocks are seen as having different criteria. The distinction is based on the idea that a petrochemical refinery is built on a small number of initial building blocks: olefins, BTX, methane, CO. Those compounds that are able to have an analogous role in a biorefinery, as primary building blocks, will be of high importance. Following these criteria, according to Bozell and Petersen, ethanol would be at the same time a good platform chemical and a primary building block. Succinic acid, however, would be a good platform chemical but a poor primary building block. Nevertheless, probably the dominant trend has been to name building blocks as platforms.

We consider in this work that the idea of platform chemicals should not be taken as simply a new way to rename the existing petrochemical building blocks. A platform is taken here as a strategy and a way to organize the business of the industry for at least two reasons. The emergence of the platform concept places the chemical industry within the open innovation[17] paradigm that currently dominates innovation strategies of the most dynamic sectors of the economy. The second reason is due to the idea of the bioeconomy as a system innovation that could lead to a transition to new socio-technical regimes departing from the fossil-based one. This transition[18] can be transformative for the current regime and open opportunities to new forms of business organization.

We suggest then that the so-called platform chemicals are particular cases of technological platforms whose backbone takes the form of a chemical compound. This means that platform chemicals are chemical intermediates capable of giving rise to a wide range of derivatives with diverse final applications. These final products result from specific physical and chemical transformations and are consumed in different markets.

Platform chemicals are able to create great commercial value. However, as intermediates in the value chain, in order to actually create that value these chemicals need to be properly transformed into end products and these products need to be adopted by end users. This adoption[19] may depend only on cost competitiveness, in the case of the drop in substitutes for conventional petrochemicals. But if the platform chemical leads to new molecules, a new plastic for example, the innovation adoption depends on market development, which can be very challenging. The platform leader must face not only technological difficulties in order to transform an intermediate molecule into different final products, but also challenges in developing new market applications. The set of technological and managerial competencies that must be mobilized exceeds what is held by a single company, not to mention the financial resources. Except for some fully vertically integrated companies that hold all stages of the production chain and the necessary specific skills and technologies, in general, it is necessary to take advantage of external capabilities. That is, in general, this challenge also involves the need for engagement of complementors, developers and contributors from different and diverse industries. These complementors, as downstream chain actors, are responsible for driving innovation produced by the platform leader to a final use.

Thus, a bio-based platform chemical can be considered an industry platform immersed in an innovation ecosystem that, in this case, is a set of value chains. These value chains are diverse in the sense that the nature of the participating complementors can be distinct, particularly in the extreme downstream positions. Value chains with different end users may demand very different approaches, which means that the innovation ecosystem to be built may be very complex.

The innovation ecosystem, which is so diverse and complex, has to be coordinated by the platform leader for two reasons: to facilitate the value creation and at the same time to find a favorable position for value capture. In this extremely competitive and dynamic environment, focal firms are vulnerable to external innovation efforts and therefore have to deal with technological and behavioral uncertainties.

Considering the identification of these particularities and challenges in the development of chemical platforms and taking them as specific cases for product technological platforms, it is possible to identify the dimensions that will allow us to characterize a given product as a potential platform chemical. In this way, products characterized as a platform chemical necessarily:

(I) are an intermediate in the value chain;

(II) have a flexible chemical structure with functional groups that allow the generation of a set of derivatives;

(III) have interfaces, with varying degrees of openness, that allow their transformation into a wide range of derivatives at competitive cost. These interfaces can be seen as chemical and physical transformation processes that can be integrated by the focal company or performed by external agents;

(IV) are connected to innovation agents with varying levels of competence and diverse interests located at different positions of the value chain and towards different end products, so that they are structured within innovation ecosystems formed by a set of production chains;

(V) are regulated by control and command mechanisms (governance) that vary according to the organizational context. The governance mechanisms should allow a favorable position for value capture by the platform leader;

(VI) create value through scope and scale economies.

The structuring process is challenging as we can see by the conditions presented above. Even if the platform molecule fulfills the technological conditions, the platform leader faces some strategic dilemmas.

Thus, similar to technological platforms, the platform chemicals, in order to be commercially effective, need to overcome some dilemmas that correspond to challenges to be faced by the focal companies, namely:

(I) degree of openness *versus* value appropriation: the platform leader must be able to manage the interfaces in such a way as to allow the generation of value with external collaboration in an open innovation mode, but, at the same time, to prevent this value from being totally captured by these agents;

(II) innovation/collaboration *versus* competition: the platform leader should create relationships of trust and collaboration with the complementors and end users. This is essential for forming an innovative environment. However, the platform leader must be attentive to the opportunistic behavior of other agents;

(III) large-scale production and market application/development: the platform leader must be able to deal with this egg–chicken problem in which network effects define interdependence between these factors. It is necessary to achieve production on a large scale to enter new markets. However, new markets need an application development to persuade end users to adopt the innovation. At the same time, end users usually are not ready to adopt the innovation if large scale production is not available yet.

Comparing biosuccinic acid, biobutanol and farnesene as potential platform chemicals

Biosuccinic acid, biobutanol and farnesene were chosen as cases to apply our framework and discuss the structuring process for new bio-based platform chemicals. They are representative of the new generation of bio-based molecules and have already begun to undergo commercial production, but they are quite diverse concerning their technological and strategic aspects (information on the products is provided by a database organized by the authors, compiling data from company websites, specialized sites and conference presentations).

Succinic acid has been produced only through a petrochemical route. Due to the cost limitations of the petrochemical process, its market is limited to no more than 40 000 tons per year. Bio-based production, potentially cost competitive, could open up the opportunity to develop a wide number of derivatives. There are four important producers currently: Bioamber, Myriant, Reverdia and Succinity. The first two are startups created to explore the opportunity. Reverdia is a joint venture between DSM and Roquette, and Succinity is a joint venture between BASF and Corbion (former Purac).

Butanol has been produced through a petrochemical route and also through a biochemical one, called ABE. The petrochemical process was able to be used to obtain an important market of more than 3 million tons per year, essentially for chemical uses such as solvents and plasticizers. The bio-based production is aimed at drop-in substitutions but also new markets, to provide renewable fuels for gasoline, diesel and jet fuel applications, and new bio-based chemicals such as *p*-xylene, used as a PTA precursor for PET production. There are three important players: Gevo, Butamax and Green Biologics. Gevo and Green Biologics are startups while Butamax is a joint venture between Du Pont and BP.

Farnesene was not produced at a commercial level until an innovative biochemical route was proposed by Amyris, a startup created to explore the potential uses of farnesene and other similar isoprenoids. Farnesene itself has no market, but its derivative applications are quite diverse: from drop-in diesel and jet fuels to use in a variety of specialty markets within the chemical, cosmetic and food ingredient industries. Amyris is the only farnesene producer.

How are these molecules performing as potential platform chemicals? Starting with the basic condition – to be an intermediate –, it is fulfilled by the three molecules. Concerning the chemical structure, which must be adequate for performing transformations to a wide range of derivatives, important differences can be found among the three molecules. This issue is normally considered in publications on promising bio-based molecules. Chemical functionality can be based on the number of potential derivatives that can be synthesized through chemical and biological transformations. Simply, a candidate with one functional group will have a limited potential for derivatives, while candidate molecules with multiple functional groups will have a much larger potential for derivatives and new families of useful molecules. We propose having at least two functional groups as the basis for a platform molecule.

A platform chemical should have at least two functional groups. The nature of the functional groups – equal or different, easy or difficult to access – can improve the quality of the starting molecule. A diacid, such as succinic acid, fulfills this condition with two equal functional groups. Farnesene has four double bonds, which open up transformation possibilities. But butanol, with only one functional group, provides considerably less options. It should be converted to butene in order to increase its functionality and open up possibilities for new derivatives.

The derivatives have to be cost competitive. This is probably a very difficult evaluation to conduct mainly in the first stage of development, where the learning curve is still in its first stages. Many factors have to be considered to give a platform candidate a good position for cost competitiveness. We propose that a first proxy could be the biomass utilization efficiency,[20] which can be evaluated using the theoretical yield of sugar. This theoretical yield should be compared with the industrial yield. But as a general rule it is possible to consider that 90% should be the limit achieved at the mature industrial scale.

Succinic acid has a clear advantage at this point: with a CO_2 absorbing fermentation the theoretical yield of glucose is 100%. Butanol has a theoretical yield of 41% and that of farnesene is around 30%.

The availability of derivatives at a competitive cost depends also on the number of steps needed to obtain the final product. Each step represents not only additional capital expenditure (Capex) but also a lower final yield. The productivity measured in g per liter per hour is another factor that can strongly influence the Capex cost of fermentative processes.

Accumulated knowledge on the molecule could also contribute to its competitiveness. Taking the number of papers cited on Web of Science as an indicator, butanol is by far the molecule with more accumulated knowledge. There are almost 25 000 papers cited on Web of Science. Succinic acid has less than 8 000 and farnesene has less than 1 500 papers cited on the same database. Trying to overcome this situation, Amyris started to sell *via* the internet small quantities of farnesene to promote more research and experiments using the molecule. The same strategy to increase knowledge on succinic acid molecules has been proposed by Reverdia.

The ability to create value through scope and scale economies is directly related to the number of derived products (economy of scope) and their market volume (economy of scale). A platform chemical will be successful only if its final markets are a combination of commodity and specialty applications. If there are only a couple of commodities, the value chain tends to involve very small numbers of actors, probably with vertical integration by commodity producers. On the other hand, if there are only specialty applications, the lack of scale makes it difficult to produce the platform molecule at a cost low enough to multiply the applications.

The two final dimensions in our framework are strategic ones, which depend on management approaches: the innovation ecosystem and governance.

The term innovation ecosystem[21] refers to the set of innovative actors – upstream suppliers, buyers and downstream complementors – normally organized into a network. This set of actors provides products and services in order to create value and enable market diffusion of an innovation produced by a central organization called the leader or focal firm. The challenge for the platform chemical leader is to identify the set of actors and to mobilize their capabilities to contribute to the platform development. As long as the platform leader is able to combine the existing capabilities and induce the development of new ones, the innovation ecosystem works in an open innovation mode through a high level of collaboration among the actors. In this context, it is required[22] that the leader has the ability to organize and participate in an innovation process that combines a very diverse set of actors performing at the various stages of this innovation process.

Nonetheless, to develop a platform, the ecosystem must be coordinated. That is, some kind of governance must be in place. The term innovation ecosystem refers to the comprehensiveness of the competencies that have to be mobilized; the ecosystem must involve the complete set of competencies necessary to develop the platform innovation. So, an innovation ecosystem is critical to achieve innovation. On the other hand, governance refers to the way that relationships are organized and how the focal firm manages these relationships. An innovation ecosystem is essential for creating value and governance is critical for enhancing the value creation but also for defining the value capture, which means that these dimensions are interrelated.

It is possible to identify five types of global value chain governance[23] – hierarchy, captive, relational, modular, and market – which range from high to low levels of explicit coordination and power asymmetry. These types are used to try to categorize the variety of network forms and particularly those between the traditional poles: hierarchy and market coordination. Platform development involves very intensive coordination efforts in order to enable contact among the producers, a quite varied range of complementors, and end users or brand owners. This coordination aims at the same time to diffuse information and knowledge about the product, to acquire knowledge from the network and to combine this knowledge to create or co-create product applications to be adopted by end users. At the present stage, bio-based platform chemicals are still immature and under development. Thus the governance is essentially relational.

Relational value chains[24] "arise when product specifications cannot be easily codified, products are complex and supplier capabilities are high; this leads to frequent communication between buyers and suppliers within the framework of a certain degree of mutual dependence, which may be regulated through reputation, social ties and/or spatial proximity". Sturgeon[25] emphasizes that the access to expertise and competencies is a strong motivation for developing and maintaining external relationships.

Previous stories of the introduction of new molecules to the market can illustrate the nature of the relational competencies that innovating firms had to develop to get new applications adopted by end users. This was the case, for example, with GE Plastics in the 1980s when the firm innovated in the engineering plastics segment to provide a series of new applications.[26,27] However it is important to understand that although relational governance prevails in the development phase, it will probably evolve later to market or hierarchical governance, depending on the nature of the value chain. In that situation, the platform chemical producer could become a simple commodity supplier with low, if any, value capture from the derived products. At this point, it is important to distinguish another possible outcome: a wide range of new applications is developed based on the bio-based molecule but at the end of the day the molecule producers do not have a relevant position for value capture. In other words, value created by the innovation system may, depending on the governance and coordination mechanisms, be captured by other actors and not by the producer at the platform origin. Thus, it is critical that molecule producers, aiming to profit from a platform chemical, are involved in the platform structuring process, particularly in the innovation ecosystem organization and in the governance coordination.

How are biosuccinic acid, biobutanol and farnesene producers performing at the innovation ecosystem and governance level?

Farnesene has only one producer, Amyris. The ecosystem is vast and involves actors from very different industries and knowledge bases: chemistry, fuels, lubricants, cosmetics, flavors and fragrances. There is apparently a perception that the way to value farnesene for the market depends strongly on the nature of each market. Amyris has different business models from a joint venture with minority participation (with Total for biofuels) to a joint venture with equal participation (with Cosan for lubricants), a vertical integration from farnesene to a final market in cosmetics, and R&D cooperation with Braskem and Michelin. These different business models could be interpreted as a strategy to adapt the value creation and value capture to the particularities of each business.

Succinic acid is a very interesting case due to the strategic diversity of the players. Bioamber and Myriant, both startups with no capabilities other than succinic acid technology, have been trying to diversify their relationships in order to build an innovation ecosystem. But it is less clear whether the producers have been able to coordinate the ecosystem in their favor. Reverdia, a joint venture between DSM and Roquette, combines a strong chemical company, able to provide R&D and production expertise, with Roquette which has expertise in the supplying and processing of renewable feedstock. At the same time, Reverdia has been partnering with players in the downstream side of the value chain. Through this action, Reverdia shows a clear strategy to build a governance form in its favor. There is explicitly a proposal to license the technology to key users of succinic acid that has the potential to generate captive consumers. But the licensee cannot commercialize any succinic acid if the production is in excess of the captive use. The fourth player, Succinity, a joint venture between BASF and Corbion, has a very different strategy. Even if the company suggests a strategy for platform building, no alliance or relationships have been announced up to now. It is possible to suppose in this case that Succinity, considering that the competencies of BASF and Corbion are sufficient to develop the succinic acid platform, has decided to develop a product platform and not an industry platform.

Butanol is a particular case due to two important aspects: there is an efficient petrochemical production route and the technological barrier for entry to bio-based production is not high. As a result, many companies have been trying to enter the business. Three of them – Gevo, Butamax and Greenbiologics – are the most important. Gevo and Greenbiologics are startups and Butamax is a joint venture between Du Pont and BP. The platform vision of these companies seems less developed compared to farnesene and succinic acid. Their focus is apparently on butanol production, and less on the creation of new applications. Butamax has an explicit strategy to license its technology to potential producers of biofuels, particularly within the gasoline market. Gevo and Greenbiologics have some initiatives for building a platform vision. Gevo has been building an innovation ecosystem with a focus on jet fuels and resins (a renewable *p*-xylene for the production of drop-in PET in cooperation with Coca-cola). Some commercial partnerships are also in place. Greenbiologics' targets are solvents for various types of paints and adhesives, flavorings, fragrances, cosmetics and personal care. Its network is not very well developed.

Concluding remarks

We have explored the concept of technological platforms, which is currently applied in many industries, in the particular case of bio-based platform chemicals. Platform chemicals have been frequently mentioned within the specialized literature as a key development, being one of the most promising innovations for the bioeconomy. Taking the literature on technological platforms as a reference and the particular aspects involved in platform chemicals, we have proposed an analytical framework with six dimensions that we present as a tool for better understanding of the platform structuring process. The key dimensions, involving technological and strategic issues, are that the chemical should: be an intermediate molecule, have a flexible structure to make a wide range of derivatives possible, be cost competitive at the level of the platform molecule and at the level of the derivatives, be capable of generating scale and scope economies in the value chain, be organized within an innovation ecosystem and have associated well-developed mechanisms of governance. Through applying the framework to three bio-based platform chemicals – succinic acid, farnesene and butanol – we have discussed their current status and possible outcomes. Two of these molecules – farnesene and succinic acid – adhere to most of the conditions required for structuring a platform chemical. However, the innovation ecosystem is not complete and the governance mechanisms are still under development, so it is not clear if they will be capable of allowing a favorable position for value capture by the platform leader. Butanol structuring for a platform does not seem promising. The potential of the molecule is apparently not high and the strategic initiatives are in general not focused on innovation ecosystem structuring.

The framework should be further elaborated, particularly concerning the definition of clearer and more precise criteria for each dimension. This is a development that should be made so that the analytical framework may possibly be useful to help with the decision making of research and business personnel.

References

1 J. V. Bomtempo and F. C. Alves, *Chem. Biol. Technol. Agric.*, 2014, **1**, 19.
2 J. Bozell and R. Petersen, *Green Chem.*, 2010, **12**(4), 525–728.
3 J. Becker, A. Lange, J. Fabarius and C. Wittmann, *Curr. Opin. Biotechnol.*, 2015, **36**, 168–175.
4 D. P. Mcintyre and A. Srinivasan, *Strat. Manag. J.*, 2017, **38**, 141–160.
5 A. Gawer and M. A. Cusumano, *J. Prod. Innovat. Manag.*, 2014, **31**(3), 417–433.
6 M. Araujo, J. V. Bomtempo, F. Alves and F. Oroski, *ALTEC 2015*, 2015, http://www.altec2015.org/anais/altec/papers/608.pdf.
7 A. Gawer and M. A. Cusumano, *International Encyclopedia of the Social & Behavioral Sciences*, 2nd edn, 2015, vol. 3, DOI: 10.1016/B978-0-08-097086-8.73012-1.
8 G. G. Parker, M. W. Van Alstyne and S. P. Choudary, *Platform revolution*, W W Norton & Company, 2016.
9 D. S. Evans and R. Schmalensee, *Matchmakers: The New Economics of Multisided Platforms*, Harvard Business Review Press, 2016.
10 A. Gawer, *DRUID Summer Conference 2010*, 2010.
11 A. Gawer, *Res. Pol.*, 2014, **43**(7), 1239–1249.
12 A. Brandenburger and B. J. Nalebuff, *Coopetition*, Currency Doubleday, New York, 1997.
13 S. R. O. Weastra, *WP 8.1. Determination of market potential for selected platform chemicals: itaconic acid, succinic acid, 2,5-furandicarboxylic acid*, Report for BioConSepT FP7 project, 2013, http://www.bioconsept.eu/wp-content/uploads/BioConSepT_Market-potential-for-selected-platform-chemicals_report1.pdf.
14 E4tech, RE-CORD and WUR, *From the sugar platform to biofuels and biochemicals*, Final report for the European Commission, contract No. ENER/C2/423-2012/SI2.673791, 2015.
15 Department of Energy, *Top Value Added Chemicals from Biomass Volume I—Results of Screening for Potential Candidates from Sugars and Synthesis Gas*, 2004.
16 J. S. Luterbacher, D. Martin Alonso and J. A. Dumesic, *Green Chem.*, 2014, **16**, 4816.
17 H. Chesbrough, *Open Innovation: the New Imperative for Creating and Profiting from Technology*, Harvard Business School Press, Boston, MA, 2003.
18 F. W. Geels, F. Kern, G. Fuchs, N. Hindererc, G. Kungl, J. Mylan, M. Neukirch and S. Wassermann, *Res. Pol.*, 2016, **45**, 896–913.
19 F. Oroski, J. V. Bomtempo and F. Alves, *J. Bus. Chem.*, 2014, **11**(1), 131–138.
20 K. Iffland, J. Sherwood, M. Carus, A. Raschka, T. Farmer and J. Clark, *nova paper#8 on biobased economy*, 2015, http://www.biobased.eu/novapapers.
21 R. Adner and R. Kapoor, *Strat. Manag. J.*, 2010, **31**(3), 306–333.
22 J. West and M. Bogers, *J. Prod. Innovat. Manag.*, 2014, **31**(4), 814–831.

23 G. Gereffi, J. Humphrey and T. Sturgeon, *Rev. Int. Polit. Econ.*, 2005, **12**(1), 78–104.

24 P. Gibbon, J. Bair and S. Ponte, *Econ. Soc.*, 2008, **37**(3), 315–338, DOI: 10.1080/03085140802172656.

25 T. Sturgeon, *Industry Studies Association Working Paper Series*, 2008, From Commodity Chains to Value Chains: Interdisciplinary Theory Building in an Age of Globalization.

26 J. V. Bomtempo, D Sc thesis, CERNA, École des Mines de Paris, 1994.

27 J. T. Coe, *Unlikely Victory: How General Electric Succeeded in the Chemical Industry*, American Institute of Chemical Engineers, New York, 2000.

Faraday Discussions

DISCUSSIONS

Bio-based chemicals: general discussion

Harry Bitter, James Clark, Gadi Rothenberg, Avtar Matharu, Claudia Crestini, Dimitris Argyropoulos, Carlos I. Cabrera-Rodríguez, Bruce E. Dale, Christian Stevens, Assunta Marrocchi, Ines Graca, Hui Luo, Deepak Pant, Karen Wilson, Douwe Sjirk Zijlstra, Florence Gschwend, Xindong Mu, Long Zhou, Changwei Hu, Alexei Lapkin, Mark Mascal, Vitaliy Budarin, Andrew Hunt, Keith Waldron, Fang Zhang, Anna Zhenova, Joseph Samec, George Huber, Marta Coma, Xiaoming Huang and José-Vitor Bomtempo

DOI: 10.1039/C7FD90048A

George Huber opened the discussion of the paper by Xiaoming Huang: Where does the ash go? Does it go into the methanol solution? How much ash ends up on the catalyst *versus* the methanol *versus* the cellulose? Is the methanol being converted?

Xiaoming Huang answered: We did not trace the fate of the ash, and therefore we cannot give a firm answer to this question. Nevertheless, I speculate that a small part of the acid-soluble ash might end up in the methanol solution, and most of the ash should remain in the biomass matrix. A more dedicated study is needed. The conversion of methanol solvent is very small under these relatively mild reaction conditions (180 °C). A small amount of DME was produced due to dehydration. A small amount of methanol reacts with lignin monomers and xylose by methylation. In addition, a small amount of methanol reacts with hemicellulose-derived acetic acid to form methyl acetate. The GC peaks of these side products are typically smaller than the peak of the internal standard (22.5 mg *n*-dodecane).

Mark Mascal said: If you cling to the "lignin-first" approach, you will be left with the catalyst, cellulose, some hemicellulose, and almost certainly some inorganics in the solid after processing the lignin. How are you going to recover the catalyst in this case? Also, do you have an analysis of the residual cellulose? It must be at least partially degraded.

Xiaoming Huang responded: Indeed, one challenge of this lignin-first approach is that the metal catalyst ends up in the solid residue. There are several possible solutions available for tackling this separation issue. It has been demonstrated that the use of a magnetic catalyst such as Ni/C and RANEY® Ni allows successful separation of the metal catalyst from the biomass residue.

Alternatively, it is possible to trap the metal catalyst in a microporous cage or a basket before and during the reaction. The lignin could be extracted by the acidic solvent, followed by entering the basket for further depolymerization. The solid–solid interaction between the metal catalyst and biomass is not required. This strategy has been demonstrated by the groups of Mahdi Abu-Omar and Bert Sels. The alternative solution proposed by us is to integrate the lignin release step and hydrogenolysis steps in a two-stage flow reactor. Lignin can be extracted from the biomass in the first-stage reactor. The lignin–methanol solution could be immediately pumped through the metal catalyst bed immobilized in the second-stage reactor for depolymerization. In this way, there is no need to separate the catalyst and biomass in the downstream process. We have sent out some of our samples to third parties to analyze the residual cellulose fraction. It was found that the average fiber length of this cellulose fiber is slightly shorter than those obtained from the traditional Kraft pulping process. XRD analysis showed that the crystallinity of this cellulose residue is higher than that of the raw biomass. SEM analysis did not show much difference in terms of morphology.

Harry Bitter commented: Thank you for the interesting paper on using catalysis in a lignin-first approach. In your approach you used a Pd/C catalyst under acidic conditions and in the presence of H_2. The acidic conditions you used might lead to Pd oxidation and leaching. On the other hand, the oxidation can be overcome to a certain extent by keeping the catalyst reduced as we have shown for Ni in water – here the H_2 helps.[1] Can you comment on the stability of your Pd catalyst under the chosen reaction conditions?

1 T. van Haasterecht, C. C. I. Ludding, K. P. de Jong and J. H. Bitter, *J. Catal.*, 2014, **319**, 27–35.

Xiaoming Huang replied: We also realized that the presence of acid might cause catalyst leaching of the Pd metal catalyst. To verify this, we have performed model reactions in the absence of biomass, so that we can recover the Pd/C catalyst for ICP-AES elemental analysis. We found that only HCl caused significant leaching – only 78% of Pd could be recovered after the reaction in methanol at 180 °C for 2.5 h in the presence of 30 bar H_2. About 92% of Pd could be recovered when H_3PO_4 acid was used as the co-catalyst. When Al triflate and H_2SO_4 were used as acid co-catalysts, the leaching was negligible as about 96% of Pd could be recovered, which is nearly the same as the acid-free case (95%). The more pronounced leaching in the case of HCl and H_3PO_4 might be due to the addition of water as 37 wt% (in water) of HCl and 85 wt% (in water) of H_3PO_4 were used for the reactions. Based on these experiments, we believe that the Pd catalyst is rather stable under the chosen reaction conditions in the cases of Al triflate and H_2SO_4. I think the H_2 and water-free atmosphere helps to avoid leaching.

Changwei Hu asked: Is there a difference between the conversion of converted biomass component and the yield of monomers? Has all been converted to small molecules? My concern is that there could be some oligomers forming in the solution.

Xiaoming Huang answered: The reactions were performed in methanol solvent at high temperature. Lignin could be selectively extracted and converted into small molecules. In a typical reaction, about 90 wt% of lignin was extracted from the biomass and mainly converted to monomers. The monomer yield is about 50 wt% based on the total lignin content in the biomass, which is close to the theoretical maximum value. The rest is mainly dimers and oligomers, which are linked by C–C bonds and cannot be depolymerization under these conditions. However, we could not quantify the yield of these dimers and oligomers. The hemicellulose is also extracted from the biomass to a large extent. About half of the hemicellulose was removed from the biomass matrix and mainly converted into methyl xylosides. The cellulose fraction is largely untouched because of its highly crystalline nature.

Christian Stevens enquired: Can you comment on why the acid you use performs better than some others? You would expect that the pK_a of the acid is playing a more important role.

Xiaoming Huang replied: Yes, the pK_a value of the Brönsted acid is playing a very important role. For the strong acids like H_2SO_4 and HCl, a very small amount of acid is able to effectively promote the catalytic performance. For a medium strength acid like H_3PO_4, a large amount of acid is needed. For the weak organic acids like acetic acid, even a large amount of acid does not enhance the catalytic performance. Therefore, the right concentration of strong or medium strength acids is preferred for this process.

Joseph Samec queried: Can you comment on the hemicellulose – what happens to it in your system?

Xiaoming Huang responded: Hemicellulose is always released in the solution together with lignin because they are directly linked with each other. In a typical experiment, about half of the hemicellulose was removed from the biomass matrix and it is mainly converted into methyl xylosides.

Andrew Hunt asked: My understanding is that the catalyst ends up in your solid material at the end, have you tried to recover the catalyst and reuse it?

Xiaoming Huang responded: It is difficult to separate the catalyst from the biomass residue in the current process. In one of our experiments, we managed to recover some of the Pd/C catalyst. When wood chips were used as the feedstock, we can separate a large amount of Pd/C simply by sieving based on the particle size difference. The recovered Pd/C was subjected to a new reaction in order to test its reusability. We found that the catalyst maintained about 80% of its initial activity. The slight activity drop could be partially due to the loss of Pd/C catalyst during recovery, because some of the Pd/C was trapped in the porous biomass structure and could not be recovered completely in this way.

Xindong Mu said: Combining hydrogenation with an acid catalyst would be an effective system to remove oxygen. Can you determine any aromatics, such as benzene, as products in your system? Do all of your products contain oxygen?

Also, the use of acid will cause metal leaching into the liquid phase, can you prove if it is a homogeneous or a heterogeneous catalytic system?

Xiaoming Huang answered: No, we did not detect any deoxygenated aromatics. Indeed, the combination of metal and acid catalysts is typically used for removing oxygen from biomass-derived molecules by the hydrodeoxygenation reaction. However, this type of reaction typically occurs at higher reaction temperatures (above 200 °C) in non-alcoholic solvents, such as water and alkanes. In our case, we used methanol as the solvent and operated the reaction at 180 °C. This catalytic system is not effective for removing oxygen. It was reported that methanol interacts with the Pd metal catalyst surface, hindering the adsorption of lignin molecules on the Pd catalyst and reducing the Pd catalyst activity.[1] We observed severe leaching of Pd metal caused by HCl and a small amount of leaching caused by H_3PO_4. For the Al triflate and H_2SO_4 acids, the leaching of Pd is not noticeable as 96% of Pd could be recovered in the solid catalyst, which is nearly the same as for the acid-free case. Therefore, we believe that the Pd catalyst mainly functioned as a heterogeneous catalyst when Al triflate and H_2SO_4 acids were used as co-catalysts.

1 X. Y. Wang and R. Rinaldi, *ChemSusChem*, 2012, **5**, 1455–1466.

Long Zhou said: You mentioned different types of acids. Organic acids do not have good results at a low concentration. Sulfuric acid is better as the concentration can be lower. Can you comment on which is better and more sustainable, a small amount of sulfuric acid or a large amount of organic acid?

Xiaoming Huang replied: Based on the catalytic performance, a small amount of strong acid is preferable over a large amount of organic acid. However, I cannot say for sure which acid is better and more sustainable, because we also need to take many other aspects into account. This includes economic, safety and environmental issues.

Douwe Sjirk Zijlstra queried: You use methanol as a solvent for your tandem reactions. Did you try any other solvents?

Xiaoming Huang responded: We also tested other solvents such as ethanol, 2-propanol, tetrahydrofuran (THF), water and *n*-hexane under the same conditions. When ethanol was the solvent, the monomer yield was similar to that obtained with methanol. The use of 2-propanol solvent gave poor results with a lignin monomer yield of 33 wt% after 2 h of reaction. The use of *n*-hexane as the solvent led to the lowest monomer yield, which is likely due to the low solubility of lignin fragments and monomers in this apolar solvent. The limited solubility of these products in water also caused a low monomer yield. Although THF is a very good solvent for lignin and its products, the lignin monomer yield was still much lower than that obtained in alcoholic solvents. We believe that the main reason for this is the absence of transetherification and transesterification in this solvent, which play important roles in the delignification. Therefore, alcohols are the preferred solvents in this process.

Changwei Hu commented: Biorefinery faces the problem of the starting materials being in the solid state, and one of the problems is how to dissolve the solid. If it can be dissolved, then the obtained fluid could be treated in more traditional ways. How did your solid catalyst activate the solid state starting material?

Xiaoming Huang answered: It is true that one of the problems for biorefinery is how to effectively liquefy the biomass components. In my opinion, one promising approach is to selectively remove the lignin fraction from the biomass matrix by solvolysis using cheap and green solvents, like alcohols or their mixtures with water. I do not think that solid catalysts will play much role in activating this delignification process due to the insufficient solid–solid interaction. However, the addition of a solid catalyst can play a crucial role in stabilizing the reactive intermediates formed during the solvolysis process. The traditional solvolysis process typically results in the formation of reactive intermediates that contain aldehydes, ketones and other unsaturated double bonds. These intermediates will transform into condensed products like humins. The addition of a metal catalyst and hydrogen allows quick elimination of these reactive intermediates and conversion into more stable small molecules. In this way, the overall efficiency of the fractionation process could be improved.

Mark Mascal opened the discussion of the paper by Andrew Hunt: The main thing that stands out to me is the 233 °C boiling point. This is going to involve a substantial energy burden in the recycling of the solvent, can you comment on this? Also, chloroform and DCM are hydrolytically stable, but there are sensitive functionalities (acetal, ester) in MCC, so it is not a one-for-one replacement for halogenated hydrocarbons as such.

Andrew Hunt responded: Frequently, bio-based solvents have higher boiling points and are more functionalised compared to traditional petrochemical-derived solvents. This is a significant challenge that must be addressed, as the use and subsequent removal of higher boiling point solvents may require significant energy input. The development of new sustainable separation systems for bio-based solvents that avoid evaporation or distillation is one approach that could be utilised to resolve this issue. Separations for bio based molecules to retain functionality are also an important part of a holistic biorefinery concept. It is agreed that methyl(2,2-dimethyl-1,3-dioxolan-4-yl)methyl carbonate (MMC) has sensitive functionalities and as such would not be a one-for-one replacement for halogenated hydrocarbons. The fact that MMC has a boiling point of 233 °C, means it failed an AMES test and cannot be used as a one-for-one replacement for halogenated hydrocarbons. This demonstrates the shortcomings of this bio-based molecule as a potential solvent, but highlights the importance of a step-wise method for developing bio-based solvents.

Gadi Rothenberg remarked: It is legitimate to use the Hansen parameters, however, when you have a multi-parameter analysis, you will, at the end, face the problem that the solvent is objectively okay but does not fit to your application. You need to have axes that fit to your application when doing analysis. Dipolar aprotic solvents are normally either hydrophilic/lipophilic or donor/acceptors –

you could use these parameters for analysis. This gives a simple solvent map that you can use (see Fig. 2 in ref. 1). In general, solvent optimisation is per application, so the conditions of the application should be included in the optimisation parameters.

1 E. Burello and G. Rothenberg, *Adv. Synth. Catal.*, 2003, **345**, 1334–1340.

Andrew Hunt answered: Many thanks for the comment. I agree it is legitimate to use the Hansen parameters and that solvents can have a wider range of properties that are application specific. One challenge that is faced in the development of replacements for traditional solvents (that are being restricted due to their toxicity, under legislation such as REACH) is to produce candidates with properties that enable their use as drop-in alternatives for a wide variety of applications. These new candidates must also be non-toxic, efficient and cost effective. The method set out in this work highlights how important it is to have toxicity or mutagenicity testing early on in the solvent development process. When developing solvents for specific applications, your suggestion of adding axes (or set of parameters) that fit to an application will be important when assessing the new solvent's performance. Similarly to your suggestion, in this current work, we utilised solvent maps based on the Kamlet–Abboud–Taft (KAT) solvatochromic polarity scale to place our new solvents in context with the current solvent space. The KAT scale can be used to assess the potential properties of a solvent, including the solvent dipolarity (π^*), the hydrogen bond accepting ability (β) and the hydrogen bond donating ability (α).

Bruce Dale said: I am afraid that we are often too "pure" in our approach to bio-based chemicals. It becomes more of a religion than an approach to economy. The key objective is to make real progress toward sustainability by introducing more circular flows of materials. So if we are "greenish-brown" in some of our approaches, but make real progress, I do not think that we should make the perfect the enemy of the good.

Andrew Hunt replied: It is important to make real progress toward sustainability and certainly the development of circular economies is a key objective that we are all working towards. There are many shades of green and moving towards these greener shades (from brown to green) are positive steps forward, even if these changes are preserved to be small at the present time.

Avtar Matharu asked: In light of your talk and paper, what would now be the best operational order? Toxicology after step 6?

Andrew Hunt answered: The authors believe that the operational order that has been suggested in this work makes sense in terms of solvent development. Firstly, it is important to identify the key properties of the solvent that is to be replaced. These are then used in the selection of potential candidates, and further *in silico* modelling is used to narrow down the potential number of candidates. Adopting a green synthetic approach early on in the development process to optimise the system ensures that safety, sustainability and the principles of green chemistry are at the heart of the solvent development process and that enough solvent can be produced to be used for property or toxicity testing. It is agreed that steps 6 (Testing physical properties) and 7 (Assessing performance and toxicity) can take place in parallel.

Xindong Mu enquired: Have you ever checked the stability – the thermal stability and the stability in acidic and alkaline solutions? Also, for industrial applications, is the solvent combustible? What is the flash point of your solvent?

Andrew Hunt responded: The flash point, auto-ignition temperature, thermal stability and stability in acids or bases are important parameters that should be tested for new solvents. In this current work, steps 6 (Testing physical properties) and 7 (Assessing performance and toxicity) were conducted in parallel. The testing of methyl(2,2-dimethyl-1,3-dioxolan-4-yl)methyl carbonate (MMC) was halted when the solvent failed the AMES test. It would be a normal requirement for the auto-ignition temperature, thermal stability and stability in acids tests to be undertaken, and other examples where we have conducted such tests can be found in the literature, for example, ref. 1.

1 F. Byrne, B. Forier, G. Bossaert, C. Hoebers, T. J. Farmer, J. H. Clark and A. J. Hunt, *Green Chem.*, 2017, **19**, 3671–3678.

Christian Stevens asked: Why did you select the Diels–Alder reaction as a model reaction, since it is very rarely used in industry? There are very few examples of the Diels–Alder reaction in commercial applications.

Your synthesis of a green solvent uses dimethylcarbonate as the starting material, however the solvent characteristics are similar to dimethylcarbonate. So, would it be more logical to use dimethylcarbonate as a green solvent?

Andrew Hunt replied: For some time it has been known that Diels–Alder reactions can be strongly influenced by solvent effects. Such reactions are used in industry, including in the flavour and fragrance sector. This reaction was therefore selected to probe the solvent effects of methyl(2,2-dimethyl-1,3-dioxolan-4-yl)methyl carbonate (MMC) and set this solvent's properties in context with other commercially available traditional solvents. The other example in this paper is the Friedel–Crafts acylation reaction, which is widely used throughout industry.

The method set out in this paper provides a concise route to develop new safer solvents. As highlighted in the introduction discussion, the case study on methyl(2,2-dimethyl-1,3-dioxolan-4-yl)methyl carbonate (MMC) presents some of the challenges faced when developing new solvents. In this case, it was found that the MMC solvent failed an AMES test, thus indicating its potential mutagenic properties. This demonstrates that the approach set out in this work is a valid method to not only identify promising candidates but also eliminate those with issues, such as MMC. The solvent properties of MMC differ from dimethylcarbonate (DMC), demonstrating a higher solvent dipolarity (π^*) based on the Kamlet–Abboud–Taft (KAT) solvatochromic polarity scale, and in fact MMC outperforms DMC in both the Friedel–Crafts acylation and Diels–Alder reactions. However, both solvents are in the same region of the solvent space (please see Fig. 2, DOI: 10.1039/C7FD00049A). Therefore, due to the issues of the toxicity of

MMC and the relative proximity of DMC in the solvent space, in this case it would be recommended to utilise dimethylcarbonate.

Assunta Marrocchi communicated: What about a comparison between the toxicity of MMC and that of the traditional solvents most commonly used for the Friedel–Crafts or Diels–Alder transformations? In other words, wouldn't MMC be a better solution anyway, if it has a reduced toxicity?

Andrew Hunt communicated in reply: In the methodology set out in this paper, methyl(2,2-dimethyl-1,3-dioxolan-4-yl)methyl carbonate (MMC) was not taken to full toxicological testing as it failed the AMES test. Ideally, we would like to replace a problematic solvent with one that is safe, cost effective and ideally bio-derived, however, this is not always possible. Based on the route for solvent development as discussed in this paper, it would be recommended that the MMC solvent should not be taken further or advocated for use, but there are many shades of green on the way to a sustainable bio-economy, thus moving towards those that are greener will be a positive a step forward.

Alexei Lapkin remarked: In your methodology, a life cycle assessment (LCA) is added at the end as a "good to have" thing. However, LCA is the only methodology that allows the identification of any system-level problems that relate to the overall chemical route, as well as product use. LCA is no longer that hard to do, with LCAs on complex reactions now possible. The use of a LCA at an early chemical development stage has also been advanced and LCA as a design tool is becoming rather powerful. Hence, looking at a LCA as early as possible may prevent many errors at the late process development stage.

Andrew Hunt replied: This is a good point. As outlined in the paper, steps 1–7 of the solvent development methodology can be undertaken over a relatively short time frame, thus enabling the collection of essential data required for a full LCA. Some aspects of the LCA could be brought forward to an earlier stage in the process and may lead to further impact in improving the process or solvent. However, a full LCA should take place alongside acquiring the data from steps 1–9.

James Clark opened the discussion of the paper by Marta Coma: what is the effect of algae on the plastic conversion? Do we know much about the "catalysis", optimum amounts, *etc.*?

Marta Coma answered: The plastic does not degrade sufficiently under these conditions on its own, and we have demonstrably proven that plastic is incorporated into the final products. In addition, the yields are higher with plastic than with the biomass alone. This suggests that the algae (or at least the ash) have a catalytic effect on this reaction. We are currently undertaking a much bigger study on a range of seaweeds and different types of plastics, and will look into the mechanism of reaction here. From the literature though, it does look necessary to generate free radicals to help break down the plastics. This brings in an interesting factor, that some biological degradation prior to the HTL process might be beneficial to helping generate these. Of course, free radical generators (such as peroxide) would also be an interesting supplement to the reaction, as would a catalyst that can aid this process. We have had some success in the past with simple iron salts, and this could be a potential route for this system.

Karen Wilson said: You are looking at hydrothermal processing of polyethylene and polypropylene with algae – I assume they are pure and not waste plastics that you are working with? Do you know if cross-linking of the polymer influences the ability to degrade the polymer? Regarding the mechanism, are you planning to use model systems, using, for example, organic acids as possible degradation products of algal oil to help understand how polyethylene and polypropylene degrade during co-processing?

Marta Coma responded: We used pure plastics sourced from Sigma. I think the cross-linking question is fascinating. As we are assuming a radical mechanism, then branching in the polymer should have a reasonably large effect with the sp^2 hybridisation. This is certainly not something we considered, but would make a very interesting aim for future work on this. I think the model compound approach is also a good one. We would use radical initiators, organic acids, amines and alternative oxygenates. This would also benefit from a firm analysis of the original bio-crudes by GC to try to isolate some of the plastic breakdown products. This is something that, with polypropylene at least, should be relatively easy to spot due to the branching.

Carlos I. Cabrera-Rodríguez asked: 1. You mentioned that during the anaerobic fermentation the pH goes below 6, what exactly is the obtained pH?

2. I was glad to see you got 50% propionic acid production, which is different from the normal case in which most is acetic acid, can you elaborate on this? Is this because of the type of organic waste used?

Marta Coma replied: 1. pH indeed decreases below 6 if it is not controlled, but it stabilises around pH 5.5. When fermentation takes over, volatile fatty acids (VFA) are accumulated and these have a pK_a of around 4.7, which would be the minimum pH achieved. However, the buffer capacity of the media, mostly carbonates ($pK_{a1} = 6.3$) or less probably phosphates ($pK_{a2} = 7.21$), will have an influence on the equilibration of the final pH, thus bringing it over the VFA equilibrium point.

2. The tests presented in the manuscript were carried out over 7 days of fermentation. From our experience, propionic acid appears when overloading a non-acclimatised inoculum. As it can be seen in Fig. 2 (DOI: 10.1039/C7FD00070G), a higher proportion of propionic acid is obtained when the food to microorganisms (F/M) ratio is larger. When we have used mid-conditions (F/M of 5), the proportion of propionic acid within the total VFA keeps on decreasing until it stabilises.

Avtar Matharu enquired: Fig. 6 (DOI: 10.1039/C7FD00070G) – why don't the C : N ratios change more when you add plastics? You seem to have the same level of nitrogen all through the experiments? Unless I am reading the figure wrong, the nitrogen ratio seems to stay the same.

Marta Coma responded: Yes, the N content in the bio-oil does not reduce dramatically as would be expected if you were just liquefying the plastic and depositing it there. However, this needs to be looked at in context of the N content of the solids. This reduces dramatically, suggesting two things – that we are, in the presence of plastic, making more oil soluble nitrogen containing compounds, effectively depositing them from the solids to the oils – but also that we are not reacting all the plastic under these conditions, and so are, in effect, diluting the solids with a non-N containing species.

Andrew Hunt queried: You mentioned economic analysis in your paper, are you looking to do a full economic analysis or just an economic analysis of certain parts of the process?

Marta Coma answered: A comparison of alternative products and routes will need a full economic analysis, including capital and operating costs of the whole process comprising reactors, separations and purification of the products. This will be done once the main reaction conditions are optimised experimentally.

Xindong Mu said: The picture you presented was taken in the city where I come from. Some local universities have worked on the treatment of algae. Algae contain a lot of sulfur and so they generate a lot of bad smelling gases. In your system, where does the sulfur go?

Marta Coma replied: The sulfur cycle tells us that sulfur can be in the solid phase as elemental sulfur (easy to be oxidised), sulfate, sulfite, *etc.*, in the aqueous phase, and sulfide (lethal at high concentrations) or sulfur oxides in the gas phase. Not many sulfur gases are produced in HTL. Sulfide is usually produced by the reduction of sulfate; hydrogen formed in anaerobic fermentation could enhance this process and generate such toxic and corrosive gases. Therefore, we are now looking at the elemental composition of each stream to see how the integration of the processes could be affected and the sulfur composition is an important element to take into account.

Christian Stevens commented: The goal of your work is to produce alcohols *via* fatty acids that are being produced by fermentation. Is there a specific reason from an economical point of view why you do not try to isolate the fatty acids and stop the process at this stage?

Marta Coma responded: The goal of the work is indeed to produce fatty acids instead of methane, not alcohols. However, ethanol, lactic acid and hydrogen are produced in the first stage of fermentation and the accumulation of these and short chain fatty acids promote the reversed β-oxidation and thus, the generation of medium chain fatty acids with higher values, which are easier to extract.

Hui Luo asked: Regarding the aqueous phase produced in your hydrothermal liquefaction, have you done any characterization?

Marta Coma answered: Yes, we did characterise the aqueous phase and found suitable C, P and N deposition in the phase (although we did not characterise for all metals). It seemed not particularly dependent on the plastic concentration, so we felt that there were more interesting aspects of this system to present in the space that we had available.

Changwei Hu remarked: The selectivity of the products is a very important concern, otherwise the products cannot be used in the manner that we use chemicals from petrol.

Marta Coma replied: The selectivity of the products is indeed a key point in the process. The technology is still under study, but we observed that by simply modifying some operational parameters the selectivity towards a single product increases. Furthermore, as these chemicals are dissolved in water, extraction techniques must be applied. By generating compounds with higher hydrophobicity (*i.e.* medium chain fatty acids), the extraction from aqueous media could be facilitated and this would improve the selective production of chemicals.

Deepak Pant opened a general discussion of the papers by Xiaoming Huang, Andrew Hunt and Marta Coma: In her presentation, Dr Coma mentioned the term "organic waste" and also algal bloom was mentioned as being waste. However, the term "waste" has a negative connotation to it. The point is, not everything that has been mentioned is really waste, and some of it is currently being used for some other applications. The moment a utilization value is ascribed to something, it stops being waste. So somebody's waste may be a resource for someone else. Similarly, algal bloom is biomass and although it has a nuisance value, it is a biomass source nevertheless. So my question to the esteemed audience is – is it not better to call such materials "residues", "byproducts" or "biomass" rather than "waste"?

Changwei Hu answered: I think "biomass" is suitable.

Gadi Rothenberg replied: "Biomass" is a suitable term because it is general enough to encompass many activities (including biomass waste recycling and reuse), yet excludes non-biomass waste.

Christian Stevens asked: Which boundaries should be chosen for a LCA study on green solvents?

Andrew Hunt responded: It can be hard to define system boundaries for a LCA study on green solvents. However, in our current work (DOI: 10.1039/C7FD00049A), we have suggested that "at the same time as REACH registration is being sought, a full cradle to grave LCA of the solvent will provide a holistic assessment of the solvent, thus providing investors and end users with the confidence to commercialise the process or utilise the solvent in their processes."

Anna Zhenova commented: While LCA is a priority for us at the Green Chemistry Centre of Excellence, it is not practical to do it as one of the first stages in solvent discovery. Even predictive LCA consumes considerable time and resources, so it can only be done when there are a few final candidates. The first stage has to be narrowing down the library of potential solvents *via* solubility parameters, COSMO-RS, and other important characteristics.

Deepak Pant said: Most speakers showed us how to use waste and bioresources in a better and more economic way, however, currently most of industry is based on fossils, so how can we better quantify the impact of bioresources so that people can understand in numbers how much of an impact it will have? This is important if we talk about moving from a fossil-based economy to a bioresources-based bioeconomy.

Bruce Dale answered: I cited some of these data in my Introductory Lecture (DOI: 10.1039/C7FD00173H). Ref. 1 is a bit old but the lecture has the most recent data. I think the energy = prosperity relationship should resonate with most people.

1 B. E. Dale and R. G. Ong, *Biotechnol. Prog.*, 2012, **28**, 893–898.

Changwei Hu replied: A systematic consideration of the biomass-based industry should be established. At present, only separated products are reported, and the competitiveness is not as strong as the fossil-based industry.

James Clark commented: Brown chemistry may not be a term that kicks off, perhaps different shades of green would be better. The theme here is bio-based chemicals and we have heard a lot about solvents and aromatics, both in very large volumes. How quickly can we expect bio-based alternatives to replace traditional petro-based products? Are we going to use the same amount of aromatics in the future? Do we have particular targets? I realise that we are subject to many factors, often outside of our control.

Changwei Hu responded: Perhaps we might also change the ways that we use chemicals in the future, because of the bio-based products obtained.

Andrew Hunt opened the discussion of the paper by Changwei Hu: I see lipid extraction gives you advantages, including higher yields of the bio-oil, but also leads to enhanced degradation of proteins and production of certain compounds such as indoles. Can you give us some insight into the mechanism and why lipid removal is enhancing the pyrolysis?

Changwei Hu replied: The main cause is the fact that after lipid extraction, the pyrolytic process is simplified. Because of the interaction between the intermediates from the pyrolysis of lipids and the proteins, the products obtained before lipid extraction include a certain amount of amides. After lipid extraction, the formation of amides is significantly reduced. Thus, the extraction of lipids on the one hand breaks the interaction between the lipids and proteins, making the starting material simplified; on the other hand, the pyrolysis process is also simplified, and then the product distribution is changed.

Gadi Rothenberg asked: The yield of bio-oil is 28–38%, is that based on dry mass? This is an impressive number, higher than I have seen in other reports for algae. What is the leftover/what is the residue? Is it cellulose? Also, can you comment on the energy input necessary for reaching this yield?

Changwei Hu answered: This yield is based on the dry mass of algae used. Of course, because of the pre-extraction of lipids or saccharides, the amount extracted was added to the total yield. We also got residue. Based on our analysis, the residue might be heat resistant protein or char formed. Considering the ratio of energy input and output, we have analyzed the microwave assisted process, and found a highest energy gain of 216%.[1] We have not yet analyzed the thermal process.

1 R. Zhang, L. Li, D. Tong and C. Hu, *Bioresour. Technol.*, 2016, **212**, 311–317.

Ines Graca enquired: What are the main differences between bio-oil from algae and bio-oil from biomass? If you want to use this bio-oil in the transportation sector, can it be directly applied? Or is further upgrading/processing needed?

Changwei Hu replied: The differences between the bio-oil from algae and that from lignocellulosic biomass come from the differences in their composition. The lignocellulosic biomass samples are usually composed of hemicellulose, cellulose, and lignin, while algal samples are usually composed of lipids, saccharides, and proteins. Thus, the most obvious difference in the composition of these bio-oils is the nitrogen contents. The bio-oil obtained directly from the pyrolysis of algae could not be used directly for the transportation sector, it must be upgraded.

Dimitris Argyropoulos said: My experience and my understanding of bio-oils, is that they are extremely reactive. They polymerize if left at room temperature on the bench. You use GPC and GC-MS, do you do any stabilization prior to these analyses? This could be extremely acute of a problem, especially when injecting such bio-oils in GC-MS when the injector temperature is well above 150–180 °C and the column temperature gradients are at similar elevated levels. Therefore, in view of the extreme reactivity of bio-oils, are the data that you received representative?

Changwei Hu responded: Thank you very much for your comments and question. It is true that bio-oils obtained from pyrolysis are not stable and polymerization may occur if left even at room temperature on the bench. Considering this aspect, we kept the time interval between the analysis and pyrolysis the same for different samples to ensure the same systematic errors. Based on the fact that the injector temperature is lower than the pyrolytic temperature, we neglected the possible variation of the samples in the injector. Therefore, we considered these data to be representative for a comparison study. Of course, the analysis of pyrolytic bio-oils is still a challenge due to their complexity, including their instability.

Vitaliy Budarin remarked: One of the major products of biomass pyrolysis is anhydrosugars (levoglucosan, levoglucosenone). These sugars are volatile. Have you seen any of these sugars in GC?

Changwei Hu answered: Yes, we have observed different pyrolytic products of carbohydrates from Taihu Lake algae and lipid extracted algae.

Christian Stevens said: The algae are first sun dried. If the process was performed on a large scale, how difficult would it be to prevent rotting of the biomass?

Changwei Hu replied: The scale up of the pyrolysis process could be realized, and then the process could be performed on a large scale. The algal samples should be dried to avoid rotting. We have first dried the sample in the sunshine.

Fang Zhang enquired: Why choose such a short treatment time of 2.0 h? If we use a lower temperature, could we use a longer time. Does 3 or 4 h alter the results?

Changwei Hu responded: We have evaluated the effect of treatment time and found that for the amount of sample treated, 2 h is sufficient. Temperature is another factor for the efficiency of the extraction. We have optimized these parameters and those reported in the present paper (DOI: 10.1039/C7FD00065K) are the most effective. Perhaps other options are also possible.

Xindong Mu said: Concerning the solid residue from low temperature pyrolysis, we can get some biocrude. Have you ever determined the composition and characteristics (*e.g.* surface area) of the solid residue? Also, during the extraction, could some of the protein be extracted? And will this cause a difference in the next step reaction?

Changwei Hu answered: We have analyzed the composition of the solid residue from our pretreatment. The results showed that in the extraction process, small portions of protein could also be extracted, and this will affect the next step reaction.

Mark Mascal remarked: You have chosen a pyrolytic method to process your algae after extraction of the lipids and you obtain a modest yield of bio-oil as a complex mixture of compounds. If you instead chose a chemo-catalytic method, you could valorize your carbohydrate fraction to a much greater extent. So why choose the thermal route?

Changwei Hu replied: Thank you very much for your comment. In this work, we want to show the importance of fractional conversion of biomass materials. As you pointed out, the pyrolysis of biomass would give us bio-oil as a complex mixture of compounds. Furthermore, even if we use a chemo-catalytic method, as actual algae are used as the starting material, the selectivity of the products would still be low because of the complexity of the starting material. Thus, fractionation of the raw material is important. Of course, after fractionation, the use of a chemo-catalytic method to convert the simplified component will get higher selectivity to the target products. In the present work (DOI: 10.1039/C7FD00065K), we just want to show that different conversion ways could provide different products, and the selectivity of the target products is controllable, while fractionation provides an efficient option.

George Huber asked: Have you done a nitrogen balance? Do you know how much nitrogen ends up in the solid *versus* the liquid *versus* the gas? Do you get any ammonia?

Changwei Hu responded: We have not done nitrogen balance measurements and the corresponding calculation. We just analyzed the nitrogen content in the liquid products. There is ammonia produced in the pyrolysis process. The reaction of the ammonia formed with the lipids to form further amides has also been observed. So, the formation of ammonia in the pyrolysis is certain. The denitrogenation of the liquid products also seems very important for the sake of using the bio-oil as fuel.

James Clark opened the discussion of the paper by José-Vitor Bomtempo: One of the central premises I am struggling with is that you propose having at least two functional groups for a platform molecule, but some single functional group molecules, *e.g.* methanol and butanol, have been used for years. Your method would rule these out. Are you looking at this differently? Or is it a case of terminology and definitions? I think it is not realistic to dismiss for example, butanol, which may be a large volume, low cost bio-based chemical that can be converted into many useful chemical products.

José-Vitor Bomtempo replied: I take the idea of platform chemicals in the sense of technological platforms as they are used in many industries. I agree that it is possible, as in the case of methanol and butanol, that a molecule with one functional group could be converted into many useful chemicals. However, if the molecule producer wishes to create and capture value, as a platform leader, the molecule should be easily transformed into many final applications. Having only one functional group does not exclude a molecule as a platform, I agree. The point I have intended to stress is that having two or more functional groups would give the molecule producer a better position to manage the value chain and a significant share of the value created. The DOE Report from 2004 is a classic reference on the identification of promising bio-based molecules.[1] The idea of more than one functional group is in the report, which presented twelve high-value bio-based chemicals as building blocks: "Building block chemicals, as considered for this analysis, are molecules with multiple functional groups that possess the potential to be transformed into new families of useful molecules". The point deserves more development, I agree.

1 Department of Energy, *Top Value Added Chemicals from Biomass Volume I— Results of Screening for Potential Candidates from Sugars and Synthesis Gas*, 2004.

Gadi Rothenberg said: This is a very interesting approach that differs from the traditional "chemical" way of looking at bioresources, yet is still very important. Regarding the three so-called "strategic factors" in your paper, factor 5 says that one condition is that the molecule should be regulated by governance mechanisms. Can you explain more about this? As far as I know, the existing platform compounds (such as ethylene, propylene and BTEX) did not come with governance – they simply were used on a large scale and the rules evolved with them (or more likely after them). Why do you say that the governance mechanisms must come first?

José-Vitor Bomtempo answered: The vision we have been trying to develop in our research on platform chemicals is not based on the established petrochemical industry but on the possibility of new business models similar to the technological platforms we see now in many industries.

If a producer intends to explore its potential platform molecule and to capture value from the final products, it has to coordinate a quite complex development. As a platform leader, it is required that the molecule producer has the ability to organize and participate in an innovation process that combines a very diverse set of actors performing at the various stages of this innovation process. As the governance refers to the way that relationships are organized and how the focal company manages these relationships, it would be critical to the molecule producer to participate in the governance development. This is not to say that the governance mechanisms must come first. But the platform leader must follow and coordinate these governance mechanisms with appropriate timing. If not, the governance in place may be unfavorable to the molecule producer.

Florence Gschwend commented: There are two hurdles that have to be overcome for a new platform chemical to be viable. We have to develop technology and value chain to produce the platform chemical in large enough amounts but also the technology and value chain to make the final products from it. Do you have any ideas how to do both in the right time-frame? Are there any strategies that can be followed? It seems like a chicken and egg problem.

José-Vitor Bomtempo responded: You are right. There is a chicken and egg problem in platform development. How could companies manage this situation? The first point is to be aware of the problem. The second is to communicate as soon as possible with all the production chain, especially with complementary and downstream agents. It is particularly important to establish some links with end users as early as possible. Anyway, this a key challenge in the development of platform chemicals.

Christian Stevens enquired: When you are talking about chemical platform molecules, your starting point seems to be that the same processes should be used as for petrochemicals. The question is, do we really have to put all our efforts into producing bulk chemicals first? Is the best strategy to produce bio-based bulk chemicals? Would it be more valuable to look first for high value niche applications in order to introduce bio-based products more quickly into the market?

José-Vitor Bomtempo replied: I completely agree that it is certainly more valuable to look for high value niche applications in order to introduce bio-based molecules into the market. Our reference is not the established petrochemical industry. The central point of our platform vision is a strategy of a (bio)chemical company that develops (or organizes the development of) a portfolio of products based on a platform molecule. This portfolio must combine niche applications and at least one or two applications on a larger scale, which are critical to opening a kind of virtuous cycle linking specialities and commodities. If the innovating company intends to build a platform, it could start with some niche applications. However, to explore the platform potential and expand the portfolio, a volume application is desirable.

Fang Zhang said: Usually a platform chemical comes from crude oil, not from complex materials. Perhaps bioresources are only usable for small amounts, for fine chemicals, and are not suitable for platform chemicals?

José-Vitor Bomtempo responded: The vision we have been trying to develop in our research on platform chemicals is not based on the established petrochemical industry but on the possibility of new business models similar to the technological platforms we see now in many industries. In this vision, a platform chemical based on bio-based molecules is not organized in the same way as the traditional petrochemical industry. It would be a virtuous combination of niche (specialities) and volume applications. The key point is the company's ability to organize the platform in an innovative way that allows the value creation and capture.

Keith Waldron opened a general discussion of the papers by Changwei Hu and José-Vitor Bomtempo: We have successfully used FTIR spectroscopy and Partial Least Squares modelling for the rapid screening of chemical (sugars, lignin) and tissue compositions of large sets of biomass samples.[1] The approach requires a high degree of calibration using reference samples but once set up is extremely rapid.

1 S. R. A. Collins, N. Wellner, I. Martinez Bordonado, A. L. Harper, C. N. Miller, I. Bancroft and K. W. Waldron, *Biotechnol. Biofuels*, 2014, 7, 121

Gadi Rothenberg remarked: Regarding the use of FTIR and other spectroscopic methods for biomass and bio-oil identification, in general, people want to use FTIR as it is a non-invasive and non-destructive method. Chemists often miss out the mathematics that can be used. It is possible to identify and quantify single components from complex mixtures by using multivariate techniques. There are research groups that develop the maths for these programs, chemists should ask these people for help. Typically, Matlab toolboxes are used, sometimes even with graphical user interfaces (GUIs), which are accessible to chemists.

Claudia Crestini commented: It is advisable that newcomers to the biomass utilisation area become acutely aware of the numerous and extensive previous efforts of this community in standardising material and analytical protocols.

Deepak Pant replied: I fully agree with this and having an informed opinion is a must. Bioenergy and biofuels have attracted a lot of bad press in recent years. I came across an interview with Prof. André Faaij from the University of Groningen in the Netherlands, which I find very interesting and it clears up several issues related to this topic.[1] It covers almost all the topics that were discussed at this *Faraday Discussions* meeting.

1 http://www.theenergycollective.com/karelbeckman/2407088/interview-bio-energy-expert-andre-faaij-much-nonsense-told-high-time-real-story.

Bruce Dale said: There has been a lot of enthusiasm for algae as a means of producing biofuels and bioproducts. The motivation for using algae seems to be the idea that they do not require much in the way of land and other resources to produce. When I look closely at algae, however, I come to a very different conclusion. To grow algae on a commercial scale, you need to assemble the following resources in the same location: (1) lots of flat land for the raceways, (2) lots of concentrated carbon dioxide and sunlight to feed the algae, and (3) lots of water to replace the water evaporated from the open raceways. Algae in open raceways are also prone to contamination by other algal species. Every approach that I have seen for processing algae in bioreactors looks horribly capital intensive to me. I cannot even come close to making this work economically. Will someone please tell me where I am mistaken?

Changwei Hu answered: That is why we chose natural algae as our starting materials for the conversion study. We want to develop conversion methods to treat the naturally formed algae, which would be less capital intensive. Of course, the collection of natural algae might not be cheap.

Christian Stevens enquired: When trying to stimulate the use of renewable resources, it is important to ask the question about if we should focus first on bulk chemicals or if it would be better to focus on high value niche applications. Which choice will lead to a faster adoption of renewable resources in the production industry?

José-Vitor Bomtempo replied: There are indeed many dilemmas when choosing which kind of product should be the strategic focus in the bio-based industry. One of them is the choice between bulk or niche applications. I agree that looking for a niche application first makes sense. However, if we have a long term vision, it is important to explore the potential of volume applications. Niche and volume applications may be combined in a virtuous way. A niche application in this strategy is not only the first market but a learning space in order to attain productivity and reduce costs. It is interesting to note that initially the industry focus was on bulk products, such as biofuels and bulk polymers. Apparently, this has been changing over the last few years.

Christian Stevens asked: What do you think about the economical viability and sustainability of growing algae in special plastic bags on top of landfill sites using the purified carbon dioxide produced in the landfill?

Bruce Dale answered: I do not know. I have not seen the economic analysis of such a project and I do not want to prejudge without some data. However, every one of these systems I have looked at has the same problems: (1) lots of surface area needed because of light attenuation, (2) very dilute nature of the product leading to large separation costs, (3) costly harvesting and breaking of the cells for intracellular products, and (4) costly concentration of carbon dioxide as a feed.

Bruce Dale commented: We have so many challenges to meet in renewable energy and bioproducts, and so little time to meet them in if we are going to avoid drastic changes in our way of life and standard of living. I hope everyone attending this conference will ask themselves this question, and then act on the answer. Am I doing something that I really have thought carefully about and have decided will make a significant difference if I am successful? Or is this just another job? I shared the ideas of "Sobering Numbers"[1] in my presentation and my exhortation for all of us to try to really make a difference. Hopefully this short paper captures some of these ideas.

1 B. E. Dale, *Resource: Engineering and Technology for Sustainable World*, 2014, **21**, 25.

Bruce Dale remarked: I asked Dr Waldron the question "why duckweed" to highlight the importance of being very careful and self-aware about the biomass feedstocks we choose to work with. Mostly, this is an issue of scale and logistics. Dr Waldron explained to me that duckweed has very little lignin and lots of starch, but primarily he is interested in it as a means of remediating water. I think this is a valuable feature of duckweed and may provide some scale. We have to look for biomass materials that are appropriate for the location and scale we are considering, and then be very severe with ourselves and self-skeptical until we arc convinced that the biomass material has real promise.

Faraday Discussions

PAPER

New catalytic strategies for α,ω-diols production from lignocellulosic biomass†

Jiayue He,[a] Kefeng Huang,[a] Kevin J. Barnett,[a] Siddarth H. Krishna,[a] David M. Alonso,[a] Zachary J. Brentzel,[a] Samuel P. Burt,[ab] Theodore Walker,[a] Williams F. Banholzer,[a] Christos T. Maravelias,[a] Ive Hermans,[ab] James A. Dumesic[a] and George W. Huber*[a]

Received 26th January 2017, Accepted 9th March 2017

DOI: 10.1039/c7fd00036g

Catalytic strategies for the synthesis of 1,5-pentanediol (PDO) with 69% yield from hemicellulose and the synthesis of 1,6-hexanediol (HDO) with 28% yield from cellulose are presented. Fractionation of lignocellulosic biomass (white birch wood chips) in gamma-valerolactone (GVL)/H_2O generates a pure cellulose solid and a liquid stream containing hemicellulose and lignin, which is further dehydrated to furfural with 85% yield. Furfural is converted to PDO with sequential dehydration, hydration, ring-opening tautomerization, and hydrogenation reactions. Acid-catalyzed cellulose dehydration in tetrahydrofuran (THF)/H_2O produces a mixture of levoglucosenone (LGO) and 5-hydroxymethylfurfural (HMF), which are converted with hydrogen to tetrahydrofuran dimethanol (THFDM). HDO is then obtained from hydrogenolysis of THFDM. Techno-economic analysis demonstrates that this approach can produce HDO and PDO at a minimum selling price of $4090 per ton.

1 Introduction

α,ω-Diols, chemicals such as 1,4-butanediol (BDO) and 1,6-hexanediol (HDO), which have a linear carbon chain and OH groups at both terminal positions, are high value commodity chemicals with market prices of $2000 per ton and $4400 per ton, respectively.[1] α,ω-Diols are monomers used predominantly for the production of polyesters and polyurethanes (Fig. 1). HDO is primarily used as a polyester or polyurethane component to increase flexibility, hydrolytic stability, and adhesion.[2] 1,5-Pentanediol (PDO) currently has a higher market price ($6000 per ton) than HDO but a smaller production level (3000 tons per year) due to limited readily accessible C5 petroleum feedstocks.[2] Polymers produced from PDO can include polyesters and polyurethanes. These polymers have a low

[a]Department of Chemical and Biological Engineering, University of Wisconsin-Madison, Madison, WI 53706, USA. E-mail: gwhuber@wisc.edu

[b]Department of Chemistry, University of Wisconsin-Madison, Madison, WI 53706, USA

† Electronic supplementary information (ESI) available. See DOI: 10.1039/c7fd00036g

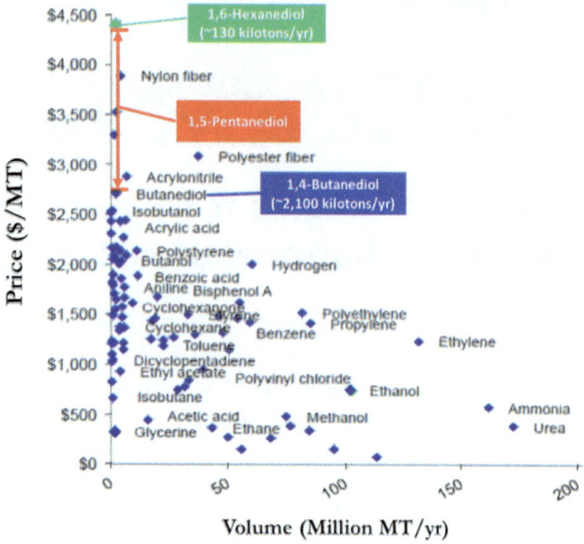

Fig. 1 Market price of commodity chemicals in 2013.[8]

viscosity, low glass transition temperature, and good flexibility.[2] In 2014, the HDO global market value was $607 million with an annual market size of 138 000 tons.[1] BDO has a market size of 2 500 000 tons per year with an annual growth rate of 5.1% from 2014 to 2020. PDO could most likely be used as a direct replacement for HDO or BDO in several applications (*e.g.* polycarbonate diol, urethane, and polyetherpolyol).[2] Polyesters made from PDO have different physical properties compared to polyesters made from BDO (*e.g.* the melting point of C5 polyester is lower than that of PET, PTT and PBT, and the biodegradability of C5 polyesters is worse than that of C4 polyesters (PBS)). However, polyesters produced from PDO are a safer additive than polyesters produced from BDO, especially in children's toys,[3] since BDO in the body is rapidly absorbed and metabolized into γ-hydroxybutyric acid (GHB),[4–6] commonly referred to as the date-rape drug because of its ability to induce deep sleep.[7]

The current industrial process for the production of HDO relies on the catalytic reduction of adipic acid (AA) at 170–240 °C and 15–30 MPa H_2 over Cu-, Co- or Mn-catalysts or its esters.[2] The most common industrial route for the production of AA from benzene includes three steps:[9] (i) hydrogenation of benzene to cyclohexane, (ii) oxidation of cyclohexane to KA oil (a mixture of cyclohexanol and cyclohexanone),[10–12] and (iii) oxidation of the KA oil to AA using an excess of HNO_3 in the presence of copper(II) and ammonium metavanadate catalysts.[13–15] Current production of PDO is by hydrogenation of glutaric acid (GA) or its esters.[16] The industrial production of GA is accomplished *via* a four-step approach: (i) conversion of acetylene with two equivalents of formaldehyde to form 1,4-butynediol, which is known as "Reppe chemistry", (ii) hydrogenation of 1,4-butynediol to 1,4-butanediol, (iii) dehydration of 1,4-butanediol to γ-butyrolactone,[17] and (iv) ring-opening of butyrolactone with potassium cyanide followed by hydrolysis to form GA.[18] Both HDO and PDO synthesis routes above suffer from low conversion reactions, many reaction steps (lower yields), homogeneous chemistry involving toxic chemicals (HNO_3 or KCN), and nonrenewable, petroleum feedstocks which contribute to climate change. It would be preferable to develop production routes for HDO and PDO from renewable biomass feedstocks.

Several approaches for the production of HDO and PDO from biomass feedstocks have been reported in the literature. Previous research efforts to synthesize PDO from lignocellulosic biomass-derived chemicals have focused on the selective hydrogenolysis of tetrahydrofurfuryl alcohol (THFA), a hydrogenation product from furfural with Cu- or Ni-catalysts,[19–21] into PDO over bimetallic noble metal catalysts, which include a reducible noble metal, including Rh,[22–26] Ru,[27] and Ir,[28] with an oxophilic second metal, including Re[22,24–26,29] and Mo.[22–24] Multiple hypotheses exist as to the mechanism for the enhancement in the rate. Tomishige suggested that the oxophilic promoter induces heterolytic cleavage of the hydrogen on itself and the adjacent noble metal.[30] The Dumesic group has demonstrated with computational calculations that the secondary metal strongly bonds with oxygen atoms on the surface, leading to the formation of Brønsted acid hydroxyl group sites on the secondary metal oxide.[29] The acidity assists in the formation of carbenium ion intermediates. This metal and acid interfacial site are critical for the selective hydrogenolysis of the alpha alcohol ether bond. However, one of the challenges of this approach is the low catalyst activity and hence high catalyst cost.

In 1983, Faber patented a three-step process for HDO production.[31] This process includes: (i) hydrolysis/dehydration of biomass to HMF with dilute acid, (ii) hydrogenation of HMF to tetrahydrofuran dimethanol (THFDM) with RANEY®-nickel, and (iii) hydrogenolysis of THFDM to HDO with a copper chromite catalyst.[32] Buntara *et al.* presented a process to convert HMF to HDO which also involves hydrogenation of HMF to THFDM, ringing opening of THFDM into 1,2,6-hexanetriol (HTO), and hydrogenolysis of HTO into HDO (selectivity to HDO was 73% with a 17% conversion).[33] The authors reported a one-pot reaction for converting THFDM into HDO by using a combination of $RhRe/SiO_2$ and Nafion SAC-13 with a selectivity to HDO of up to 86%. The drawbacks are low catalyst activity, and high solubility of the catalyst support (SiO_2) in hot water.[34,35] Xiao *et al.* presented a direct synthesis of HDO from HMF over double-layered catalysts of Pd/SiO_2 and $Ir\text{-}ReO_x/SiO_2$ in a solution of water and THF with a yield to HDO of 57.8%.[36] Chia *et al.* and Chen *et al.* have demonstrated that ReO_x-promoted Rh is selective for the hydrogenolysis of THP-2M into HDO in the aqueous phase.[29,37] By combining a highly reducible metal with an oxophilic metal, the ring-opening of THP-2M and subsequent dehydration were facilitated with stabilized oxocarbenium ion intermediates by the Brønsted acidity of the oxophilic promoter. A DuPont patent reported on a $PtWO_x/TiO_2$ catalyst for the conversion of THFDM to HDO with 83% yield from levoglucosenone (LGO) derived THFDM over a Pt/C catalyst at 160 °C and 5.5 MPa H_2 in the aqueous phase.[38]

In this paper we demonstrate a multi-step catalytic process for the production of PDO and HDO starting from white birch, as shown in Scheme 1. First, lignocellulosic biomass is converted into cellulose, furfural and solid lignin *via* the TriVersa Process™ developed by Glucan Biorenewables LLC. Furfural is converted into PDO by four catalytic steps, as shown in Schemes 1 and 2:[40] (i) furfural is hydrogenated into THFA, (ii) THFA is dehydrated to dihydropyran (DHP), (iii) DHP is hydrated to 2-hydroxytetrahydropyran (2-HY-THP) and 2-HY-THP dimers, and (iv) 5-hydroxyvaleraldehyde (5-HY-Val) is hydrogenated into PDO. This

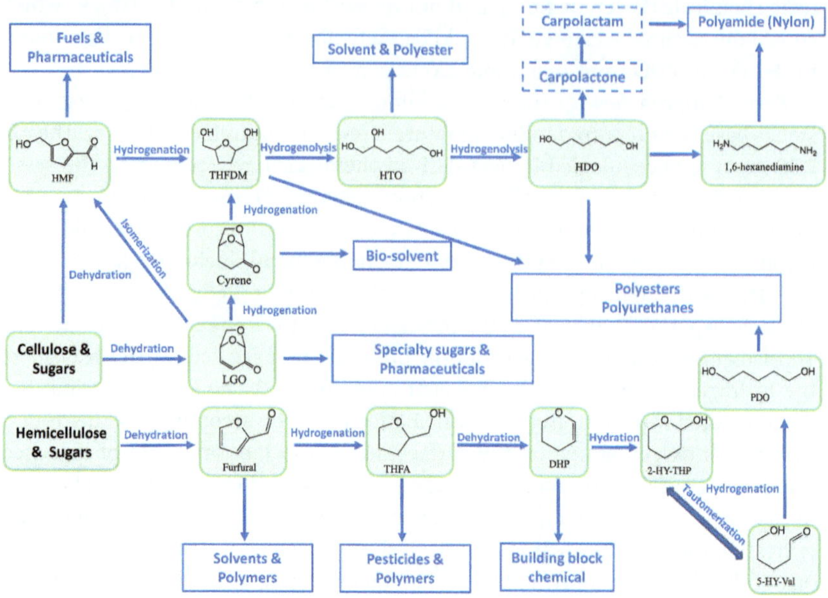

Scheme 1 Potential products produced from cellulose and hemicellulose.[39]

pathway will be referred to as the dehydration, hydration, and hydrogenation (DHH) route (see Section 3.2).

Cellulose was converted into HDO by a three step approach, as shown in Schemes 1 and 3: (i) LGO and HMF were obtained with high yields from cellulose using a mixture of water and a polar-aprotic solvent such as tetrahydrofuran (THF).[41] (ii) THFDM was produced by hydrogenation and hydrogenolysis of LGO and HMF with Ni-based and Pd-based catalysts,[42] and (iii) HDO was produced by hydrogenolysis of THFDM over a Pt-based catalyst (see Section 3.3).

The intermediates in this process have many potential applications, as shown in Scheme 1. LGO can be used as a precursor to produce cyrene, THFDM, HTO and HDO. Cyrene is a new bio-based solvent that has been reported as a safe replacement for *N*-methyl-2-pyrrolidone (NMP) and dimethylformamide (DMF).[43] These polyhydric alcohols (THFDM, HTO) are key intermediates in the industrial production of 1,6-hexanediamine, caprolactone, and caprolactam, which are used for the manufacturing of polyesters, polyamides, and polyurethanes, representing multimillion ton operations.[32,38,44] LGO can also be isomerized into HMF, another precursor for both pharmaceuticals and fuels.[45] The chiral nature of LGO also lends itself to the synthesis of natural products. Furfural, a platform chemical, is used as a bio-solvent and polymer precursor.[46] THFA and DHP can be produced from furfural. THFA is used for manufacturing pesticides and intermediates to produce DHP, a building block chemical that can be hydrogenated to bio-solvent tetrahydropyran (THP), converted to γ-valerolactone, or used as an intermediate in pesticide and pharmaceutical production.[47]

2 Experimental

2.1 Fractionation of lignocellulosic biomass

Biomass fractionation experiments were carried out in 60 mL high pressure glass reactors (Chemglass) placed in an oil bath.[48] In a typical experiment, 5 g of ground white birch and 28 g of a liquid solution of GVL/water 80/20 by weight and 0.1 M sulfuric acid were added to the reactor and heated at 130 °C. At the end of the reaction, the liquid was separated from the solids using a centrifuge filter or vacuum filtration. The cellulose was analyzed to determine its chemical composition following NREL/TP-510-42618.[49] The liquid fraction was analyzed by HPLC using a HPLC Waters 2695 system with a Bio-Rad Aminex HPX-87H column, and using a RI 410 detector for carbohydrates, organic acids, and GVL, and a UV detector for furfural and HMF. In addition, to convert soluble oligomeric carbohydrates into monomers, a protocol similar to NREL/TP-510-42618 was followed.[49] A portion of the liquid was diluted 10 times with 4.4 wt% sulfuric acid and heated at 120 °C for 1 hour. The sample was analyzed by HPLC as mentioned above.

2.2 Conversion of hemicellulose to furfural

Experiments to convert the soluble hemicellulose into furfural were performed in a continuous up-flow reactor. A $\frac{1}{4}''$ tubular stainless steel reactor was filled with SiO_2 to fill the reactor volume and improve heat and mass transfer. The reactor was mounted inside an aluminum block heated by heating tape. The reaction temperature was monitored at the reactor wall by a Type K thermocouple (Omega) mounted within the aluminum block and controlled by a Series 4B temperature controller (Love Controls). The pressure was kept at 2.06 MPa to prevent water vaporization. The liquid feed was introduced in the reactor using an HPLC pump (Lab Alliance Series I). The feedstock utilized was the liquid obtained after the biomass fractionation step without any purification or treatment step. The effluent was analyzed by HPLC (Waters 2695 system with a Bio-Rad Aminex HPX-87H column) using a RI 410 detector for carbohydrates, organic acids, and GVL, and a UV detector for furfural and HMF.

2.3 THFA dehydration

The dehydration studies of THFA were performed in a fixed bed catalytic reactor in an up-flow configuration at 375 °C and 0.1 MPa with γ-Al_2O_3. The catalyst was loaded into a $\frac{1}{4}''$ OD stainless steel tube and secured in place by alternating layers of quartz wool and granular silicon dioxide. The γ-Al_2O_3 dehydration catalyst was crushed and sieved below 100 mesh. The silicon dioxide was used to decrease the void volume and increase heat transfer. The quartz wool layers served as a buffer between the catalyst and silicon dioxide layers. The tube was secured in a furnace comprising aluminum blocks, heating tape (Briskheat, 135W), and insulation. The catalyst was pretreated in flowing He at 400 °C for 1 hour prior to the reaction. The vapor phase reactor effluent was passed through an ice bath condenser. Liquid samples were collected from the condensed effluent every 2 to 8 hours. The vapor phase was sampled *via* online gas chromatography (Shimadzu GC2010 equipped with a flame ionization detector (FID) and a RT-Q-Bond column). Liquid samples were analyzed by gas chromatography (Shimadzu GC2010 equipped with a flame ionization detector (FID) and a DB-5MS capillary column). The γ-A_2O_3 catalyst was pretreated in flowing He at 400 °C for 1 hour prior to the reaction. The reactor was then cooled to a reaction temperature of 375 °C. The flowing He was then used as a sweep gas at the effluent of the reactor. Neat THFA was pumped into the reactor with an HPLC pump (Lab Alliance Series I).

2.4 DHP hydration

The catalysts were loaded into Hastelloy autoclaves (Parr Instrument, Multi Reactor System 5000). The autoclaves were seated in their housing and purged 3 times with 2.0 MPa of Ar and then 3 times with 2.0 MPa of H_2. The catalysts were reduced at 200 °C for 15 minutes, with a 4 °C min^{-1} ramp. 1 wt% DHP feeds in deionized water were prepared and pumped into each autoclave once cooled to 27 °C. The autoclaves were stirred with magnetic stir bars at 500 rpm. The autoclaves were then pressurized to 6.5 MPa with H_2. The autoclaves were heated to 120 °C (6 °C min^{-1}). The autoclaves were held at 120 °C for 1 to 3 hours. After cooling, the samples were analyzed by both gas chromatography (Shimadzu GC2010 equipped with a flame ionization detector (FID) and a RTX-VMS column) and high-performance liquid chromatography (HPLC) (Waters Alliance 2695 system equipped with a Waters 410 differential refractometer and Aminex HPX-87H ion exchange column) using a 5 mM H_2SO_4 mobile phase.

2.5 Hydrogenation of 2-HY-THP

The hydrogenation studies were performed in a fixed bed catalytic reactor in an up-flow configuration at 4.5 MPa and 393 K with Ru/TiO_2. The catalyst was loaded into a $\frac{1}{4}''$ OD stainless steel tube and secured in place by alternating layers of quartz wool (Grace) and granular silicon dioxide. Silicon dioxide was used to decrease the void volume and increase heat transfer. The quartz wool layers served as a buffer between the catalyst and silicon dioxide layers. The tube was secured in a furnace comprising aluminum blocks, heating tape (Briskheat, 135W), and insulation. Liquid samples were collected from the condensed effluent every 2 to 24 hours. Liquid samples were analyzed by both gas chromatography (Shimadzu GC2010 equipped with a flame ionization detector (FID) and an RTX-VMS column) and high-performance liquid chromatography (HPLC) (Waters Alliance 2695 system equipped with a Waters 410 differential refractometer and Aminex HPX-87H ion exchange column) using a 5 mM H_2SO_4 mobile phase.

2.6 Dehydration of cellulose to LGO and HMF

The reactions were carried out in a 100 mL autoclave (Parr Instrument Company, series 4560, Hastelloy (C-276)). The vessel and head were dried overnight at 70 °C to remove residual water prior to each reaction. Cellulose, THF, and sulfuric acid were added into the autoclave sequentially, at which point the vessel was purged five times with helium and charged to 3.4 MPa He. The vessel was then heated to the desired reaction temperature and pressurized to a final pressure of 6.9 MPa. Zero time was defined as the point at which the heating was started, except where noted. The temperature and stirring maintained at 700 rpm were controlled by a Parr 4848 Controller. Samples (1 mL) were periodically taken through a dip tube during the reaction. The reactor was repressurized with helium to 6.9 MPa after each sampling. The samples were quenched in a dry ice bath and then filtered with a 0.22 μm syringe filter (Restek, PTFE (polytetrafluoroethylene)) prior to analysis.

LGO and HMF standards were analyzed using a high-performance liquid chromatograph (HPLC; Shimadzu, LC-20AT) equipped with UV (UV-vis; SPD-20AV) and RI (RID-10A) detectors. Separation was achieved using a Biorad Aminex HPX-87H column at 30 °C with 5 mM H_2SO_4 as the mobile phase, flowing at a rate of 0.6 mL min^{-1}. The injection volume was 1 μL in each analysis.

2.7 Hydrogenation/hydrogenolysis of LGO, HMF

The reactions were carried out in 45–75 mL Parr high-pressure batch reactors. The catalyst, feedstock, and a magnetic stir bar were added to the reactor. The reactor was purged four times with H_2 prior to the reaction, pressurized to the desired H_2 pressure, stirred at 750 rpm, and heated to the reaction temperature. After the reaction, the reactor was cooled in ice water and depressurized. The product mixture was then filtered using a 0.2 μm PTFE (polytetrafluoroethylene) syringe filter.

The products were quantified using a Shimadzu Gas Chromatograph incorporating a Flame Ionization Detector (FID) with liquid injection. A Restek RTX-VMS capillary column (length: 30 m, ID: 0.25 mm, film thickness: 1.4 μm) was used. Lgol was assumed to have a GC-FID sensitivity equal to the average sensitivity of LGO and cyrene. Because *trans*-THFDM was overlapped by one of the stereoisomers of THP2M5H in the GC, the concentration of *trans*-THFDM relative to *cis*-THFDM was calculated using quantitative ^{13}C NMR spectroscopy. Then, the concentration of the overlapped stereoisomer of THP2M5H was back-calculated using the GC peak area, assuming that THP2M5H has a GC sensitivity equal to that of THFDM.

Quantitative ^{13}C NMR spectra were collected on a Bruker Avance 500 MHz spectrometer at room temperature, using an inverse-gated decoupling pulse sequence with a 30° pulse. 256 scans were used, with an acquisition time of 1 s and a relaxation delay of 15 s. Samples were prepared for analysis by evaporating the THF solvent from 5 mL of sample, then adding 1 g of d_6-DMSO. Further details about the analytical methods for this section are provided in our recent work on LGO hydrogenation.[42]

2.8 Hydrogenolysis of THFDM in a continuous flow reactor

THFDM hydrogenolysis was carried out in a stainless-steel tubular flow reactor (30 cm long, 6.35 mm outer diameter), arranged in an up-flow configuration and heated using a tube furnace (Thermcraft Inc. No. 114-12-1ZH). The Pt-based catalyst (0.10 g) was added into the reactor with quartz wool and quartz beads packed on both sides without dilution. The catalyst was reduced under a H_2 flow (100 mL min^{-1}) at 250 °C for 2 h before reaction. The reactor was cooled to 160 °C and pressurized to 3.4 MPa H_2. An aqueous 1 wt% THFDM solution was pumped to the reactor using an Agilent Prostar HPLC pump with co-fed H_2 (40 mL min^{-1}). Two stainless steel tanks (150 mL each) were used to accumulate the liquid products at the reactor outlet. The accumulated liquid product was drained periodically into a collector container, filtered through a 0.22 μm PES syringe filter and then analyzed by GC. The gaseous products continued to flow through a back pressure regulator to maintain the reaction pressure. Both an excess flow shut-off valve and a pressure relief valve were also installed in the reactor system to ensure safe operation. The feed solutions and the liquid products were analyzed using a gas chromatograph (Shimadzu, GC-2010) equipped with a flame ionization detector (FID) and a Restek RTX-VMS capillary column. The gas effluent from the continuous flow reactor was analyzed using a gas chromatograph (Shimadzu, GC-2014) equipped with an FID and a Restek RT-Q-Bond capillary column as well as a thermal conductivity detector and a ShinCarbon ST micropacked column.

The chemicals, catalyst preparations, THFDM synthesis, batch reaction studies of 2-HY-THP and THFDM, catalyst characterizations are included in the ESI.†

3 Results and discussion

3.1 Production of furfural and cellulose from lignocellulosic biomass

White birch was converted into a cellulose solid and liquid stream (containing hemicellulose and lignin) by the TriVersa Process™. This process uses GVL to fractionate biomass.[50–52] The GVL solvent minimizes hemicellulose and lignin degradation reactions since the hydrolysis reaction is 100 times faster in GVL than in water.[53] The reactor also operates at a low temperature (<130 °C). Finally, because lignin is highly soluble in GVL/water mixtures, it is possible to work at high biomass loadings (>15 wt%).

3.1.1 Fractionation of lignocellulosic biomass. Fig. 2 shows the hemicellulose and cellulose yields from white birch in the TriVersa Process (80/20 GVL/water, 0.1 M sulfuric acid (SA), at 130 °C). At shorter reaction times (45 min), more than 90% of the hemicellulose is removed while 42% of the original biomass is solid cellulose with a 90% purity. Increasing the reaction time to 60 min increases the hemicellulose extraction to 95% and the cellulose purity to 92.5%; however, the overall yields decrease to 38% in part due to the cellulose hydrolyzing to sugars. Further increasing the reaction time leads to the highest cellulose purity

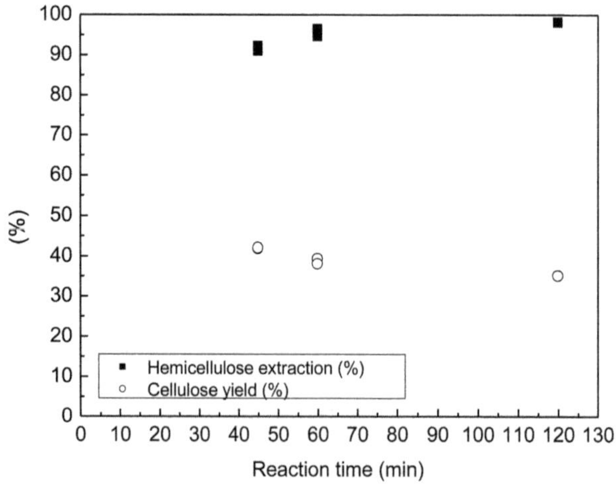

Fig. 2 White birch fractionation at 15 wt% biomass loading in 80/20 GVL/water, 0.1 M SA and 130 °C.

(>94.5%) and hemicellulose extraction (>98%), but the solid cellulose yield decreased to 35% and >30% of the hemicellulose was converted to furfural.

3.1.2 Production of furfural and cellulose. The liquid and solid phases (cellulose) are separated, and the liquid fraction obtained from the fractionation reaction is directly utilized to produce furfural in a continuous flow reactor without further purification. Furfural yields over 85% are obtained at 225 °C and less than 30 seconds residence time (Fig. S1†). The furfural is then purified by neutralizing the sulfuric acid and separating the furfural from the GVL by distillation in two distillation columns. The final step is the separation of the lignin from the GVL, which involves evaporating the GVL. The produced lignin has a similar structure of lignin to that in the native biomass and is of high purity (low sugar and low ash content).[48,54] This lignin could potentially be used to produce lignin monomers,[54] carbon foams,[48] or battery anodes.[55]

3.2 Synthesis of PDO from furfural *via* the DHH route

The furfural derived from xylose can be converted to THFA *via* complete hydrogenation over monometallic catalysts such as nickel. THFA is converted to PDO *via* the dehydration–hydration–hydrogenation (DHH) pathway in 84% overall yield (Scheme 2). In this 3-step approach, PDO is produced *via* vapor-phase dehydration of THFA to DHP, homogeneous hydration of DHP to 2-HY-THP in water, and hydrogenation of the ring-opened tautomer of 2-HY-THP, 5-HY-Val, to PDO. During the hydration step, the monomers undergo etherification to their respective dimers. These dimers, however, are reconverted back into their monomer forms and further converted to PDO in the final hydrogenation step.

3.2.1 THFA dehydration into DHP. THFA was dehydrated over a γ-Al_2O_3 catalyst in the vapor-phase at 375 °C. Dehydration of neat THFA in the vapor phase gave 87% overall yield of DHP (Scheme 2). This result was consistent with the work of Schniepp and Geller.[56]

3.2.2 DHP hydration into 2-HY-THP. DHP (95%, AK Scientific) was added to DI water to 20 wt% and hydrated homogeneously in water from 70 to 160 °C in a 45 mL Parr batch reactor in a He atmosphere (1.7 MPa). Hydration of the DHP resulted in the formation of the main 2-HY-THP hydration product and dimers formed *via* etherification (as shown in Table 1). Nearly quantitative yields to PDO precursors (2-HY-THP + dimers) were achieved at 100–130 °C. The main dimer products formed were 5-tetrahydropyranyl-oxypentanal (THP-oxypentanal) and 2-

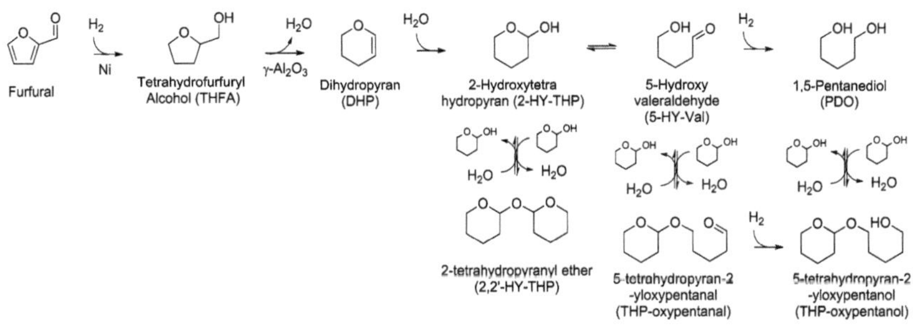

Scheme 2 Reaction pathway for the dehydration, hydration, and hydrogenation (DHH) of THFA into PDO.[40]

Table 1 DHP dehydration without catalysts[40]

Temperature (°C)	Reaction time (h)	Yield (%) 2-HY-THP	THP-oxypentanal	2,2′-HY-THP	Overall yield (%)
70	4	84.5	5.3	1.9	91.7
100	4	92.0	6.6	1.2	99.8
130	4	89.5	6.4	1.0	96.9
160	4	65.4	4.2	1.3	70.9

tetrahydropyranyl ether (2,2′-HY-THP). 2,2′-HY-THP and THP-oxypentanal were formed *via* the etherification of 2-HY-THP with itself and with 5-HY-Val, respectively. At higher temperatures (160 °C), self-condensation of the alcohols was observed, leading to solid polymer formation and lower product yields.

3.2.3 Hydrogenation of 2-HY-THP into PDO. An aqueous solution of 2-HY-THP (no dimers) was hydrogenated into PDO. A variety of monometallic catalysts were studied for their activity and selectivity to PDO, as shown in Table 2. Of the six monometallic catalysts studied, Ni had the highest turnover frequency (TOF), 0.3 s^{-1}. Pt and Rh were the next most active, with TOF values of 0.1 and 0.07 s^{-1}, respectively. All catalysts were greater than 99% selective to PDO and dimers of 2-HY-THP. The number of surface sites of these catalysts was measured by CO chemisorption.

The underlying mechanism for the quantitative conversion of 2-HY-THP into PDO over monometallic catalysts was then probed, given that THFA is unable to be ring-opened to PDO under the same conditions with the same catalysts. Furthermore, 2-HY-THP was found to be 80 times more active over a traditional bimetallic hydrogenolysis catalyst, RhRe, than THFA. ^{13}C NMR spectroscopy[57,58] (Fig. 3A), attenuated total reflectance Fourier transform infrared spectroscopy (ATR-FTIR),[40] and aldehyde hydrogenation studies have demonstrated that the high reactivity of 2-HY-THP is a result of the quasi-equilibrated ring-opening tautomerization to 5-HY-Val. PDO is formed *via* the hydrogenation of 5-HY-Val, with the hydrogenation being the rate determining step.

Hydrogenation of the product solution from DHP hydration was performed over supported Ru catalysts. This hydrogenation resulted in >96% yields of PDO

Table 2 Monometallic catalyst screening results

Catalyst	Surface site density (μmol g^{-1})	Reaction time (h)	Conversion (%)	PDO selectivity (%)	Turnover frequency (TOF) (s^{-1})
3 wt% Ni/SiO_2	34	1	6.3	>99	0.30
5 wt% Pt/C	120	3	30.7	>99	0.10
5 wt% Rh/C	117	3	21.8	>99	0.07
5 wt% Co/SiO_2	56	4	10.5	>99	0.01
50 wt% $Cu/\gamma\text{-}Al_2O_3$	—	6	75	>99	0.0003[a]

[a] TOF is normalized to total metal sites since a surface site density was not obtained.

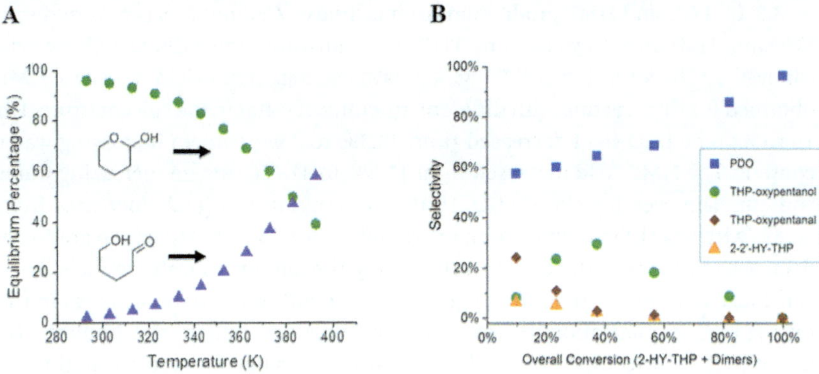

Fig. 3 (A) Equilibrium percentage of 2-HY-THP and 5-HY-Val from 25 to 120 °C based on quantitative ^{13}C variable temperature NMR results.[40] (B) Product selectivities over 1 wt% Ru/TiO_2 in a continuous flow reactor (40 mg of $Ru-TiO_2$ diluted 10× in silica gel, $T = 120$ °C, $P = 4.5$ MPa H_2, H_2 flowrate = 40 mL min^{-1}).[40]

from 2-HY-THP and the dimers. At low conversions, the PDO selectivity is <60% because a shift in equilibrium towards the dimers occurs due to hydrogenation of THP-oxypentanal to THP-oxypentanol (Scheme 2 and Fig. 3B). At higher conversions, the dimers are hydrolyzed back to their monomers, which are then hydrogenated to PDO, resulting in near quantitative yields of PDO from 2-HY-THP and the dimers. Ru-based catalysts were used due to their high hydrogenation activity and relative stability compared with other monometallic catalysts.

3.3 Synthesis of HDO from cellulose

The catalytic conversion of cellulose into HDO includes three steps: (i) hydrolysis/dehydration of cellulose into LGO and HMF,[41] (ii) hydrogenation/hydrogenolysis of LGO and HMF into THFDM,[42] and (iii) hydrogenolysis of THFDM into HDO (Scheme 3). In the second step, LGO was hydrogenated to Lgol, followed by Lgol hydrogenolysis to THFDM, tetrahydropyran-2-methanol-5-hydroxyl (THP2M5H) and 2-methyl-tetrahydrofurfurylalcohol (2MTHFA) over a Pd-based catalyst.

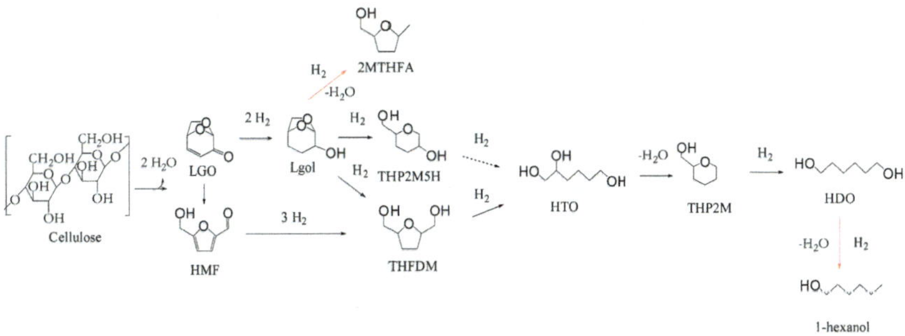

Scheme 3 Reaction network for 1,6-HDO production from cellulose.[59] The red arrows indicate undesired products. The dotted arrows indicate hypothetical reactions.

3.3.1 LGO and HMF production from cellulose. Cellulose was converted into LGO and HMF in 48% yield using THF–water mixtures and a dilute sulfuric acid catalyst, as shown in Fig. 4.[41,60] Fig. 4 shows the maximum yield of LGO + HMF obtained for this reaction with different amounts of water in the solvent (from 0 to 10 wt%). The LGO yield decreased from 32.3% to 3.0% with an increasing water content. The HMF yield increased from 13.2% to 41.6% with an increasing water content. The overall yield of LGO, HMF and levoglucosan (LGA) increases from 46.6% with 0% H_2O to a maximum of around 50% with 2.5% H_2O, then gradually decreases to 44.6% with 10% H_2O. Increasing the amount of water in the solvent can cause both (a) changes in the glucose/LGA equilibrium or (b) changes in the rate of LGO isomerization to HMF (a water-mediated reaction).[41,45] Similarly, De Bruyn *et al.* reported that LGO could be obtained from unconverted saccharides in a lignin-rich biorefinery waste stream in yields up to 8 wt%.[39]

3.3.2 Hydrogenation/hydrogenolysis of LGO and HMF into THFDM. Our upgrading strategy to produce HDO (Scheme 3) involves the upgrading of cellulose derived LGO and HMF in the same reactor. The upgrading pathways for LGO and HMF involve similar chemistry: lower-temperature hydrogenation of C=C and C=O bonds followed by higher temperature hydrogenolysis of C–O bonds. HMF hydrogenation to THFDM has been reported in near-quantitative yields over supported metal catalysts.[32,61] This reaction proceeds *via* initial hydrogenation of the aldehyde group to furandimethanol, followed by hydrogenation of the furan rings to form THFDM.[32] HMF hydrogenation to THFDM results in predominantly the *cis*-THFDM isomer, most likely because both C=C bonds in the furan ring are hydrogenated from the same side of the adsorbed reactant.[62]

In contrast, there are relatively few reports of LGO hydrogenation to THFDM in the literature. A DuPont patent reports 84% yield of THFDM from LGO over a Pt/C catalyst using a low temperature (60 °C) hydrogenation step to produce Lgol, followed by a higher temperature (180 °C) hydrogenolysis step to produce THFDM.[38] We recently investigated the reaction network for LGO hydrogenation over supported Pd catalysts.[42] At low temperature (40 °C) over a Pd/Al_2O_3 catalyst,

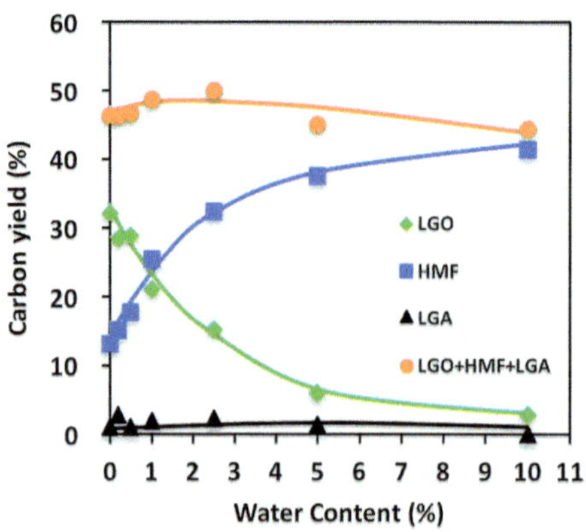

Fig. 4 Impact of initial water content on the product distributions from cellulose. Reaction conditions: cellulose (3 wt%), THF (60 mL), H_2SO_4 (64 μL, conc., 20 mM), 6.9 MPa He, 210 °C, 700 rpm. The reaction times for various water contents to reach the maximum yield of LGO and HMF are summarized in Table S1.†[41]

the C=C bond of LGO was selectively hydrogenated to cyrene.[43] Cyrene can be further hydrogenated to Lgol over the same catalyst at intermediate temperatures (100 °C). Using a bifunctional Pd/Si–Al catalyst, Lgol is hydrogenolyzed to THFDM at higher temperature (150 °C) with 58% selectivity. Side-products produced in parallel with THFDM include THP2M5H (a likely HDO precursor) and 2MTHFA (an undesired dehydration product). The THFDM *cis*/*trans* ratio in this reaction was approximately 2.5, distinctly different from the *cis*/*trans* ratio observed during HMF hydrogenation. We hypothesize that this ratio is influenced by steric hindrance during the ring rearrangement of Lgol to THFDM.

To improve our co-upgrading strategy, we carried out an integrated two-step approach to convert a mixture of LGO and HMF to THFDM. First, we hydrogenated an LGO/HMF mixture (in THF) at a lower temperature (100 °C) over a Ni/SiO_2 catalyst. This non-noble metal catalyst was selected to reduce the catalyst cost compared to the Pd/Al_2O_3 catalyst used for LGO hydrogenation in our previous work.[42] THF was selected as the reaction solvent because it is used in the upstream cellulose dehydration reaction.[60] As shown in Table 3, the reaction displayed near-quantitative yields of LGO hydrogenation to Lgol and HMF hydrogenation to THFDM. ICP analysis of the reaction products suggested that leaching of the Ni/SiO_2 catalyst was negligible (<0.1%).

Next, we further upgraded the product mixture from the hydrogenation reaction over a Pd/Si–Al catalyst at a higher temperature (150 °C) in which Lgol was hydrogenolyzed to THFDM. Lgol was converted to THFDM with approximately 63% selectivity, in agreement with the selectivity of pure Lgol conversion to THFDM (58%).[42] The overall THFDM *cis*/*trans* ratio of the final product was 4.1, consistent with a high *cis*/*trans* ratio for HMF hydrogenation to THFDM (6.3) combined with a lower *cis*/*trans* ratio of Lgol hydrogenolysis to THFDM (~2.2). Because *trans*-THFDM overlaps with one of the isomers of THP2M5H in the GC, product quantification was carried out using a combination of quantitative ^{13}C NMR spectroscopy and GC analysis, as described in our previous work.[42]

Importantly, the THFDM generated during the hydrogenation step was not degraded during the Lgol hydrogenolysis step. The THFDM yields attainable from the co-upgrading of LGO and HMF are nearly identical to those obtained when these species are upgraded separately. The overall selectivity to THFDM from LGO and HMF was 83%. Because we expect that THP2M5H can also be upgraded to HDO through ring-opening to HTO (Scheme 2), the overall selectivity from LGO and HMF to HDO precursors was 94%.

Table 3 Hydrogenation of LGO and HMF to THFDM[a]

Reaction step	Temp. (°C)	Cat.	Lgol (mM) (*exo* : *endo*)	THFDM (mM) (*cis* : *trans*)	THP2M5H (mM)	2MTHFA (mM)
Hydrogenation	100	64% Ni/SiO_2	52.6 (1.9 : 1)	50.7 (6.3 : 1)	0	0
Hydrogenolysis	150	1% Pd/Si–Al	14.1 (1.9 : 1)	76.4 (4.1 : 1)	10.1	7.0

[a] Reaction conditions: feed: LGO (55 mM) + HMF (51 mM) in THF, 3 h reaction time, 3.45 MPa H_2, 100 mg of cat.

3.3.3 Conversion of THFDM to HDO. THFDM first undergoes hydrogenolysis to HTO, followed by conversion of HTO into HDO over a Pt-based catalyst at 160 °C and 5.5 MPa H_2 in a batch reactor.[59] A HDO yield of >70% could be attainable in a one-pot reaction. The major side product is 1-hexanol, which is produced from HDO hydrogenolysis. The other minor products include 1,2-hexanediol and 1,5-hexanediol. As reported in previous studies,[26,37] the hydrogenolysis over RhRe-based catalysts requires the presence of water to generate the Brønsted acidity of the oxophilic promoter; however, it was found that the hydrogenolysis of THFDM could be carried out in the absence of H_2O, in pure THF. This behavior indicates that there is some difference in the hydrogenolysis mechanism between Pt-based catalysts and RhRe-based catalysts.[59] A total selectivity to HDO (60%) and HTO (20–30%) of 80–90% with 30–40% THFDM conversion was obtained at 160 °C and 3.4 MPa H_2 in a continuous flow reactor.[59]

3.4 Process modeling

3.4.1 Process overview. Fig. 5 shows the overall process diagram for the conversion of biomass into PDO and HDO. Raw biomass feedstock is stored and handled so that it flows into the process at a constant, continuous rate. In area A200, the hemicellulose is hydrolyzed into soluble sugars and the lignin is solubilized in a mixture of water and GVL with a sulfuric acid catalyst. The remaining solid is primarily cellulose, which is separated from the liquid fraction for further processing. In area A300, GVL is evaporated so it can be then condensed and recycled for reuse. Upon separation of the GVL, the lignin becomes insoluble and is sent to a combustor to produce process steam and electricity (area A900). The process steam and electricity are also assumed to be available to purchase if the energy generated *via* burning lignin cannot satisfy the overall process consumption. The remaining liquid is heated to convert the xylose to furfural. Furfural is recovered *via* distillation and the stillage (mainly water) is sent to wastewater treatment (A700).

Area A400 contains the conversion of xylose-derived furfural to PDO. First furfural is hydrogenated into THFA over a nickel-based catalyst, which is then

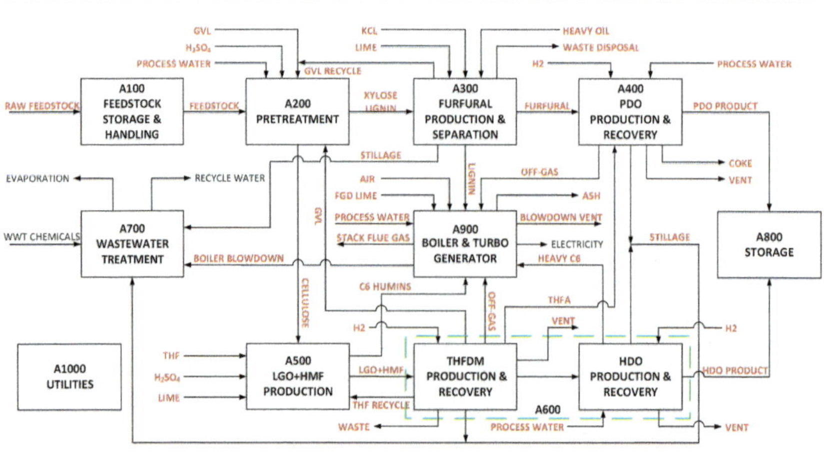

Fig. 5 Overall process diagram.

converted into PDO with high yields in the three step DHH process. PDO and any minor side products are separated from the water and the PDO is purified using distillation. The purified PDO is stored in area A800.

The concentrated cellulose stream from area A200 is converted to LGO and HMF in area A500 using THF as a solvent with a 0.25% sulfuric acid catalyst. Reaction byproducts include LGA, glucose, HMF, furfural, formic acid and humins. The resulting LGO and HMF are converted to HDO *via* THFDM as an intermediate in area A600 where the resulting HDO is separated from the water solvent and other organic materials. The purified HDO is stored in area A800.

3.4.2 Techno-economic analysis. A process simulation model was developed in Aspen Plus (V8.8 Aspen Technology) based on experimental data. The size and cost of all the equipment in areas A200–A700 were estimated using Aspen Process Economic Analyzer (V8.8 Aspen Technology). The equipment cost of the areas A100 and A800–A1000 was estimated using an exponential scaling expression for the equipment size and cost data in NREL's design.[63] The plant size is 1000 dry metric tons per day of white birch. The plant produces 23 000 tons per year of HDO and 40 000 tons per year of PDO. This amount is 34% of the 2019 projected market for HDO if PDO is considered as a direct substitute for HDO. The feedstock cost was assumed to be $80 per dry short ton, which is consistent with several NREL reports.[63,64] Table 4 summarizes the economic analysis for this design. The total installed equipment cost is estimated to be $361.4 million, where the equipment cost for the LGO and HMF production area is the largest component (29.5% of the total installed equipment cost). The total capital investment, which includes other direct (warehouse, site development and additional piping) and indirect (*e.g.* project contingency, construction, field expense, and engineering) costs, is $810.4 million. It should be noted that a relatively high project contingency (40%) is used.

The minimum selling price (MSP) for the diols (*i.e.* the price that leads to zero net present value) is $4089 per ton, which is similar to the current market price of HDO. Fig. 6 summarizes the cost drivers and yields for the catalytic conversion of lignocellulosic biomass to α,ω-diols. The most expensive area of the process is the LGO + HMF production area ($889 per ton of HDO or 22% of the MSP) because of the high capital cost for the cellulose dehydration reactor. This is driven by the low cellulose loading in THF (3 wt%). This dilute system also leads to a high cost for the separation and recovery (THFDM and HDO recovery) in the subsequent area (A600), which is due to the purchased steam cost for the recovery of THF solvent from THFDM and water solvent from HDO since the energy generated *via* burning lignin cannot satisfy the overall process consumption. The next most expensive area is the pretreatment area (A200), contributing $481 per ton to the MSP (12% of the total). This section includes the biomass digester, C5/C6 filters, and GVL evaporator. Details for the individual parameters affecting the overall MSP are presented *via* a single-point sensitivity analysis or tornado chart, as shown in Fig. 7. Reasonable minima and maxima for each variable were chosen to understand and quantify the resulting cost impact on the overall MSP (ΔMSP in Fig. 7 is defined as the change of the MSP from the base case). Each variable was changed to its maximum and minimum value with all other parameters held constant.

From a process improvement point of view, the THF and water solvent systems (production of LGO and HMF from cellulose in A500 and production of HDO from

Table 4 Economic summary for base case process (all values in 2015$)

Minimum Selling Price (MSP)	**$4089 per ton**
HDO production	22 743 tons per year
PDO production	40 291 tons per year
HDO yield	0.063 tons per dry U.S. ton feedstock
PDO yield	0.111 tons per dry U.S. ton feedstock
Feedstock + handling cost	$80.00 per dry U.S. ton feedstock
Internal rate of return (after tax)	10%
Equity percent of total investment	40%
Installed equipment costs	
Pretreatment	$75 600 000
Furfural production and separation	$18 700 000
PDO production and recovery	$12 300 000
LGO + HMF production	$106 600 000
HDO production and recovery	$37 600 000
Wastewater treatment	$37 300 000
Storage	$3 600 000
Boiler and turbo generator	$60 300 000
Utilities	$9 400 000
Total installed equipment cost	$361 400 000
Added direct + indirect costs	$449 000 000
(% of TCI)	55%
Total capital investment (TCI)	$810 400 000
Installed equipment cost/annual ton	$15 891
Total capital investment/annual ton	$35 634
Loan rate	8.0%
Term (years)	10
Capital charge factor (computed)	0.137
Specific operating conditions	
Excess electricity (kW h per kg HDO)	2.5
Plant electricity use (kW h per kg HDO)	3.5
Manufacturing costs ($ per ton HDO)	
Feedstock + handling	$460
GVL solvent	$59
THF solvent	$199
Hydrogen	$153
Catalyst recycling	$89
Other raw materials	$172
Purchased steam	$1025
Waste disposal	$18
Net electricity	-$130
Fixed costs	$282
Capital depreciation	$408
Average income tax	$244
Average return on investment	$1111
Manufacturing costs ($ per year)	
Feedstock + handling	$29 000 000
GVL solvent	$3 700 000
THF solvent	$12 500 000
Hydrogen	$9 600 000
Catalyst recycling	$5 600 000

Table 4 (*Contd.*)

Other raw materials	$10 800 000
Purchased steam	$64 600 000
Waste disposal	$1 100 000
Net electricity	-$8 200 000
Fixed costs	$17 800 000
Capital depreciation	$25 700 000
Average income tax	$15 400 000
Average return on investment	$74 000 000

THFDM in A600, respectively) have the largest impact on the MSP. The economics of this process can be improved by increasing the concentration of both cellulose (from 3% to 10%) and THFDM (from 5% to 15%) in the dehydration and hydrogenolysis of A500 and A600, respectively, which could highly reduce both the size of the equipment and the energy requirement for the solvent recovery by $1207 per ton. For instance, increasing the concentration of THFDM would reduce the water requirement and thus reduce the produced wastewater since the largest source of wastewater is A600 due to the low concentration of THFDM in water. This also leads to a lower utility consumption for the subsequent separation and recovery steps, which would reduce the size of the cooling water system and thus lower the amount of wastewater from the cooling water system that is the second largest source of wastewater. Since the major expense for A700 is the capital cost for infrastructures, decreasing the water usage by increasing the reactant loading in water would reduce the scale and cost of A700. Additionally, increasing the HDO yield from 25% to 50% will decrease the MSP of HDO by $1091 per ton. This analysis indicates that further improving the catalytic reaction processes to produce higher yields of HDO with higher reactant loadings could decrease the MSP of the products by over 50%.

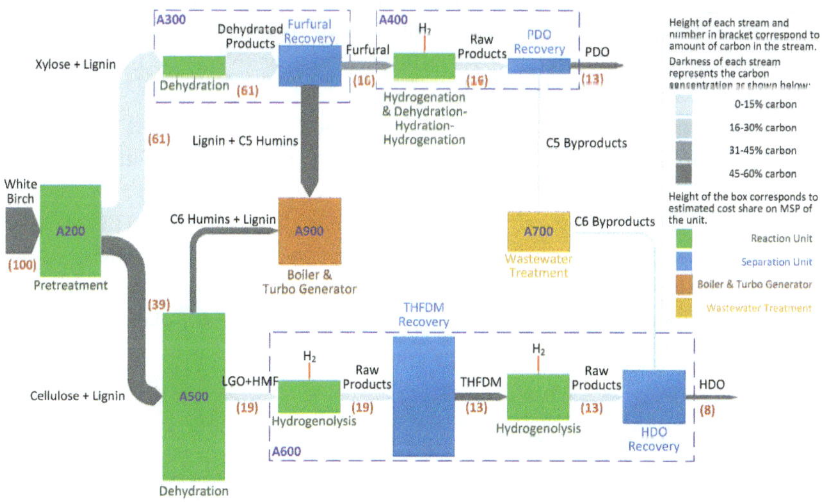

Fig. 6 Sankey diagram on cost contribution and carbon flow for each process area.

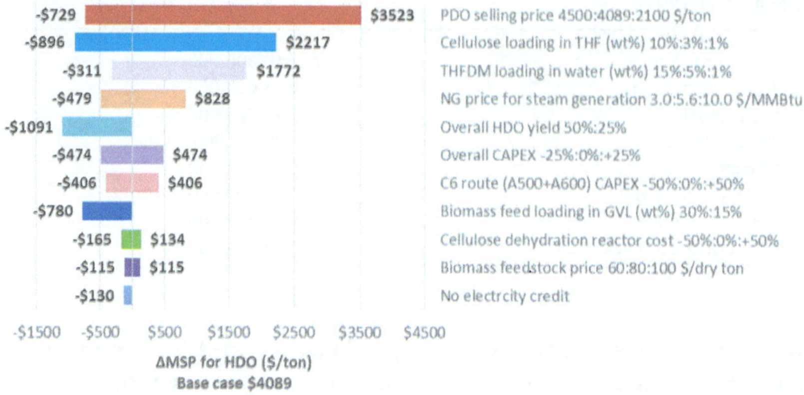

Fig. 7 Single-point sensitivity tornado chart for the benchmark process.

4 Conclusions

We have demonstrated a pathway to synthesize PDO and HDO from lignocellulosic biomass. In the TriVersa Process™, a white birch biomass feedstock was solubilized by dilute sulfuric acid to produce a high purity solid cellulose (90–94.5% purity and 35–42% yield), a soluble hemicellulose (>90% of hemicellulose in biomass feedstock), and a lignin stream in GVL/H_2O (80/20) solvent. The soluble hemicellulose was converted into furfural at 85% yield in less than 30 seconds. Furfural has been shown to form THFA in 96% yield by vapor phase hydrogenation.[20] THFA was converted into PDO in 84% overall yield *via* the three-step DHH process. THFA was converted into DHP in 87% yield by vapor phase dehydration with γ-Al_2O_3. Uncatalyzed DHP hydration in water produced 2-HY-THP and 2-HY-THP dimers in 99% yield. Ru catalysts hydrogenated 2-HY-THP to PDO in 97% yield. This conversion represents an 80% yield of PDO from furfural. HDO was produced from the cellulose by first dehydrating cellulose into LGO and HMF (48% yield) in a THF/H_2O mixture with acid catalysts. This mixture was then hydrogenated using Ni/SiO_2, followed by hydrogenolysis over Pd/SiO_2–Al_2O_3 with 83% overall selectivity to THFDM. A 70% yield to HDO from THFDM was obtained with Pt-based catalysts.

A process model was developed to estimate the economics of this process for a 1000 dry metric tons white birch per day plant. The minimum selling price for the production of HDO and PDO is $4090 per ton, which is similar to the market price of HDO derived from petroleum. Our analysis suggests that two parameters that appear to be the most important are the concentration of cellulose and THFDM and the HDO yield. In addition, further process integration and continuous flow studies are also required to evaluate the impact of contaminants and the catalyst stability.

Acknowledgements

This material is based upon work supported by the Department of Energy, Office of Energy Efficiency and Renewable Energy (EERE), under Award Number DE-

EE0006878. Any opinions, findings, and conclusions or recommendations expressed in this material are those of the authors and do not necessarily reflect the views of the Department of Energy. S. H. K. acknowledges that this material is based upon work supported by the National Science Foundation under Grant No. DGE-1256259. We thank Q. Rashke and M. J. Liu for assistance with experiments.

References

1 Markets and Markets, *1,6-Hexanediol Market by Application Polyurethanes, Coatings, Acrylates, Adhesives, Unsaturated Polyester Resins, Plasticizers, and Others*, Trends and Forecasts to 2019, 2014.

2 P. Werle, M. Morawietz, S. Lundmark, K. Sörensen, E. Karvinen and J. Lehtonen, in *Ullmann's Encyclopedia of Industrial Chemistry*, 2008, pp. 263–281.

3 T. L. Brown, H. E. LeMay Jr, B. E. Bursten, C. Murphy, P. Woodward, S. Langford, D. Sagatys and A. George, *Chemistry: The Central Science*, Pearson Higher Education AU, 2013.

4 M. A. M. Carai, G. Colombo, R. Reali, S. Serra, I. Mocci, M. P. Castelli, G. Cignarella and G. L. Gessa, *Eur. J. Pharmacol.*, 2002, **441**, 157–163.

5 O. Snead, R. Furner and C. Liu, *Biochem. Pharmacol.*, 1989, **38**, 4375–4380.

6 M. Perry, J. Pomfret and R. Crabb, *Australia bans China-made toy on toxic drug risk*, Reuters, 2007.

7 Z. Németh, B. Kun and Z. Demetrovics, *J. Psychopharmacol.*, 2010, **24**, 1281–1287.

8 Lux Research, *Bio-based Materials and Chemicals Intelligence Service*, 2013.

9 S. Van De Vyver and Y. Román-Leshkov, *Catal. Sci. Technol.*, 2013, **3**, 1465.

10 T. F. Cavani Fabrizio, *Centi Gabriele, Perathoner Siglinda, Sustainable Industrial Chemistry: Principles, Tools and Industrial Examples*, Wiley-VCH Verlag GmbH & Co. KGaA, 2009.

11 I. V. Gürsel, V. Hessel, Q. Wang, T. Noël and J. Lang, *Green Process. Synth.*, 2012, **1**, 315–336.

12 V. Hessel, I. Vural Gürsel, Q. Wang, T. Noël and J. Lang, *Chem. Eng. Technol.*, 2012, **35**, 1184–1204.

13 C. Aellig, C. Girard and I. Hermans, *Angew. Chem., Int. Ed.*, 2011, **50**, 12355–12360.

14 C. Aellig, U. Neuenschwander and I. Hermans, *ChemCatChem*, 2012, **4**, 525–529.

15 C. Aellig, D. Scholz and I. Hermans, *ChemSusChem*, 2012, **5**, 1732–1736.

16 J. Scarlett, P. Willett and M. A. Wood, European Patent Office, EP 1 048 638 B1, 2003.

17 W. Schwarz, J. Schossig, R. Rossbacher and H. Höke, in *Ullmann's Encyclopedia of Industrial Chemistry*, John Wiley, 2000.

18 G. Paris, L. Berlinguet and R. Gaudry, *Org. Synth.*, 1963, **4**, 496.

19 S. Sitthisa, T. Sooknoi, Y. Ma, P. B. Balbuena and D. E. Resasco, *J. Catal.*, 2011, **277**, 1–13.

20 Y. Nakagawa, H. Nakazawa, H. Watanabe and K. Tomishige, *ChemCatChem*, 2012, **4**, 1791–1797.

21 T. P. Sulmonetti, S. H. Pang, M. T. Claure, S. Lee, D. A. Cullen, P. K. Agrawal and C. W. Jones, *Appl. Catal., A*, 2016, **517**, 187–195.

22 S. Koso, N. Ueda, Y. Shinmi, K. Okumura, T. Kizuka and K. Tomishige, *J. Catal.*, 2009, **267**, 89–92.

23 L. F. Han, W. Zhang, M. H. Lu, M. S. Li and Y. H. Shan, *Chin. J. Inorg. Chem.*, 2015, **31**, 297–302.

24 Y. Nakagawa, M. Tamura and K. Tomishige, *Catal. Surv. Asia*, 2015, **19**, 249–256.

25 S. Koso, I. Furikado, A. Shimao, T. Miyazawa, K. Kunimori and K. Tomishige, *Chem. Commun.*, 2009, 2035–2037.

26 M. Chia, B. J. O'Neill, R. Alamillo, P. J. Dietrich, F. H. Ribeiro, J. T. Miller and J. A. Dumesic, *J. Catal.*, 2013, **308**, 226–236.

27 Y. Nakagawa and K. Tomishige, *Catal. Today*, 2012, **195**, 136–143.

28 K. Chen, K. Mori, H. Watanabe, Y. Nakagawa and K. Tomishige, *J. Catal.*, 2012, **294**, 171–183.

29 M. Chia, Y. J. Pagán-Torres, D. Hibbitts, Q. Tan, H. N. Pham, A. K. Datye, M. Neurock, R. J. Davis and J. A. Dumesic, *J. Am. Chem. Soc.*, 2011, **133**, 12675–12689.

30 S. Koso, Y. Nakagawa and K. Tomishige, *J. Catal.*, 2011, **280**, 221–229.

31 M. Faber, Process for Producing Adipic acid from biomass, *US Pat.*, 4400468, 23 Aug 1983.

32 T. Buntara, S. Noel, P. H. Phua, I. Melián-Cabrera, J. G. De Vries and H. J. Heeres, *Angew. Chem., Int. Ed.*, 2011, **50**, 7083–7087.

33 T. Buntara, S. Noel, P. H. Phua, I. Melián-Cabrera, J. G. De Vries and H. J. Heeres, *Top. Catal.*, 2012, **55**, 612–619.

34 I. Gunnarsson and S. Arnórsson, *Geochim. Cosmochim. Acta*, 2000, **64**, 2295–2307.

35 R. Siever, *J. Geol.*, 1962, **70**, 127–150.

36 B. Xiao, M. Zheng, X. Li, J. Pang, R. Sun, H. Wang, X. Pan, A.-Q. Wang, X. Wang and T. Zhang, *Green Chem.*, 2016, **18**, 2175–2184.

37 K. Chen, S. Koso, T. Kubota, Y. Nakagawa and K. Tomishige, *ChemCatChem*, 2010, **2**, 547–555.

38 A. M. Allgeier, W. I. N. De Silva, E. Korovessi, C. A. Menning, J. C. Ritter, S. K. Sengupta and C. S. Stauffer, Process for preparing 1,6-hexanediol, *US Pat.*, US 8865940 B2, 21 Oct 2014.

39 M. De Bruyn, J. Fan, V. L. Budarin, D. J. Macquarrie, L. D. Gomez, R. Simister, T. J. Farmer, W. D. Raverty, S. J. McQueen-Mason and J. H. Clark, *Energy Environ. Sci.*, 2016, **9**, 2571–2574.

40 Z. J. Brentzel, K. J. Barnett, K. Huang, C. T. Maravelias, J. A. Dumesic and G. W. Huber, *ChemSusChem*, 2017, **10**(7), 1351–1355.

41 J. He, M. Liu, K. Huang, T. W. Walker, C. T. Maravelias, J. A. Dumesic and G. W. Huber, *Green Chem.*, 2017, DOI: 10.1039/c7gc01688c.

42 S. H. Krishna, D. J. McClelland, Q. A. Rashke, J. A. Dumesic and G. W. Huber, *Green Chem.*, 2017, **19**, 1278–1285.

43 J. Sherwood, M. De bruyn, A. Constantinou, L. Moity, C. R. McElroy, T. J. Farmer, T. Duncan, W. Raverty, A. J. Hunt and J. H. Clark, *Chem. Commun.*, 2014, **50**, 9650–9652.

44 A. M. Allgeier, N. Desilva, E. Korovessi, C. Menning, J. C. Ritter and S. K. Sengupta, *US Pat.*, US 2013/0172629 A1, 2013.

45 S. H. Krishna, T. W. Walker, J. A. Dumesic and G. W. Huber, *ChemSusChem*, 2017, **10**, 129–138.

46 R. Mariscal, P. Maireles-Torres, M. Ojeda, I. Sádaba and M. López Granados, *Energy Environ. Sci.*, 2016, **9**, 1144–1189.
47 IHS Markit, *Furfuryl Alcohol and Furan Resins-Chemical Economics Handbook*, 2016.
48 D. M. Alonso, S. H. Hakim, S. Zhou, W. Won, O. Hosseinaei, J. Tao, V. Garcia-Negron, A. H. Motagamwala, M. A. Mellmer, K. Huang, C. J. Houtman, N. Labbé, D. P. Harper, C. Maravelias, T. Runge and J. A. Dumesic, *Sci. Adv.*, 2017, DOI: 10.1126/sciadv.1603301.
49 A. Sluiter, B. Hames, R. Ruiz, C. Scarlata, J. Sluiter, D. Templeton and D. Crocker, *Technical Report NREL/TP-510-42618*, 2008.
50 D. M. Alonso, S. G. Wettstein, M. A. Mellmer, E. I. Gurbuz and J. A. Dumesic, *Energy Environ. Sci.*, 2013, **6**, 76–80.
51 D. M. Alonso, S. G. Wettstein and J. A. Dumesic, *Green Chem.*, 2013, **15**, 584.
52 W. Fang and H. Sixta, *ChemSusChem*, 2015, **8**, 73–76.
53 M. A. Mellmer, D. Martin Alonso, J. S. Luterbacher, J. M. R. Gallo and J. a. Dumesic, *Green Chem.*, 2014, **16**, 4659–4662.
54 J. S. Luterbacher, A. Azarpira, A. H. Motagamwala, F. Lu, J. Ralph and J. a. Dumesic, *Energy Environ. Sci.*, 2015, **8**, 2657–2663.
55 V. García-Negrón, N. D. Phillip, J. Li, C. Daniel, D. Wood, D. J. Keffer, O. Rios and D. P. Harper, *Energy Technol.*, 2016.
56 L. E. Schniepp and H. H. Geller, *J. Am. Chem. Soc.*, 1946, **68**, 1646–1648.
57 C. D. Hurd and W. H. Saunders, *J. Am. Chem. Soc.*, 1952, **74**, 5324–5329.
58 T. Shono, M. Godo and K. Shinra, *Jpn. Anal.*, 1971, **20**, 1183.
59 J. He, S. P. Burt, L. Zhang, M. Ball, I. Hermans, G. W. Huber and J. A. Dumesic, in preparation.
60 F. Cao, T. J. Schwartz, D. J. McClelland, S. H. Krishna, J. A. Dumesic and G. W. Huber, *Energy Environ. Sci.*, 2015, **8**, 1808–1815.
61 R. Alamillo, M. Tucker, M. Chia, Y. Pagán-Torres, J. Dumesic, Y. Pagan-Torres and J. Dumesic, *Green Chem.*, 2012, **14**, 1413–1419.
62 X. Hu, R. J. M. Westerhof, L. Wu, D. Dong and C.-Z. Li, *Green Chem.*, 2015, **17**, 219–224.
63 R. Davis, L. Tao, C. Scarlata, E. C. D. Tan, J. Ross, J. Lukas and D. Sexton, *Process Design and Economics for the Conversion of Lignocellulosic Biomass to Hydrocarbons: Dilute-Acid and Enzymatic Deconstruction of Biomass to Sugars and Catalytic Conversion of Sugars to Hydrocarbons NREL/TP-5100-62498*, 2015.
64 R. Davis, *Process Design and Economics for the Conversion of Lignocellulosic Biomass to Hydrocarbons-Dilute-Acid and Enzymatic Deconstruction of Biomass to Sugars and Biological Conversion of Sugars to Hydrocarbons NREL/TP-5100-60223*, 2013.

Faraday Discussions

PAPER

The influence of the explosive decompression in steam-explosion pretreatment on the enzymatic digestibility of different biomasses†

Christoph-M. Seidel, *[a] Thomas Pielhop,[a] Michael H. Studer[b] and Philipp Rudolf von Rohr[a]

Received 14th February 2017, Accepted 13th March 2017
DOI: 10.1039/c7fd00066a

For the production of second generation biofuels from lignocellulosic biomass, pretreatment of the biomass feedstock is necessary to overcome its recalcitrance in order to gain fermentable sugars. Due to many reasons, steam-explosion pretreatment is currently the most commonly used pretreatment method for lignocellulosic biomass on a commercial scale [S. Brethauer and M. H. Studer, *CHIMIA*, 2015, **69**, 572–581]. In contrast to others, we showed that the explosive decompression at the end of this pretreatment step can have a positive influence on the enzymatic digestibility of softwood, especially in combination with high enzyme dosages [T. Pielhop, *et al.*, *Biotechnology for Biofuels*, 2016, **9**, 152]. In this study, the influence of the explosive decompression on the enzymatic digestibility of hardwood and herbaceous plants was systematically studied. Beech and corn stover were pretreated under different pretreatment conditions and enzymatically hydrolysed with different enzyme dosages. The maximum enhancement of the digestibility of corn stover was 16.53% after a 2.5 min pretreatment step at 15 barg steam pressure. For beech, a maximum relative enhancement of 58.29% after a 10 min pretreatment step at 15 barg steam pressure could be reached. With this, we show that the explosive decompression can also enhance the enzymatic cellulose digestibility of hardwood and herbaceous plants.

1 Introduction

Lignocellulosic biomass like wood, energy crops or agricultural waste is the most abundant raw material on earth[1] and a potential carbon-neutral source for the production of liquid fuels. A very promising approach is the production *via*

[a]*Institute of Process Engineering, ETH Zurich, Sonneggstrasse 3, 8092 Zurich, Switzerland. E-mail: seidel@ipe.mavt.ethz.ch; vonrohr@ipe.mavt.ethz.ch*

[b]*School of Agricultural, Forest and Food Sciences, Bern University of Applied Sciences, Länggasse 85, 3052 Zollikofen, Switzerland*

† Electronic supplementary information (ESI) available. See DOI: 10.1039/c7fd00066a

enzymatic hydrolysis of cellulose and hemicellulose, sugar fermentation and ethanol purification.[2,3] The lignin is separated and usually burned for energy production.

The hydrolysis of the complex network is challenging since the entanglement of the lignin with cellulose and hemicellulose protects the biomass against enzymatic hydrolysis by limiting the accessibility of enzymes to cellulose and non-productive binding of enzymes.[4] This makes pretreatment an essential step to overcome the biomass recalcitrance by disrupting the cell walls, which improves the access of hydrolytic enzymes to the polysaccharides. Autohydrolysis pretreatments are very attractive due to their cost-saving potential,[2] since no additional chemicals like acids, bases or organic solvents are required. This simplifies the whole biorefinery process, because less neutralisation or removal of organic solvents is necessary, which can inhibit cellulase enzymes or fermentative microorganisms.

Steam-explosion pretreatment is also a very flexible pretreatment method for the feedstock.[5] It has low investment costs, energy requirements and environmental impact.[6–8] Furthermore, steam-explosion pretreatments allow for high biomass loadings and can deal with large biomass particles due to the particle size reducing the effect of the explosion.[9–11] For these reasons, steam-explosion pretreatment has become the dominating biomass pretreatment technology on a commercial scale.[10]

During steam-explosion pretreatment, the biomass is treated with saturated steam at a certain temperature for a certain time. Due to fast diffusion, the steam accesses the inner "capillary-like" structures, where it condenses and heats up the biomass.[12] An acidic milieu (pH 3 to 4) is created by acids released from the hemicellulose,[13] which leads to hydrolysation of the hemicellulose and cleavage of ether bonds in the lignin fraction.[14,15] In this way, the enzymatic digestibility is enhanced. The last step of the steam-explosion pretreatment is the explosion, which means that the pressure is abruptly released. In batch pretreatments this is done by opening a valve and in continuous treatments by a small slit valve to a tank or pipe usually with atmospheric conditions. The water inside the biomass capillaries was in equilibrium with the high temperature and pressure during the pretreatment. A flash evaporation of the liquid water inside the biomass occurs due to the pressure drop, resulting in an explosion or extensive defibriation of the biomass.[12]

Steam-explosion pretreatment has been studied for a long time, but very little information on the effect of the explosive decompression at the end of the pretreatment is currently available. Several studies state that steam-explosion is an efficient way to decrease the particle size[7,16] and increase the accessibility of carbohydrates to enzymes.[17,18] Only two systematic studies investigated the impact of the explosive decompression on the enzymatic cellulose digestibility of wood. Brownell *et al.*[19] identified no influence on the digestibility of aspen wood chips, although the enzymatic dosage was not varied. We showed that for spruce wood chips there is an influence at higher enzyme dosages, while at lower dosages the digestibility is similar.[5] Altogether, there is still a big gap in the general knowledge of the influence of the explosion at the end of steam pretreatment on lignocellulosic biomass. In both studies only wood as one type of lignocellulosic biomass was investigated and the results of the two studies are inconsistent.

Summarised, for a biorefinery concept, the only proven positive effect of the explosion is the particle size reduction. Regarding the enhancement of the enzymatic cellulose digestibility, inconsistent studies are available, which is why in this study the influence on the enzymatic digestibility of hardwood (beech) and herbaceous plants (corn stover) was evaluated.

2 Experimental

2.1 Biomass

Corn stover was harvested in November 2015 in Bärfischenhaus (canton of Bern, Switzerland) and chopped through a 30 mm sieve. After air drying for four days, the corn stover had a dry matter content of 93.63 ± 0.67% and the composition was determined to be: glucan 31.88 ± 0.61%, xylan 17.25 ± 0.26%, acid-soluble lignin (ASL) 4.27 ± 0.15%, acid-insoluble lignin (AIL) 11.09 ± 0.20%, lignin ashes 0.43 ± 0.07%, and alcohol and water extractives 32.97% (total 97.73%).

Beech was cut in August 2016 in Messen (canton of Solothurn, Switzerland) from a roughly 40 year old tree and chopped through a 30 mm sieve. After air drying for one day and oven drying at 40 °C for two days, the beech had a dry matter content of 91.74 ± 0.21% and the composition was determined to be: glucan 41.34 ± 0.97%, xylan 20.50 ± 0.41%, acid-soluble lignin (ASL) 10.20 ± 0.13%, acid-insoluble lignin (AIL) 23.21 ± 0.67%, lignin ashes 9.19 ± 1.8%, and extractives 4.07 ± 0.15% (total 96.25%).

2.2 Pretreatment

The pretreatment experiments were carried out in a steam gun set up by the Industrieanlagen Planungsgesellschaft (IAP, Graz, Austria). A process scheme and a detailed description can be found elsewhere.[5] The boiler of the steam gun provided saturated steam of 32 bar. The pretreatment experiments were carried out at 170.5, 201.4, and 230.1 °C, corresponding to a pressure of 7, 15, and 27 barg in the reactor. Before the experiments, the reactor was preheated to 20 °C less than the pretreatment temperature to reduce the steam condensation during the experiments. After preheating, the steam and condensate of the reactor were released again and the reactor was filled with the biomass. The reactor was filled with 100 g of corn stover or 500 g of beech. This corresponds to 2.5 l or filling the reactor halfway. During the first 10 s, the reactor was flushed to remove entrapped air. The pretreatment time was started when 99% of the target temperature was reached. After the predefined pretreatment time of 2.5 to 20 min, the biomass was discharged from the reactor to the blow tank through a ball valve. If lower explosion pressures than the pretreatment pressure were investigated, the pressure was released by an exhaust valve at the top of the reactor. The obtained biomass was weighed and filtered, recording the volume, weight and pH of the filtrate. The dry matter of the filter cake was recorded in triplicate. Experiments carried out with an overpressure of 3 bar, which was necessary to empty the steam gun, are referred to without explosion. The different process conditions of the different experiments can be found in Tables 1 and 2. The dimensionless pretreatment severity used was defined analogously to Overend[20] as follows:

$$R_0 = \frac{t}{[\text{min}]} e^{\frac{T - 100\ ^\circ\text{C}}{14.75\ ^\circ\text{C}}} \tag{1}$$

Table 1 Overview of the experiments and experimental conditions for corn stover

T [°C]	p [barg]	t [min]	$\log R_0$ [—]	Δp_{expl} [bar]
143.7	3	5	1.87	3
170.5	7	2.5	2.46	7
170.5	7	2.5	2.46	3
170.5	7	5	2.76	7
170.5	7	5	2.76	3
170.5	7	10	3.06	7
170.5	7	10	3.06	3
170.5	7	20	3.38	7
170.5	7	20	3.38	3
201.4	15	2.5	3.34	15
201.4	15	2.5	3.34	3
201.4	15	5	3.64	15
201.4	15	5	3.64	3
201.4	15	10	3.98	15
201.4	15	10	3.98	3
201.4	15	20	4.24	15
201.4	15	20	4.24	3
230.1	27	5	4.52	27
230.1	27	5	4.52	3

where T represents the pretreatment temperature and t represents the pretreatment time.

2.3 Compositional analysis of biomass and pretreatment liquor

The analysis of the dry matter content, the biomass composition and the sugar content in the pretreatment and enzymatic hydrolysis liquor were done according to the laboratory methods from the National Renewable Energy Laboratory (NREL).[21–25] All biomass and pretreatment liquor analyses were performed in triplicate and duplicate, respectively. Mean values with single standard deviations are reported in this work.

Table 2 Overview of the experiments and experimental conditions for beech

T [°C]	p [barg]	t [min]	$\log R_0$ [—]	Δp_{expl} [bar]
170.5	7	5	2.76	7
170.5	7	10	3.06	7
170.5	7	10	3.06	3
201.4	15	10	3.98	15
201.4	15	10	3.98	7
201.4	15	10	3.98	3
230.1	27	15	5.00	27
230.1	27	15	5.00	20
230.1	27	15	5.00	13
230.1	27	15	5.00	7
230.1	27	15	5.00	3

2.4 Enzymatic hydrolysis

The enzymatic hydrolysis of the biomass was conducted according to the NREL standard procedure using a cellulose concentration of 1% w/w.[26] The following changes were made: sodium azide with a final concentration of 0.01 g l^{-1} was used instead of antibiotics and the pH was adjusted to 5.0. Accellerase 1500 (Genencor; batch number 4902 279 303) with an activity of 45 filter paper units (FPU) per ml (ref. 27) was used with different concentrations between 2.5 and 60 FPU per g cellulose in the sample preparation. The samples were incubated in a shaker (Multitron; Infors-HT) with a shaking throw of 50 mm at 50 °C and 200 rpm for 120 h and then the supernatant was analysed for sugars. All hydrolysis experiments were carried out in triplicate and single standard deviations are reported with the mean.

2.5 Sugar analysis

The sugar analysis was performed according to the laboratory method from the NREL.[23] A Waters 2695 separation module equipped with a Bio-Rad Aminex HPX-87H column with a precolumn and a Waters 410 differential refractometer were used for the analysis. The column was operated at 60 °C with a flow of 0.6 ml min^{-1} with 0.005 M sulfuric acid as the eluent.

2.6 Wet sieving

The raw biomass and selected pretreated biomass samples were analysed by wet sieving analysis with a vibratory sieve shaker (AS200, Retsch, Germany). All samples were sieved for 5 min at a sieving amplitude of 1.1 mm and a water flowrate of 45 l h^{-1}. Samples were not dried and were adjusted to contain 10 g of dry biomass. The used mesh sizes were 6700, 4000, 2000, 1000, 500, 355, 180 and 25 μm. After the sieving procedure, the sieves with the remaining biomass were dried at 80 °C (maximum tolerated temperature of the sieves) overnight. For the investigation of the enzymatic digestibility of different particle sizes, the biomass samples were prepared without drying.

3 Results and discussion

3.1 Influence of the explosive decompression on the enzymatic digestibility

3.1.1 Corn stover. Corn stover has a comparatively high digestibility of 32.31 ± 1.18% already under low pretreatment conditions of 5 min at 143.7 °C ($\log R_0 = 1.97$). Under these conditions, an investigation of the explosion effect could not be done, since the pretreatment pressure of 3 bar was the minimum required pressure to empty the steam gun.

The pretreatments at medium temperature were carried out at 170.5 °C (corresponding to a vapor pressure of 7 barg). The time for the experiments was varied between 2.5 and 20 min so that pretreatment severities between $\log R_0 = 2.46$ and $\log R_0 = 3.36$ could be realised. These pretreatments led, without explosion, to digestibilities between 41.33 ± 1.38% (2.5 min) and 48.88 ± 2.90% (20 min) (see Fig. 1). With explosions, the digestibility could be increased in all experiments by around 4–8%, with the maximum increase after 20 min (a relative increase of 16.18%). Presumably, the explosion pressure difference of only 4 bar between the

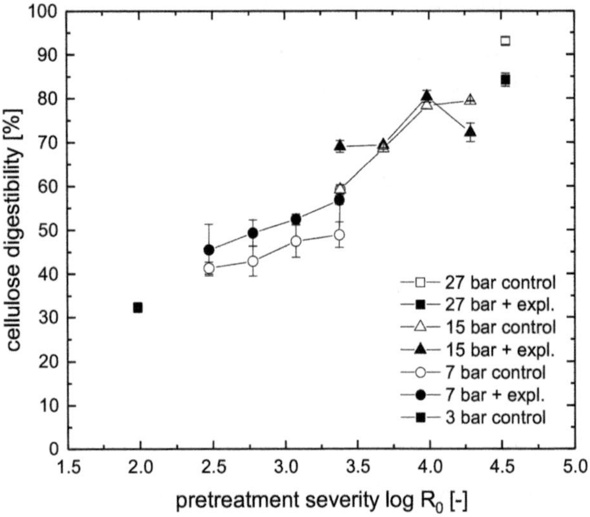

Fig. 1 Enzymatic cellulose digestibility of corn stover after pretreatment at different severities (steam pressures of 3, 7, 15 and 27 bar, time varied between 2.5 and 20 min) with and without explosion. Hydrolysis conditions: 1% w/w cellulose, 60 FPU per g cellulose.

samples pretreated with explosion and the samples without explosion is not high enough to defibrate the biomass.

The pretreatments at higher temperature were performed at 201.4 °C (corresponding to a vapor pressure of 15 barg). The times were varied between 2.5 and 20 min so that pretreatment severities between $\log R_0 = 3.38$ and $\log R_0 = 4.29$ could be realised. With these pretreatments, digestibilities between 59.21 $\pm$ 1.08% and 80.43 $\pm$ 1.39% could be reached. A beneficial effect of the explosion could only be observed at 2.5 min. The digestibility could be increased from 59.21 $\pm$ 1.08% to 69.01 $\pm$ 1.34%. For the other samples, the explosive decompression showed no effect on the digestibility or was even negative (see Fig. 1). The difficulty with the pretreatment severity factor (assuming a first-order reaction) can be seen here. The sample pretreated at a pressure of 15 barg for 2.5 min has a digestibility that is more than 10% higher than that of the sample pretreated at 7 bar for 20 min, although the severity for both was the same ($R_0 = 3.38$).

3.1.2 Beech. Beech was pretreated under harsher conditions compared to corn stover. The results are shown in Fig. 2. After 10 min pretreatment at 7 bar ($\log R_0 = 3.07$) steam pressure, the cellulose digestibility with and without explosion is below 10%. After 10 min pretreatment at 15 barg ($\log R_0 = 3.98$) steam pressure, the cellulose digestibility is 42.87 $\pm$ 1.93% without explosion and 67.86 $\pm$ 1.84% with explosion, showing that the explosion could increase the enzymatic digestibility by 58.29% (relative). After 15 min pretreatment at 27 barg ($\log R_0 = 5.00$), the cellulose of the exploded biomass is completely digestible (99.14 $\pm$ 0.66%), which is 15.60% (relative) higher than that for the non-exploded biomass (85.76 $\pm$ 2.63%). This shows that the explosive decompression can enhance the enzymatic digestibility of hardwood (in contrast to the study of Brownell *et al.*[19]).

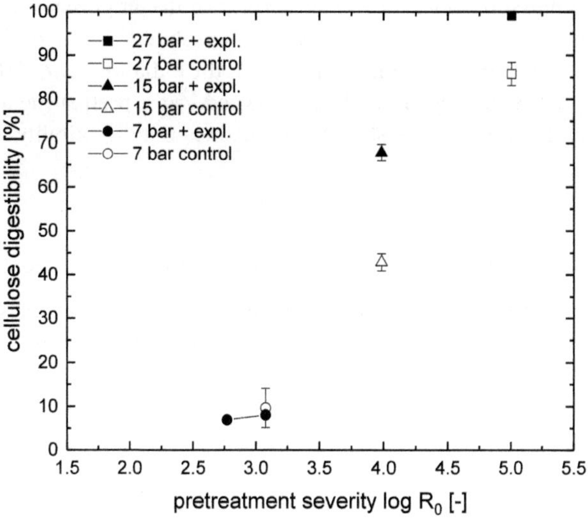

Fig. 2 Enzymatic cellulose digestibility of beech after pretreatment at different severities with and without explosion. Hydrolysis conditions: 1% w/w cellulose, 60 FPU per g cellulose.

Since the explosion showed a strong influence on the enzymatic digestibility, different explosion pressure differences were evaluated for conditions that allowed for a pronounced effect of the explosion ($\log R_0 = 3.98$ and $\log R_0 = 5.0$). The results are shown in Fig. 3. It can be seen that the cellulose digestibility is distinctly increasing with an increased pressure difference, especially at a pretreatment severity of $\log R_0 = 3.98$ where it is increased by 72%. With an

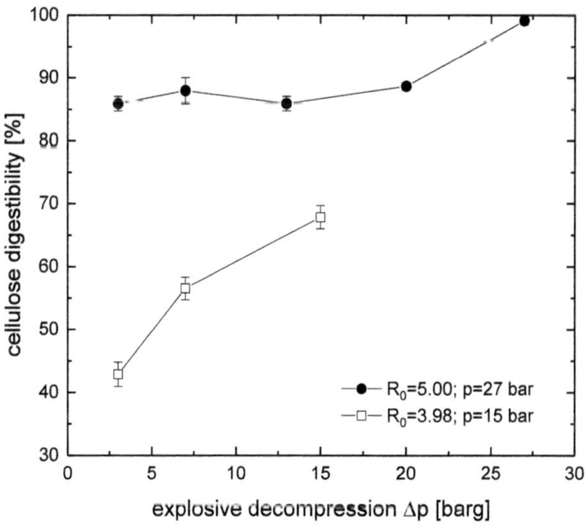

Fig. 3 Enzymatic cellulose digestibility of beech depending on the pressure difference of the explosion at the end of the pretreatment step. Hydrolysis conditions: 1% w/w cellulose, 60 FPU per g cellulose.

intermediate pressure difference of the explosive decompression of 7 bar, the enhancement is reduced. This shows that the absolute pressure of the explosive decompression also plays an important role in the enhancement of the cellulose digestibility, as we already showed for softwood.[5] At a pretreatment severity of $\log R_0 = 5.00$, the influence is slightly lower, since the digestibility is already at 85.76%.

3.2 Influence of the explosive decompression on the particle size

3.2.1 Corn stover. The particle size distribution was compared between samples of exploded and non-exploded biomass pretreated at 15 barg for 2.5 min ($\log R_0 = 3.34$) and 10 min ($\log R_0 = 3.98$). The results can be seen in Fig. 4. As can be seen from the cumulative particle size, the number of smaller particles after the pretreatment is increased by both the pretreatment severity, *i.e.* time, and by the explosion. The samples that were pretreated for a shorter time show a much larger difference between the exploded and non-exploded biomass. For samples that were pretreated for a longer time, the influence of the explosion on the particle size reduction, as well as on the enzymatic digestibility, is very low. This suggests that the particle size has a great influence on the enzymatic digestibility.

3.2.2 Beech. The particle size distribution was compared between the samples pretreated at 27 barg and explosively discharged at pressure differences of 27, 20, 13, 7 and 3 bar. The cumulative particle size distribution can be found in Fig. 5. The largest difference is between the sample discharged without explosion (3 bar) and the other samples. The sample with no explosion has larger particles and, as can be seen in Fig. 3, also has the lowest enzymatic digestibility. As already seen for corn stover in the last section, this leads to the same suggestion that the particle size has a great influence on the enzymatic digestibility.

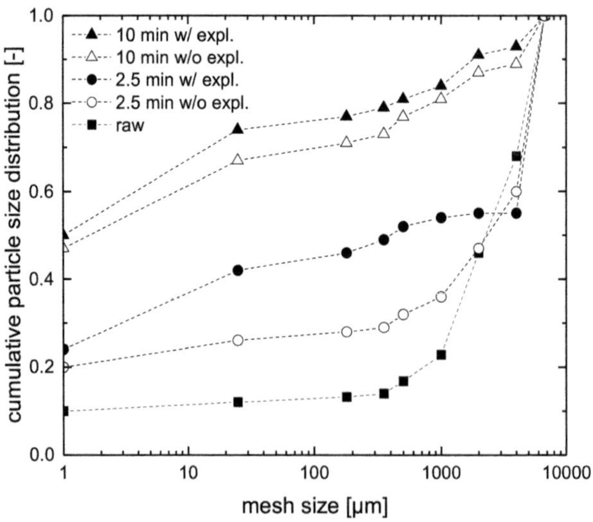

Fig. 4 Cumulative particle size distribution of corn stover pretreated at 15 barg for 2.5 and 10 min ($\log R_0 = 3.34$ and $\log R_0 = 3.98$) with and without explosive decompression at the end of the pretreatment step.

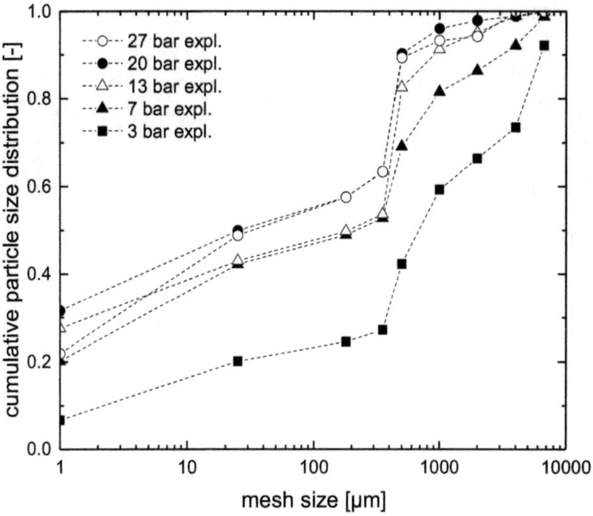

Fig. 5 Cumulative particle size distribution of beech pretreated at 27 barg for 15 min ($\log R_0 = 5.00$) with different explosion pressures at the end of the pretreatment step.

To determine whether the enzymatic digestibility is only increased by the particle size reduction or if other alterations were provoked by the explosive decompression, the samples exploded with 27 and 3 bar were fractionated by sieving to obtain samples of different particle sizes and the individual fractions were enzymatically hydrolysed. As Fig. 6 shows, the exploded biomass does not exhibit a higher digestibility than the sample prepared without explosion. Cellulose digestibilities above 100% for the small particles might be due to sugar

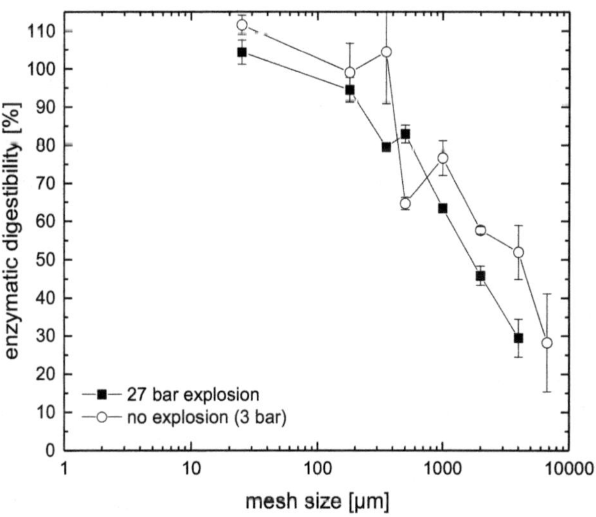

Fig. 6 Enzymatic cellulose digestibility of biomass fractions of different particle sizes from beech treated for 15 min at 27 barg with and without explosion. Hydrolysis conditions: 1% w/w cellulose, 60 FPU per g cellulose.

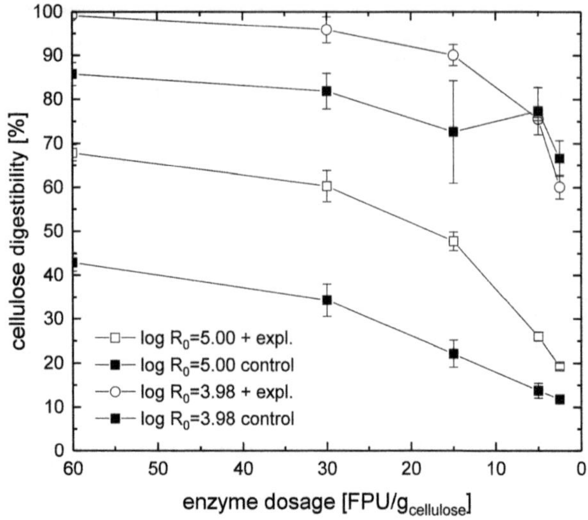

Fig. 7 Enzymatic cellulose digestibility of exploded and non-exploded beech pretreated with log $R_0 = 3.98$ and log $R_0 = 5.00$. Hydrolysis conditions: 1% w/w cellulose, 60, 30, 15, 5 and 2.5 FPU per g cellulose.

degradation during the cellulose determination. During the procedure, sugar recovery standards are only used for the second dilute acid hydrolysis in the autoclave and not for the first hydrolysis step where concentrated sulfuric acid is used at 30 °C.

Since Brownell *et al.* did not vary the enzyme dosage and we showed that the enzymatic digestibility of softwood is only enhanced when high enzyme dosages are applied, in an additional study, the influence of different enzyme dosages for the enzymatic hydrolysis was investigated. Biomass from all beech experiments where the explosion had an influence was taken and the enzymatic dosage was reduced in 4 steps to 2.5 FPU per g cellulose (Fig. 7). The enhancement of the cellulose digestibility shrinks with decreasing enzyme dosages. At an enzyme dosage of 2.5 FPU per gram cellulose, almost no difference for the biomass pretreated with log $R_0 = 5.00$ and little influence for the biomass pretreated with log $R_0 = 3.98$ could be found between the exploded and non-exploded biomass. This result suggests that the explosive decompression enhanced the enzymatic digestibility mostly by an increased accessibility of cellulose, since yields at low enzyme dosages are often limited by enzyme deactivation on the biomass.[3]

4 Conclusion

This study shows that the explosive decompression at the end of a steam-explosion pretreatment step can enhance the enzymatic digestibility of corn stover. It also shows that the explosion can greatly enhance the digestibility of beech wood chips as we already showed for spruce wood chips[5] and which is in contrast to the study of Brownell *et al.*[19] Low enzyme dosages lead to a disappearance of the enhancement, indicating that the enhancement is mainly provoked by an increased accessibility of the cellulose for the hydrolytic enzymes. Furthermore, it could be shown that the enhancement occurs mainly due to particle size reduction, since particles fractionated by size from exploded and non-exploded beech wood show similar glucose yields from enzymatic hydrolysis at an enzyme loading of 60 FPU per gram cellulose.

Studying the explosion effect on other pretreatment methods, like acid catalysed methods, might give more insight into the effect of the explosive decompression. We have already started preliminary studies with the steam pretreatment of acid impregnated spruce and the explosion seems to also have a strong influence on the enzymatic digestibility.

References

1 Ó. J. Sánchez and C. A. Cardona, *Bioresour. Technol.*, 2008, **99**, 5270–5295.

2 P. Alvira, E. Tomás-Pejó, M. Ballesteros and M. Negro, *Bioresour. Technol.*, 2010, **101**, 4851–4861.

3 T. Pielhop, G. O. Larrazábal, M. H. Studer, S. Brethauer, C.-M. Seidel and P. Rudolf von Rohr, *Green Chem.*, 2015, **17**, 3521–3532.

4 A. R. Esteghlalian, V. Srivastava, N. Gilkes, D. J. Gregg and J. N. Saddler, in *An Overview of Factors Influencing the Enzymatic Hydrolysis of Lignocellulosic Feedstocks*, ed. M. E. Himmel, J. O. Baker and J. N. Saddler, American Chemical Society, 2000, ch. 6, pp. 100–111.

5 T. Pielhop, J. Amgarten, P. Rudolf von Rohr and M. H. Studer, *Biotechnol. Biofuels*, 2016, **9**, 152.

6 Y. Kang, P. Bansal, M. J. Realff and A. S. Bommarius, *Biotechnol. Prog.*, 2013, **29**, 909–916.

7 M. Shafiei, R. Kumar and K. Karimi, *Lignocellulose-Based Bioproducts*, Springer, 2015, pp. 85–154.

8 Z.-H. Liu, L. Qin, F. Pang, M.-J. Jin, B. Z. Li, Y. Kang, B. E. Dale and Y.-J. Yuan, *Ind. Crops Prod.*, 2013, **44**, 176–184.

9 M. T. Holtzapple, *Energy Requirements for the Size Reduction of Poplar and Aspen Wood*, D. of Chemical Engineering, T. A. University, C. Station and Texas, 1989.

10 S. Brethauer and M. H. Studer, *Chimia*, 2015, **69**, 572–581.

11 B. C. Vidal, B. S. Dien, K. C. Ting and V. Singh, *Appl. Biochem. Biotechnol.*, 2011, **164**, 1405–1421.

12 E. Chornet and R. P. Overend, Phenomenological kinetics and reaction engineering aspects of steam/aqueous treatments, in *Process Fundamentals*, ed. A. Focher, B. ans Marzetti and V. Crescenzi, Goran and Breach Science Publishers, New York, 1991, pp. 21–58.

13 N. Mosier, C. Wyman, B. Dale, R. Elander, Y. Lee, M. Holtzapple and M. Ladisch, *Bioresour. Technol.*, 2005, **96**, 673–686.

14 M. G. Chua and M. Wayman, *Can. J. Chem.*, 1979, **57**, 1141–1149.

15 K. Lundquist and R. Lundgren, *Acta Chem. Scand.*, 1972, **26**, 2005–2023.

16 W. Kim, M. Yahaya and M. Goto, *Grassl. Sci.*, 2005, **51**, 139–144.

17 P. Kumar, D. M. Barrett, M. J. Delwiche and P. Stroeve, *Ind. Eng. Chem. Res.*, 2009, **48**, 3713–3729.

18 S.-L. Sun, J.-L. Wen, M.-G. Ma and R.-C. Sun, *Appl. Energy*, 2014, **136**, 519–526.

19 H. H. Brownell, E. K. C. Yu and J. N. Saddler, *Biotechnol. Bioeng.*, 1986, **28**, 792–801.

20 R. P. Overend, E. Chornet and J. Gascoigne, *Philos. Trans. R. Soc. London, Ser. A*, 1987, **321**, 523–536.

21 B. Hames, R. Ruiz, C. Scarlata, A. Sluiter, J. Sluiter and D. Templeton, *Preparation of Samples for Compositional Analysis*, National Renewable Energy Laboratory techreport NREL/TP-510-42620, 2008.

22 A. Sluiter, B. Hames, D. Hyman, C. Payne, R. Ruiz, C. Scarlata, J. Sluiter, D. Templeton and J. Wolfe, *Determination of Total Solids in Biomass and Total Dissolved Solids in Liquid Process Samples*, National Renewable Energy Laboratory techreport NREL/TP-510-42621, 2008.

23 A. Sluiter, B. Hames, R. Ruiz, C. Scarlata, J. Sluiter, D. Templeton and D. Crocker, *Determination of Structural Carbohydrates and Lignin in Biomass*, National Renewable Energy Laboratory techreport NREL/TP-510-42618, 2008.

24 A. Sluiter, R. Ruiz, C. Scarlata, J. Sluiter and D. Templeton, *Determination of Extractives in Biomass*, National Renewable Energy Laboratory techreport NREL/TP-510-42619, 2005.

25 A. Sluiter, B. Hames, R. Ruiz, C. Scarlata, J. Sluiter and D. Templeton, *Determination of Sugars, Byproducts, and Degradation Products in Liquid Fraction Process Samples*, National Renewable Energy Laboratory techreport NREL/TP-510-42623, 2006.

26 M. Selig, N. Weiss and Y. Ji, *Enzymatic Saccharification of Lignocellulosic Biomass*, National Renewable Energy Laboratory techreport NREL/TP-510-42629, 2008.

27 B. Adney and J. Baker, *Measurement of Cellulase Activities*, National Renewable Energy Laboratory techreport NREL/TP-510-42628, 1996.

Faraday Discussions

PAPER

Sustainable sources need reliable standards

Maxim V. Galkin,[a] Davide Di Francesco,[a] Ulrica Edlund*[b] and Joseph S. M. Samec*[a]

Received 30th January 2017, Accepted 9th February 2017
DOI: 10.1039/c7fd00046d

This review discusses the challenges within the research area of modern biomass fractionation and valorization. The current pulping industry focuses on pulp production and the resulting cellulose fiber. Hemicellulose and lignin are handled as low value streams for process heat and the regeneration of process chemicals. The paper and pulp industry have therefore developed analytical techniques to evaluate the cellulose fiber, while the other fractions are given a low priority. In a strive to also use the hemicellulose and lignin fractions of lignocellulosic biomass, moving towards a biorefining concept, there are severe shortcomings with the current pulping techniques and also in the analysis of the biomass. Lately, new fractionation techniques have emerged which valorize a larger extent of the lignocellulosic biomass. This progress has disclosed the shortcomings in the analysis of mainly the hemicellulose and lignin structure and properties. To move the research field forward, analytical tools for both the raw material, targeting all the wood components, and the generated fractions, as well as standardized methods for evaluating and reporting yields are desired. At the end of this review, a discourse on how such standardizations can be implemented is given.

Introduction

A renaissance of the research area of biomass valorization has taken place during the past decade for platform chemicals, materials and biofuels. Probably because of a combination of factors such as a surge in oil prices, environmental awareness of greenhouse gas emissions, a decline in the discovery of giant oil fields, human catastrophes in shale water and oil related conflicts, and maybe also a surge in the desire of researchers from new research fields to find new challenges. There are several reasons why the interest in this research area will not be temporary, as it was during the 1970s energy crisis (Fig. 1). As humans currently use only 3.5% of the net growth of biomass, a more bio-based industry and economy would be

[a]*Department of Organic Chemistry, Stockholm University, SE-106 91, Stockholm, Sweden. E-mail: joseph.samec@su.se*

[b]*Fiber and Polymer Technology, Royal Institute of Technology (KTH), Teknikringen 56, SE-100 44 Stockholm, Sweden. E-mail: edlund@polymer.kth.se*

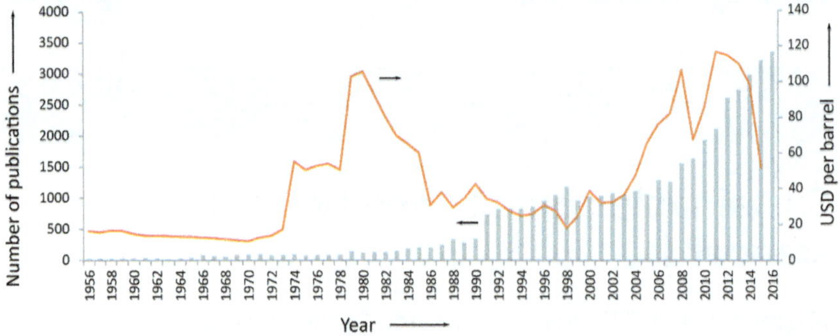

Fig. 1 The number of publications (grey bars) indexed by Web of Science containing the word "lignin", as an example of a biomass component, and real crude oil prices (in orange).

a reasonable and sustainable pathway forward. Corma and Huber wrote a series of inspiring review articles that motivated researchers outside of the traditional biomass research community to start their journey into biomass valorization a decade ago.[1] Promising examples from the organometallic chemistry community emerged, with reports from renowned research groups such as those of Bergman, Toste, Hartwig, and Silks on lignin depolymerization.[2] These studies within the field were performed mainly utilizing model substances to represent one or two biomass constituents, especially where the focus of the research was on lignin depolymerization.[2] For current progress made in the field, usually real biomass is used instead of model compounds. Many research groups including those of Zhang, Rinaldi, Ford, Abu-Omar, and Sels have revived the approach of fractionating wood, in which the lignin is depolymerized in parallel with organosolv pulping. The research interest in hemicellulose fractionation and valorization is also rapidly increasing, both within academia and industry.

This critical review discusses the challenges of wood fractionation with lignin depolymerization, including biomass valorization as such. There are major challenges for the community both in terms of raw material analysis as well as in reporting yields. One important aspect of moving the research area forward will be to systemize these two aspects of the research field. This review, to start with, will briefly explain current biomass fractionation accounting for the majority of paper and pulp production, then proceed with examples of the modern fractionation approach described above and then cover current analytical tools as well as recent developments and also further needs. A discussion on how to progress and mature the research field is given at the end of this review.

Traditional pulping techniques

The main goal of current lignocellulosic biomass fractionation processes is to produce a strong cellulose fiber with tailored delignification and a degree of retention of the hemicellulose for further use in the manufacturing of paper, packaging materials, and fabrics. The rest of the biomass (lignin and hemicellulose) is handled as low value side streams.[3] Parts of these streams are used for the regeneration of process chemicals, and the production of process heat and electricity at low efficiencies. Growing demands in pulp and paper,[4] low oil prices, and the public unawareness of the negative side effects of fossil fuels consumption during the 20th century favored the current situation. Nevertheless, the lignocellulosic biomass constitutes 50% or less of the cellulose by weight and this means that the efficiency of biomass intake by the present-day fractionation technologies is limited by 50% (Fig. 2). Calculations of calorific values for the biomass constituents result in an even worse situation: energy-wise cellulose constitutes less than 25% of the feed-stock.[5]

There are several chemical and mechanical fractionation methods used for wood delignification, the separation of fibers, and discoloration. All pulping processes involve the same general steps for the manufacture of pulp and paper: raw material preparation (*e.g.*, debarking and chipping); either mechanical, chemical or a combined separation of the wood fibers (grinding, digestion, *etc.* to dissolve the lignin and extractives); removal of the coloring agents (primarily residual lignin) by bleaching; and the final product formation (paper, fiber, *etc.*).[6,7] Chemical pulping is the main current lignocellulosic biomass fractionation process, where sulphite and kraft processes account for approximately 70% of today's global pulp production, with kraft pulping dominating the industry.[8–10]

During the kraft process, biomass is treated with a sodium hydroxide and sodium sulfide water solution at 423–453 K. That results in partial hemicellulose degradation and lignin polymer solubilization due to cleavage of the interunit ether linkages (mainly β-O-4′). The liberated lignin fragments are reactive and condense to form new strong C–C bonds.[11] The dissolved organic matter (lignin, part of extractives, and hemicelluloses) serves as a fuel for the operation of a kraft mill and as a reductant in the regeneration of process chemicals.[12] In the kraft process, a highly-condensed lignin may be isolated by precipitation from the pulping liquor by lowering the pH.[13] Only a few examples of valorized products from this type of lignin have been reported[14] due to its high heterogeneity, low purity, and partial involvement in electricity generation (the largest share of biofuel in the Finnish and Swedish energy matrices *etc.*).[15]

The sulfite pulping operates at similar temperatures (403–423 K) to kraft pulping, however, it can tolerate a broader pH range (2 to 12), and sulfurous acid salts are used as a pulping reagent instead of sulfides.[16] By partial lignin degradation and introduction of sulfonate groups to the lignin skeleton, the lignin water solubility increases. The liquor obtained with the dissolved organics is

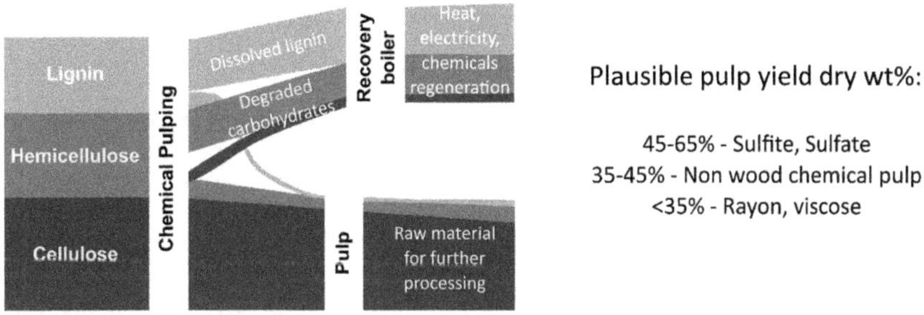

Fig. 2 Traditional pulp mills focus on generating a strong fiber. The yield of the valorized fraction is usually below 50%.

incinerated in a recovery boiler to produce process heat and regenerate process chemicals. About 2 wt% of the generated lignosulfonates are isolated and available commercially as dispersants, emulsifiers, and binders.[17–19]

In view of the above, modern pulping techniques only utilize part of the biomass at a high quality, namely to provide cellulose and certain extractives (crude tall oil),[20] while leaving half of the initial biomass amount behind. The reason is that harsh conditions and long reaction times are needed to produce high quality cellulose, consequently resulting in degradation of the lignin and the hemicellulose fractions on the molecular level.[11,21,22]

Besides limited examples of technical lignin valorization as macromolecules,[23] there are a few examples of successful valorization of technical lignin to small molecules. The challenge in utilizing technical lignin can be illustrated from industrial examples: the Noguchi process 1959–1966, Crown-Zellerbach process 1966, and Lignol process 1960–1983 once started were all discontinued due to a number of drawbacks making them cost-ineffective.[24–27] The following studies address issues such as a lower lignin conversion and char formation due to recondensation reactions.[28,29] One such approach in technical lignin depolymerization is the use of so-called capping agents, such as alcohols, that can enable O-alkylation and C-alkylation reactions that hinder repolymerization of the phenolic monomers formed during lignin disassembly.[30–34] Kumar *et al.* demonstrated that it is possible to perform kraft lignin depolymerization under solvent free conditions.[35,36] An interesting approach was proposed by Löfstedt *et al.* It was shown that when esterified with tall oil fatty acids, kraft lignin could undergo hydroprocessing to yield alkylbenzenes as the major products.[37]

Likewise, there are few examples of the valorization of hemicelluloses isolated from the pulping process liquors. Due to the harsh conditions applied during the digestion stage of the pulping process, the hemicelluloses released into the spent cooking liquor are often more or less degraded to oligomeric residuals with mechanical properties inferior to those of a hemicellulose with a molecular weight closer to the native state. Isolation and valorization of such hemicellulose-rich fractions is still possible and of commercial interest if cost-effective processes are developed and the yields are sufficient.[38–40] The worldwide production of dissolving pulp is rapidly increasing. In the past decade, several large pulp mills converted their production from paper pulp to dissolving pulp for textile production. Dissolving pulp plants may include a prehydrolysis unit where wood is hydrothermally pretreated prior to cooking. The prehydrolysis liquor contains dissolved and dispersed hemicelluloses that could be recovered and used as a material or chemical resource.[41]

Fractionation by organosolv pulping in tandem with transition metal catalysis

An alternative approach to lignocellulosic biomass valorization was pioneered in the 1950s with significant contributions from Hibbert, Pepper and Lee.[42–49] In this approach, an organosolv pulping to fractionate the woody components was combined with a hydrogenolysis reaction to depolymerize and reduce the lignin to yield a lignin-derived bio-oil.

Starting from 2008, different research groups (Kou, Scion R&D, Zhang, Sels, Rinaldi, Abu-Omar, *et al.*)[50–56,66] have revived the field. Earlier than this, there were numerous reports, however, the goal of these earlier reports was often to study the biomass and not to develop fractionation methodologies. A few different terms have been coined to name this process such as "Lignin-first",[53] "Early-stage Catalytic Conversion of Lignins" (ECCL)[54] or "Catalytic Upstreaming Biorefining" (CUB),[55] or "Catalytic Reductive Fractionation" (CRF). It has been demonstrated that this approach is advantageous over processes that attempt valorization of the lignin after fractionation (pulping) has been finished.[50,51,56–58] A rationale for the success of this methodology is that native lignin contains weak ether bonds that are readily cleaved through fractionation to generate very reactive monomeric species that react to form strong carbon–carbon (C–C) bonds through condensation reactions. The formed condensed lignin is not susceptible to further valorization. In the present approach, the lignin-containing biomass in the solvent, under elevated temperatures, is treated in the presence of a transition metal catalyst under reducing conditions (Fig. 3). Depolymerization is achieved through the cleavage of weak C–O interunit linkages in the lignin, and the formed molecules are stabilized against repolymerization through reduction reactions.

There are many important examples of this emerging methodology, which have been intensely covered and discussed within two recent reviews to which we would like to draw the reader's attention.[11,21] The progress in this research field, from the very first examples of how lignin could be more efficiently depolymerized in wood to a promising technology that can be put into practice for a bio-economy, is demonstrated. We would like to exemplify this further using recent reports not covered in the previous reviews.

Organosolv pulping in tandem with lignin hydrogenolysis – selected examples

Recent publications (2015–2017) differ significantly from the previously reported examples: the recent examples demonstrate the possibility to gain lignin monomers in yields close to the theoretical maximum, where the overall mass balance is close to 90 wt% and all parts of the biomass are valorized in one or another way.

Considerable improvements were achieved through Lewis and Brønsted acid catalyzed lignin depolymerization processes coupled with organosolv pulping. Hensen's group demonstrated an acid-catalyzed process combined with metal-catalyzed hydrogenolysis. The process separates the woody biomass into a few

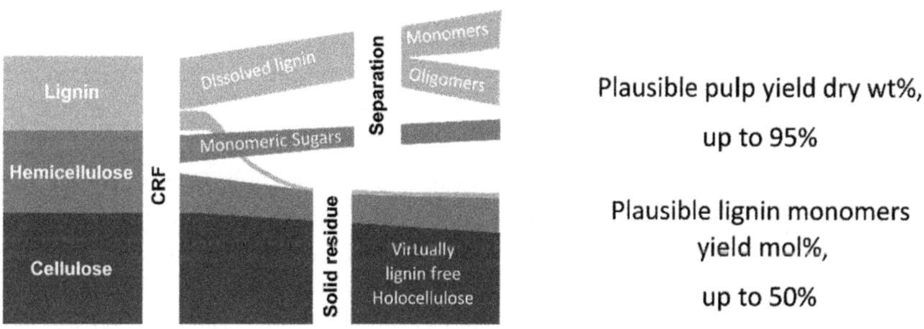

Fig. 3 Fractionation using organosolv pulping in tandem with transition metal catalysis.

lignin-derived alkylmethoxyphenols (36–55 wt%, lignin base) and cellulose under tandem metal triflate and Pd/C catalysis.[59,60] Deuss *et al.* showed a difference in the reactivity between metal triflates and triflic acid, which favors metal triflates.[61] Bruijnincx's group reported that it is possible to control selectivity during similar processes by varying the amount and strength of the Lewis (metal triflate) acid catalyst.[62]

The influence of pH on the catalytic reductive fractionation was investigated by the Sels group. Both H_3PO_4 and NaOH catalysis was studied and their effects on the degree of holocellulose fractionation were shown. Lignin monomers were obtained in near to theoretical maximum yields.[63] In another contribution, the group studied the role of the solvent (composition: MeOH/water and EtOH/water blends) in catalytic reductive fractionation (CRF). When the water concentration was 30 vol% or less, most of the carbohydrates were preserved in the solid residue. Contrarily, upon processing with water-rich mixtures (70 vol% or more), most of the hemicellulose fraction was removed, while the cellulose fraction was left untouched.[64]

Samec's research group performed organosolv pulping in tandem with a Pd/C catalyzed transfer hydrogenolysis, where no hydrogen was needed for hydrogenolysis to proceed. It was proposed that the hemicellulose was acting as an internal source of hydrogen.[57] The group later reported that the degree of lignin monomer backbone unsaturation could be controlled by changing the reaction conditions (time and temperature). When a fully reduced and deoxygenated propyl chain was obtained, the hemicellulose fraction was lost. The cellulose was of a high purity and susceptible to enzymatic hydrolysis.[65]

A series of reports from different laboratories have demonstrated an applicability for CRF processes using catalysts based on earth-abundant Ni, under relatively mild conditions. Abu-Omar's group obtained a yield of aromatic monomers of over 68% from catalytic reductive fractionation of *Miscanthus* biomass. The carbohydrate residue was treated with iron chloride to give furfurals and levulinic acid. Thereby the biomass fractionation gave a total mass balance of valorized products of 98%.[66] When corn stover was utilized as a substrate by Román-Leshkov's group, a moderate yield of lignin oil (55%) and sugar retention values higher than 90% were obtained.[67] Xu's group transformed lignin from woody biomass, under similar conditions, into natural phenolic alcohols (39 wt% yield).[68] The influence of hemicellulose derived sugars on lignin monomer hydrogenolysis was studied by Rinaldi's group. It was demonstrated that hemicellulose sugars decrease the activity of the Ni based catalyst (RANEY® nickel) and, therefore, significantly inhibit hydrogenation of the aromatic species composing the lignin stream.[69]

An outstanding discovery was made on the border of two different strategies – the use of capping agents and CRF processing. Luterbacher's group demonstrated a possibility to stabilize the native β-O-4′ motif within lignin in the wood, before the pulping step, by treating the lignocellulosic biomass with an acidified formaldehyde solution. The formaldehyde prevented lignin condensation by forming 1,3-dioxane structures with the lignin side-chain and by reacting with guaiacol based lignin fragments at the fifth position. The subsequent hydrogenolysis step resulted in 47 mol% Klason lignin for beech and 78 mol% for a high-syringyl transgenic poplar. The process is very efficient overall and when depolymerization of cellulose, hemicelluloses, and lignin was performed separately, monomer yields between 76 and 90 mol% were obtained.[70]

Major challenges in biomass fractionation

This young research area has started to bloom (Fig. 1). To progress the overall research field, this area has to mature. The timing is right as the research field has a solid ground with many active research groups and the right momentum with an increasing number of publications. In the following section, different challenges within this young research field are described. We will start with describing the natural variations in the feedstock, then describe and discuss the analysis of the feedstock. After this, a discussion of the variations in transforming the feedstock and reporting yields is given.

Natural variations in the raw material

Lignocellulose is the dry matter of all (vascular) higher plants. Among all vascular plants, three groups are of interest for biorefining: hardwoods (dicotyledons), softwoods (conifers), and grasses (monocotyledons). Each group represents a range of families and numerous species, for example, the hardwoods group includes more than 200 families: birch, willow (aspen, poplar, and populus), myrtle, sweetgum, soapberry (maple), *etc.* Taking birch as an example that is widespread in the northern hemisphere, there are more than 30 species. Within one species, there are structural and compositional variations depending on a variety of different factors: longitude, height, water supply, nutrient supply, microbial interaction, and age of harvest. Even within the same tree, the lignocellulose will be different depending on what part of the wood is utilized (*e.g.* heart wood or sap wood). Finally, there will, of course, be batch-to-batch differences in the handling and storing of the biomass before processing. All the mentioned factors will influence the raw material (variation in the lignocellulose composition for biomass of a different origin, as an example, is depicted in Fig. 4).

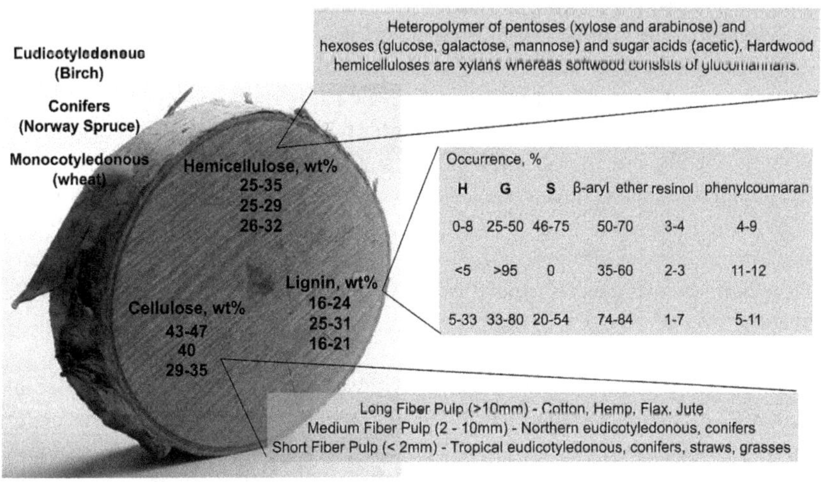

Fig. 4 Natural variations in the raw material.

At the moment, most research groups will report the type of lignocellulosic biomass that has been used, for example birch or corn stover. Yet, statistically, it would be pure luck if one research group obtained a similar distribution of products to another, given the factors influencing the raw material. This is a significant challenge with lignocellulosic bio-refining and one important aspect that needs to be addressed.

Analysis of the lignocellulosic feedstock

The standardized techniques developed by the Technical Association of Pulp and Paper Institute (TAPPI) are optimized for pulp and paper production with the focus on generating a strong fiber with a controlled degree of delignification. With the ambition to develop biorefineries, new standards that take all fractions of the biomass into account are needed. The standardized methodologies that are used today are disruptive techniques where the biomass is degraded prior to or while analyses are performed. A conventional and industrially established technique of lignin content analysis is Klason lignin determination.[71] This methodology is probably still the most commonly used method to determine the lignin content. Klason lignin refers to acid-insoluble lignin, which is typically determined gravimetrically,[72] while the part of the lignin that escapes gravimetric determination (acid-soluble lignin) can be quantified spectrophotometrically.[73] Advantages of Klason lignin determination include a wide-spread acceptance, a straight-forward procedure, and that it is relatively easy to perform. Disadvantages are that it operates better with softwood than hardwood biomass, and you need to complement it with UV spectroscopy of the acid soluble lignin fractions for the latter. Cellulose has been quantified and qualified by determining the fiber length and strength as well as the degree of delignification using the kappa number determination.[74] Hemicellulose determination constitutes an additional challenge, given the structural diversity of hemicelluloses. A conventional strategy for hemicellulose characterization involves acid hydrolysis followed by monosaccharide content determination using ion chromatography (IC).[75] IC is time consuming and requires adequate reference calibration to provide accurate results, but it will indicate the relative occurrence of constituent sugars in the analyzed biomass sample, which may be helpful for distinguishing one type of hemicellulose from another or the purity of a given polysaccharide. IC fails to distinguish polysaccharides composed of the same monosaccharide units, for instance amylose from cellulose, which are both homopolysaccharides of glucose. As mentioned before, the variability of the substrate does not allow researchers to compare their results, since every research group is studying a different source of biomass or at least not the same sample. The other main cause of the inability to compare data is the different analytical methods used. Standardization in terms of the biomass samples and analytical methods is evidently necessary in order to create a web of research groups able to collaborate directly and share their progress in that field.

Recently, an effort has been made in this direction. Thanks to the use of standardized biomass samples from NIST,[76] Templeton *et al.* have been able to evaluate the different parameters of biomasses from different natural sources.[77] The same substrates were sent to several laboratories and the data was collected and compared. The authors also compared their results with a previous attempt dating back to 1992. For many of the parameters that were analyzed, results with low deviations were obtained from the different laboratories. However, there were also inconsistencies between the laboratories. The reason for this, as proposed by the author, can be differences in the techniques used, in instrumentation, and in performing the analysis. This example illustrates both the challenges and the feasibility of uniform analysis of biomass.

Within the field of analytical methods applied to biomass, procedures used at the National Renewable Laboratory of the U.S. Department of Energy are considered as referential methods. In the last 20 years, they have published several modifications to previous methods taking into account all fractions of the biomass. The current NREL/TP-510-42618 (ref. 78) method is a 40 step procedure including pretreatment, hydrolysis, burning, autoclaving, and HPLC analyses. To date, this is the best method for quantifying the carbohydrates and lignin in lignocellulosic biomass. However, the method is time-consuming, and requires specific equipment and preparation of standards. Therefore, most of the research groups will probably not use it in their laboratories since a huge effort required. Even from an industrial point of view, the lack of standardized and universal analytical procedures is demonstrated by the presence of only four methods concerning the whole biomass in the ASTM catalogue.[79–82]

Almost every analytical technique has been applied to biomass and its byproducts, but only a few of those are used in routine analyses.[83,84] Examples are HPLC, GC, and IC methods for carbohydrate determination,[85,86] size exclusion chromatography (SEC) for analyzing the molecular weight distribution,[87,88] and NMR for obtaining structural information.[89–91] These techniques are all able to provide fundamental information on the chemical composition of the relative biomass fraction. The methods listed above have found recognition amongst non-specialists and are used separately or in combination on a daily basis, but still use of the same analytical parameters and sample preparation procedures is missing, complicating accurate comparison of the data.

The complexity in the biomass composition has historically separated the analytical research field in terms of the function of the particular biomass component studied. As there are research groups specialized in cellulose chemistry and lignin chemistry, many of the available analytical tools are designed to analyze mainly one biomass component at a time. The current trend is to develop a comprehensive method to qualify and quantify as many components of the biomass as possible using the same method, with the same instrument and, preferably, within the same sample run or stream. Several analytical techniques have been investigated for this purpose and are still under development including Near Infrared Spectroscopy (NIR),[92,93] Diffusion Ordered Spectroscopy NMR (DOSY),[94] and solid-state NMR spectroscopy.[91] The main advantage with these techniques is that they are non-disruptive. This enables study of the structural features in addition to quantifying the different fractions, even if they may require a tough sample preparation or isolation. Near infrared spectroscopy was recently reviewed[95] and has many advantages in terms of accuracy and ease of operation. The main disadvantage is that the obtained data should be processed using chemometric techniques, which still rely on wet chemistry procedures.[96] The utilization of standardized biomass model samples as references would perhaps overcome this obstacle.[97] Diffusion ordered spectroscopy (DOSY) NMR was recently proposed to be a good alternative for multi-compound analysis of lignocellulosic biomass,[94] as it is able to quantify several classes of compound within the same experiment. Since this method only works with solutions, the main disadvantages are: the sample preparation and the restriction on the molecular weight of the analyzed compound (not higher than 2–3 kDa).[98]

Solid-state NMR spectroscopy using cross polarization magic-angle spinning (CP-MAS) has been applied for determination of lignocellulosic feedstocks.[99] This technique, like NIR spectroscopy, allows determination of the structural features of different biomass components. Such equipment is very expensive and, moreover, the methodology still relies on wet chemistry standards.[91,99]

Luminescent conjugated oligothiophenes have recently been shown to offer rapid and highly selective detection of cellulose in lignocellulosic biomass matrices.[100] This method is straight-forward, needs very little sample preparation and hence does not result in artefacts in the analysis results due to changes in the structure of the analytes during sample preparation.

Further development will hopefully lead in the future to a non-disruptive technique that can analyze the biomass without the need to prepare standards. This would be a great advancement and would mature the research field significantly. Several reviews on analysis have recently been published and we recommend these to the interested reader for further reading.[83,101,102]

Pretreatment of lignocellulose

Both for fractionation and analysis of lignocellulosic biomass, a pretreatment is usually performed. Different research groups pretreat the biomass using various different procedures.[103,104] The pretreatments can be divided into physical (drying,[105,106] comminution,[107] *etc.*), chemical,[108,109] biological, and multiple or combinatorial pretreatments (like dewaxing[110]). The pretreatment will affect the chemical composition and the structure of the biomass,[111] creating another variable within the whole process. Among the pretreatment methods, chemical pretreatments are certainly the most expensive and produce the most waste.[111] It is necessary to report the efficiency of the method applied by considering the recyclability of the solvents and chemicals used, in addition to quantification of the extractives. Pretreatment usually is considered as a variable in the screening of the efficiency of the later valorization steps.[112] However, there are a few studies on how different pretreatments affect the different fractions on a molecular level.[109,111–114] More studies would be valuable for understanding and designing more efficient methodologies.

Processing of biomass

Many research groups target one of the biomass streams when fractionating lignocellulosic biomass. This is natural when the research is focused on one particular aspect of valorization or upgrading. Taking paper mills as an example, with few exceptions the extraction of cellulose has been the only purpose of these industries for about a century, while ignoring the possibility to develop a process able to valorize other fractions of the lignocellulosic biomass. During the 21st century, a growing interest in the valorization of lignin and hemicellulose has emerged.[115,116] There are some good examples of research groups that handle more than one of the lignocellulosic streams.[117–120] If this research area can gain trustworthiness in biorefining, it would be natural to study how a procedure affects the multiple fractions of the raw material. There are of course exceptions, for example, when starting from an isolated fraction such as lignosulfonate from the sulfite process.[121] In this case, the added procedure would convert an existing pulping process towards a biorefinery.

In order to progress the research area, it would also be desirable to report mass balances for all the reagents, including solvents used for purification, besides providing incentives for catalyst recycling, new green catalyst development, greener solvents, more benign reaction conditions and purification systems.

Analysis of the fractions from biomass fractionation

Cellulose. Cellulose is probably the fraction for which there are the most analytical methods available to determine the structure, purity, and performance as a material resource. The cellulose fraction of biomass is the main constituent of different products made from pulp (newsprint, packaging, *etc.*) and viscose products. This is the fraction of biomass that is further transformed through hydrolysis to glucose and then enzymatically transformed into ethanol for biofuels. Current methods for analyzing the cellulose fraction mostly address the purity and fiber quality.

In the pulping industry, purity most often equates to the degree of delignification. While research groups working on biomass fractionation usually assess Klason lignin, the paper and pulp industry heavily depend on the kappa number to determine the degree of delignification. These methods provide an estimation of the fraction of remaining lignin in the cellulose product but do not assess the absence or presence of other components. Fiber quality assessments include assessment of the mechanical properties (*e.g.* wet and tear strength, ductility), physicochemical properties (*e.g.* wettability, surface charge), and morphology (*e.g.* fiber dimensions and roughness, degree of crystallinity). These properties can typically be assessed using traditional analytical methods and are intimately associated with the type and origin of the biomass and the process parameters by which it was refined.

Since the 1980's, the production of nano-sized fibrils and particles from cellulose[122,123] has attracted ever-increasing attention with a concomitant increasing need for analytical techniques able to assess the mechanical and compositional properties of samples of minute dimensions and with unprecedented cellulose surface-to-bulk ratios. Implementation of nano-sized cellulosic materials in commodity products is expected in the near future and this also raises the need for further analysis of the potential health issues and tissue interactions of such nano-sized items.

To be viable for a broader range of biomasses in future biorefineries, new and selective analytical methods are needed that can assess the composition and purity of cellulose-containing biomass matrices, also when these contain complex mixtures of native components. Conventional methods are typically based on hydrolysis pretreatments followed by traditional analytical methods for structural determination such as NMR, IC, GC, and mass spectrometry.[124,125] A limiting factor in adapting conventional wet chemistry analyses to cellulose samples is their poor solubility in water and a vast majority of all organic solvents. IC, as mentioned previously, detects the presence of other mono-, oligo-, and polysaccharides but will not be selective for cellulose if one or several other glucose-based polysaccharides are co-existing in the sample. A couple of exploratory, new techniques for selectively assessing the presence and purity of cellulose in biomass matrices have been reported in the literature as discussed above. NIR has been shown to be viable for the detection of lignin, hemicellulose, and cellulose components in lignocellulosic biomass,[93,126] and luminescent conjugated oligothiophenes provide highly selective detection of cellulose.[100]

Hemicelluloses. Hemicelluloses belong to a family of structurally diverse heteropolysaccharides that may be more or less branched and are composed of one or several of a range of monosaccharide units, the most commonly occurring ones being mannose, xylose, glucose, galactose, and arabinose as well as some uronic acids.[127,128] Hemicelluloses typically have a much lower degree of polymerization (DP) than cellulose. While abundant in nature, hemicelluloses have not found extensive use as raw materials for platform chemicals, fuels or materials. Some hemicellulose-rich biomass fractions, *e.g.* from seeds, are used in the food industry as consistency regulator additives (for instance guar gum and locust bean gum), otherwise broad commercial introduction of hemicellulose based materials is still to be realized. In the scientific literature, hemicellulose-based films and coatings have been demonstrated to possess excellent oxygen barrier properties and have thus been proposed as bio-based packaging materials.[129–134] Also, hemicelluloses have been proposed as suitable raw materials for hydrogels[135–138] and wound dressings,[139] given their inherent hydrophilicity. Alternatively, hydrolysis can be performed to generate carbohydrate degradation products such as glucose, xylose, furfural, or levulinic acid, which can be further modified into commodity chemicals such as xylitol or ethanol.[140]

In analogy to cellulose, the bio-refining process conditions, the means of recovery and purification will significantly affect the composition[141] of the isolated hemicellulose fraction. Furthermore, it is inherently difficult to manage a complete separation of the wood components given their strong intermolecular interactions. It has been suggested that hemicelluloses are partly covalently bonded to lignin, thus referred to as lignin–carbohydrate complexes (LCCs).[142] For valorization of hemicelluloses from lignocellulosic bio-refining, determination of the structural composition and purity is important. Mostly, the same methods that are applied to cellulose are used, for example IC or GC-MS, with the same pros and cons as previously discussed. Hemicelluloses are typically much more susceptible to depolymerization reactions than cellulose and may degrade more or less significantly during processing and purification. For this reason, molecular weight assessments are often desirable as the molecular weight is a factor strongly affecting the mechanical integrity and performance of the resulting material.

Lignin. The approach where a catalytic procedure is combined with organosolv fractionation has recently attracted considerable attention. This approach allows conversion of reactive lignin fragments released through pulping into nonreactive products to prevent recondensation reactions. Many reports use reductive conditions such as an atmosphere of hydrogen gas or the addition of a hydrogen donor. Depending on the reaction conditions, different products will be generated. Mild reductive conditions will give guaiacol- or syringolpropanols. Using harsher conditions, fully deoxygenated and dearomatized cyclohexanes will be formed as the major products. A challenge in comparing different methodologies is how different research groups report yields. Using the mass yield for lignin derived monomers will favor a low degree of deoxygenation as the products containing oxygen weigh more than the corresponding products from deoxygenation. An alternative method based on mol% or more specifically the C-9 yield is significantly better and more transparent for comparison (see Sels' reports), and will reveal the atom-economy of the process with regard to the starting material used. In general, further understanding of the structure and fundamental chemical reactivity of real lignin streams is still needed. Accumulation of raw analytical data is still needed and besides the currently used 2D-NMR techniques such as HSQC, other techniques including ^{31}P and long-range correlation (HMBC) NMR spectroscopy, SEC, ICP, IR, and two-dimensional gas chromatography need to be used more frequently for elucidation of both the raw lignin structure and the products formed. This knowledge will allow evaluation of the reactivity of various lignins from both different natural sources and different pretreatment techniques. In total, this will help to change the current situation where it is hard to compare the results of different research groups, especially as each lignin resource has an inherent variability of the chemical composition and pretreatment may often make it even more complex. As was already mentioned, the field needs standard protocols for lignin manipulation, and comparison of the materials, catalysts, and analysis tools, to make it possible to compare the results on-the-fly. As a guideline, the following critical reviews and protocols of biomass analysis are recommended.[78,83,113,124]

Discussion

To mature the research field of lignocellulosic biorefining, it is desirable that a more unified way of performing the research is implemented. The timing is good as the research field is expanding. The question is how should this be performed? Taking a look at the paper and pulp industry, the Technical Association of the Pulp and Paper Industry (TAPPI) has unified standardizations for the entire industry. As mentioned earlier, the biorefining research field can only adopt standardizations for the cellulose fraction from the paper and pulp industry. The current biorefining industry mostly consists of either bioethanol or biodiesel, which have their own standards, and there are few incentives for the development of standards for lignocellulosic biorefining and the generated products. The problem is that there are no real stakeholders at the moment. Future stakeholders will be found in forestry, agriculture waste handling, pulp producers and converters, and biorefining towards chemicals, fuels, materials, and food. The National Renewable Laboratory (NREL) has taken important steps toward the standardization of biomass analysis and this is very encouraging. However, the methodologies are complicated and few laboratories will be able to adopt them. The NREL also provides a service for analysis of samples, however, their throughput cannot meet the global demand for analysis in this growing field.

The NREL is a good example of how the government, in this case the US Department of Energy, has realized that the future stakeholders are not mature enough to take responsibility yet. A suggestion would be that regional areas open institutes, similar to the NREL, to handle the analysis of biomass from regional researchers. For instance, the European Union could open an institute similar to the North American one, specialized in analysis of lignocellulosic biomass from forestry and agriculture in Europe. Likewise, Asia and South America should open regional institutes. These institutes should strive to develop methodology for biomass analysis, agree globally on standardizations, and provide a service for researchers. A further ambition should be to provide standardized samples of lignocellulosic biomass so that researchers can benchmark their methodology. In the future, commercial institutes could take over the service from the regional institutes. In this case, quality assurance through standardization and also a control function is necessary.

Hopefully, the academic research groups that are performing research in this area will mature from showing proof of concept towards more systematically studying what factors influence the outcome of the reactions. We would welcome researchers studying how different factors within one species affect the fractionation and valorization, for example, aging (both age of tree and aging of the biomass stored), latitude of growth, sap *versus* heart wood, and woody parts from the tops and roots *versus* pulp or sawmill parts. In addition, it should be called upon that researchers report mass balances for both the biomass samples and also the reagents and solvents used in the transformations. Development of analytical tools is another research area that needs more attention. It would be desirable to have non-disruptive, reliable, simple, fast, and cost efficient techniques for the analysis of biomass. These future techniques could also provide a better understanding of the native biomass.

Conclusions

Current technologies for utilizing lignocellulosic biomass for pulp production were developed at a time when petroleum was inexpensive and available, and the negative effects were unknown. Therefore, there were low incentives during the second half of the 20th century for pulp producers to produce anything but high quality pulp for paper and packaging. This has changed drastically as the negative effects of fossil fuel use have been uncovered. One of major challenges for the 21st century is, therefore, to substitute petroleum with a renewable and sustainable source. Lignocellulosic biomass has the potential to serve as a raw material for the production of chemicals, materials, and fuels. Methodology that combines organosolv fractionation with catalysis to yield valorized fractions of cellulose, hemicellulose, lignin, and extractives is a very promising pathway to reach these goals. In order to do this, the research field needs to mature. Regional institutes such as the NREL could bridge the current development gap with future stakeholders. There is an immense need for reference feedstocks of lignocellulose so that researchers can compare and benchmark their methodologies. If reference feedstocks were available, analysis would also be much more convenient with respect to standards for the different analytical tools. It is also important to develop more robust, faster, and convenient analytical procedures for biomass. Research groups should become more unified in terms of how to report mass balances and yields. Given the potential of biomass valorization, it is worthwhile to take these important steps so that these studies can mature to innovations and furthermore into future sustainable industries.

Acknowledgements

We would like to acknowledge the Swedish Energy Agency for financial support of our work.

Notes and references

1 (*a*) G. W. Huber, S. Iborra and A. Corma, *Chem. Rev.*, 2006, **106**, 4044–4098; (*b*) A. Corma, S. Iborra and A. Velty, *Chem. Rev.*, 2007, **107**, 2411–2502; J. N. Chheda and G. W. Huber, *Angew. Chem., Int. Ed.*, 2007, **46**, 7164–7183.

2 (*a*) J. M. Nichols, L. M. Bishop, R. G. Bergman and J. A. Ellman, *J. Am. Chem. Soc.*, 2010, **132**, 12554–12555; (*b*) S. Son and F. D. Toste, *Angew. Chem., Int. Ed.*, 2010, **49**, 3791–3794; (*c*) S. K. Hanson, R. Wu and L. A. P. Silks, *Angew. Chem., Int. Ed.*, 2012, **51**, 3410–3413; (*d*) A. G. Sergeev and J. F. Hartwig, *Science*, 2011, **332**, 439–443.

3 H. Zhu, W. Luo, P. N. Ciesielski, Z. Fang, J. Y. Zhu, G. Henriksson and L. Hu, *Chem. Rev.*, 2016, **116**, 9305–9374.

4 Food and Agriculture organization of the United Nations, http://www.fao.org/faostat/en/#data/FO, accessed January 2017.

5 L. D. Shackford, *Presented in part at 36th International Pulp and Paper Congress and Exhibition*, Sao Paulo, Brazil, October, 2003.

6 G. Gellerstedt, in *Pulp and Paper Chemistry and Technology*, ed. M. Ek, G. Gellerstedt and G. Henriksson, De Gruyter, Berlin, 2009, vol. 2, pp. 92–119.

7 F. F. Wangaard, *Wood: Its Structure and Properties*, Pennsylvania State University, University Park, PA, 1981, vol. 1, p. 335.

8 Z. Strassberger, S. Tanase and G. Rothenberg, *RSC Adv.*, 2014, **4**, 25310–25318.

9 H. Sixta, in *Handbook of Pulp*, ed. H. Sixta, Wiley-VCH Verlag GmbH, Weinheim, Germany, 2006, vol. 1, pp. 2–9.

10 A. Tejado, C. Peña, J. Labidi, J. M. Echeverria and I. Mondragon, *Bioresour. Technol.*, 2007, **98**, 1655–1663.

11 R. Rinaldi, R. Jastrzebski, M. T. Clough, J. Ralph, M. Kennema, P. C. A. Bruijnincx and B. M. Weckhuysen, *Angew. Chem., Int. Ed.*, 2016, **55**, 8164–8215.

12 M. Ragnar, G. Henriksson, M. E. Lindstrçm, M. Wimby, J. Blechschmidt and S. Heinemann, in *Ullmann's Encyclopedia of Industrial Chemistry*, Wiley-VCH, Weinheim, 2014.

13 W. Zhu, *Equilibrium of Lignin Precipitation – The Effects of pH, Temperature, Ion Strength and Wood Origins*, Chalmers University of Technology, Gothenburg, Sweden, 2013, pp. 6–10.

14 G. Gellerstedt, *Ind. Crops Prod.*, 2015, **77**, 845–854.

15 J. Jçnsson, K. Pettersson, T. Berntsson and S. Harvey, *Int. J. Energy Res.*, 2013, **37**, 1017–1035.

16 B. Saake and R. Lehnen, in *Ullmann's Encyclopedia of Industrial Chemistry*, Wiley, 2007, vol. 21, pp. 21–35.

17 J. E. Holladay, J. F. White, J. J. Bozell and D. Johnson, *Top Value Added Chemicals from Biomass*, Pacific Northwest National Laboratory, Richland, WA, 2007.

18 P. Sannigrahi and A. J. Ragauskas, *J. Biobased Mater. Bioenergy*, 2011, **5**, 514–519.

19 Borregaard LignoTech, http://www.lignotech.com/Industrial-Applications, accessed January 2017.
20 L.-H. Norlin, in *Ullmann's Encyclopedia of Industrial Chemistry*, Wiley-VCH, Weinheim, 2002.
21 M. V. Galkin and J. S. M. Samec, *ChemSusChem*, 2016, **9**, 1544–1558.
22 S. Constant, H. L. J. Wienk, A. E. Frissen, P. de Peinder, R. Boelens, D. S. van Es and J. N. Saddler, *Green Chem.*, 2016, **18**, 2651–2665.
23 B. M. Upton and A. M. Kasko, *Chem. Rev.*, 2016, **116**, 2275–2306.
24 D. W. Goheen, Hydrogenation of lignin by the Noguchi process, in *Lignin Structure and Reactions, Advances in Chemistry*, ed. J. Marton, American Chemical Society, Washington, D.C., 1966, vol. 59, pp. 205–225.
25 H. H. Nimz, in *Alternative Uses for Agricultural Surpluses*, ed. W. F. Raymond and P. Larvor, Elsevier, London, 1986, pp. 67–68.
26 D. W. Goheen, Process of hydrogenating sulfonated lignin to phenolic substances, US3253044, 1966.
27 H. L. Chum, D. K. Johnson, S. Black, M. Ratcliff and D. W. Goheen, in *Liquid Fuels from Lignins, Annual report*, ed. H. L. Chum and D. K. Johnson, 1986, pp. 63–87, can be found under http://www.nrel.gov/docs/legosti/old/2907.pdf, accessed January 2017.
28 J. A. Barrett, Y. Gao, C. M. Bernt, M. Chui, A. T. Tran, M. B. Foston and P. C. Ford, *ACS Sustainable Chem. Eng.*, 2016, **4**, 6877–6886.
29 S. K. Singh and J. D. Ekhe, *Catal. Sci. Technol.*, 2015, **5**, 2117–2124.
30 R. Ma, W. Hao, X. Ma, Y. Tian and Y. Li, *Angew. Chem., Int. Ed.*, 2014, **53**, 7310–7315.
31 A. Narani, R. Kumar Chowdari, C. Cannilla, G. Bonura, F. Frusteri, H. J. Heeres and K. Barta, *Green Chem.*, 2015, **17**, 5046–5057.
32 X. Huang, T. I. Korányi, M. D. Boot and E. J. M. Hensen, *Green Chem.*, 2015, **17**, 4941–4950.
33 F. Yan, R. Ma, X. Ma, K. Cui, K. Wu, M. Chen and Y. Li, *Appl. Catal., B*, 2017, **202**, 305–313.
34 L. Chen, T. I. Korányia and E. J. M. Hensen, *Chem. Commun.*, 2016, **52**, 9375–9378.
35 C. R. Kumar, N. Anand, A. Kloekhorst, C. Cannilla, G. Bonura, F. Frusteri, K. Barta and H. J. Heeres, *Green Chem.*, 2015, **17**(11), 4921–4930.
36 S. Agarwal, R. K. Chowdari, I. Hita and H. J. Heeres, *ACS Sustainable Chem. Eng.*, 2017, **5**, 2668–2678.
37 J. Löfstedt, C. Dahlstrand, A. Orebom, G. Meuzelaar, S. Sawadjoon, M. V. Galkin, P. Agback, M. Wimby, E. Corresa, Y. Mathieu, L. Sauvanaud, S. Eriksson, A. Corma and J. S. M. Samec, *ChemSusChem*, 2016, **9**, 1392–1396.
38 T. Persson, A.-K. Nordin, G. Zacchi and A.-S. Jönsson, *Appl. Biochem. Biotechnol.*, 2007, **137–140**, 741–752.
39 G. Schild, H. Sixta and L. Testova, *Cellul. Chem. Technol.*, 2010, **44**, 35–45.
40 M. Jansson, S. Danielsson, S. Saadatmand, U. Edlund and A.-C. Albertsson, *Cellulose*, 2014, **21**, 2045–2062.
41 O. Dahlman, M. Söderqvist Lindblad, J. Parkås, A.-C. Albertsson and U. Edlund, Utilization of a wood hydrolysate, PCT/EP2008/066148, WO 2009/068525 A1, 2009.
42 M. Pepper and H. J. Hibbert, *J. Am. Chem. Soc.*, 1948, **70**, 67–71.

43 J. M. Pepper, C. J. Brounstein and D. A. Shearer, *J. Am. Chem. Soc.*, 1951, **73**, 3316–3319.
44 J. M. Pepper and D. C. Hagerman, *Can. J. Chem.*, 1954, **32**, 614–627.
45 I. Sobolev, H. G. Arlt and C. Schuerch, *Ind. Eng. Chem.*, 1957, **49**, 1399–1400.
46 J. M. Pepper and Y. W. Lee, *Can. J. Chem.*, 1969, **47**, 723–727.
47 J. M. Pepper and P. Supathna, *Can. J. Chem.*, 1978, **56**, 899–902.
48 J. M. Pepper and R. W. Fleming, *Can. J. Chem.*, 1978, **56**, 896–898.
49 L. S. Nahum, *Ind. Eng. Chem. Prod. Res. Dev.*, 1965, **4**, 71–74.
50 N. Yan, C. Zhao, P. J. Dyson, C. Wang, L.-t. Liu and Y. Kou, *ChemSusChem*, 2008, **1**, 626–629.
51 K. M. Torr, D. J. van de Pas, E. Cazeils and I. D. Suckling, *Bioresour. Technol.*, 2011, **102**, 7608–7611.
52 F. H. M. Graichen, W. J. Grigsby, S. J. Hill, L. G. Raymond, M. Sanglard, D. A. Smith, G. J. Thorlby, K. M. Torr and J. M. Warners, *Ind. Crops Prod.*, 2016, DOI: 10.1016/j.indcrop.2016.10.036.
53 S. Vanden Bosch, W. Schutyser, R. Vanholme, T. Driessen, S. F. Koelewijn, T. Renders, B. De Meester, W. J. J. Huijgen, W. Dehaen, C. M. Courtin, B. Lagrain, W. Boerjan and B. F. Sels, *Energy Environ. Sci.*, 2015, **8**, 1748–1763.
54 P. Ferrini and R. Rinaldi, *Angew. Chem., Int. Ed.*, 2014, **53**, 8634–8639; *Angew. Chem.*, 2014, **126**, 8778–8783.
55 P. Ferrini, C. A. Rezende and R. Rinaldi, *ChemSusChem*, 2016, **9**, 3171–3180.
56 C. Li, M. Zheng, A. Wang and T. Zhang, *Energy Environ. Sci.*, 2012, **5**, 6383–6390.
57 M. V. Galkin and J. S. M. Samec, *ChemSusChem*, 2014, **7**, 2154–2158.
58 C. S. Lancefield, G. M. M. Rashid, F. Bouxin, A. Wasak, W.-C. Tu, J. Hallett and T. D. H. Bugg, *ACS Sustainable Chem. Eng.*, 2016, **4**, 6921–6930.
59 X. Huang, J. Zhu, T. I. Korányi, M. D. Boot and E. J. M. Hensen, *ChemSusChem*, 2016, **9**, 3262–3267.
60 X. Huang, O. M. Morales Gonzalez, J. Zhu, T. I. Korányi, M. D. Boot, E. J. M. Hensen and W. W. L. Lam, *Green Chem.*, 2017, **19**, 175–187.
61 P. J. Deuss, C. Lahive, C. S. Lancefield, N. J. Westwood, P. C. J. Kamer, K. Barta and J. G. de Vries, *ChemSusChem*, 2016, **9**, 2974–2981.
62 R. Jastrzebski, S. Constant, C. S. Lancefield, N. J. Westwood, B. M. Weckhuysen and P. C. A. Bruijnincx, *ChemSusChem*, 2016, **9**, 2074–2079.
63 T. Renders, W. Schutyser, S. Van den Bosch, S.-F. Koelewijn, T. Vangeel, C. M. Courtin and B. F. Sels, *ACS Catal.*, 2016, **6**, 2055–2066.
64 T. Renders, S. Van den Bosch, T. Vangeel, T. Ennaert, S.-F. Koelewijn, G. Van den Bossche and B. F. Sels, *ACS Sustainable Chem. Eng.*, 2016, **4**, 6894–6904.
65 M. V. Galkin, A. T. Smit, E. Subbotina, K. A. Artemenko, J. Bergquist, W. J. J. Huijgen and J. S. M. Samec, *ChemSusChem*, 2016, **9**, 3280–3287.
66 H. Luo, I. M. Klein, Y. Jiang, H. Zhu, B. Liu, H. I. Kenttämaa and M. M. Abu-Omar, *ACS Sustainable Chem. Eng.*, 2016, **4**, 2316–2322.
67 E. M. Anderson, R. Katahira, M. Reed, M. G. Resch, E. M. Karp, G. T. Beckham and Y. Román-Leshkov, *ACS Sustainable Chem. Eng.*, 2016, **4**, 6940–6950.
68 J. Chen, F. Lu, X. Si, X. Nie, J. Chen, R. Lu and J. Xu, *ChemSusChem*, 2016, **9**, 3353–3360.
69 C. Chesi, I. B. D. de Castro, M. T. Clough, P. Ferrini and R. Rinaldi, *ChemCatChem*, 2016, **8**, 2079–2088.

70 L. Shuai, M. T. Amiri, Y. M. Questell-Santiago, F. Héroguel, Y. Li, H. Kim and J. S. Luterbacher, *Science*, 2016, **354**, 329–333.
71 TAPPI T 249 cm-00, Carbohydrate composition of extractive-free wood and wood pulp by gas-liquid chromatography, Technical Association of the Pulp and Paper Industry, 2009, http://www.tappi.com, accessed January 2017.
72 TAPPI T 222 om-02, Acid-insoluble lignin in wood and pulp, Technical Association of the Pulp and Paper Industry, 2006, http://www.tappi.com, accessed January 2017.
73 TAPPI UM 250, Acid-soluble lignin in wood and pulp, Technical Association of the Pulp and Paper Industry, 2000, http://www.tappi.com, accessed January 2017.
74 ISO 302:2015, Pulps – Determination of Kappa number, 2015, International Organization for Standardization, http://www.iso.com, accessed January 2017.
75 SCAN-CM 71:09, *Pulps – Carbohydrate composition, Scandinavian Pulp, Paper and Board Testing Committee*, Stockholm Sweden, 2009.
76 The National Institute of Standards and Technology, https://www.nist.gov, accessed January 2017.
77 D. W. Templeton, E. J. Wolfrum, J. H. Yen and K. E. Sharpless, *BioEnergy Res.*, 2016, **9**, 303–314.
78 A. Sluiter, B. Hames, R. Ruiz, C. Scarlata, J. Sluiter, D. Templeton and D. Crocker, *Determination of Structural Carbohydrates and Lignin in Biomass, Laboratory Analytical Procedures (LAP) for Standard Biomass Analysis*, National Renewable Energy Laboratory (NREL), Golden, CO, 2011, NREL/TP-510-42618, http://www.nrel.gov/biomass/analytical_procedures.html, accessed January 2017.
79 ASTM E1756-08(2015), *Standard Test Method for Determination of Total Solids in Biomass*, ASTM International, West Conshohocken, PA, 2015, http://www.astm.org, accessed January 2017.
80 ASTM E1721-01(2015), *Standard Test Method for Determination of Acid-Insoluble Residue in Biomass*, ASTM International, West Conshohocken, PA, 2015, http://www.astm.org, accessed January 2017.
81 ASTM E1757-01(2015), *Standard Practice for Preparation of Biomass for Compositional Analysis*, ASTM International, West Conshohocken, PA, 2015, http://www.astm.org, accessed January 2017.
82 ASTM E1690-08(2016), *Standard Test Method for Determination of Ethanol Extractives in Biomass*, ASTM International, West Conshohocken, PA, 2016, http://www.astm.org, accessed January 2017.
83 J. S. Lupoi, S. Singh, R. Parthasarathi, B. A. Simmons and R. J. Henry, *Renewable Sustainable Energy Rev.*, 2015, **49**, 871–906.
84 D. Harris, V. Bulone, S.-Y. Ding and S. DeBolt, *Plant Physiol.*, 2010, **153**, 420–426.
85 S. Van Wychen and L. M. L. Laurens, *Determination of Total Carbohydrates in Algal Biomass: Laboratory Analytical Procedure (lap)*, National Renewable Energy Laboratory (NREL), Golden, CO, 2013.
86 ASTM E1758-01(2015), *Standard Test Method for Determination of Carbohydrates in Biomass by High Performance Liquid Chromatography*, ASTM International, West Conshohocken, PA, 2015, http://www.astm.org, accessed January 2017.

87 A. Tolbert, H. Akinosho, R. Khunsupat, A. K. Naskar and A. J. Ragauskas, *Biofuels, Bioprod. Biorefin.*, 2014, **8**, 836–856.

88 A. Potthast, S. Radosta, B. Saake, S. Lebioda, T. Heinze, U. Henniges, A. Isogai, A. Koschella, P. Kosma, T. Rosenau, S. Schiehser, H. Sixta, M. Strlič, G. Strobin, W. Vorwerg and H. Wetzel, *Cellulose*, 2015, **22**, 1591–1613.

89 C. G. Yoo, Y. Pu, M. Li and A. J. Ragauskas, *ChemSusChem*, 2016, **9**, 1090–1095.

90 T. Mori, Y. Tsuboi, N. Ishida, N. Nishikubo, T. Demura and J. Kikuchi, *Sci. Rep.*, 2015, **5**, 11848.

91 Y. Pu, B. Hallac and A. J. Ragauskas, in *Aqueous Pretreatment of Plant Biomass for Biological and Chemical Conversion to Fuels and Chemicals*, ed. C. E. Wyman, John Wiley & Sons, Ltd, 2013, pp. 369–390.

92 L. Fu, S. A. McCallum, J. Miao, C. Hart, G. J. Tudryn, F. Zhang and R. J. Linhardt, *Fuel*, 2015, **141**, 39–45.

93 X. Li, C. Sun, B. Zhou and Y. He, *Sci. Rep.*, 2015, **5**, 17210.

94 W. Ge, J. H. Zhang, C. M. Pedersen, T. Zhao, F. Yue, C. Chen, P. Wang, Y. Wang and Y. Qiao, *ACS Sustainable Chem. Eng.*, 2016, **4**, 1193–1200.

95 L. Xiao, H. Wei, M. E. Himmel, H. Jameel and S. S. Kelley, *Front. Plant. Sci.*, 2014, **5**, 388.

96 A. Ferrer, C. Alciaturi, A. Faneite and J. Ríos, in *Analytical Techniques and Methods for Biomass*, ed. S. Vaz Jr, Springer International Publishing, 2016, pp. 45–83.

97 C. E. Payne and E. J. Wolfrum, *Biotechnol. Biofuels*, 2015, **8**, 43.

98 A. J. Simpson, *Magn. Reson. Chem.*, 2002, **40**, S72–S82.

99 O. D. Bernardinelli, M. A. Lima, C. A. Rezende, I. Polikarpov and E. R. deAzevedo, *Biotechnol. Biofuels*, 2015, **8**(110), 1–11.

100 F. X. Choong, M. Bäck, S. E. Steiner, K. Melican, K. P. R. Nilsson, U. Edlund and A. Richter-Dahlfors, *Sci. Rep.*, 2016, **6**, 35578.

101 J. B. Sluiter, R. O. Ruiz, C. J. Scarlata, A. D. Sluiter and D. W. Templeton, *J. Agric. Food Chem.*, 2010, **58**, 9043–9053.

102 S. Vaz, *Anal. Methods*, 2014, **6**, 8094–8105.

103 M. J. Taherzadeh and K. Karimi, *Int. J. Mol. Sci.*, 2008, **9**, 1621–1651.

104 J. D. McMillan, in *Enzymatic Conversion of Biomass for Fuels Production*, American Chemical Society, 1994, vol. 566, pp. 292–324.

105 A. Barakat, H. de Vries and X. Rouau, *Bioresour. Technol.*, 2013, **134**, 362–373.

106 X. Wang, H. Chen, K. Luo, J. Shao and H. Yang, *Energy Fuels*, 2008, **22**, 67–74.

107 A. Barakat, C. Mayer-Laigle, A. Solhy, R. A. D. Arancon, H. de Vries and R. Luque, *RSC Adv.*, 2014, **4**, 48109–48127.

108 S. Behera, R. Arora, N. Nandhagopal and S. Kumar, *Renewable Sustainable Energy Rev.*, 2014, **36**, 91–106.

109 M. Badiei, N. Asim, J. M. Jahim and K. Sopian, *APCBEE Proc.*, 2014, **9**, 170–174.

110 G. Qi, F. Peng, L. Xiong, X. Lin, C. Huang, H. Li, X. Chen and X. Chen, *Prep. Biochem. Biotechnol.*, 2016, 1–6, DOI: 10.1080/10826068.2016.1224246.

111 S. Singh, G. Cheng, N. Sathitsuksanoh, D. Wu, P. Varanasi, A. George, V. Balan, X. Gao, R. Kumar, B. E. Dale, C. E. Wyman and B. A. Simmons, *Front. Energy Res.*, 2015, **2**, 62, DOI: 10.3389/fenrg.2014.00062.

112 P. Kumar, D. M. Barrett, M. J. Delwiche and P. Stroeve, *Ind. Eng. Chem. Res.*, 2009, **48**, 3713–3729.

113 K. Karimi and M. J. Taherzadeh, *Bioresour. Technol.*, 2016, **200**, 1008–1018.
114 V. B. Agbor, N. Cicek, R. Sparling, A. Berlin and D. B. Levin, *Biotechnol. Adv.*, 2011, **29**, 675–685.
115 A. Dieste, L. Clavijo, A. I. Torres, S. Barbe, I. Oyarbide, L. Bruno and F. Cassella, *Energy Fuels*, 2016, **30**, 10494–10498.
116 L. Kouisni, A. Gagné, K. Maki, P. Holt-Hindle and M. Paleologou, *ACS Sustainable Chem. Eng.*, 2016, **4**, 5152–5159.
117 C. Li, M. Zheng, A. Wang and T. Zhang, *Energy Environ. Sci.*, 2012, **5**, 6383–6390.
118 A. Corma, S. Iborra and A. Velty, *Chem. Rev.*, 2007, **107**, 2411–2502.
119 C. O. Tuck, E. Pérez, I. T. Horváth, R. A. Sheldon and M. Poliakoff, *Science*, 2012, **337**, 695–699.
120 W. Lan, C.-F. Liu and R.-C. Sun, *J. Agric. Food Chem.*, 2011, **59**, 8691–8701.
121 R. N. Miller and W. H. Swanson, *Ind. Eng. Chem.*, 1925, **17**, 843–847.
122 Y. Habibi, L. A. Lucia and O. J. Rojas, *Chem. Rev.*, 2010, **110**, 3479–3500.
123 D. Klemm, F. Kramer, S. Moritz, T. Lindström, M. Ankerfors, D. Gray and A. Dorris, *Angew. Chem., Int. Ed.*, 2011, **50**, 5438–5466.
124 J. B. Sluiter, R. O. Ruiz, C. J. Scarlata, A. D. Sluiter and D. W. Templeton, *J. Agric. Food Chem.*, 2010, **58**, 9043–9053.
125 D. W. Templeton, C. J. Scarlata, J. B. Sluiter and E. J. Wolfrum, *J. Agric. Food Chem.*, 2010, **58**, 9054–9062.
126 F. Xu, J. Yu, T. Tesso, F. Dowell and D. Wang, *Appl. Energy*, 2013, **104**, 801–809.
127 A. Ebringerova, *Macromol. Symp.*, 2006, **232**, 1–12.
128 A.-C. Albertsson, U. Edlund and I. K. Varma, in *Biopolymers–New Materials for Sustainable Films and Coatings*, ed. D. Plackett, 2011, ch. 7, pp. 133–150.
129 N. M. L. Hansen and D. Plackett, *Biomacromolecules*, 2008, **9**, 1493–1505.
130 A.-C. Albertsson and U. Edlund, Barrier layers for packaging laminates and packaging laminates comprising such barrier layers, WO 2011005181 A1, 2011.
131 U. Edlund, Y. Z. Ryberg and A.-C. Albertsson, *Biomacromolecules*, 2010, **11**, 2532–2538.
132 M. Gröndahl, L. Eriksson and P. Gatenholm, *Biomacromolecules*, 2004, **5**, 1528–1535.
133 K. S. Mikkonen, S. Heikkinen, A. Soovre, M. Peura, R. Serimaa, R. A. Talja, H. Helen, L. Hyvonen and M. Tenkanen, *J. Appl. Polym. Sci.*, 2009, **114**, 457–466.
134 E. I. Goksu, M. Karamanlioglu, U. Bakir, L. Yilmaz and U. Yilmazer, *J. Agric. Food Chem.*, 2007, **55**, 10685–10691.
135 M. Söderqvist Lindblad, E. Ranucci and A.-C. Albertsson, *Macromol. Rapid Commun.*, 2001, **22**, 962–967.
136 X.-W. Peng, L.-X. Zhong, J.-L. Ren and R.-C. Sun, *J. Agric. Food Chem.*, 2012, **60**, 3909–3916.
137 X.-F. Sun, H.-H. Wang, Z.-X. Jing and R. Mohanathas, *Carbohydr. Polym.*, 2013, **92**, 1357–1366.
138 W. Zhao, K. Odelius, U. Edlund, C. Zhao and A.-C. Albertsson, *Biomacromolecules*, 2015, **16**, 2522–2528.
139 L. L. Lloyd, J. F. Kennedy, P. Methacanon, M. Paterson and C. Knill, *Carbohydr. Polym.*, 1998, **37**, 315–322.

140 M. J. Taherzedeh and K. Karimi, *BioResources*, 2007, **3**, 472–499.

141 A. I. Yaich, U. Edlund and A.-C. Albertsson, *Biomacromolecules*, 2012, **13**, 466–473.

142 M. Lawoko, G. Henriksson and G. Gellerstedt, *Biomacromolecules*, 2005, **6**, 3467–3473.

Faraday Discussions

PAPER

A study of biocatalysts based on glucose oxidase

E. P. Golikova,[a] N. V. Lakina,[a] O. V. Grebennikova,[a] V. G. Matveeva *[ab] and E. M. Sulman[ab]

Received 30th January 2017, Accepted 7th March 2017
DOI: 10.1039/c7fd00042a

During this work, we studied the possibility of glucose oxidase (GOx) covalent immobilization on a modified inorganic support. A series of GOx-based biocatalysts was synthesized by crosslinking the enzyme to a surface of modified silica or alumina. Polyelectrolyte layers were used as modifiers for the silica and alumina surfaces. These layers promote tight binding of the GOx to the support. The biocatalyst's activity and stability were studied using an oxidation reaction of D-glucose to D-gluconic acid. It was found that GOx immobilized on the modified SiO_2 using glutardialdehyde as a crosslinking agent was the most active and stable catalytic system, showing an 85% yield of gluconic acid. A study of the synthesized biocatalyst structure using FTIR spectroscopy showed that the enzyme was covalently crosslinked to the surface of an inorganic support modified with chitosan and glutardialdehyde. In the case of SiO_2, the quantity of the immobilized enzyme was higher than in the case of Al_2O_3.

Introduction

Monosaccharide oxidation reactions are the foundation for various modern methods of obtaining biologically active compounds. One of such reactions is D-glucose oxidation to D-gluconic acid, which pertains to the group of aldonic acids and is widely used in pharmaceutical and food industries. For instance, in the food industry gluconic acid is known as food additive E574 and is used as an acidity regulator and leavening agent. Besides, it can play the role of an acidulent and sequestrant as well as a potentiating antioxidant. In the pharmaceutical industry gluconic acid is successfully used for the synthesis of such important drugs as sodium, calcium or ferrous gluconates and glucono-delta-lactone.

Microbiological methods can be applied for highly selective oxidation of monosaccharides, including D-glucose, under mild conditions. For this purpose, different microbial strains such as *Gluconobacter* spp., *Aspergillius niger*, *Tricholoma robustum*, *Tricholoma bakamatsutake*, *Gluconobacter oxydans*, *Zymomonas mobilis*, *Penicillium variabile*, *Acetobacter methanolicus*, *Acetobacter diazotrophicus*

[a]Tver State Technical University, A.Nikitin emb., 22, Tver, 170026, Russia. E-mail: matveeva@science.tver.ru
[b]Tver State University, Zhelyabova st., 33, Tver, 170100, Russia

and *Acetobacter suboxydans* as well as their isolated enzymes are widely used.[1] At present, methods for enzymatic production of gluconic acid are of great interest. One of these methods is based on D-glucose enzymatic oxidation in the presence of native glucose oxidase (GOx) at a temperature of 25–35 °C and within the pH range 5.0–7.5, whilst maintaining a D-glucose concentration of 40–60% (the scheme of this reaction is presented in Fig. 1). In this case, the gluconic acid yield with respect to the initial glucose concentration was 95%.[2]

Glucose oxidase (GOx EC 1.1.3.4) catalyzes the oxidation of D-glucose to gluconic acid by utilizing molecular oxygen as an electron acceptor, with simultaneous production of hydrogen peroxide (H_2O_2).[3] GOx has been used in several commercial applications including glucose removal from dried egg; improvement of the color, flavor, and shelf life of food materials; oxygen removal from fruit juices, canned beverages and from mayonnaise to prevent rancidity.[3a] It has also been used in an automated glucose assay kit in conjunction with catalase[3b] and chiefly in biosensors for the detection and estimation of glucose in industrial solutions and in body fluids such as blood and urine.[3b,4] Besides, enzyme systems (GOx and catalase) that are free and/or immobilized on different membranes are also used for the development of bioreactor systems.[5]

There are two ways to improve the existing enzyme process. Firstly, combining the properties of catalytic systems and GOx allows exclusion of the use of catalase and makes the process more feasible. Recent investigations[6] have shown that combinations of GOx and metal-containing catalysts supported on different oxides provide a "synergetic effect". In this case, the side product (H_2O_2) decomposition takes place in the presence of metallic catalysts and does not require the addition of catalase. Secondly, optimization of the GOx immobilization methods leads to an increase in the enzyme life-time and biocatalyst stability. In this regard, we proposed to optimize GOx-based biocatalyst synthesis using modified alumina or silica as a support. The developed method combines both approaches: silica and alumina catalyze the decomposition of the side product – hydrogen peroxide, and polyelectrolytes as modifying agents allow tighter binding of the enzyme.

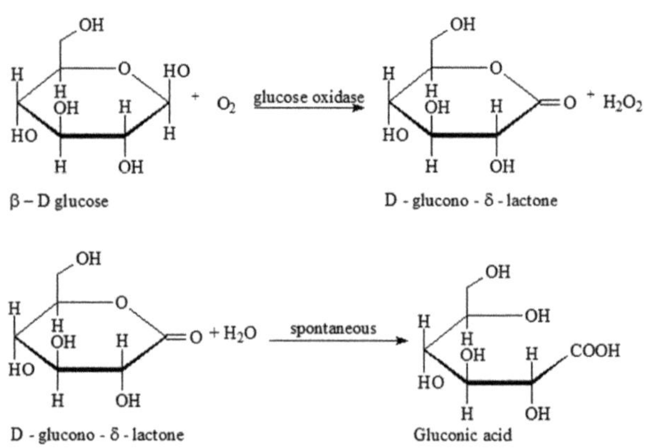

Fig. 1 Scheme of the enzymatic oxidation of D-glucose to gluconic acid in the presence of native glucose oxidase.

In order to combine the advantages and exclude the limitations of traditional catalytic methods, the interest of researchers is focused on the synthesis of immobilized enzymatic catalytic systems. According to the literature, data on glucose oxidase immobilization and the application of immobilized enzymes in the food and pharmaceutical industries have been obtained. Enzyme immobilization has attracted interest for use in a wide range of areas from fundamental academic research to various industrial applications. Up to now, several immobilized enzyme-based processes have been proven to be economical. They have been implemented on a larger scale, mainly in the food industry, where they replace free enzyme catalyzed processes, and in the production of fine chemicals and pharmaceuticals.[3a,7]

A variety of supports have been used for the immobilization of enzymes such as cellulose, solid glass particles, porous glass particles, and nickel screens. Because of the advantages for catalytic activity offered by materials with relatively high surface areas, porous glass and cellulose have been the most popular supports. For example, for preparation and improvement of the storage stability of GOx on porous glass, γ-amino-propyltriethoxysilane (APTES) has been used by various researchers for silanization of the glass prior to immobilization of GOx. GOx immobilization on both solid and porous glass has also been performed by some researchers.[3a,8] In addition, a nickel oxide screen can also be silanized and GOx can be coupled using a thiophosgene method, but the screen offers little surface area.[6b] GOx is difficult to crosslink with an agent like glutaraldehyde, but it crosslinks rather easily in the presence of certain proteins that have a high number of reactive amino groups such as polyethyleneimine.[8a]

In the work of Jagdish and Neelam,[9] the purified enzyme GOx was immobilized using covalent and gel entrapment methods. The former was carried out with oxirane beads, while the latter was carried out in sodium alginate and polyacrylamide gels. Covalent immobilization showed a higher stability for glucose removal in comparison with the entrapment methods. The enzymes immobilized through entrapment methods were stable at temperatures up to 50–55 °C for one hour and above this temperature there was a negligible decrease in the enzyme activity. The thermal stability of the enzyme immobilized in polyacrylamide gel was higher compared to free and sodium alginate immobilized systems. The glucose oxidation rates were 600 and 500 mg h^{-1} with acrylamide and sodium alginate immobilized enzymes (20 IU) at an optimum pH and temperature of 6 and 30 °C, respectively. For kinetic characterization of the immobilized enzyme, various concentrations of glucose (1–40 mM) were used. Kinetic constants K_m and V_{max} were calculated using a Lineweaver–Burk plot. The Michaelis constant for the immobilized enzymes was 12.0 and 8.0 mM respectively for sodium alginate and acryl amide.[9]

MacAodha *et al.*[10] immobilized GOx on multiwalled carbon nanotubes with osmium redox polymers under pseudo-physiological conditions of 5 mM glucose, 150 mM NaCl and 37.8 °C. Glucose oxidation current densities above 800 μA cm^{-2} were obtained from films containing the $[Os(4,4'\text{-dimethyl-}2,2'\text{-bipyridine})_2(\text{polyvinylimidazole})_{10}Cl]^+$redox polymer, with a redox potential 0.1 V *vs.* Ag/AgCl, and GOx. Such a result was promising for the development of glucose-oxidising enzymatic fuel cells. Besides, the enzyme-based redox systems showed good stability over 24 hours during the glucose oxidation.[10] The immobilization methodology of an enzyme and redox complex on an electrode surface can have an impact on the magnitude and stability of the amperometric current response, with implications for application as a biosensor and fuel cell enzyme electrodes. Kumar and Leech[11] have reported an investigation of carboxymethyl dextran (CMD) and polyacrylic acid polymers, bearing carboxylic functional groups, as chemical supports for the immobilization of amine-containing osmium redox complexes and enzymes on electrode surfaces. Crosscoupling of GOx using a carbodiimide reagent to the CMD polymer support, the [Os(2,2′-bipyridine)$_2$(4-aminomethyl pyridine)Cl]PF$_6$ redox complex and multiwall carbon nanotubes (MWCNT) provides a 3-dimensional matrix for the catalytic electrooxidation of glucose. Similar enzyme electrodes, thus, can be used as low potential glucose oxidizing biosensors and as anodes for *in vivo* enzymatic fuel cells for power generation.

Interesting results were obtained during the work of F. M. Bautista and coworkers.[12] These authors studied the properties of GOx covalently immobilized on an $AlPO_4$-based clay mineral sepiolite system in an oxidation reaction of D-glucose to D-gluconic acid. It was shown that the use of $AlPO_4$ as an enzyme support allowed exclusion of the use of catalase; in that case, the activity of the developed system was 75% of the activity of native GOx.

In this work, we have developed a design methodology for biocatalysts based on GOx covalently immobilized on modified inorganic supports of Al_2O_3 and SiO_2. To modify the surface of the support, we proposed the use of polyelectrolyte layers[13] (poly(sodium-4-styrene sulfonate) as a polyanion and chitosan as a polycation). Chitosan contains easily-accessible amino groups, which can be used for intermolecular crosslinking with GOx. Glutardialdehyde was used for enforced binding of the enzyme to the support.

Testing of the obtained biocatalysts was performed using an oxidation reaction of D-glucose to D-gluconic acid.

Experimental

Materials

In this work, the following materials were used as received: alumina (γ-Al_2O_3, $d <$ 60 μ, sieved) (Reachim, Russia); silica (SiO_2, $d <$ 60 μ, sieved) (Reachim, Russia); poly(sodium-4-styrene sulfonate) (PSS, MW 70 000) (SIGMA-ALDRICH, USA), chitosan (viscosity 20–200 cP for 1% solution in 1% acetic acid, 75–85% deacetylation) (SIGMA-ALDRICH, USA), glutardialdehyde (Fluka, Switzerland), glucose oxidase from Aspergillus niger (lyophilized lot result: 174.9 U mg^{-1}) (SIGMA-ALDRICH, USA).

Method of enzyme immobilization

In this work, alumina and silica were used as inorganic supports to immobilize GOx. The support samples (mass of 1 g) were calcined at 300 °C for three hours, and then the sample surfaces were modified and activated. For surface modification, 10 mL of a poly(sodium-4-styrenesulfonate) (PSS) solution (0.1 g L^{-1} concentration) was added to 1 g of alumina or silica and the mixture was stirred for 60 minutes, and then the modified support was washed with distilled water to reach pH = 7 and dried under vacuum at 60 °C for 24 hours. Then, the samples were treated with a solution of chitosan (chit) (0.1 g L^{-1}) in 0.01 M acetic acid by

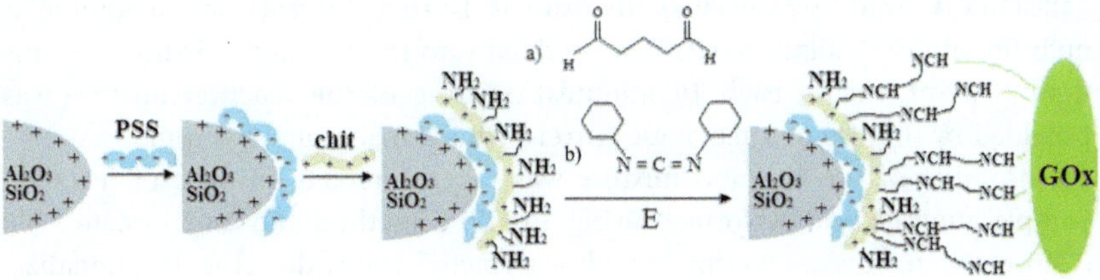

Fig. 2 Scheme of the biocatalyst synthesis using crosslinking agents: (a) glutardialdehyde; (b) carbodiimide.

stirring for 60 minutes, followed by filtration, washing with distilled water and drying under vacuum at 60 °C. The obtained samples were mixed with 50 mL of a glutardialdehyde solution (glut) of 0.1 g L^{-1} concentration for 60 minutes. Chitosan application during the modification of the support surface is due to the presence of amino groups, which allow its use in covalent enzyme immobilization. The use of glutardialdehyde is based on the formation of azomethine bonds (-CH=N-) in weakly alkaline medium during the reaction of glutardialdehyde with amino groups of the proteinic part of the enzyme and chitosan coupled to the cationic surface through a donor–acceptor mechanism.

After modification, the support was treated with 10 mL of an enzyme (GOx) solution of 0.015 g L^{-1} concentration for 60 minutes, filtered, washed and dried under vacuum at 25 °C. Thus, the following samples of biocatalysts were obtained: SiO_2/PSS/chit/glut/GOx and Al_2O_3/PSS/chit/glut/GOx. A scheme of the biocatalyst synthesis is shown in Fig. 2(a). Besides glutardialdehyde, carbodiimide (carb) was also used as a crosslinking agent (Fig. 2(b)). These biocatalysts (SiO_2/PSS/chit/carb/GOx and Al_2O_3/PSS/chit/carb/GOx) were synthesized using the same techniques as those used in the case of glutardialdehyde.

Biocatalyst analysis

Investigation of the structure of the multicomponent biocatalysts was performed using an FTIR-spectrophotometer IRPrestige 21 equipped with a diffusion reflection device, which allows analysis of homogeneous samples with high accuracy in the spectral range 350–7800 cm^{-1}. High sensitivity was provided by a special ceramic emitting source, an interferometer with gold-filled mirrors and a thermostabilized detector DLATGS for the medium infrared region. Switching of the operating range of the infrared emission (NIR-MIR-FIR) was accomplished by the installation of program-controlled split mirrors with automatic change of the emission sources and detectors. The spectrum of a KBr sample crushed to a particle size of less than 2 μm was used as the background spectrum. The spectra were recorded over a range of 250–4000 cm^{-1} with an interval of 4 cm^{-1}.

D-Glucose oxidation and product analysis

The D-glucose oxidation process was carried out at atmospheric pressure and a temperature between 30–60 °C. The reaction temperature was kept constant by supplying a heating medium into the reactor jacket. After reaching the required temperature, 12 mL of a 4 wt% aqueous solution of D-glucose and 0.025 g of the biocatalyst were put into the glass reactor. Molecular oxygen with a feeding rate

440–450 mL min^{-1} was used as the oxidant. During the reaction, an equimolar quantity of an alkalizer $NaHCO_3$ was added into the reaction mixture over one hour (0.5 mL within each 10 minutes). Stirring of the reaction mixture was provided by the use of a magnetic stirrer with a stirring rate 800 rpm.

A sample of the reaction mixture was taken after 60 minutes of reaction. Sample analysis was performed using an HPLC method and an UltiMate 3000 (ChromaTech, Russia) equipped with a refractive index detector. Fractionalization of the reaction mixture was carried out using a Reprogel-H column (500 × 10 mm, NTP 160000). Chromatographic separation was conducted under the following conditions: eluent – 9 mM H_2SO_4 solution; eluent consumption – 0.5 mL min^{-1}; eluent pressure – 65 atm; column temperature – 250 °C; duration of analysis – 30 min. The product identification was carried out using standard samples of pure compounds.

Results and discussion

Comparison of biocatalyst activity

To choose the most effective biocatalyst, experiments to investigate the influence of the biocatalyst synthesis method on the enzyme activity were carried out. The catalyst activity was estimated based on the D-gluconic acid content (wt%) of the reaction mixture after 60 minutes of oxidation. A comparison of the biocatalyst activities is presented in Fig. 3.

The yields of D-gluconic acid obtained at the end of the D-glucose oxidation process were as follows: 70% – using Al_2O_3/PSS/chit/glut/GOx; 85% – for the case of SiO_2/PSS/chit/glut/GOx; 64% – for Al_2O_3/PSS/chit/carb/GOx; 78% – in the presence of SiO_2/PSS/chit/carb/GOx. It can be noted that the biocatalysts based on silica showed a decrease in activity compared to the native enzyme of 15 and 22% for SiO_2/PSS/chit/glut/GOx and SiO_2/PSS/chit/carb/GOx respectively, while the alumina-based samples showed activity decreases of 30 and 36%. As for the influence of the crosslinking agent, the use of glutardialdehyde resulted in an increase in the gluconic acid yield of 16–17% in comparison with carbodiimide. Thus, based on these results, it can be concluded that the most effective support

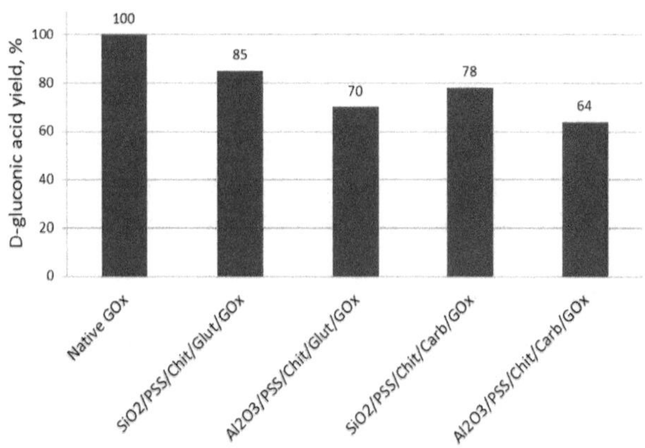

Fig. 3 Comparison of biocatalyst activity in terms of the gluconic acid yield.

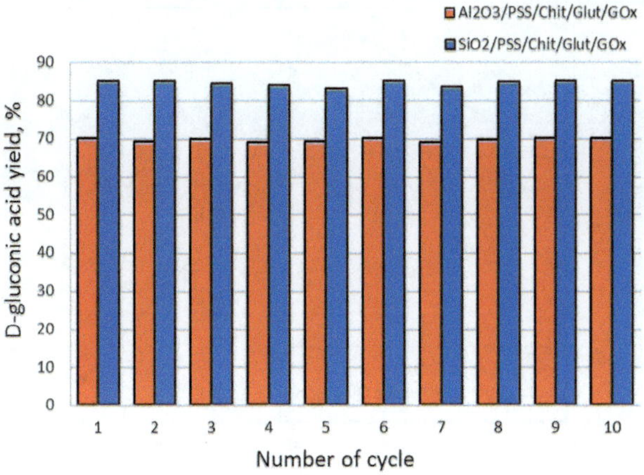

Fig. 4 A study of biocatalyst stability over 10 cycles.

for GOx immobilization is modified SiO_2 treated with glutardialdehyde as a crosslinking agent.

In order to estimate the stability of the biocatalysts synthesized using glutardialdehyde, 10 successive experiments of D-glucose oxidation were carried out. The results obtained from these experiments, in terms of the D-gluconic acid content in the reaction mixture after the reaction, are shown in Fig. 4. As can be seen, both the biocatalysts did not show a significant decrease in the activity.

Based on the presented results, the biocatalyst SiO_2/PSS/chit/glut/GOx was chosen as the most active and stable enzymatic system for D-glucose oxidation to gluconic acid.

Oxidation parameter influence

Further experiments to investigate the optimal reaction parameters were performed for the selected SiO_2/PSS/chit/glut/GOx biocatalyst. A study of the effect of the pH on the D-glucose oxidation in the presence of the native and synthesized

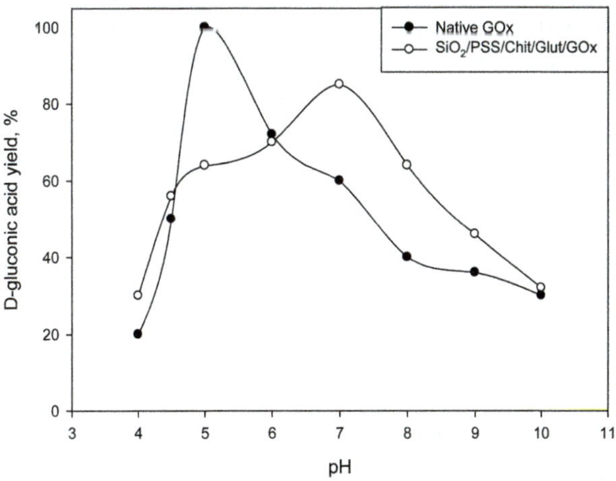

Fig. 5 pH influence on D-glucose oxidation in the presence of the biocatalyst.

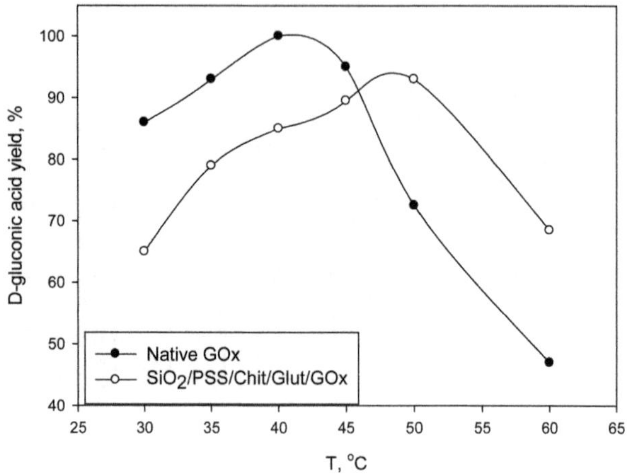

Fig. 6 Temperature influence on D-glucose oxidation in the presence of the biocatalyst.

biocatalysts was performed under a temperature of 40 °C for one hour. The influence of the pH on the yield of gluconic acid is shown in Fig. 5.

It can be seen that the maximum product yield for the immobilized GOx was obtained at pH = 7, while for the native enzyme the optimum pH was 5. It is notable that the shift in the optimum pH does not lead to a change in the enzyme activity. Thus, enzyme immobilization allows conduction of the D-glucose oxidation process in a neutral or slightly basic medium. Moreover a slight increase in the activity of the immobilized enzyme at pH 4–4.5 was observed, which can be connected with the presence of polyelectrolytes on the biocatalyst surface.

To study the influence of the temperature on D-glucose oxidation in the presence of the native and synthesized biocatalysts, experiments were performed at various temperatures within the range of 30–60 °C at pH = 5 for one hour. The dependence of the gluconic acid yield on the temperature is shown in Fig. 6.

The native enzyme showed a maximum yield at 40 °C, then a sharp decrease in the product yield was observed indicating loss of the enzyme activity due to temperature-related destruction of the enzyme structure. In the case of the immobilized GOx, there was a decrease in the gluconic acid yield at temperatures of 30–40 °C compared to the native GOx. Further increase of the temperature led to an increase in the product yield for the immobilized enzyme. The optimal temperature for D-glucose oxidation in the presence of SiO_2/PSS/chit/glut/GOx was 50 °C. It is notable that use of the immobilized GOx allows increase of the product yield by 1.2–1.4 times at temperatures higher than 50 °C, due to improvement of the enzyme's thermal stability.

Thus, it was shown that GOx immobilization on the surface of modified silica leads to a shift in the optimal pH and temperature in comparison with the native enzyme. The optimal conditions for D-glucose oxidation were found to be pH = 7, $t = 50$ °C.

FTIR study of the biocatalysts

To study the formation of covalent grafting between the enzyme and the support, the obtained biocatalyst samples were investigated using diffuse reflection

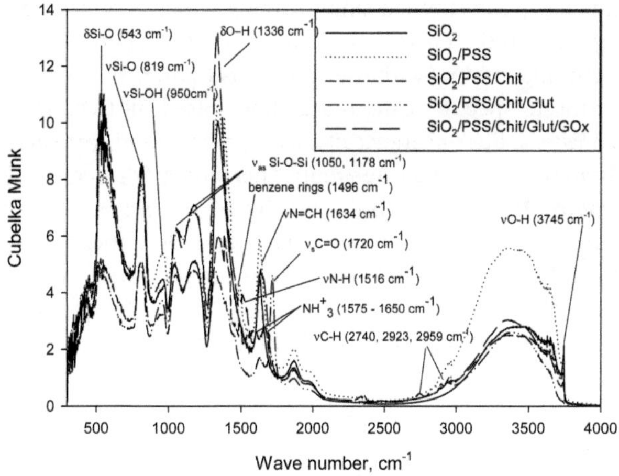

Fig. 7 IR spectra of the silica-based biocatalyst samples.

infrared Fourier transform (DRIFT) spectroscopy. IR spectra were collected at all of the stages of the biocatalyst synthesis. Fig. 7 shows the data for the support surface modification study for the biocatalysts based on GOx immobilized on SiO_2.

The IR-spectra (Fig. 7) show that the SiO_2 sample has the following relevant absorption bands: 3745 cm^{-1} corresponding to the stretching vibrations of O–H bonds; a wide band in the range of 3000–3600 cm^{-1}, which also refers to O–H stretching vibrations; the band at 1336 cm^{-1}, corresponding to νO–H vibrations; absorption bands at 1050 and 1178 cm^{-1} relating to asymmetric Si–O–Si stretching vibrations; a band at 950 cm^{-1} corresponding to νSi–OH (OH group vibrations to Si); νSi–O (819 cm^{-1}); deformation Si–O vibrations at 543 cm^{-1}; a band at 1870 cm^{-1} corresponding to combined and overtone vibrations of the silica skeleton.[14] After treatment of the silica support with PSS, a characteristic band at 2923 cm^{-1} corresponding to the stretching vibrations of C–H bonds of the modifying agent was observed in the spectra. The intensity of the absorption bands relating to O–H bond vibrations increased after the surface treatment. This allows us to suggest that modification of the support with PSS leads to an increase in the quantity of OH-groups on the silica surface. Thus it can be concluded that PSS does not form any covalent bonds with the support, as it is evidently held at the silica surface by electrostatic forces. After treatment of the SiO_2/PSS sample with chitosan, besides the absorption bands characteristic of the initial sample, bands corresponding to deformation (in the range of 1575–1650 cm^{-1}) and stretching (3030–3130 cm^{-1}) vibrations of the NH_3^+ groups of chitosan were observed. After the sample treatment with chitosan, a decrease in the intensity of the OH-group characteristic bands was noticed. The IR-spectra of the sample SiO_2/PSS/chit/glut show that it has a new azomethine bond, which vibrates at 1634 cm^{-1} (νN=CH), due to interaction of the glutardialdehyde with the chitosan amino groups. As a result of this interaction, the characteristic bands corresponding to chitosan NH_3^+ group deformation vibrations decreased. Besides, new absorption bands relating to free carbonyl groups (νC=O) at 1720 cm^{-1} were observed. When using glutardialdehyde as a crosslinking agent, the enzyme binds with the modified support through amino groups, which is preferable as there are more amino groups in the enzyme than there are carboxyl ones in the support, and thus the enzyme binding with the support is tighter.

Analysis of the IR-spectra of the biocatalysts based on GOx immobilized on SiO_2 (Fig. 7) shows the disappearance of an absorption band corresponding to the stretching vibrations of an aldehyde group. This indicates its conversion to other functional groups. In addition, the intensity of the νN=CH vibration band increases, indicating that the enzyme was covalently bonded to the support through glutardialdehyde. The following bands are indicative of the presence of the enzyme: δN–H (1516 cm^{-1}); ν-CO_2^- (1570 cm^{-1}). The peptide bonds are identified by the band due to $\delta_{NH} + \nu_{CN}$ (1559 cm^{-1}).

The second sample studied was the biocatalyst based on GOx immobilized on Al_2O_3 (Fig. 8). Analysis of the IR-spectra of Al_2O_3 showed the presence of the following characteristic absorption bands: 400–800 cm^{-1} corresponding to νAl–O–Al; 1645 cm^{-1} relating to νAl–O and 1068 cm^{-1} corresponding to Al–OH stretching vibrations.[15] After treatment of the alumina with PSS, bands typical of polystyrenesulfonic acid (νC–H (2933 cm^{-1}), benzene ring (1500 cm^{-1}), νR-SO_2–OH (1070 cm^{-1})) were observed. New adsorption bands corresponding to covalent interactions of Al_2O_3 and PSS did not form. PSS is held at the alumina surface by electrostatic interactions. After treatment of the sample with a chitosan solution, absorption bands relating to the deformation vibrations of NH_3^+ groups typical of this polysaccharide (in the range of 1575–1650 cm^{-1}), as well as δC–H (1456 cm^{-1}), δO–H (1420 cm^{-1}) and νC–N (1341 cm^{-1}) were observed in the IR-spectra. Glutardialdehyde forms azomethine bonds at the surface of the modified Al_2O_3 support, which was confirmed by the presence of an absorption band at 1644 cm^{-1} (νN=CH). The formation of azomethine bonds can also be verified by the decrease in the intensity of the bands corresponding to chitosan amino groups (1575–1650 cm^{-1}). Analysis of the IR-spectra of the Al_2O_3/PSS/chit/glut/GOx biocatalyst allows us to conclude that successful grafting of the enzyme occurred. This is confirmed by the increase in intensity of the absorption bands relating to azomethine bonds and the disappearance of the band corresponding to the free

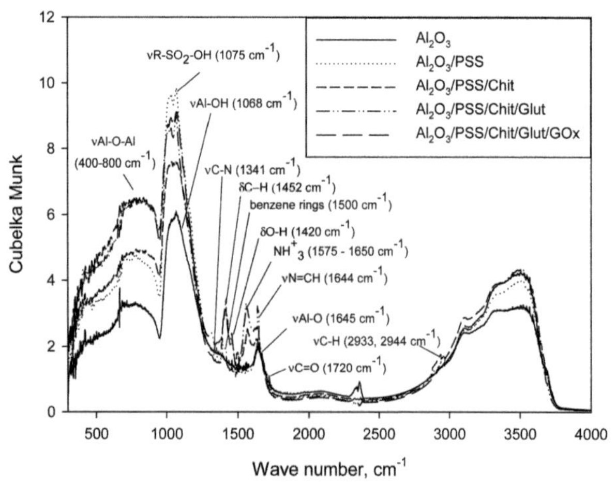

Fig. 8 IR spectra of the alumina-based biocatalyst samples.

carbonyl groups of glutardialdehyde, attributed to interaction with the GOx amino groups. Thus, based on the IR-spectrophotometry analysis, it can be concluded that GOx was successfully covalently bonded to the surface of the modified support. It is necessary to note that in the case of alumina the quantity of the immobilized enzyme was less than that of silica, which allows us to suggest that the activity of the silica-based biocatalyst must be higher compared to the alumina-based system.

Conclusions

Based on the experiments, it can be concluded that the proposed method of D-glucose oxidation to gluconic acid in the presence of a biocatalyst immobilized on a modified inorganic support can be an alternative to the existing methods for gluconic acid production. Biocatalytic methods for organic compound oxidation meet with the stringent requirements of GMP regarding the presence of toxic impurities in pharmaceutical products.

A novel biocatalyst design was presented. Alumina and silica modified with polyelectrolytes (*i.e.* chitosan) were used for GOx immobilization. The application of chitosan is due to the presence of amino groups in its structure. This allows use of such molecules for the covalent immobilization of enzymes. FTIR-spectroscopy data confirmed the presence of covalent bonds between the polyelectrolyte layers, crosslinking agent and enzyme. The most effective biocatalytic system SiO_2/PSS/chit/glut/GOx showed 85% activity compared with the native enzyme in the D-glucose oxidation to gluconic acid.

A higher thermostability (optimal temperature 50 °C) of the designed biocatalysts compared to native GOx and immobilized systems presented in the literature[12] (optimal temperature 30–40 °C) was shown.

Thus, the combination of the properties of the support, modifier, crosslinking agent and GOx can be considered as having a "synergetic effect" which allows obtainment of a high activity and stability of the biocatalysts based on immobilized enzymes. The use of supports based on silica and alumina, which catalyze decomposition of the side product (H_2O_2), allows exclusion of the addition of catalase and consequently the stage of its immobilization.

Acknowledgements

This work was supported by the Ministry of Science and Education of the Russian Federation (grant RFMEFI58615X0024).

References

1 (*a*) M. Yoshimoto, Y. Miyazaki, Y. Kudo, K. Fukunaga and K. Nakao, *Biotechnol. Prog.*, 2006, **22**, 704; (*b*) L. Betancor, F. Lopez-Gallego, A. Hidalgo, N. Alonso-Morales, G. Dellamora-Ortiz, J. M. Guisan and R. Fernandez-Lafuente, *J. Biotechnol.*, 2006, **121**, 284; (*c*) S. Anastassiadis and I. G. Morgunov, *Recent Pat. Biotechnol.*, 2007, **1**, 167; (*d*) S. Ramachandran, P. Fontanille, A. Pandey and C. Larroche, *Lett. Appl. Microbiol.*, 2007, **44**, 155; (*e*) F. Pezzotti and M. Therisod, *Carbohydr. Res.*, 2006, **341**, 2290.

2 S. D. Varfolomeev, *Chemical enzymology*, Academia, Moscow, Russia, 2005.

3 (*a*) S. B. Bankar, M. V. Bule, R. S. Singhal and L. Ananthanarayan, *Biotechnol. Adv.*, 2009, **27**, 489; (*b*) D. G. Hatzinikolaou and B. J. Macris, *Enzyme Microb. Technol.*, 1995, **17**, 530.

4 M. Petruccioli, F. Federici, C. Bucke and T. Keshavarz, *Enzyme Microb. Technol.*, 1999, **24**, 397.

5 S. T. Morthensen, A. S. Meyer, H. Jorgensen and M. Pinelo, *Biochem. Eng. J.*, 2017, **117**, 41.

6 (*a*) A. Taketoshi, S. Takenouchi, T. Takei and M. Haruta, *Appl. Catal., A*, 2014, **474**, 257; (*b*) P. N. R. Vennestrøm, E. Taarning, C. H. Christensen, S. Pedersen, J.-D. Grunwaldt and J. M. Woodley, *ChemCatChem*, 2010, **2**, 943.

7 B. Krajewska, *Enzyme Microb. Technol.*, 2004, **35**, 126.

8 (*a*) J. Bouin, M. Atallah and H. Hultin, *Biotechnol. Bioeng.*, 1976, **18**, 179; (*b*) W. Herring, R. Laurexce and J. Icittrell, *Biotechnol. Bioeng.*, 1972, **14**, 975; (*c*) P. Sreedivya, D. Savitri and C. Mitra, *J. Biosci.*, 1998, **23**, 131; (*d*) M. Wasilewska, J. Luterek and A. Leonowicz, *Biotechnol. Bioeng.*, 1987, **32**, 507; (*e*) B. Wassrman and H. Hultin, *Biotechnol. Bioeng.*, 1980, **22**, 271.

9 S. Jagdish and V. Neelam, *Adv. Appl. Sci. Res.*, 2013, **4**(3), 250.

10 D. MacAodha, P. O. Conghaile, B. Egan, P. Kavanagh, C. Sygmund, R. Ludwig and D. Leech, *Electroanalysis*, 2013, **25**(1), 94.

11 R. Kumar and D. Leech, *J. Electrochem. Soc.*, 2014, **161**(13), 3005.

12 F. M. Bautista, J. M. Campelo, A. García, A. Jurado, D. Luna, J. M. Marinas and A. A. Romero, *J. Mol. Catal. B: Enzym.*, 2001, **11**, 567.

13 (*a*) T. Swrizawa, M. Yamaguchi and M. Akashi, *Macromolecules*, 2002, **35**, 8656; (*b*) I. B. Tsvetkova, L. M. Bronstein, S. N. Sidorov, O. L. Lependina, M. G. Sulman, P. M. Valetsky, B. Stein, L. Zh. Nikoshvili, V. G. Matveeva, A. I. Sidorov, B. B. Tikhonov, G. N. Demidenko, L. Kiwi-Minsker and E. M. Sulman, *J. Mol. Catal. A: Chem.*, 2007, **276**(1–2), 116.

14 G. Katumba, B. W. Mwakikunga and T. R. Mothibinyane, *Nanoscale Res. Lett.*, 2008, **3**, 421.

15 (*a*) Z. Ghezelbash, D. Ashouri, S. Mousavian, A. Hossein Ghandi and Y. Rahnama, *Bull. Mater. Sci.*, 2012, **35**(6), 925; (*b*) G. D. Chukin, *Structure of alumina and desulfurization catalysts. Reaction mechanisms*, Printa, Moscow, Russia, 2010; (*c*) K. Nakamoto, *Infrared and Raman Spectra of Inorganic and Coordination Compounds: Part A: Theory and Applications in Inorganic Chemistry*, John Wiley & Sons Inc., New York, USA, 6th edn, 2008; (*d*) K. Nakanishi, *Infrared Absorption Spectroscopy: Practical*, Holden-Day, Tokyo, Nakokodo, 1962.

Faraday Discussions

PAPER

Cellulose nanocrystals by acid vapour: towards more effortless isolation of cellulose nanocrystals

Marcel Lorenz,[a] Stefan Sattler,[b] Mehedi Reza,[c] Alexander Bismarck[ab] and Eero Kontturi *[abc]

Received 31st January 2017, Accepted 27th February 2017
DOI: 10.1039/c7fd00053g

Cellulose nanocrystals (CNCs) are topical in materials science but their full potential is yet to be fulfilled because of bottlenecks in the production: the process consumes huge amounts of water, recycling the strong acid catalyst is difficult, and purification steps are cumbersome, particularly with lengthy dialysis. Production of CNCs with HCl vapour overcomes many of these difficulties but the dispersion of CNCs from the already hydrolysed fibre matrix is a formidable challenge. This study is a fundamental effort to explore very basic means to facilitate CNC dispersion from cotton linter fibres (filter paper), hydrolysed to levelling off degree of polymerization by HCl vapour. The introduction of carboxylic groups on the cellulose crystal surface proved the most efficient method to alleviate dispersion with good yields (*ca.* 50%) and a provisional possibility to tune the CNC length. By contrast, attempts to directly disperse untreated hydrolysed fibres in various organic solvents and aqueous surfactant solutions were unsuccessful. The results showed that hydrolysis of native cellulose fibres by HCl vapour is indeed a viable method for producing CNCs but it has more potential as a pre-treatment step rather than a full-fledged process on its own.

Introduction

Cellulose is a polysaccharide that is responsible for the structural integrity of all plant fibres. While cellulose fibres have been traditionally applied in paper, cardboard and textile applications, the isolation of nanosized cellulosic objects from a plant-based matrix has become increasingly popular.[1] These nanocellulose species possess potential in applications such as strong renewable composites,[2–4] biomedical devices,[5] chiral templates,[6] and responsive materials.[7]

[a]*Polymer and Composite Engineering (PaCE) Group, Department of Chemical Engineering, Imperial College London, SW7 2AZ, UK*

[b]*Polymer and Composite Engineering (PaCE) Group, Institute of Materials Chemistry and Research, University of Vienna, Währinger Strasse 42, A-1090, Vienna, Austria*

[c]*Department of Bioproducts and Biosystems, School of Chemical Engineering, Aalto University, P.O. Box 16300, FI-00076 Aalto, Finland. E-mail: eero.kontturi@aalto.fi; Tel: +358 503442978*

Nanocellulose comes in two categories: nanofibres and nanocrystals. Cellulose nanofibres (often dubbed nanofibrils) are essentially isolated microfibrils, which are the smallest supramolecular units in the cell walls of plants. Cellulose nanocrystals (CNCs), on the other hand, are prepared by controlled hydrolysis of cellulose microfibrils.[8,9] These microfibrils are semi-crystalline, consisting of highly ordered crystalline domains that are interrupted by short disordered regions in the longitudinal direction.[10] Specifically, cellulose chains are cleaved by acid catalysed hydrolysis selectively on the disordered domains of cellulose microfibrils, reaching the so-called levelling off degree of polymerization (LODP),[11] which means that the crystallites are left intact, that is, CNCs are formed (Fig. 1a).[12] The main distinction between cellulose nanofibres and CNCs lies in their aspect ratio and shape.[1] Their thickness is nearly identical but the length of the nanofibres runs up to several micrometres whereas most cellulose sources yield CNCs of around 50–300 nm in length after acid hydrolysis. Moreover, nanofibres are flexible, which allows entanglement at relatively low concentrations, leading to gel formation in water.[13] CNCs, by contrast, are straight and stiff, forming liquid crystals in concentrated suspensions.[14]

Nearly all literature accounts utilise *ca.* 65% (w/w) sulphuric acid at moderately elevated temperatures (45–60 °C) for CNC production.[8,9] The reaction is quenched with 10-fold dilution and the CNCs are subsequently purified by centrifugation

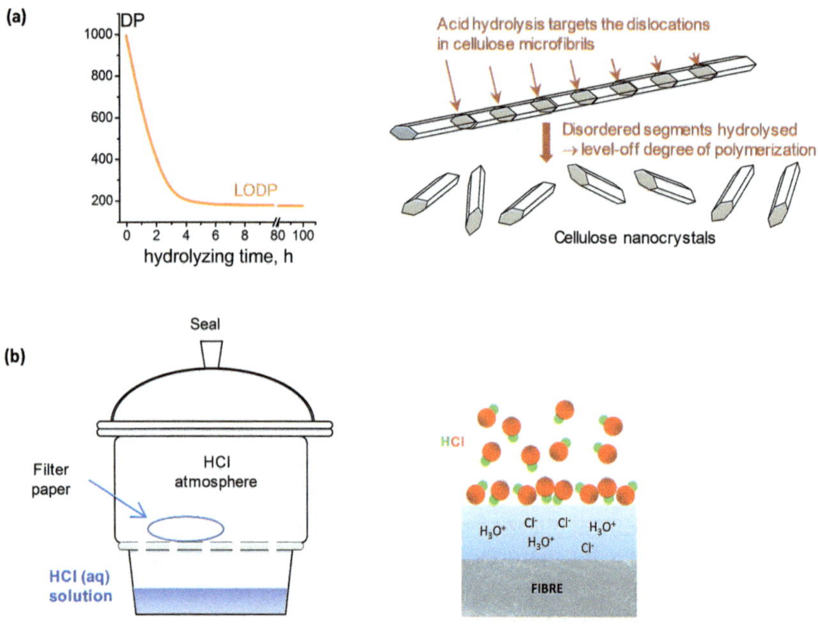

Fig. 1 (a) The principle behind CNC preparation: disordered segments can be selectively hydrolysed by aqueous acid, resulting in fast hydrolysis of dislocations or disordered segments within cellulose microfibrils. The levelling off degree of polymerization (LODP) denotes the DP where only the recalcitrant crystalline segments are left and the hydrolysis nearly halts. (b) An alternative pathway of cellulose hydrolysis by hydrogen chloride vapour: HCl fumes and the vapour adsorb on the fibres, which are always covered by a thin water layer where the HCl can dissociate and catalyse the hydrolysis of cellulose, reaching the LODP.

and dialysis. The yields are generally not very high (between 15–50%), the procedure consumes a lot of water, and it is impossible to fully recycle the acid from the mixture if upscaling is desired.[15–18] The production with sulphuric acid, however, bears a significant asset: concomitant with hydrolysis, the sulphate groups of the acid form half-esters with the hydroxyl groups of cellulose, thus imparting charge on the resulting CNC surfaces, greatly facilitating their dispersion in water.[14,15] Alternative methods to sulphuric acid hydrolysis do exist, namely with oxidation,[19–22] esterification,[23,24] and enzymatic procedures,[25,26] but these are usually odd accounts, not further investigated, and none of them have been widely accepted as a viable replacement.

Recently, one of the authors (EK) pioneered a method to produce CNCs by exposing cellulosic fibres to HCl vapour (Fig. 1b).[27] HCl adsorbs on the fibres, which are partially covered by a thin water layer under ambient conditions. Subsequently, HCl dissociates and the commencing hydrolysis quickly degrades the cellulose down to the LODP value. Unfortunately, reaching the LODP is not the only prerequisite for CNC isolation. One must also disperse the CNCs from the fibre matrix, which is a major challenge, taking into account the strong, hierarchical structure of native cellulose fibres, holding the CNCs strongly together inside the fibre matrix and the lack of charge after HCl-based hydrolysis of the fibres – in contrast to the one with sulphuric acid. The previous effort featured a proof-of-concept of dispersing the CNCs from the fibres hydrolysed in HCl vapour: a prolonged 72 h sonication in concentrated formic acid.[27] Obviously, this is not a viable process for large-scale production and it is a cumbersome undertaking even on the laboratory scale.

Since the DP of cellulose is easily reduced to the LODP stage in HCl vapour hydrolysis,[27] the isolation of CNCs is essentially a question of dispersing them from the fibre matrix. Indeed, the CNCs already exist within the hydrolysed fibres as degraded microfibrils. In this effort, we focus on exploring various methods for CNC dispersion from a cellulose fibre substrate that has been hydrolysed to LODP with HCl vapour. The methods involve: oxidation mediated by (2,2,6,6-tetramethylpiperidine-1-yl)oxyl radical (TEMPO), probing different organic solvents as dispersion media, and utilizing surfactants to assist the aqueous dispersion of CNCs. The motivation is to further explore the possibilities of utilising HCl vapour hydrolysis as a genuine alternative to the current paradigm of CNC preparation.

Materials and methods

Materials

Whatman 1 filter paper (from cotton linters) was obtained from GE Healthcare, hydrochloric acid (37% stock solution, exact concentration checked with titration against NaOH solution) was obtained from Sigma Aldrich, and KOH and NaOH pellets were analytical grade and were obtained from VWR Chemicals. NaBr, TEMPO, dimethylacetamide, and sodium dodecyl sulfate (SDS) were analytical grade from Sigma Aldrich. Cetyltrimethylammonium bromide (CTAB) of >96% purity and hypochlorite, a 10–15% aqueous solution (exact concentration checked with titration against NaOH), were from Sigma Aldrich. Formic acid was reagent grade (>95%, Sigma Aldrich). All other solvents were analytical grade from Sigma Aldrich. Water was HPLC grade from VWR unless otherwise mentioned.

Hydrolysis with HCl vapour

The hydrolysis with HCl vapour was performed in a vacuum desiccator with 35% HCl solution poured on the bottom, as shown in Fig. 1b. The desiccator valve was left open for >5 days so that the HCl atmosphere would completely substitute the air inside the vessel. Subsequently, filter paper was placed inside the desiccator and the hydrolysis was allowed to proceed for 12 h, after which the hydrolysed filter paper was placed in water to rinse away the excess HCl. The filter paper was then left to dry in a fumehood overnight.

TEMPO-mediated oxidation

TEMPO-mediated oxidation was performed according to the method of Saito *et al.*[28] Filter paper (hydrolysed or unhydrolysed) was suspended in water (375 mL) containing TEMPO (0.0125 g) and sodium bromide (0.125 g). Subsequently, 13.4 g of 9% NaClO solution was added slowly to the cellulose slurry, and the pH of the mixture was maintained at 10.5 at room temperature by adding 0.5 M NaOH. When no more decrease in pH was observed, the pH was adjusted to pH 7. After the oxidation, the sample was washed, dispersed by sonication (10 min), and centrifuged (8000 rpm for 1 h). The supernatant, consisting of CNCs, was collected for further analysis.

Dispersion attempts in organic solvents

1% of the dried paper was tested. The paper was balanced in a glass vial and the necessary amount of solvent (~5 mL) was then added. The paper was cut in the vial with scissors to break it into smaller pieces. Then, the vial was placed in an ice bath and the tip sonicated for 10 min (UP100H Ultrasonic processor, cycle: 100%, amplitude: 1/1), until the remaining particles appeared to have a similar size and shape in all solvents (as dusty and powder-like as possible).

Dispersion attempts in aqueous surfactant solutions

1% HCl vapour hydrolysed filter paper was mixed with an aqueous surfactant solution (either 0.73 mM SDS or 0.47 mM CTAB, roughly half of their critical micelle concentrations at room temperature). Sonication on the mixture was performed for 50 min.

Gel permeation chromatography (GPC)

Determination of the molecular weight and degree of polymerization was done using a RI-101 detector (Shodex) coupled with an UltiMate 3000 LC system (Dionex). A 50 $\pm$ 5 mg sample was weighed into a tube and activated overnight by adding 4 mL of HPLC water. The water was removed by rinsing the sample in 2 mL of acetone and subsequent activation in 4 mL of acetone for 12 h. The last activation was done in DMAc (*N*,*N*-dimethylacetamide) overnight. The activated cellulose was finally dissolved in 5 mL of LiCl/DMAc (90.0 g L^{-1}) at room temperature under magnetic stirring. After dissolution, 0.5 mL of the sample was diluted in 4.5 mL of pure DMAc and stirred for 2 h. This gave a sample concentration of 1.0 mg mL^{-1} and a LiCl concentration of 9.0 g L^{-1}. The diluted samples were transferred into vials using 0.2 μm syringe filters.

Conductometric titration

The carboxylate content of the nanocrystals was determined following a procedure by Saito and Isogai.[29] A dried sample of 0.3 g was dispersed into water (100 mL) and 0.01 M NaCl solution (10 mL). Subsequently, the pH was set to 2.5 by adding 0.1 M HCl to the slurry. In the case of the samples that were first oxidized by TEMPO and then hydrolysed, the titration was carried out using a 0.1 M KOH solution. The titrations for the samples that were first hydrolysed and subsequently TEMPO-oxidized were carried out using 0.05 M KOH solution. The titration was stopped once a pH of 11 was achieved.

Scanning electron microscopy (SEM)

Pristine and HCl vapour hydrolysed filter paper fibres were imaged with a Zeiss Sigma VP SEM at the accelerating voltages of 3–5 kV. Filter paper specimens were attached to Al studs with double-sided carbon tapes and sputtered with Au/Pd on an Emitech K100X sputtering device before imaging.

Cryoscopic transmission electron microscopy (cryo TEM)

The CNC dispersions were characterized using a JEOL JEM-3200FSC field emission cryo-transmission electron microscope operated at 300 kV accelerating voltage. Vitrified specimens were prepared with an FEI Vitrobot device using Quantifoil holey carbon copper grids with a hole size of 3.5 μm. The grids were plasma cleaned prior to use on a Gatan Solarus 9500 plasma cleaner and then transferred into an environmental chamber of FEI Vitrobot at room temperature and 100% humidity. Subsequently, a drop of 3 μL of sample dispersion was applied on the grid and was blotted once for 1 second and then vitrified in a 1/1 mixture of liquid ethane and propane at a temperature of −180 °C. The grid with the vitrified sample dispersion was maintained at liquid nitrogen temperature and then cryo-transferred into the microscope. Images were taken in bright-field mode and using zero loss energy filtering (Omega type) with a slit width of 20 eV. Micrographs were recorded using a Gatan Ultrascan 4000 CCD camera. The specimen temperature was maintained at −187 °C during the imaging.

Results

Hydrolysis by HCl vapour

Hydrolysis of Whatman 1 filter paper substrates with HCl vapour was performed and the LODP was comfortably reached (see Fig. 4) over the hydrolysis period. The SEM images in Fig. 2 show that no morphological alterations of the cellulose fibre substrate occurred during the hydrolysis. It is essentially a solid/gas system where HCl adsorbs on the solid fibres. The resulting hydrolysed filter paper is still a macroscopic object where the integrity of the fibres and the whole paper have been preserved – although the filter paper is admittedly fairly brittle after the cellulose chains have been cleaved to their LODP. The acid can be washed away by simple rinsing of the substrate instead of laborious centrifugation and dialysis steps.

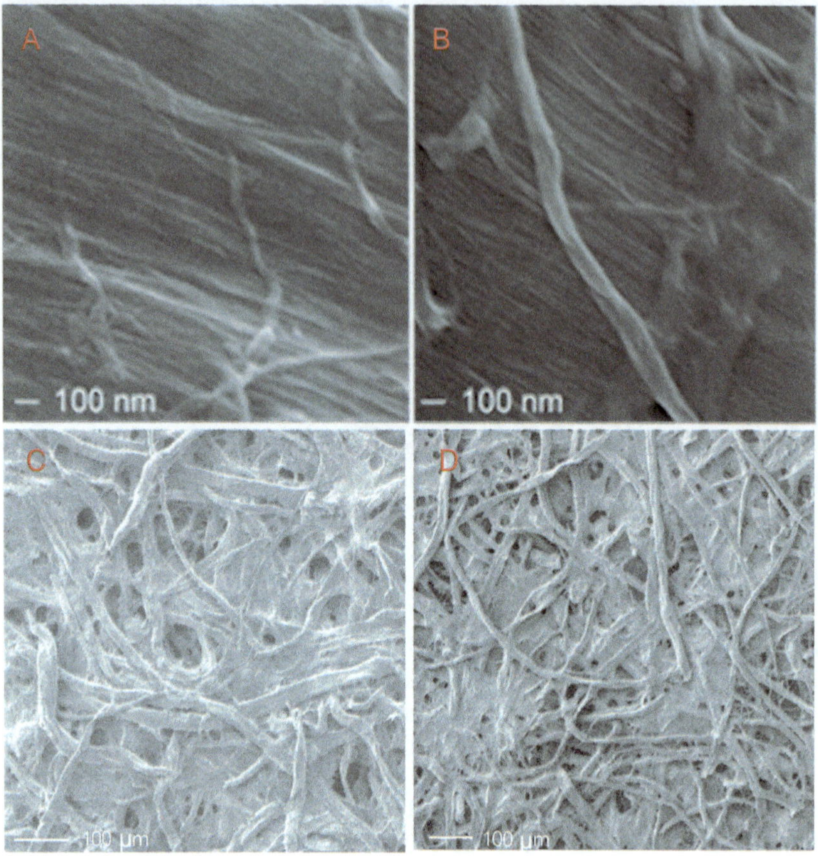

Fig. 2 Scanning electron microscopy images of pristine (A and C) and hydrolysed (B and D) filter paper fibres. (A) and (B) are high magnification images, focussed on a single fibre, whereas (C) and (D) are low magnification images, exposing the full fibre network.

TEMPO-mediated oxidation before and after hydrolysis by HCl vapour

TEMPO-mediated oxidation introduces carboxylic groups on the surface of cellulose microfibrils by exclusively oxidizing the primary alcohols of cellulose into carboxylic groups *via* an aldehyde intermediate (Fig. 3a).[28,29] The procedure is utilised for isolating microfibrils as nanofibrils from a native fibre matrix because it introduces sufficient charge on the fibril surface in order to facilitate effortless dispersion of the fibrils, but it is still mild enough not to break the integrity of the microfibrils or the subsequent nanofibrils.[28,30] The established method of TEMPO-oxidation under alkaline conditions (pH 10) as described by Saito *et al.*[28] was exploited for Whatman 1 filter paper substrates using two different experimental approaches: (i) the substrate was first hydrolysed in HCl vapour, followed by TEMPO-oxidation, and (ii) the substrate was TEMPO-oxidized to begin with, and hydrolysed in HCl vapour afterwards (Fig. 3b). Fig. 4 displays the molecular weight distributions as determined by GPC in a cellulose solvent system for the untreated substrate, hydrolysed substrate, TEMPO-oxidized substrate, and for a sample that was first TEMPO-oxidized followed by hydrolysis. HCl vapour

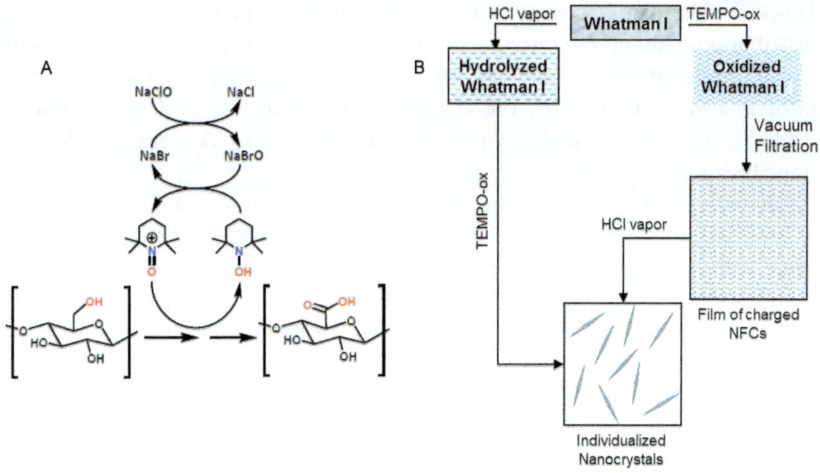

Fig. 3 (A) Schematic of the TEMPO-mediated oxidation of cellulose: exclusively the primary alcohols are oxidized into carboxylic acids *via* an aldehyde intermediate. (B) Experimental scheme for the two TEMPO-oxidations performed in this study: (i) the filter paper is hydrolysed by HCl vapour and subsequently TEMPO-oxidized or (ii) the filter paper is first TEMPO-oxidized and subsequently hydrolysed by HCl vapour.

hydrolysis of a pristine Whatman 1 fibre substrate decreases the DP of cellulose down to the LODP value, as reported previously, and the shape of the M_w distribution is similar to that of fibres that have been hydrolysed down to their LODP, either by vapour or by a liquid acid.[27] It is also evident that TEMPO-oxidation alone greatly reduces the DP, although not quite to the LODP level. This phenomenon has been ascribed to β-elimination and/or generation of hydroxyl radicals upon the oxidation.[31] Interestingly, when the substrate has been first

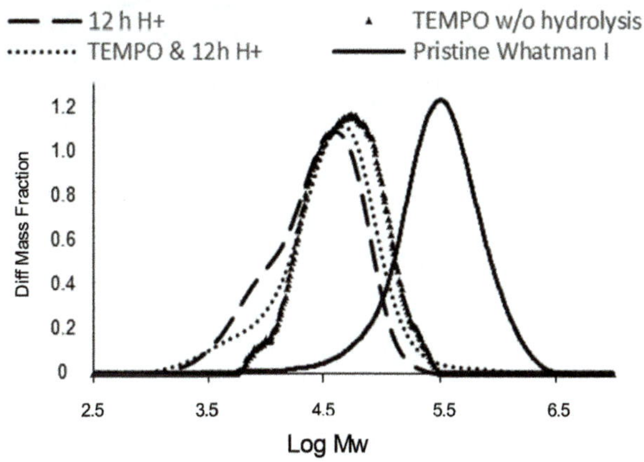

Fig. 4 Gel permeation chromatograms of untreated filter paper cellulose (Pristine Whatman 1), filter paper hydrolysed for 12 h (12 h H^+), TEMPO-oxidized filter paper (TEMPO w/o hydrolysis), and TEMPO-oxidized and subsequently hydrolysed filter paper (TEMPO & 12 h H^+).

TEMPO-oxidized and subsequently subjected to HCl vapour, the shape of the M_w distribution is different from just the hydrolysed filter paper: the shoulder with low M_w components is far smaller when the substrate has been TEMPO-oxidized prior to hydrolysis. Clearly, the carboxylic groups on the surface of TEMPO-oxidized microfibrils slightly impede acid hydrolysis. The reasons for the impediment are unclear.

The charge, as determined by conductometric titration, was 1.36 ± 0.02 meq g^{-1} for both the hydrolysed and the non-hydrolysed samples. This is exactly the same value as measured by Okita *et al.* for cotton-based microfibrils whose surfaces were completely TEMPO-oxidized[32] and it shows that pre-hydrolysis by HCl vapour does not affect the behaviour of the cellulose upon oxidation.

Fig. 5 shows the TEM images of the supernatants of the samples sonicated and fractionated by centrifugation. In a control experiment (Fig. 5a), sonication was performed on untreated, pristine filter paper, showing that the sonication itself does little more than small breakages of the fibres. Fig. 5b shows a sample that

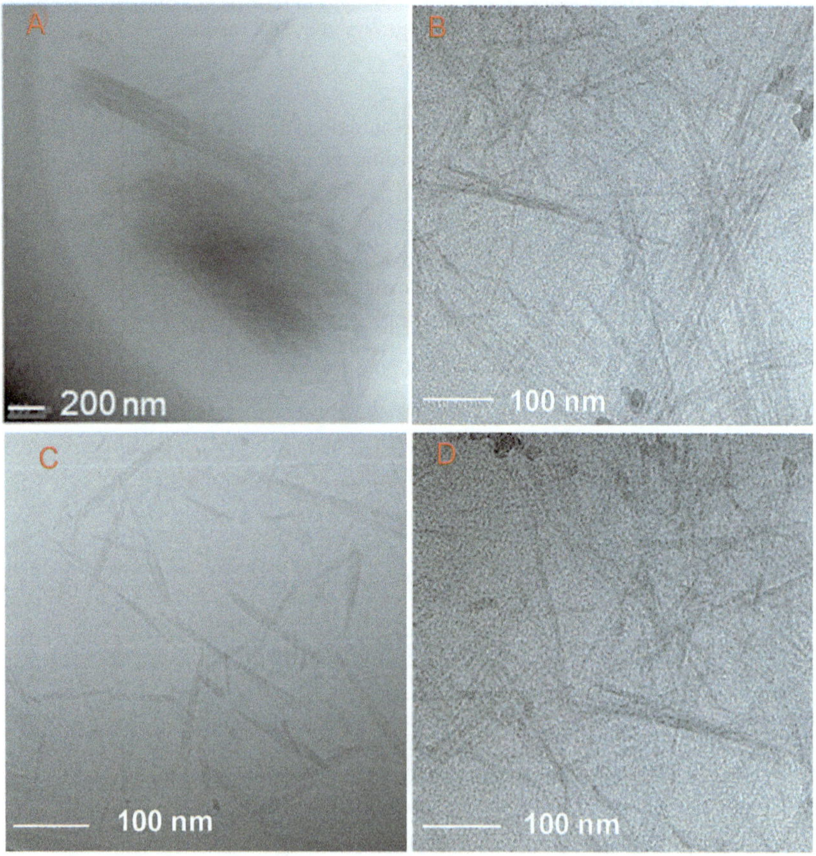

Fig. 5 TEM images of (A) pristine filter paper, treated with sonication, (B) TEMPO-oxidized filter paper (not hydrolysed), resulting in nanofibres as reported by Saito *et al.*, (C) filter paper hydrolysed by HCl vapour and subsequently TEMPO-oxidized, and (D) filter paper TEMPO-oxidized first and subsequently hydrolysed by HCl vapour.

has been TEMPO-oxidized but not hydrolysed, corresponding well to the previously published images on nanofibrils prepared by TEMPO-oxidation.[28,30] Fig. 5c shows a pre-hydrolysed filter paper that has been TEMPO-oxidized afterwards, clearly demonstrating the formation of CNCs. Likewise, Fig. 5d illustrates that CNCs can be produced from TEMPO-oxidized filter paper that has been subsequently hydrolysed. Here, however, the CNCs are clearly longer, well in-line with the GPC results that suggest an impediment to hydrolysis caused by the preceding TEMPO-oxidation.

The properties of the CNCs isolated by TEMPO-oxidation are gathered in Table 1. The width/length dimensions of the sample that was first hydrolysed, then TEMPO-oxidized, are similar to those of the CNCs produced by conventional sulphuric acid hydrolysis of cotton linter fibres (*i.e.* filter paper).[33] By contrast, the sample that has first been TEMPO-oxidized, then hydrolysed, results in CNCs that are much longer, in-line with the molecular weight data. The CNC yields (40–54%) are somewhat better than those of the CNCs produced with conventional sulphuric acid hydrolysis, which usually results in yields of around 20–40%, although higher yields have occasionally been reported.[16–18]

Solvent trials for the dispersion of CNCs from the filter paper substrate hydrolysed by HCl vapour

Among the most straightforward ways of dispersing the CNCs from a fibre matrix hydrolysed down to LODP would obviously be direct exposure to a dispersing solvent. Regrettably, the lack of charge with the HCl hydrolysed CNCs renders water an impracticable solvent and other common organic solvents are not really applicable either. To investigate whether the isolation of CNCs to any realistic degree is possible with a simple organic solvent, we tried to sonicate the hydrolysed filter paper in the following solvents: ethanol, DMSO, toluene, dichloroethane, 1,4-dioxane, dimethoxymethane, and pyridine, with water as a control. It is evident from the photographs in Fig. 6 that none of the solvents worked the way they were desired to function: sedimentation occurs inevitably after sonication stopped. 1,4-Dioxane yields a marginally better result for dispersion but even there the sediment dominates over the dispersion.

The results presented in Fig. 6 largely concur with the seminal study by van den Berg *et al.*[34] where several organic solvents were trialled for dispersing already prepared CNCs, both charged and uncharged. It was this particular publication

Table 1 Properties of the CNCs isolated from the TEMPO-oxidized samples: surface charge (from the carboxylic groups), average M_w as determined by GPC, average dimensions from the TEM images, and yield as determined from the share of solids in the supernatant after centrifugation

Sample	Charge [μeq g^{-1}]	Average M_w [g mol^{-1}]	Width [nm]	Length [nm]	CNC yield [%]
Hydrolysis succeeded by TEMPO	1.37	37 500	12 ± 5	102 ± 38	40 ± 27%
TEMPO succeeded by hydrolysis	1.36	57 900	11 ± 5	216 ± 75	54 ± 22%

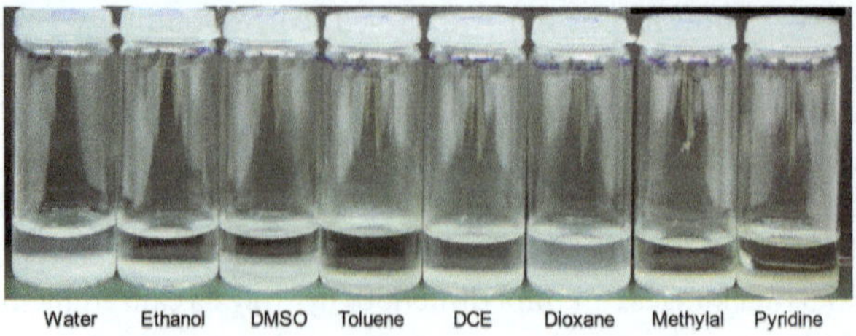

Fig. 6 Photographs of the attempts to isolate CNCs from a filter paper matrix hydrolysed by HCl vapour in different solvents. The dispersions were subjected to ultrasonication for 10 min, after which they were left for 19 h untouched. DMSO stands for dimethylsulfoxide, DCE for dichloromethane, and methylal for dimethoxymethane.

that established formic acid as the superior solvent for dispersing uncharged CNCs, prompting us to use formic acid in the previous proof-of-concept effort where the production of CNCs with HCl vapour was introduced. To illustrate the excellent dispersion quality of formic acid, we dispersed the CNCs again in that solvent, like in the proof-of-concept study, after which the dispersion was centrifuged (6000 rpm/10 min). Even after centrifugation, the dispersion was immaculate, virtually without any sediment (Fig. 7). It is up for discussion as to why formic acid in particular is such a powerful dispersing agent for uncharged CNCs.

Fig. 7 Photograph of a dispersion of CNCs isolated from filter paper hydrolysed by HCl vapour. The dispersion was obtained by 50 min sonication and subsequent centrifugation (6000 rpm/10 min).

Attempts to disperse isolate CNCs from HCl hydrolysed fibres with surfactants

For a colloid chemist, the use of a surfactant is perhaps the most obvious way of dispersing a colloidal object like CNC. With this in mind, the use of two common surfactants were trialled for the isolation of CNCs in water: CTAB (cationic) and SDS (anionic). Although initially a degree of dispersion was detected directly after sonicating the samples, practically complete sedimentation took place within 60 hours (Fig. 8). So far, only one specific surfactant has been reported to work for CNCs: Beycostat NA, which is a phosphoric ester of polyoxyethylene (9) nonylphenyl ether, and was successfully applied to disperse sulphated CNCs in non-polar organic solvents.[35] In conclusion, none of the quick-and-easy non-covalent methods here with different solvents or surfactants resulted in acceptable levels of CNC dispersion.

Discussion

Despite the importance of the dispersion step in CNC manufacturing, it has received relatively meagre attention in the literature. Because of the importance of sulphuric acid hydrolysis in CNC preparation at present, re-dispersion of dried, sulphated CNCs in water has been explored rather extensively and the studies stress the importance of choosing the correct counter-ion prior to drying and re-dispersing.[36] Viet *et al.*[37] managed to disperse traditionally produced sulphated

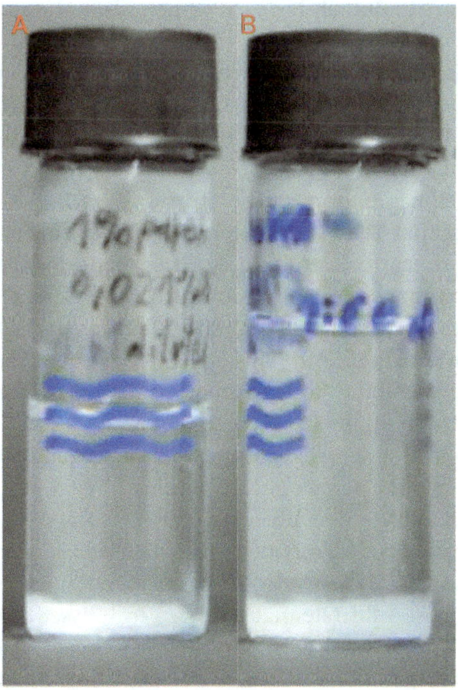

Fig. 8 Photographs from the attempts to isolate CNCs from filter paper hydrolysed with HCl vapour: aqueous dispersions with surfactants (A) SDS and (B) CTAB. The dispersions were sonicated for 50 min and left untouched to sediment for 60 h.

CNCs in some polar organic solvents, such as DMSO, but successful dispersion actually required small amounts of water in the mixture. *N*-Methyl pyrrolidone was also found to disperse sulphated CNCs, alongside formic acid and *m*-cresol.[34] For uncharged CNCs, only formic acid and *m*-cresol have proved to be satisfactory dispersion media and only after extensive (24 h) sonication.[34] Our study with a range of common solvents of differing polarities did not introduce any new possible solvents for uncharged CNC dispersion. Interestingly, Ferguson *et al.*[38] were able to disperse (uncharged) bacterial cellulose nanofibres in a number of apolar organic solvents, which highlights the differences between long nanofibres and short CNCs.

Chemical modification of CNCs is an efficient route to enable their dispersion in various solvents, although systematic studies on the explicit relationship between the nature of the covalently attached moiety and the dispersion medium are lacking. Quite expectedly, attaching hydrocarbon chains by, *e.g.*, esterification on the surface hydroxyls of CNCs allows their dispersion in non-polar media.[39] Studies on uncharged CNCs, generally prepared by HCl catalysed hydrolysis, are fewer in number but TEMPO-oxidation has been used as a post-treatment for the introduction of carboxylic groups and the subsequent dispersion in water or as a platform for further modification.[40] Otherwise, uncharged CNCs in water aggregate in a matter of seconds (see Fig. 6).

All these quoted studies, however, have focussed on already prepared CNCs and therefore the treatments, be it counter-ion exchange or covalent modification, have been imposed on already nanosized materials. In our case, we are still dealing with hierarchically organized fibres (Fig. 2) consisting of cellulose microfibrils that have essentially been cut at their dislocations by the HCl vapour hydrolysis (Fig. 1), as shown by the M_w distributions of the hydrolysed material (Fig. 4). One may consider this system to require more energy input for dispersion than is the case with already nanosized CNCs which are being re-dispersed. Yet TEMPO-oxidation for cellulose nanofibre production also leaves behind fibres with morphological integrity and they require minimal energy input for nanofibre dispersion;[30] sometimes only magnetic stirring does the trick.[28] Indeed, our results pointed out that TEMPO-oxidation is the most efficient way to disperse CNCs from the fibre matrix hydrolysed by HCl vapour (Fig. 5c). When TEMPO-oxidation was performed as a pre-treatment, by contrast, the hydrolysis was somewhat impeded and the CNCs – although efficiently dispersed – were left clearly longer than usual (Fig. 5d). This is not necessarily a disadvantage and it hints towards the possibility to do a more wide-ranged tuning of the CNC dimensions than has traditionally been possible by adjusting the conditions in the sulphuric acid hydrolysis, where tuning the length automatically means also affecting the yield and the surface charge density.[15–18,41]

Furthermore, TEMPO-oxidation worked well in terms of the CNC yield from filter papers hydrolysed by HCl vapour (Table 1). Previously, our group attempted to disperse CNCs by TEMPO-oxidizing microcrystalline cellulose (MCC), a commercial cellulose substrate that has been hydrolysed down to its LODP.[20] TEMPO-oxidation of MCC resulted in CNCs but with a scant yield of around 4%. A follow-up study showed that sonication was generally an inadequate way of dispersing CNCs from MCC which consists of fibre fragments because sonication requires a critical length for properly being able to enact on a substrate.[42] Hence, CNC yields from MCC were virtually never more than 20% even after optimizing all the conditions.[42] With our hydrolysed fibres which are still intact (Fig. 2), the requirement for the critical length is fulfilled – hence the good yield of CNCs, generally over 50% (Table 1). Nanofibre production from TEMPO-oxidized cellulosic fibres is even more efficient because the length is always retained in the process: the yield there is generally >90%.[28] Furthermore, our yields were inferior to the proof-of-concept study where the CNCs dispersed in formic acid from a filter paper hydrolysed in HCl vapour resulted in >97% yield.[27]

It is intriguing to see how ineffective the addition of a surfactant is for CNC dispersion (Fig. 8). In principle, a cellulose crystal has both large hydrophilic (OH decorated) and smaller hydrophobic (CH decorated) faces.[43,44] The original hypothesis was that the hydrophobic moiety of the surfactant would attach on the hydrophobic face of the CNC, adding charge or other hydrophilic functionality, thus facilitating the dispersion in water. This did not work (Fig. 8). According to our experience, previous efforts to perform CNC dispersion in a hydrophobic solvent with surfactant addition have not been successful, either. As stated earlier, only one surfactant has worked for dispersing sulphated CNCs in apolar solvents so far: Beycostat NA, a phosphoric ester of polyoxyethylene (9) nonylphenyl ether.[35] This is surprising since it would make sense that, for example, any cationic surfactant would electrostatically attach on the sulphate groups on a CNC surface and that way enable the subsequent dispersion of CNCs in hydrophobic solvents. This works well for TEMPO-oxidized nanofibres[45,46] but not for sulphated CNCs. This may well result from different charge densities since the amount of carboxylic groups on the TEMPO-oxidized nanofibre surfaces are 2–4 times higher than the amount of sulphate groups on the traditionally prepared CNC surfaces.[8,9,15,28,30,32,41] The attachment of Beycostat NA is not electrostatic. However, the hydrophobic 200 faces, although small in area, should be continuous throughout the cellulose crystal surface, independent of whether it is sulphated or not. Obviously, the amount of attached CTAB or SDS is not sufficient to impart enough charge on the cellulose crystal surface to enable dispersion in water (Fig. 8). There is little knowledge on the interactions of the hydrophobic 200 faces with other surfaces or even with each other in different media. More fundamental efforts are required in this respect.

In a previous effort where we attempted to go beyond the proof-of-concept model of dispersion in formic acid, we focused on non-covalent methods with adsorption by various biopolymers (tailored proteins and polysaccharides) on the cellulose crystal surface.[47] Notably with carboxymethyl cellulose (CMC), it was possible to efficiently isolate and disperse CNCs but only after severe mechanical treatments with ball milling and similar actions. In the present effort, we aimed at avoiding the excessive mechanical energy input and sought for easy chemical solutions instead. It transpired that only TEMPO-oxidation behaved as we had hoped for: isolation of CNCs with good yields was feasible with reasonable mechanical energy (sonication).

This study is a contribution to the current efforts and discussion on dispersing cellulosic nanomaterials, namely CNCs and cellulose nanofibres. It strengthens the view that covalent addition of charge is amongst the only straightforward routes to efficient dispersions of these nanomaterials. Therefore, the introduction of sulphate groups *via* sulphuric acid hydrolysis on CNCs and the introduction of carboxylic acids *via* TEMPO-oxidation on nanofibres have enjoyed substantial popularity amongst the nanocellulose community. Other dispersion systems than charged species in water have run into the well-known difficulties of dispersing or indeed dissolving cellulose in conventional organic or inorganic solvents. With this in mind, it is no surprise that simple trials with a number of solvents or adding a surfactant to facilitate aqueous dispersibility were not successful. Clearly, a more systematic approach is required to find suitable solvents or solvent mixtures for CNC dispersion. Since a couple of decent solvents for dispersion are already known (*m*-cresol and formic acid), a possible future study with, *e.g.*, Hildebrand parameters is likely to shed light on this matter.

Conclusions

Hydrolysis by HCl vapour appears to be a promising candidate for CNC production from native cellulosic fibres. The isolation of CNCs from the hydrolysed fibre matrix remains a significant bottleneck and this study explored various quick-and-easy methods to achieve this. Frankly, TEMPO-oxidation, already proven as a decent method for nanofibre production, was the only successful route within the trialled approaches that included simple dispersion in organic solvents and in aqueous surfactant solutions. More fundamental efforts are needed to illuminate what the actual physico-chemical bottlenecks in CNC dispersion really are. At this stage, however, it seems that hydrolysis by HCl vapour works better as a pre-treatment step to, *e.g.*, TEMPO-oxidation rather than a completely developed method for CNC production on its own.

Acknowledgements

EK acknowledges the Academy of Finland for financial support (Project 259500) and University of Vienna for funding the visiting professorship. Ms. Rita Hatakka (Aalto University) is acknowledged for helping with the GPC analyses.

References

1 D. Klemm, F. Kramer, S. Moritz, T. Lindström, M. Ankerfors, D. Gray and A. Dorris, *Angew. Chem., Int. Ed.*, 2011, **50**, 5438–5466.
2 K.-Y. Lee, Y. Aitomäki, L. A. Berglund, K. Oksman and A. Bismarck, *Compos. Sci. Technol.*, 2014, **105**, 15–27.
3 W. J. Lee, A. J. Clancy, E. Kontturi, A. Bismarck and M. S. P. Schaffer, *ACS Appl. Mater. Interfaces*, 2016, **8**, 31500–31504.
4 J. R. McKee, J. Huokuna, L. Martikainen, M. Karesoja, A. Nykänen, E. Kontturi, H. Tenhu, J. Ruokolainen and O. Ikkala, *Angew. Chem., Int. Ed.*, 2014, **53**, 5049–5053.
5 N. Lin and A. Dufresne, *Eur. Polym. J.*, 2014, **59**, 302–325.
6 J. A. Kelly, M. Giese, K. E. Shopsowitz, W. Y. Hamad and M. J. MacLachlan, *Acc. Chem. Res.*, 2014, **47**, 1088–1096.
7 J. R. Capadona, K. Shanmuganathan, D. J. Tyler, S. J. Rowan and C. Weder, *Science*, 2008, **319**, 1370–1374.
8 Y. Habibi, L. A. Lucia and O. J. Rojas, *Chem. Rev.*, 2010, **110**, 3479–3500.
9 S. J. Eichhorn, *Soft Matter*, 2011, **7**, 303–315.
10 Y. Nishiyama, U.-J. Kim, D.-Y. Kim, K. S. Katsumata, R. P. May and P. Langan, *Biomacromolecules*, 2003, **4**, 1013–1017.

11 O. A. Battista, S. Coppick, J. A. Howsmon, F. F. Morehead and W. A. Sisson, *Ind. Eng. Chem.*, 1956, **48**, 333–335.

12 B. Rånby, *Discuss. Faraday Soc.*, 1951, **11**, 158–164.

13 M. Pääkkö, M. Ankerfors, H. Kosonen, A. Nykänen, S. Ahola, M. Österberg, J. Ruokolainen, J. Laine, P. T. Larsson, O. Ikkala and T. Lindström, *Biomacromolecules*, 2007, **8**, 1934–1941.

14 J.-F. Revol, H. Bradford, J. Giasson, R. H. Marchessault and D. G. Gray, *Int. J. Biol. Macromol.*, 1992, **14**, 170–172.

15 X. M. Dong, J. F. Revol and D. G. Gray, *Cellulose*, 1998, **5**, 19–32.

16 Q. Wang, X. Zhao and J. Y. Zhu, *Ind. Eng. Chem. Res.*, 2014, **53**, 11007–11014.

17 L. Chen, Q. Wang, K. Hirth, C. Baez, U. P. Agarwal and J. Y. Zhu, *Cellulose*, 2015, **22**, 1753–1762.

18 W. Y. Hamad and T. Q. Hu, *Can. J. Chem. Eng.*, 2010, **88**, 392–402.

19 A. C. W. Leung, S. Hrapovic, E. Lam, Y. Liu, K. B. Male, K. A. Mahmoud and J. H. T. Luong, *Small*, 2011, **7**, 302–305.

20 J. Peyre, T. Pääkkönen, M. Reza and E. Kontturi, *Green Chem.*, 2015, **17**, 808–811.

21 M. Visanko, H. Liimatainen, J. A. Sirviö, J. P. Heiskanen, J. Niinimäki and O. Hormi, *Biomacromolecules*, 2014, **15**, 2769–2775.

22 H. Yang, D. Chen and T. G. van de Ven, *Cellulose*, 2015, **22**, 1743–1752.

23 L. Tang, B. Huang, N. Yang, T. Li, Q. Lu, W. Lin and X. Chen, *Green Chem.*, 2013, **15**, 2369–2373.

24 L. Chen, J. Y. Zhu, C. Baez, P. Kitin and T. Elder, *Green Chem.*, 2016, **18**, 3835–3843.

25 P. B. Filson, B. E. Dawson-Andoh and D. Schwegler-Berry, *Green Chem.*, 2009, **11**, 1808–1814.

26 P. Satyamurthy, P. Jain, R. H. Balasubramanya and N. Vigneshwaran, *Carbohydr. Polym.*, 2011, **83**, 122–129.

27 E. Kontturi, A. Meriluoto, P. A. Penttilä, N. Baccile, J.-M. Malho, A. Potthast, T. Rosenau, J. Ruokolainen, R. Serimaa, J. Laine and H. Sixta, *Angew. Chem., Int. Ed.*, 2016, **55**, 14455–14458.

28 T. Saito, S. Kimura, Y. Nishiyama and A. Isogai, *Biomacromolecules*, 2007, **8**, 2485–2491.

29 T. Saito and A. Isogai, *Biomacromolecules*, 2004, **5**, 1983–1989.

30 A. Isogai, T. Saito and H. Fukuzumi, *Nanoscale*, 2011, **3**, 71–85.

31 R. Hiraoki, Y. Ono, T. Saito and A. Isogai, *Biomacromolecules*, 2015, **16**, 675–681.

32 Y. Okita, T. Saito and A. Isogai, *Biomacromolecules*, 2010, **11**, 1696–1700.

33 S. Elazzouzi-Hafraoui, Y. Nishiyama, J.-L. Putaux, L. Heux, F. Dubreuil and C. Rochas, *Biomacromolecules*, 2008, **9**, 57–65.

34 O. van den Berg, J. R. Capadona and C. Weder, *Biomacromolecules*, 2007, **8**, 1353–1357.

35 L. Heux, G. Chauve and C. Bonini, *Langmuir*, 2000, **16**, 8210–8212.

36 S. Beck, J. Bouchard and R. Berry, *Biomacromolecules*, 2012, **13**, 1486–1494.

37 D. Viet, S. Beck-Candanedo and D. G. Gray, *Cellulose*, 2007, **14**, 109–113.

38 A. Ferguson, U. Khan, M. Walsh, K.-Y. Lee, A. Bismarck, M. S. Schaffer, J. N. Coleman and S. D. Bergin, *Biomacromolecules*, 2016, **17**, 1845–1853.

39 S. Eyley and W. Thielemans, *Nanoscale*, 2014, **6**, 7764–7779.

40 J. Araki, M. Wada and S. Kuga, *Langmuir*, 2001, **17**, 21–27.

41 E. Kontturi and T. Vuorinen, *Cellulose*, 2009, **16**, 65–74.
42 R. Salminen, M. Reza, T. Pääkkönen, J. Peyre and E. Kontturi, *Cellulose*, 2017, **24**, 1657–1667.
43 Y. Nishiyama, J. Sugiyama, H. Chanzy and P. Langan, *J. Am. Chem. Soc.*, 2003, **125**, 14300–14306.
44 A. N. Fernandes, L. H. Thomas, C. M. Altaner, P. Callow, V. T. Forsyth, D. C. Apperley, C. J. Kennedy and M. C. Jarvis, *Proc. Natl. Acad. Sci. U. S. A.*, 2011, **108**, E1195–E1203.
45 Y. Okita, S. Fujisawa, T. Saito and A. Isogai, *Biomacromolecules*, 2011, **12**, 518–522.
46 M. Shimizu, T. Saito and A. Isogai, *Biomacromolecules*, 2014, **15**, 1904–1909.
47 W. Fang, S. Arola, J.-M. Malho, E. Kontturi, M. B. Linder and P. Laaksonen, *Biomacromolecules*, 2016, **17**, 1458–1465.

Faraday Discussions

PAPER

Effect of pretreatment severity on the cellulose and lignin isolated from *Salix* using ionoSolv pretreatment†

Lisa Weigand,[a] Shahrokh Mostame,[a] Agnieszka Brandt-Talbot,[a] Tom Welton[b] and Jason P. Hallett *[a]

Received 7th February 2017, Accepted 10th March 2017
DOI: 10.1039/c7fd00059f

The ionoSolv pretreatment is a new technique employing protic low-cost ionic liquids and has previously been applied to successfully fractionate switchgrass and the grass *Miscanthus giganteus*. This study investigates the effect of using the protic ionic liquid solution $[N_{2220}][HSO_4]_{80\%}$ with two different acid/base ratios (1.02 and 0.98) at 120, 150 and 170 °C on the pretreatment outcome of the hardwood willow. The ionic liquid solution was able to fractionate willow, and a pulp and lignin fraction were recovered after treatment. The pretreatment success was determined *via* enzymatic hydrolysis of the pulp, which showed that the ionoSolv pretreatment was able to increase enzymatic glucose yields compared to untreated willow biomass. The pretreatment produced a cellulose-rich pulp with high hemicellulose and lignin removal. The pulp composition and glucose yield after saccharification were greatly influenced by the acidity of the ionic liquid solution, temperature and pretreatment time. The extracted lignin was analysed *via* 2-D HSQC NMR spectroscopy and GPC to investigate the changes in the lignin structure induced by the pretreatment severity. The lignin structure (in terms of inter-unit linkages and S/G ratio) and molecular weight varied significantly depending on the pretreatment conditions used.

Introduction

Using fossil fuels such as oil and coal for the production of electricity and chemicals generates CO_2, which is a contributor to climate change and global warming, thus necessitating the search for an alternative carbon source.[1]

Lignocellulosic biomass (composed of cellulose, hemicellulose and lignin) is the only large-scale sustainable source of carbon on the planet, which makes it a promising alternative to fossil resources for the production of fuels and

[a]*Department of Chemical Engineering, Imperial College London, SW7 2AZ London, UK. E-mail: j.hallett@imperial.ac.uk*

[b]*Department of Chemistry, Imperial College London, SW7 2AZ, London, UK*

† Electronic supplementary information (ESI) available. See DOI: 10.1039/c7fd00059f

chemicals.[1c,2] Three groups of lignocellulosic biomass are important to the bioeconomy: grasses (annual and perennial), hardwoods and softwoods, which display differences in lignin and hemicellulose composition, biopolymer content, cell wall structure as well as hemicellulose and lignin structure.[2]

The content of the biopolymers varies with feedstocks, but typically lignocellulose consists of 35–50 wt% cellulose, 20–25 wt% hemicellulose and 15–35 wt% lignin.[2a]

The lignin content and the lignins' chemical characteristics differ between grasses, softwood and hardwood, with softwoods having the highest lignin content and grasses having the lowest.[3] It is important to note that the differences in the lignin structure and properties greatly affect the delignification chemistry and play a major role in biomass deconstruction,[4] as it leads to differences in recalcitrance towards pretreatment, with grasses being the least resistant and softwoods the most resistant to pretreatment.[2a] Lignin has been identified as a major factor for biomass recalcitrance, hindering enzymatic saccharification of the cellulose to glucose either through non-productive binding with enzymes or through acting as a physical barrier and thus restricting enzyme accessibility.[5] One way to overcome this problem is to introduce a pretreatment step that either redistributes the lignin[6] or removes the lignin from the biomass.[2a]

Ionic liquids are very promising candidates for biomass pretreatment, as they are able to partially (fractionate)[2a] or completely dissolve (disentangle)[7] lignocellulose. The ionic liquid 1-ethyl-3-methylimidazolium acetate $[C_2C_1im][OAc]$ is a prominent candidate for biomass pretreatment and has been widely studied.[2a] However, commercialisation attempts have been prevented so far due to the high cost of many ionic liquids.[8] The inexpensive ionic liquid triethylammonium hydrogen sulfate $[N_{2220}][HSO_4]$ (estimated bulk price of \$1.24 per kg)[9a] was developed as a potentially commercially viable alternative.[9b] The ionic liquid $[N_{2220}][HSO_4]$, used with a minor amount of water, which is needed for pretreatment to occur, is a non-volatile organic solvent with high polarity and acidity.[10] The solvent acidity can be controlled by adjusting the acid/base ratio (a/b) during synthesis, adding an option to tune the pretreatment severity.

The ionoSolv pretreatment selectively extracts lignin and hemicellulose from the biomass and a cellulose-rich pulp and a lignin fraction are recovered.[11] The removal of hemicellulose and lignin leads to an increase in the glucose yield of enzymatic hydrolysis of the pretreated pulp compared to the untreated biomass.[11,12]

The protic ionic liquid solution $[N_{2220}][HSO_4]_{80\%}$ (80 wt% IL and 20 wt% water) with an acid/base ratio of 1 (a/b = 1) was used for the pretreatment of switchgrass[9b] and *Miscanthus giganteus*[12] in the past, but the pretreatment technology has not yet been applied to a hardwood such as willow. The pretreatment severity can be controlled either by the pretreatment time, temperature or acidity of the pretreatment solvent. As mentioned earlier, hardwoods generally show a higher recalcitrance to pretreatment than grasses.[2a] This typically necessitates the use of more severe pretreatment conditions, which can be achieved by using an a/b of greater than 1 (having excess acid present in the ionic liquid). However, little is known about the impact of the acidity of the ionic liquid solution $[N_{2220}][HSO_4]_{80\%}$ on the pretreatment of the hardwood willow.

Results and discussion

Synthesis of $[N_{2220}][HSO_4]$ and determination of the acid/base ratio

Two ionic liquid solutions with different acid/base ratios were created by carefully mixing the correct amount of the two starting materials, triethylamine and sulfuric acid.

To verify the acid/base ratio, the synthesized ionic liquid solutions with a water content of 20 wt% were characterized by pH and density measurements. The acid/base ratio of the synthesized ionic liquid was calculated according to the method developed by M. S. Y. Jennings,[13] using the following equation:

$$y = 0.0772x + 1.1135$$

where y is the density of the ionic liquid solution and x is the acid/base ratio. Table 1 shows exemplary pH values and densities for $[N_{2220}][HSO_4]$ solutions with different acid/base ratios. The difference in the acid/base ratio of the ionic liquid is clearly reflected by the change in pH and density, with the more acidic ionic liquid solution having a density of 1.1924 g mL^{-1} and a pH of 1.52 compared to 1.1889 g mL^{-1} and a pH of 1.65 for the less acidic one.

Pulp and lignin recovery after pretreatment with $[N_{2220}][HSO_4]$

In this study, the hardwood willow (variety Endurance) is used to expand the feedstock range for the ionoSolv pretreatment. To overcome the higher recalcitrance of the hardwood cell structure, the ionic liquid solution $[N_{2220}][HSO_4]$ with an excess of 2% of sulfuric acid (a/b = 1.02) was used. However, undesirable acid catalysed cellulose hydrolysis to glucose is well known,[14] which raises the concern that the high acidity of $[N_{2220}][HSO_4]$ with a/b = 1.02 will lead to cellulose degradation in the pulp, which in turn will decrease the glucose yield after enzymatic hydrolysis. To prevent this, pretreatment with $[N_{2220}][HSO_4]$ with a/b = 0.98 was also conducted.

Willow Endurance was pretreated using the following conditions: 0.5 h, 1 h, 2 h, 4 h and 8 h at 120 °C, 0.5 h, 1 h, 2 h and 4 h at 150 °C as well as 0.3 h, 0.5 h, 0.7 h and 1 h at 170 °C. Fig. 1 shows that the physical appearance of the isolated pulp changed significantly with the severity of the pretreatment, which gives a first indication of the success of the pretreatment. It can be seen that harsher pretreatment conditions resulted in a darker pulp, and an almost char-like product for very severe treatments. The pulp and lignin recovery after pretreatment gave a second indication of the pretreatment effectiveness. See Fig. 2 for an overview of the pulp and lignin recovery after pretreatment under various conditions.

Table 1 Examples of the results from pH measurements and density measurements for the synthesized $[N_{2220}][HSO_4]$

Entry	pH	Density [g mL^{-1}]	Acid/base ratio
1	1.52 ± 0.01	1.1924 ± 0.0001	1.02
2	1.65 ± 0.01	1.1889 ± 0.0002	0.98

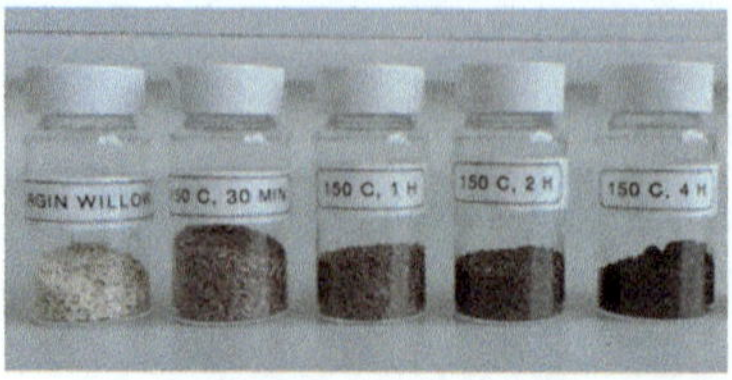

Fig. 1 Examples of the physical appearance of the pulp before and after pretreatment at 150 °C for 0.5 h, 1 h, 2 h and 4 h.

Fig. 2 shows that using $[N_{2220}][HSO_4]$ decreased the amount of biomass recovered after pretreatment compared to the starting material, consistent with the ability of $[N_{2220}][HSO_4]$ solutions to remove lignin and hemicellulose from the lignocellulose matrix as reported earlier by Gschwend *et al.*[12a] In general, the pretreatment conditions (time, temperature and acid/base ratio) dictated the amount of pulp and lignin that were recovered, with higher pretreatment severity leading to a lower pulp yield and a higher lignin yield upon precipitation. At 120 °C, an increase in pretreatment time steadily improved the removal of material from the pulp and resulted in a higher lignin recovery. It was found that using $[N_{2220}][HSO_4]$ with a/b = 1.02 overall resulted in a lower pulp yield and higher lignin yield compared to the same ionic liquid solution with a/b = 0.98 (see Fig. 2).

It was reported previously that operating at temperatures above the glass transition temperature of lignin (T_g = 130–150 °C depending on the source of lignin) might improve the delignification of biomass.[15] For more severe pretreatments at 150 °C and 170 °C, an increase in pretreatment effectiveness was discovered and more material was removed from the biomass in a shorter time compared to pretreatment at 120 °C. The most effective pretreatment condition in terms of lignocellulose fractionation was 1 hour at 170 °C ($[N_{2220}][HSO_4]$ with a/b = 1.02) with a pulp recovery of 40%. The highest lignin yield (relative to the biomass input) was 24.7 wt% when willow was pretreated for 40 min at 170 °C

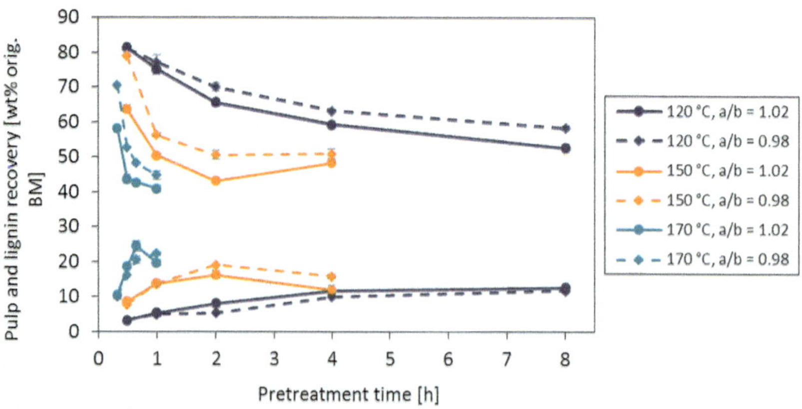

Fig. 2 Pulp and lignin recovery after pretreatment of willow with $[N_{2220}][HSO_4]$ with different acid/base ratios (a/b = 1.02 and 0.98).

($[N_{2220}][HSO_4]$ with a/b = 1.02). This indicates that lignin depolymerisation continued after extraction in the solution phase at high temperatures, increasing the amount of small water-soluble lignin fragments and hence reducing the precipitated lignin yield.

A different trend for pulp and lignin recovery observed for pretreatment at elevated temperatures compared to pretreatment at 120 °C. Initially, an increase of material removal from the biomass was found, but for longer pretreatment times (*e.g.* 4 hours at 150 °C and 1 hour at 170 °C) over-treatment of the pulp was discovered. Signs of overtreatment include the increase in pulp recovery and decrease in lignin recovery, indicating that lignin re-deposition on the pulp has occurred. This means that increasing the temperature did not only accelerate the lignin extraction but could have also increased the formation of pseudo-lignin on the pulp,[16] especially for extended pretreatment times (see section on compositional analysis for detailed discussion). At elevated temperatures, the effect of the acidity on the pretreatment performance was very prominent, with the more acidic ionic liquid solution removing more material from the biomass matrix.

The severity of the pretreatment in terms of acidity seemed to have an interesting effect on the lignin yield, as milder pretreatment conditions (*e.g.* all pretreatments at 120 °C and short pretreatment times at 150 °C and 170 °C) using $[N_{2220}][HSO_4]$ with a/b = 1.02 consistently gave a higher lignin recovery, but after 2 hours and 4 hours at 150 °C the lignin yield was found to be lower compared to using $[N_{2220}][HSO_4]$ with a/b = 0.98. This suggests that the lignin in the solution with a/b = 1.02 was more prone to condensation and re-deposition onto the pulp (rather than precipitating during the lignin isolation step) or that the lignin fragments were more water-soluble and not able to precipitate. To verify either hypothesis, compositional analysis of the pulp was performed as well as the chemical characterisation of the isolated lignin, which will be discussed later on in this publication.

Effect of $[N_{2220}][HSO_4]$ treatment on the pulp composition

Compositional analysis is a good tool for understanding the changes in biomass composition that occur during pretreatment, such as delignification and hemicellulose removal or glucan degradation. The glucan, hemicellulose and lignin contents of untreated willow and pretreated pulps are shown in Fig. 3. The composition of native willow Endurance was as follows: 42.8% glucan, 18.5% hemicellulose and 27.3% lignin (plus an additional 10.1% of extractives).

It was found that pretreatment of willow with $[N_{2220}][HSO_4]$ removed lignin and hemicellulose from the biomass matrix, leaving a cellulose-rich pulp. The effectiveness of the pretreatment and the composition of the pulp were strongly dependent on the acid/base ratio of the ionic liquid. In general, using $[N_{2220}][HSO_4]$ with a/b = 1.02 was better at removing lignin and hemicellulose from the biopolymer matrix (see Fig. 3). However, the composition of the pulp was also greatly influenced by the pretreatment temperature and time.

Pretreatment at 120 °C only partially extracted the lignin from the biomass. The highest delignification and hemicellulose removal were found for the 8 h pretreatment, which led to 44.9% and 56.6% lignin removal and 54.0% and 72.3% hemicellulose removal, using ionic liquid solutions with a/b = 0.98 and a/b = 1.02, respectively. Heteronuclear single quantum correlation (HSQC) NMR

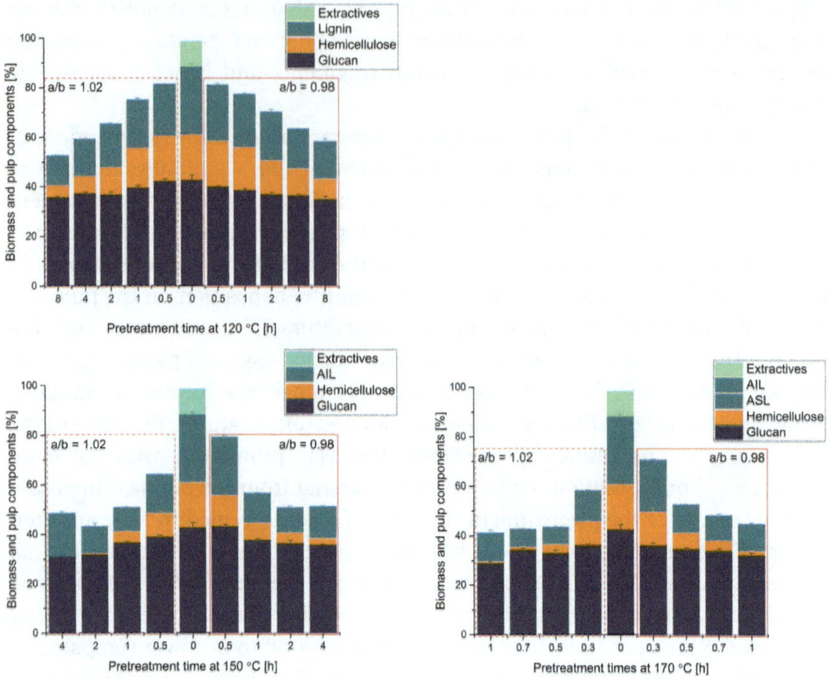

Fig. 3 Glucan, hemicellulose and lignin contents of untreated willow and pulp after pretreatment with $[N_{2220}][HSO_4]$ with two different acid/base ratios a/b = 1.02 (left hand side of graph) and a/b = 0.98 (right hand side of graph).

spectra of the isolated lignin show no carbohydrate cross-peaks, proving that hemicellulose remains in the ionic liquid liquor after pretreatment (see ESI† for HSQC spectra).

The mild pretreatment temperature (120 °C) was not the most effective in terms of delignification and hemicellulose removal but beneficial for preserving a high percentage of the glucan in the pulp, with more than 80% of the glucan remaining in the pulp after 8 hours of pretreatment, for both ionic liquid solutions used. It was also found that for mild pretreatment conditions (0.5 h, 1 h at 120 °C, 0.5 h at 150 °C and 20 min at 170 °C) around 10% of the original glucan was already extracted from the biomass, due to hydrolysis of amorphous cellulose and/or hemicellulose during the pretreatment.

Applying a more severe pretreatment by increasing the temperature to 150 °C significantly changed the composition of the pulp. The influence of the acid/base ratio on the pretreatment effectiveness and thus on the content of the pulp was marked. After only 30 min of treatment, 47.2% and 31.6% of hemicellulose were removed as well as 44.0% and 14.3% of the lignin, for $[N_{2220}][HSO_4]$ with a/b = 1.02 and a/b = 0.98, respectively. However, this significant difference in the delignification of the pulp was not reflected in the saccharification yields, with 55.9% and 44.1% glucose yields (for $[N_{2220}][HSO_4]$ with a/b = 1.02 and a/b = 0.98, respectively), which are similar. This discrepancy could be explained by the amount of glucan preserved in the pulp, namely only 91.4% for the more acidic ionic liquid compared to 99.2% for the less acidic ionic liquid (compared to the original amount of glucan).

Increasing the pretreatment time at 150 °C from 30 minutes up to 2 hours showed that the more acidic ionic liquid was better at delignifying the biomass and removing the hemicellulose. After treatment for 2 hours, 96.0% of hemicellulose and 57.8% of the lignin were removed from the pulp when $[N_{2220}][HSO_4]$ with a/b = 1.02 was used, whereas treatment with $[N_{2220}][HSO_4]$ with a/b = 0.98 extracted only 76.2% of the hemicellulose but 63.6% of the lignin. The larger amount of acid not only accelerated the lignin and hemicellulose extraction, but also sped up the formation of pseudo-lignin on the pulp. This was especially obvious for pretreatment for 4 hours at 150 °C, where the formation of pseudo-lignin was observed for pulp pretreated with the more acidic ionic liquid, whereas no formation of pseudo-lignin on the pulp was found for the ionic liquid with a lower a/b.

When comparing the delignification for the 2 hours pretreatment to the 1 hour pretreatment, it can be seen that the delignification decreased when $[N_{2220}][HSO_4]$ with a/b = 1.02 was used but increased when $[N_{2220}][HSO_4]$ with a/b = 0.98 was used, indicating that significant condensation of lignin inside the biomass matrix or lignin or humin re-deposition occurred under more acidic pretreatment conditions, with the glucan content in the pulp decreasing to levels as low as 73.1% of the original glucan in the biomass when using the more acidic ionic liquid solution compared to 83.9% for the less acidic ionic liquid solution.

The pulp isolated after pretreatment for 4 hours at 150 °C clearly showed signs of overtreatment and formation of pseudo-lignin (a higher lignin content compared to untreated biomass or pulp pretreated under milder conditions). These effects were more pronounced for the pulp pretreated with the more acidic ionic liquid, where all of the hemicellulose and 27.4% of the glucan were removed, and the delignification decreased even further from 57.8% (150 °C, 2 h) to 35.8%. The less acidic ionic liquid caused less glucan degradation and pseudo-lignin formation, only removing 16.8% of the glucan and a decrease of delignification from 63.6% (150 °C, 2 h) to 54.0%.

Pretreatment at 170 °C accelerated the lignin and hemicellulose extraction and yielded a pulp high in cellulose content after only 30 minutes (using $[N_{2220}][HSO_4]$ with a/b = 1.02). This pretreatment condition exhibited the highest delignification success, with only 23.9% of the lignin left in the pulp. Interestingly, this did not correlate with the largest amount of hemicellulose removal (found to be 96.0% after pretreatment for 2 hours at 150 °C), since only 79.1% was dissolved in the ionic liquid. This shows that, at higher temperatures, the hemicellulose and lignin extractions were not as concerted as seen for pretreatment at 120 °C but lignin removal was faster. The remaining hemicellulose may also have been hydrolysed directly from the pulp.

Elongating the pretreatment time at 170 °C to 1 hour affected the composition of the pulp significantly, which appeared to consist only of glucan, lignin/pseudo-lignin and a negligible amount of hemicellulose. However, glucan degradation was very prominent under these conditions with a glucan loss of 25% into the ionic liquid solution.

Enzymatic hydrolysis of the pulp and untreated biomass

Enzymatic hydrolysis of the pulp was performed in order to assess whether the pretreatment was successful in increasing glucose release compared to untreated biomass. It is generally believed that the lignin in the biomass/pulp hinders the cellulase enzymes from accessing the cellulose[5] and thus high delignification of the biomass is considered a key aspect of a successful fractionation pretreatment.[2a] The glucose yields after 7 days of enzymatic hydrolysis of the pulp are shown in Fig. 4, demonstrating the influence of the ionic liquid solution acidity and the deconstruction time on the glucose release.

In general, the enzymatic hydrolysis of willow Endurance resulted in a high glucose release even without a pretreatment step. The ionic liquid pretreatment of willow improved the enzymatic digestibility for certain conditions compared to the untreated biomass (see Fig. 4), but this was not universally so. Unsurprisingly, the glucose release was highly dependent on the pretreatment severity applied.

For very mild pretreatments, *e.g.* 0.5, 1 and 2 hours at 120 °C (for both a/b ratios), 0.5 hours at 150 °C (for a/b = 0.98) and 0.3 hours at 170 °C (for a/b = 0.98), the glucose yield was found to be lower than the yield from the raw biomass (46.4%).

This was also previously observed for the mild pretreatment of *Miscanthus giganteus*[12a] and might be due to acid catalysed hydrolysis of parts of the amorphous cellulose followed by dissolution into the ionic liquid. The amorphous part of cellulose is generally hydrolysed faster under acidic conditions than the crystalline part,[17] thus decreasing the glucan recovery for pulp treated with a low severity. The compositional analysis of the pulp revealed a glucan loss of around 10% even for very mild pretreatment conditions, supporting the above statement.

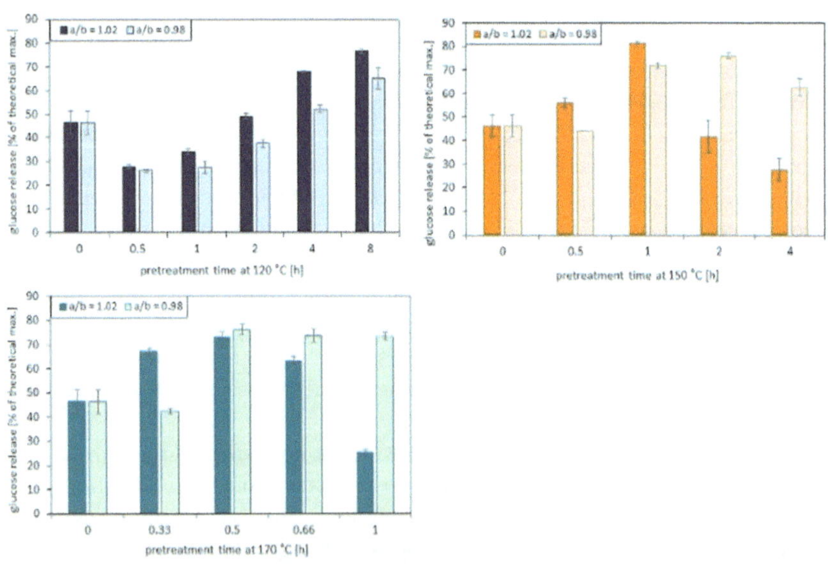

Fig. 4 Glucose yield after enzymatic hydrolysis of the pulp and raw biomass after pretreatment using $[N_{2220}][HSO_4]$ with two different acid/base ratios at 120 °C (top right), 150 °C (top left) and 170 °C (bottom left). BM = biomass.

It was found that the glucose yield after saccharification of the pulp pretreated with the more acidic solution at 120 °C was always higher compared to $[N_{2220}][HSO_4]$ with a/b = 0.98, with the highest yield being 76.5% for a pretreatment time of 8 hours. These results are in contrast to earlier findings reported by P. Verdía *et al.*[18] for the pretreatment of *Miscanthus giganteus* at 120 °C with $[C_4C_1im][HSO_4]$. In this study, the less acidic ionic liquid solution gave slightly higher glucose yields (90% and 84% for a/b = 0.99 and 1.01, respectively) after 96 hours of enzymatic hydrolysis. The contrary conclusions support the hypothesis that hardwoods are more recalcitrant towards pretreatment and harsher conditions are required for the successful break-up of the biopolymer matrix. The higher glucose yields for willow pretreated with $[N_{2220}][HSO_4]$ (a/b = 1.02) can be ascribed to the fact that the ionic liquid with an excess of sulfuric acid yields a purer pulp by removing more lignin and hemicellulose from the pulp and thus increasing the accessibility of the enzymes to the cellulose.[19]

Additionally, lignin acts as a non-active binding side for cellulases,[5] which decreases the efficiency of the enzymatic hydrolysis and explains why a pulp with higher delignification gave higher glucose yields.

The best pretreatment conditions found (in terms of highest glucose yield) were 1 hour at 150 °C with $[N_{2220}][HSO_4]$ with an a/b of 1.02, which resulted in the release of 81.8% of the glucose that was contained in the untreated willow. This shows that the ionic liquid solution $[N_{2220}][HSO_4]$ can be used to successfully pretreat the hardwood willow to improve glucose yields.

However, using very severe pretreatment conditions, such as long pretreatment times at 150 °C and 170 °C, led to overtreatment of the pulp, which dramatically decreased the enzymatic digestibility. Interestingly, using $[N_{2220}][HSO_4]$ with an a/b of 0.98 produced high glucose yields even under harsh pretreatment conditions (2 and 4 hours at 150 °C as well as 1 hour at 170 °C) compared to the same ionic liquid solution with an a/b of 1.02. For example, at 150 °C and 2 hours, a glucose yield of 76.2% (a/b = 0.98) *vs.* 41.9% (a/b = 1.02) was achieved. At 150 °C and 4 hours, the glucose yield was found to be 62.9% (a/b = 0.98) *vs.* 27.0% (a/b = 1.02).

The most significant effect of the acid/base ratio was observed for pretreatment for 1 hour at 170 °C, where the glucose yield was 73.5% (for a/b = 0.98) compared to 25.3% (for a/b = 1.02). This result is in good agreement with the glucan content of the pulps, showing that the less acidic ionic liquid reduces the propensity for severe glucan degradation.

Lignin structure

Lignin linkages in the aliphatic region. It has been shown that isolating lignin from the biomass matrix using acidic conditions, such as in the dilute acid pretreatment[20] or ionoSolv pretreatment, results in the cleavage of aryl ether C–O bonds and decreases the lignin molecular weight.[21] Previous work investigating changes in the structure of *Miscanthus gigantheus* lignin published by A. Brandt *et al.*[21] showed that pretreatment with the acidic ionic liquid 3-butylimidazolium hydrogen sulfate $[HC_4im][HSO_4]$ had a major impact.

Heteronuclear single quantum correlation (HSQC) NMR spectroscopy is a powerful two-dimensional NMR technique that resolves resonances that overlap in one-dimensional spectra and is used for semi-quantitative analysis of lignin

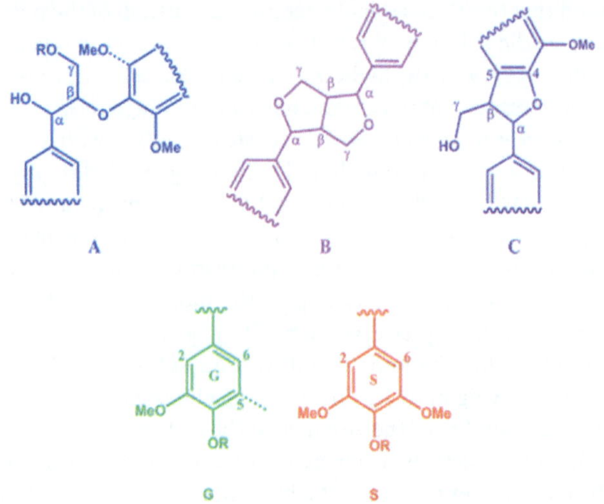

Fig. 5 Typical structural motifs found in hardwood lignin: β-*O*-4 alkyl-aryl ether (A), β-β (resinol) (B), phenylcoumaran (C), guaiacyl unit (G) and syringyl unit (S).

linkages and subunit composition. A selection of common linkages and subunits of hardwoods is shown in Fig. 5.

Several structural units of the complex lignin polymer can be detected by HSQC NMR spectroscopy, which have previously been assigned by comparison with cell wall model compounds in an NMR database.[22] Examples of common subunits and linkages that are resolved by HSQC NMR spectroscopy are β-ether (β-*O*-4, Aα at 4.87/71.8 ppm), β-β (resinol, Bα at 4.62/85.0 ppm) and phenylcoumaran (β-5, Cα at 5.45/86.9 ppm). The aromatic rings of the G (G_2 at 6.92/110.2 ppm, G_5 at 6.78/114.9 ppm and G_6 at 6.75/118.4 ppm) and S units ($S_{2,6}$ at 6.64/103.6 ppm) are also resolved. For calculation of the amount of inter-unit linkages, the α protons of β-*O*-4, β-β and phenylcoumaran were used.

HSQC NMR analysis of the lignin isolated from willow using $[N_{2220}][HSO_4]$ showed that the lignin structure (in terms of lignin linkages) was surprisingly similar for both ionic liquid solutions under mild (all time points at 120 °C and 30 min at 150 °C) or very harsh (1 hour at 170 °C) pretreatment conditions. The acidity of the ionic liquid solution seemed to have little impact on the amount of linkages present in the recovered lignin (see Fig. 6). The most prominent linkage in lignin isolated at 120 °C was the β-*O*-4 aryl ether linkage, as also described earlier by Prado *et al.*[23] It was found that the amount of the β-*O*-4 linkage in the isolated lignin strongly depended on the pretreatment severity applied. The largest amount of β-*O*-4 bonds was found for pretreatment at 120 °C for 30 min (0.66 bonds per C_9 unit for both ionic liquid solutions).

With longer pretreatment times, more aryl ether linkages had been broken and the amount of β-*O*-4 bonds dropped to 0.30 and 0.39 bonds (a/b = 1.02 *vs.* a/b = 0.98) per C_9 subunit after 8 hours. The lignin isolated at elevated temperatures contained a smaller amount of β-*O*-4 linkages compared to lignin isolated at 120 °C (*e.g.* at 30 min, 0.35–0.41 β-*O*-4 bonds per C_9 unit for lignin isolated after 150 °C and 170 °C pretreatments compared to 0.66 β-*O*-4 bonds for lignin isolated

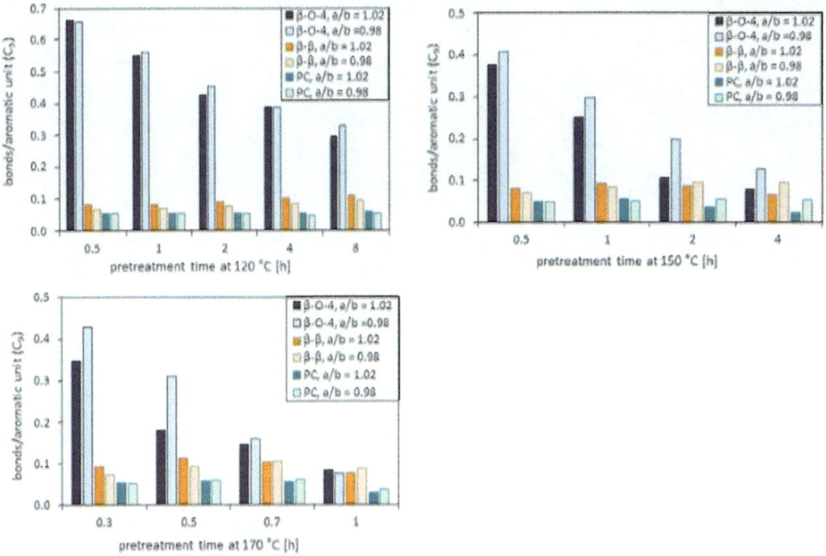

Fig. 6 Typical linkages (β-*O*-4 ether linkage, β-β (resinol) linkage and phenylcoumaran (PC) linkage) in lignin isolated after pretreatment at different temperatures with $[N_{2220}]$ $[HSO_4]$ with different acid/base ratios (a/b = 1.02 and 0.98). Top left: 120 °C, top right: 150 °C and bottom left: 170 °C.

after 120 °C pretreatment), meaning that more bonds were hydrolysed at higher pretreatment temperatures.

However, the same trend for the β-*O*-4 bond cleavage was found for lignin isolated at elevated pretreatment temperatures. Interestingly, the more acidic ionic liquid was more effective at breaking the ether bond than the less acidic solution, the phenomenon being more pronounced for pretreatments with higher severity, such as 150 °C and 170 °C. Lignin recovered after pretreatment at 150 °C for 2 hours showed a significant difference in its structure with twice as many β-*O*-4 bonds per C_9 (0.20 *vs.* 0.10) when the less acidic ionic liquid was used. A similar ratio was found for lignin isolated after 30 minutes at 170 °C, with 0.18 β-*O*-4 bonds per C_9 (a/b = 1.02) compared to 0.31 β-*O*-4 bonds per C_9 (for a/b = 0.98).

The abundance of β-β and phenylcoumaran linkages was less than 0.10 per C_9 unit for all the lignins, and the number of these linkages was largely unaffected by the pretreatment severity, with only a minor decrease in the number of β-β and phenylcoumaran linkages for very severe pretreatment conditions.

Lignin subunits in the aromatic region. The HSQC pulse sequence employed in this study is optimised for resolution and signal strength. It is important to note that this method is not a completely quantitative one, as signal relaxation following each pulse will not be complete for some correlations, especially for end groups such as PCA, which relax more slowly than the bulk.[24] Another limitation of integrating HSQC NMR spectra is the potential for the condensation reactions,[21] which replace aromatic C–H bonds with C–C bonds, particularly at the G_6 position. These C–C bonded aromatic positions do not produce cross correlations in HSQC spectra, meaning that the calculated subunit composition may be misrepresented when examining condensed lignins. With these restrictions in

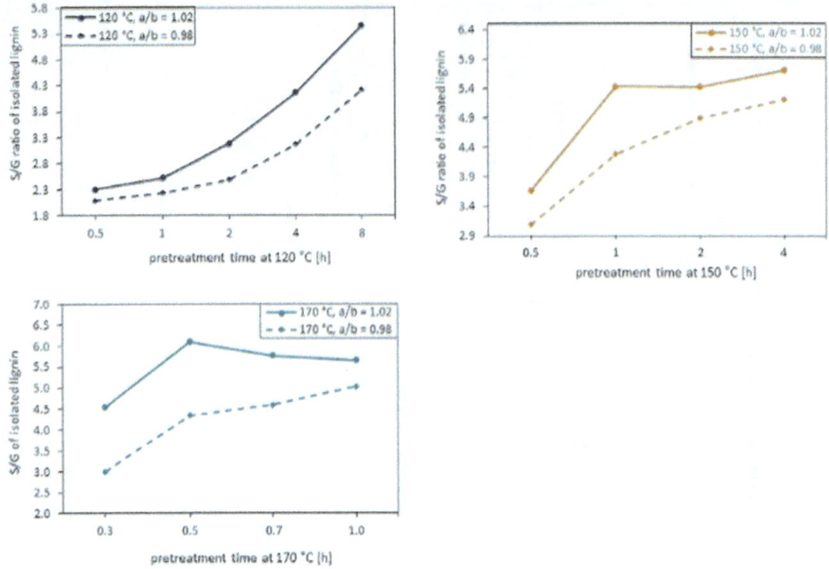

Fig. 7 S/G ratio of lignin isolated after pretreatment at different temperatures with $[N_{2220}]$ $[HSO_4]$ with different acid/base ratios (a/b = 1.02 and 0.98). Top left: 120 °C, top right: 150 °C and bottom left: 170 °C.

mind, volume integration was attempted for estimating the subunit composition (see Fig. 7). It should be noted that the G_5 and G_6 peaks cannot be relied on for quantification of the guaiacyl content because of pronounced participation in condensation reactions. Due to these reasons, the G_2 signal as well as the $S_{2,6}$ signal were selected to calculate the S/G ratio.

It was found that the S/G ratio of lignin extracted during pretreatment at 120 °C continually increased from 2.3 to 5.4 (for a/b = 1.02) and from 2.1 to 4.2 (for a/b = 0.98) with longer pretreatment times (see Fig. 7). This is in agreement with the S/G ratios found for another hardwood (poplar) after hydrothermal pretreatment with different severities.[25] As mentioned earlier, the acidity of the ionic liquid solution did not affect the composition of the lignin in terms of linkages at this temperature, but it did affect the S/G ratio. Lignin isolated with the more acidic ionic liquid solution showed a higher S/G ratio, meaning that a greater amount of S subunits was extracted from the cell wall compared to the lignin isolated with the ionic liquid with a/b = 0.98.

The S/G ratio for the lignin extracted after pretreatment at 150 °C was strongly influenced by the acidity of the ionic liquid solution. The subunit ratio for lignin isolated with the more acidic ionic liquid solution initially significantly increased from 3.7 to 5.4 (for the pretreatment times of 0.5 h and 1 h, respectively) and then reached a plateau around 5.5 for longer pretreatment times. This indicates that the pretreatment for 1 hour at 150 °C (with a/b = 1.02) extracted all the available S-lignin from the cell walls and longer pretreatment times did not change the subunit composition. Interestingly, a different trend was observed for pretreatment with the less acidic ionic liquid, where the S/G ratio gradually increased from 3.1 to 5.2, suggesting that the lower acidity of this ionic liquid solution affected the kinetics of the extraction of S-units.

Increasing the pretreatment severity by increasing the temperature to 170 °C showed the same trend for the S/G ratio as that observed for milder pretreatment conditions (when the less acidic ionic liquid was used). The subunit composition of the lignin isolated for pretreatment at 170 °C first increased from 4.5 to 6.1, reaching the highest S/G ratio observed for the lignins isolated for all the conditions studied. Interestingly, the very harsh pretreatment conditions (high temperature and high acidity of the ionic liquid solution) led to a decrease in the S/G ratio from 6.1 (for 30 min) to 5.8 (for 40 min) and 5.7 (for 1 h) of the isolated lignin for pretreatment times of more than 30 minutes. This implies that the extracted lignin got enriched in G-units over time. The data presented here might indicate that S-rich lignin may be extracted under milder reaction conditions, due to it being less cross-linked and thus more easily extracted.[26] Very harsh pretreatment conditions, such as 40 minutes and 1 hour at 170 °C using the IL solution with higher acidity, also extracted the G-rich part of the lignin and thus led to a decrease in the S/G ratio.

Molecular weight of the isolated lignin. The extracted lignin was investigated *via* GPC analysis in order to understand the effect of the pretreatment severity on the molecular weight (M_w) of the lignin. A summary of the molecular weights of the lignin using different pretreatment conditions is shown in Fig. 8 and Table 2.

The average molecular weight of the lignin changed dramatically with the pretreatment conditions applied. It was found that not only did the pretreatment time and temperature have an influence on the molecular weight, but also the acid/base ratio of the ionic liquid solution.

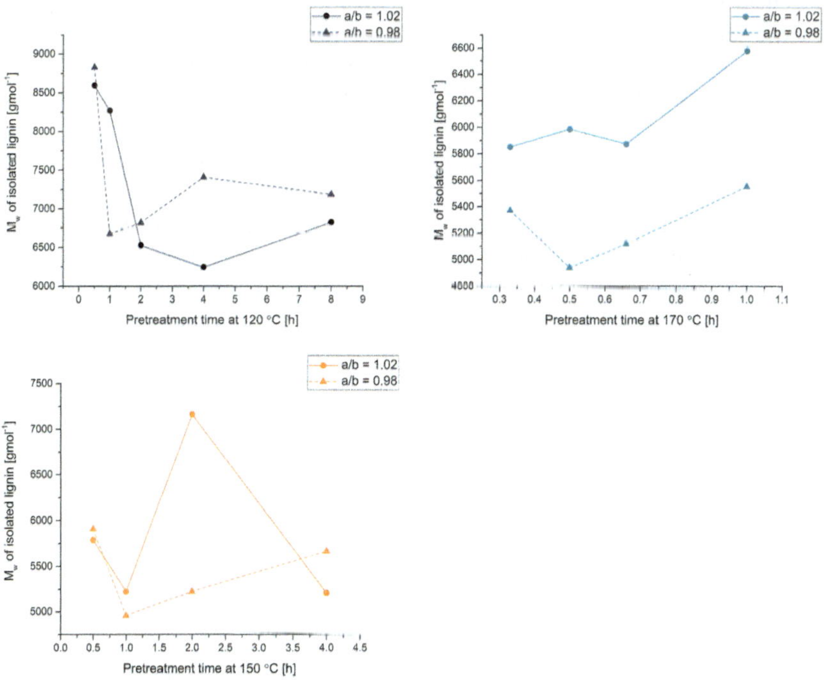

Fig. 8 Molecular weight (M_w) of the lignin isolated after pretreatment with $[N_{2220}][HSO_4]$ with different acid/base ratios.

Table 2 Polydispersity index of lignin isolated after pretreatment using $[N_{2220}][HSO_4]$ with a/b = 1.02 and a/b = 0.98

Entry	Temperature [°C]	Time [h]	PDI (a/b = 1.02)	PDI (a/b = 0.98)
1	120	0.5	7.9	7.6
2	120	1	7.5	5.5
3	120	2	5.8	5.0
4	120	4	5.8	5.4
5	120	8	5.9	5.7
6	150	0.5	5.7	4.9
7	150	1	5.6	4.6
8	150	2	6.9	5.2
9	150	4	5.8	5.9
10	170	0.3	4.9	4.6
11	170	0.5	5.7	5.0
12	170	0.7	5.8	5.3
13	170	1	7.0	5.6

The M_w of lignin isolated after pretreatment at 120 °C decreased with increasing pretreatment time from *ca.* 8500 g mol^{-1} for a pretreatment time of 30 min to *ca.* 6200 g mol^{-1} for a pretreatment time of 2 hours. This was caused by lignin depolymerisation through cleavage of β-*O*-4 bonds, as discussed in the previous section, which is likely catalysed by the acidic conditions in the ionic liquid solution.[21,23] However, no significant change in molecular weight was found after 2 hours of pretreatment and a plateau at around 6200 g mol^{-1} was reached, indicating that for longer pretreatment times at 120 °C both depolymerisation reactions and cross-linking of the lignin were equally favoured. In addition, the influence of the acidity of the ionic liquid solution appeared of little importance, which is in contrast to its effect on pulp yields, saccharification yields and the S/G ratio of the isolated lignin.

For pretreatment at higher temperatures, a different trend for the molecular weight was observed. At 150 °C and using the less acidic ionic liquid, the molecular weight first decreased from *ca.* 6000 g mol^{-1} after 30 min to *ca.* 5000 g mol^{-1} for lignin isolated after 1 hour, clearly indicating depolymerisation of the lignin. However, a slight increase in molecular weight to 5200 g mol^{-1} and 5700 g mol^{-1} was observed for longer pretreatment times (2 and 4 hours, respectively), indicating lignin condensation reactions were favoured, as suggested earlier by Brandt *et al.*[21] The acidity of the ionic liquid solution greatly impacted the molecular weight of the lignin isolated at this temperature. Using $[N_{2220}][HSO_4]$ with a/b = 1.02 for a pretreatment time of 30 minutes and 1 hour initially led to a decrease in M_w of the isolated lignin from *ca.* 5800 g mol^{-1} to 5200 g mol^{-1}, respectively, showing that depolymerisation also occurred. However, this was followed by a significant increase in M_w to *ca.* 7200 g mol^{-1} (for a pretreatment time of 2 hours), showing that lignin cross-condensation had occurred, as seen for the pretreatment at the same temperature with the less acidic ionic liquid solution. However, the molecular weight of lignin isolated after pretreatment with a/b = 1.02 increased by almost 1.5 times compared to only a slight increase for a/b = 0.98, indicating that the greater acidity of this ionic liquid solution supported the cross-condensation of the lignin. Interestingly, the conditions applied for the pretreatment for 4 hours led to a decrease in lignin molecular weight to *ca.* 5200 g mol^{-1}, which might indicate that under these conditions cleavage reactions of the lignin were favoured over cross-condensation due to the high severity of this treatment. This is supported by the HSQC NMR data showing that the amount of β-β and phenylcoumaran bonds decreased for lignin isolated under these conditions compared to shorter pretreatment times at the same temperature.

The general trend for the molecular weight of the lignin extracted after pretreatment at 170 °C showed an increase in M_w over time for both ionic liquid solutions used, indicating that the occurrence of lignin cross-condensation was favoured over lignin depolymerisation at this pretreatment temperature. However, the lignin isolated after pretreatment with the ionic liquid solution with a/b = 0.98 had a smaller molecular weight compared to the lignin extracted with $[N_{2220}][HSO_4]$ with a/b = 1.02 at 170 °C, showing that the lower acidity conditions extracted shorter lignin chains. The molecular weight increased from *ca.* 5400 g mol^{-1} and *ca.* 5900 g mol^{-1} (for a/b = 0.98 and 1.02, respectively) for lignin isolated after 30 min of pretreatment to *ca.* 5600 g mol^{-1} and 6600 g mol^{-1} (for a/b = 0.98 and 1.02, respectively) after 1 hour of pretreatment.

Experimental

Materials

Willow chips (variety: Endurance, Plot: CS689) were supplied by Rothamsted Research and the material was air-dried, ground and sieved (180–850 μm, −20 to +80 of US mesh scale) before use at Imperial College London. Triethylamine was purchased from Sigma Aldrich with a purity of >99%, 5 mM sulfuric acid and absolute ethanol were purchased from VWR. All commercial reagents and solvents were used as received unless otherwise stated.

Synthesis and characterisation of $[N_{2220}][HSO_4]$ with different acid/base ratios

Triethylamine (126.49 g, 1.25 mol) was cooled in an ice bath and sulfuric acid (256 mL, 5 mol L^{-1}, 1.28 mol for a/b = 1.02 or 246 mL, 5 mol L^{-1}, 1.23 mol for a/b = 0.98) was added drop-wise over 1 hour. The reaction mixture was stirred for 3 hours at 0 °C. Residual water was removed under vacuum from the colourless liquid until a water content of 20 wt% was achieved. The pH and density of the synthesised ionic liquid/water mixtures (containing 20 wt% water) were determined using a Jenway 3510 pH meter and a Metler Toledo DM40 density meter at room temperature. All measurements were performed in triplicates and the average pH and density values are given.

Ionic liquid pretreatment and biopolymer isolation

The biomass was pretreated following the ionoSolv protocol developed by Brandt *et al.*[12a] In brief, 2.00 g (on an oven-dried weight basis) of ground willow were weighed into a 15 mL Ace pressure tube with a screw-cap and Teflon lining, followed by the addition of 10.00 g of $[N_{2220}][HSO_4]_{80\%}$. The samples were incubated without stirring in an oven at different temperatures (120 °C, 150 °C and 170 °C) for various lengths of time. After pretreatment, the samples were cooled at room temperature and washed four times with absolute ethanol. The pulp was left to air-dry for two days and the ethanol was removed under reduced pressure and lignin was precipitated using water as the anti-solvent. The lignin was washed three times with water and dried in a vacuum oven at 40 °C over night.

Enzymatic hydrolysis of untreated biomass and pulp

The enzymatic saccharification of raw willow and pulp was performed according to the protocol "Enzymatic Saccharification of Lignocellulosic Biomass" (NREL/TP-510-42629) published by NREL[27] (National Renewable Energy Laboratory). In brief, 100 mg of untreated or pretreated biomass (oven-dried weight) were mixed with an enzyme master mix consisting of sodium citrate buffer (at pH 4.8), water, antibiotics and a commercial enzyme mixture (*Trichoderma reesei* ATCC 26921 and Cellobiase from *Aspergillus niger*) received from Novozyme. The mixture was incubated for 7 days at 50 °C under shaking at 250 rpm. The samples were analysed for monomeric sugars (glucose, xylose, arabinose, mannose and galactose) using a Shimadzu HPLC with a refractive index detector equipped with an Aminex HPX-87P column (Biorad). The mobile phase was DI water (18.8 mΩ), the flow rate was 0.6 mL min^{-1}, the column oven was set to 85 °C and the acquisition time was 20 min. Monomeric sugar yields were calculated based on the glucose content of the untreated biomass.

Compositional analysis of untreated willow and pulp

For untreated biomass, determination of extractives was included in the protocol, whereas this step was not performed for pretreated pulp due to the assumption that all extractives will be dissolved in the ionic liquid during pretreatment. For determination of the composition of biomass/pulp, the protocols "Preparation of samples for compositional analysis" (NREL/TP-510-42620), "Determination of Structural Carbohydrates and Lignin in Biomass" (NREL/TP-510-42618) and "Extractives in biomass" (NREL/TP-510-42619) published by NREL (National Renewable Energy Laboratory)[27] were followed and all experiments were performed in triplicates. In brief, 300 mg (oven dried weight) of extracted biomass or pulp were used per experiment and analysed for monomeric sugars (glucose, xylose, arabinose, mannose and galactose) using a Shimadzu HPLC with a refractive index detector equipped with an Aminex HPX-87P column (Biorad). The mobile phase was DI water (18.8 mΩ), the flow rate was 0.6 mL min^{-1}, the column oven was set to 85 °C and the acquisition time was 20 min.

NMR spectroscopy

For the NMR measurements, 20–23 mg of air dried lignin were dissolved overnight in 250 μL DMSO-d6, then transferred to a Shigemi NMR microtube. Heteronuclear single quantum correlation (HSQC) measurements were carried out on a Bruker Avance 600 MHz NMR machine at 25 °C. The acquired spectra were analysed using MestReNova and referenced to the solvent peak at 2.50/39.5 ppm (δ_H/δ_C). The S/G ratio was calculated using the signals for uncondensed and condensed S and G subunits.

Gel permeation chromatography

For the determination of the molecular weight (M_w) and polydispersity (PDI) of the isolated lignin, 20 mg of lignin were dissolved in DMF/LiBr (LiBr concentration = 1 mg mL^{-1}) and filtered using a 0.22 μm syringe filter and plastic syringe. The measurements were performed on an Agilent Technologies 1260 Infinity GPC instrument.

Conclusions

The study presented here shows that the protic ionic liquid $[N_{2220}][HSO_4]$ can be used to successfully extract lignin and hemicellulose from the hardwood willow. The amount of lignin and hemicellulose extracted were controlled by the pretreatment conditions (time, temperature and acidity of the ionic liquid) and high delignification and hemicellulose removal were achieved for pretreatments at 150 °C and 170 °C.

The pretreatment increased the digestibility of the cellulose-containing pulp during enzymatic hydrolysis. The highest glucose yield of 81.8% was found after pretreatment at 150 °C for 1 hour. However, overtreatment of the pulp (evidenced by the formation and re-deposition of pseudo-lignin on the pulp) was observed for very severe pretreatment conditions, resulting in low saccharification yields. We showed that the pretreatment severity did not only influence the pulp composition and digestibility but also the structure of the extracted lignin and its molecular weight. Lignin isolated after a mild pretreatment contained a large amount of β-*O*-4 ether linkages and a high molecular weight. More severe pretreatments changed the structure of the isolated lignin significantly, with most of the β-*O*-4 ether linkages being cleaved. Depolymerisation and re-condensation (for even more severe pretreatments) took place within the lignin, which led to a decrease in the molecular weight, followed by an increase. The S/G ratio of the isolated lignin was always higher when the more acidic IL solution was used for pretreatment and generally increased over time. An exception was observed at the highest pretreatment temperature (170 °C) using the IL with excess acid, when the S/G ratio decreased slightly at elongated treatment times (30 min or more).

A major factor that influences the price of glucose produced from lignocellulose is the cost of the pretreatment solvent, *i.e.* the ionic liquid solution used. The use of ionic liquids has hence to be weighed against the increase in glucose released from the perspective of the total carbon footprint and process costs before a full justification can be made. The cost of glucose and the environmental impact of the pretreatment can be lowered by near-quantitative recycling of the ionic liquid and reusing it for several pretreatment cycles. Further investigations into this are currently under investigation in our group.

This study demonstrated that ionoSolv pretreatment of willow has the potential to tune the properties of both fractionation products (pulp and lignin) to achieve optimised production of chemicals, materials and biofuels from willow. However, scale-up experiments are required, which are being performed in our group, to explore if the findings reported here also apply for larger scales.

Acknowledgements

Lisa Weigand gratefully acknowledges Climate-KIC for providing a scholarship and the UK Engineering and Physical Sciences Research Council for additional support.

Notes and references

1 (*a*) A. J. Ragauskas, C. K. Williams, B. H. Davison, G. Britovsek, J. Cairney, C. A. Eckert, W. J. Frederick, J. P. Hallett, D. J. Leak, C. L. Liotta, J. R. Mielenz, R. Murphy, R. Templer and T. Tschaplinski, *Science*, 2006, **311**, 484; (*b*) J. H. Clark, F. E. I. Deswarte and T. J. Farmer, *Biofuels, Bioprod. Biorefin.*, 2009, **3**, 72; (*c*) J. Pickett, *Sustainable Biofuels: Prospects and Challenge*, The Royal Society, 2008.

2 (*a*) A. Brandt, J. Gräsvik, J. P. Hallett and T. Welton, *Green Chem.*, 2013, **15**, 550; (*b*) R. D. Perlack, L. L. Wright, A. F. Turhollow, R. L. Graham, B. J. Stokes and D. C. Erbach, *Biomass as a Feedstock for a Bioenergy and Bioproducts Industry: the Technical Feasibility of a Billion-ton Annual Supply*, U.S. Department of Energy and U.S. Department of Agriculture, 2005.

3 (*a*) D. V. Evtuguin, C. Pascoal Neto, J. Rocha and J. D. Pedrosa de Jesus, *Appl. Catal., A*, 1998, **167**, 123; (*b*) X. Zhao, L. Zhang and D. Liu, *Biofuels, Bioprod. Biorefin.*, 2012, **6**, 465.

4 F. S. Chakar and A. J. Ragauskas, *Ind. Crops Prod.*, 2004, **20**, 131.

5 W. Boerjan, J. Ralph and M. Baucher, *Annu. Rev. Plant Biol.*, 2003, **54**, 519.

6 M. Li, Y. Pu and A. J. Ragauskas, *Front. Chem.*, 2016, **4**, 45.

7 T. Dutta, J. Shi, J. Sun, X. Zhang, G. Cheng, B. A. Simmons and S. Singh, Ionic Liquid Pretreatment of Lignocellulosic Biomass for Biofuels and Chemicals, in *Ionic Liquids in the Biorefinery Concept: Challenges and Perspectives*, 2016, pp. 65–94.

8 D. Klein-Marcuschamer, B. A. Simmons and H. W. Blanch, *Biofuels, Bioprod. Biorefin.*, 2011, **5**, 562.

9 (*a*) L. Chen, M. Sharifzadeh, N. Mac Dowell, T. Welton, N. Shah and J. P. Hallett, *Green Chem.*, 2014, **16**, 3098; (*b*) A. George, *et al.*, *Green Chem.*, 2015, **17**, 1728.

10 G. F. De Gregorio, C. C. Weber, J. Gräsvik, T. Welton, A. Brandt and J. P. Hallett, *Green Chem.*, 2016, **18**, 5456.

11 A. Brandt, M. J. Ray, T. Q. To, D. J. Leak, R. J. Murphy and T. Welton, *Green Chem.*, 2011, **13**, 2489.

12 (*a*) F. J. V. Gschwend, A. Brandt, C. L. Chambon, W.-C. Tu, L. Weigand and J. P. Hallett, *J. Visualized Exp.*, 2016, **114**, e54246, DOI: 10.3791/54246; (*b*) A. Brandt, J. Hallett and T. Welton, Ammonium cation ionic liquids for biomass fractionation, WO 2014/140643, 2013.

13 M. S. Y. Jennings, *New Characterization Techniques for Acidic Ionic Liquids*, Bachelor Thesis at Imperial College London, 2014.

14 A. M. Orozco, A. H. Al-Muhtaseb, A. B. Albadarin, D. Rooney, G. M. Walker and M. N. M. Ahmad, *RSC Adv.*, 2011, **1**, 839.

15 (*a*) N. J. Nagle, R. T. Elander, M. M. Newman, B. T. Rohrback, R. O. Ruiz and R. W. Torget, *Biotechnol. Prog.*, 2002, **18**, 734; (*b*) E. Kiran and H. Balkan, *J. Supercrit. Fluids*, 1994, **7**, 75.

16 P. Sannigrahi, D. H. Kim, S. Jung and A. J. Ragauskas, *Energy Environ. Sci.*, 2011, **4**, 1306.

17 (*a*) Y. H. P. Zhang and L. R. Lynd, *Biotechnol. Bioeng.*, 2004, **88**, 797; (*b*) S. D. Mansfield, C. Mooney and J. N. Saddler, *Biotechnol. Prog.*, 1999, **15**, 804.

18 P. Verdía, A. Brandt, J. P. Hallett, M. J. Ray and T. Welton, *Green Chem.*, 2014, **16**, 1617.

19 (*a*) X. Meng, M. Foston, J. Leisen, J. DeMartini, C. E. Wyman and A. J. Ragauskas, *Bioresour. Technol.*, 2013, **144**, 467; (*b*) T. Jeoh, C. I. Ishizawa, M. F. Davis, M. E. Himmel, W. S. Adney and D. K. Johnson, *Biotechnol. Bioeng.*, 2007, **98**, 112.
20 M. G. S. Chua and M. Wayman, *Can. J. Chem.*, 1979, **57**, 1141.
21 A. Brandt, L. Chen, B. E. van Dongen, T. Welton and J. P. Hallett, *Green Chem.*, 2015, **17**, 5019.
22 S. Ralph, J. Ralph, L. Landucci and L. Landucci, *NMR database of lignin and cell wall model compounds*, US Forest Prod. Lab, Madison, WI, 2004, http://ars.usda.gov/Services/docs.html.
23 R. Prado, X. Erdocia, G. F. DeGregorio, J. Labidi and T. Welton, *ACS Sustainable Chem. Eng.*, 2016, **4**, 5277.
24 S. D. Mansfield, H. Kim, F. Lu and J. Ralph, *Nat. Protoc.*, 2012, **7**, 1579.
25 S. Sun, Y. Huang, R. Sun and M. Tu, *Green Chem.*, 2016, **18**, 4276.
26 (*a*) A. Lourenço, J. Rencoret, C. Chemetova, J. Gominho, A. Gutiérrez, J. C. del Río and H. Pereira, *Front. Plant Sci.*, 2016, **7**, 1612; (*b*) J. L. Ferrer, M. B. Austin, C. Stewart Junior and J. P. Noel, *Plant Physiol. Biochem.*, 2008, **46**, 356.
27 Downloaded from: http://www.nrel.gov/bioenergy/biomass-compositional-analysis.html.

Faraday Discussions

PAPER

Fast microwave-assisted acidolysis: a new biorefinery approach for the zero-waste utilisation of lignocellulosic biomass to produce high quality lignin and fermentable saccharides†

Long Zhou,[a] Fabio Santomauro,[b] Jiajun Fan,[a] Duncan Macquarrie,[a] James Clark,[a] Christopher J. Chuck *[b] and Vitaliy Budarin*[a]

Received 10th March 2017, Accepted 29th March 2017
DOI: 10.1039/c7fd00102a

Generally, biorefineries convert lignocellulosic biomass into a range of biofuels and further value added chemicals. However, conventional biorefinery processes focus mainly on the cellulose and hemicellulose fractions and therefore produce only low quality lignin, which is commonly burnt to provide process heat. To make full use of the biomass, more attention needs to be focused on novel separation techniques, where high quality lignin can be isolated that is suitable for further valorisation into aromatic chemicals and fuel components. In this paper, three types of lignocellulosic biomass (softwood, hardwood and herbaceous biomass) were processed by microwave-assisted acidolysis to produce high quality lignin. The lignin from the softwood was isolated largely intact in the solid residue after acidolysis. For example, a 10 min microwave-assisted acidolysis treatment produced lignin with a purity of 93% and in a yield of 82%, which is superior to other conventional separation methods reported. Furthermore, py-GC/MS analysis proved that the isolated lignin retained the original structure of native lignin in the feedstock without severe chemical modification. This is a large advantage, and the purified lignin is suitable for further chemical processing. To assess the suitability of this methodology as part of a biorefinery system, the aqueous phase, produced after acidolysis of the softwood, was characterised and assessed for its suitability for fermentation. The broth contained some mono- and di-saccharides but mainly contained organic acids, oligosaccharides and furans. While this is unsuitable for *S. cerevisiae* and other common ethanol producing yeasts, two oleaginous yeasts with known inhibitor tolerances were selected: *Cryptococcus curvatus* and *Metschnikowia pulcherrima*. Both yeasts could

[a]*Green Chemistry Centre of Excellence, Department of Chemistry, University of York, York, YO10 5DD, UK. E-mail: vitaliy.budarin@york.ac.uk*

[b]*Department of Chemical Engineering, University of Bath, Claverton Down, Bath, BA2 7AY, UK. E-mail: c.chuck@bath.ac.uk*

† Electronic supplementary information (ESI) available: Proximate, ultimate and ICP analysis of feedstock; TAPPI T222 method. See DOI: 10.1039/c7fd00102a

grow on the broth, and demonstrated suitable catabolism of the oligosaccharides and inhibitors over 7 days. In addition, both yeasts were shown to be able to produce an oil with a similar composition to that of palm oil. This preliminary work demonstrates new protocols of microwave-assisted acidolysis and therefore offers an effective approach to produce high purity lignin and fermentable chemicals, which is a key step towards developing a zero-waste lignocellulosic biorefinery.

Introduction

Due to fluctuations in crude oil prices, the finite nature of the resource and the need for national energy reserves, there has been an increasing interest in developing lignocellulosic biorefineries for fuels and chemicals over the last few decades.[1–3] Lignocellulose is mainly constituted of cellulose, hemicellulose and lignin, and can be converted both chemically or biologically into a range of value added chemicals and fuels.[4–6] For example, in conventional biorefinery processes, cellulose and hemicellulose can be depolymerised to produce mono- and disaccharides such as cellobiose, glucose, xylose and furanose, which can be fermented to alcohols or other microbial products or chemically converted into furans.[7,8] However, the current utilization of lignocellulosic biomass is still relatively inefficient.[9] In industrial processes, such as paper pulping, it is common for only one product to be targeted at the expense of large waste streams.[10,11] Side products such as lignin are commonly discarded or used to provide low value energy for the process, and not valorised into suitable phenolic components,[12] as these processes are designed only to achieve high quality cellulose. The cellulose purification protocols damage the lignin structure and produce lignin with significant polysaccharide contamination. However, pure lignin with little degradation is a valuable source of aromatic functionality and could be converted into high value fuel additives or chemicals.[13] Therefore, a more selective pretreatment to efficiently cleave the bonds connecting polysaccharides and lignin is of interest. Classical lignin isolation protocols, such as the Klason method or milled wood lignin (MWL), take hours or even days according to the intensity of the process, and require the use of a range of additional acids and chemicals. As such, conventional separation methods cannot effectively isolate lignin in both a high yield and with low structure modification at the same time.

Compared with conventional thermal treatment, microwave-assisted treatment has the characteristics of high efficiency and selectivity, thus making it an efficient tool in biomass processing and solid waste recycling. Microwave heating has been widely used in pyrolysis and gasification as well as in the delignification of biomass.[14–16] Microwave heating has a dramatic effect on the reaction kinetics[17] and reduces overall reaction times substantially.[18] Cellulose and hemicellulose are more polar than lignin,[19,20] therefore they receive more energy input upon microwave irradiation. Based on this selective heating, highly efficient separation could potentially be achieved through microwave treatment, with far shorter reaction times.

However, to further develop the zero waste biorefinery concept, various products must be produced alongside lignin.[21–24] To this end, a new protocol of rapid microwave-assisted acidolysis was examined to separate lignin effectively in three types of lignocellulosic biomass. The saccharide-rich aqueous phase was then investigated as a fermentation source for two oleaginous, inhibitor-tolerant yeasts.

Experimental

Materials

Three lignocellulosic biomass samples were investigated in this research: mixed softwood pellets (MSP, UK Biochar Research Centre, School of Geosciences, University of Edinburgh), willow edging (WE, Gardman Ltd) and wheat straw (WS, Biorenewables Development Centre, University of York), which represent softwood, hardwood and herbaceous biomass respectively. The elemental and ICP analysis results are given in ESI Tables 1 and 2.†

Sulfuric acid and sodium hydroxide were purchased from Fisher Chemicals. Creosol, vanillin, 2-methoxy-phenol, glucose, fructose, rhamnose, formic acid and furfural were purchased from Sigma-Aldrich. *E*-Isoeugenol was purchased from Acros Organic. Levoglucosan and 5-hydroxymethylfurfural (HMF) were purchased from Carbosynth. Xylose was purchased from VWR. Cellobiose was purchased from Fluorochem. Lactic acid was purchased from Wardle. Acetic acid was purchased from Alfa Aesar. Levoglucosenone was purchased from Dextra.

Metschnikowia pulcherrima was isolated from the local area (named strain I1) and *Cryptococcus curvatus* (strain DSM-70022) was purchased from the German Collection of Microorganisms and Cell Cultures.

Microwave-assisted acidolysis method

All biomass was milled into 60-mesh powders using a cutting mill (Retsch SM300, Germany) in the Biorenewables Development Center (BDC), University of York. Microwave treatment was carried out using a Discovery SP microwave reactor (CEM Corporation, USA) in capped vessels. Maximum power (300 W) was applied in all of the experiments in order to achieve the holding temperature as quickly as possible. Acid solutions of different types and concentrations were applied for isolation. Processing temperatures from 170 °C to 210 °C in intervals of 10 °C were used for isolation. The holding time was 10 min. During microwave treatment, the biomass feedstock and acid solvent were heated in a capped vessel with stirring in the ratios of 0.08 g : 6 mL, 0.2 g : 6 mL, 0.4 g : 6 mL and 0.6 g : 6 mL (13, 33, 67 and 100 g L^{-1}). After microwave treatment, the residue was recovered by filtration. The residue was then washed several times with deionized water until the rinsed water was neutral. The residue was dried (105 °C, 24 h) and then weighed. All of the experiments were repeated 3 times.

Analysis of the acidolysis residue

The purity and yield were calculated using the TAPPI T222 method.[25] The method is shown schematically in ESI Fig. 1.† In this method, approximately 0.1 g of dewaxed sample was treated with 10 g of sulfuric acid (72 wt%) at 20 °C for 2 h. The solution was then diluted with deionised water to 3 wt% sulfuric acid and refluxed for 4 h. The insoluble residue (lignin) was isolated by filtration. After washing with hot water, the residue was dried at 105 °C for 24 h. This dried residue is Klason lignin (KL). The purity and yield were calculated according to the TAPPI T222 method see ESI Fig. 1.† The purity result was adjusted by subtracting the ash content measured by TG analysis.

Elemental analysis and ICP analysis data were obtained from the analytical service offered by the Department of Chemistry, University of York.

Thermogravimetric (TG) analysis was carried out using a Netzsch STA 409 analyser (Germany). The following parameters were applied: temperature ramp rate 20 K min^{-1}, final temperature 600 °C, carrier gas 50 mL min^{-1} pure nitrogen. To measure the ash content, the following parameters were applied: temperature ramp rate 20 K min^{-1}, final temperature 625 °C held for 1 h, carrier gas 50 mL min^{-1} N_2 and 100 mL min^{-1} O_2. The final mass% was used as the ash content.

FTIR data was obtained using a Perkin Elmer FTIR/FTNIR Spectrum 400 analyser (USA). The spectra were acquired between 700 cm^{-1} and 4000 cm^{-1} with a resolution of 2 cm^{-1} and a scan time of 64 s.

Solid State ^{13}C NMR spectroscopy (SSNMR) results were obtained at the EPSRC UK National Solid-state NMR Service at the University of Durham. The spectra were obtained at 100.562 MHz. The chemical shift range from 0 ppm to 240 ppm was recorded.

Py-GC/MS results were obtained from the BDC, University of York. A CDS Analytical 5250-T Trapping Pyrolysis Autosampler (UK) was used as the pyrolysis unit, an Agilent Technologies 7890B GC System (USA) as the gas chromatography unit and an Agilent Technologies 5977A MSD system (USA) as the mass spectrometer. The sample was loaded into the pyrolysis unit and pyrolyzed at 600 °C for 10 s. The volatile materials that were released were carried into the GC/MS unit by nitrogen for analysis. The following GC/MS parameters were applied: GC inlet temperature of 350 °C, initial temperature of 40 °C for 2 min, ramp rate of 10 K min^{-1} until 300 °C before being held at 300 °C for 30 min, and a split ratio of 50 : 1. The volatile compounds were identified by comparing their mass spectra with the results in the NIST Lab database. A standard sample mixture of four compounds, creosol/vanillin/2-methoxyphenol (guaiacol)/*E*-isoeugenol, was also subjected to pyrolysis and GC/MS in order to verify the mass spectral identities.

Yeast cultures

Inoculum cultures for both yeasts were prepared by culturing in 10 mL of SPME media (soy peptone: 30 g L^{-1}; malt extract: 25 g L^{-1}) and 10 mL of YPD media (yeast extract: 10 g L^{-1}; peptone: 20 g L^{-1}; glucose: 20 g L^{-1}) and were incubated at

Table 1 Purity and yield of residue lignin (wt%) obtained from different biomasses. Conditions: 0.2 M H_2SO_4 solution, 190 °C[a]

	Lignin content in the feedstock (M_{a1}/M_d)	Purity (M_{a2}/M_d)	Yield (M_{a2}/M_{a1})
MSP	39.03	92.85	82.43
WE	40.55	87.27	30.30
WS	33.83	63.12	31.23

[a] Note: all of the results are based on an extractive-basis. M_0, M_d, M_{a1}, M_i, M_{di}, M_{a2}, see ESI Fig. 1.

20 °C and 25 °C for *M. pulcherrima* and *C. curvatus*, respectively, with an agitation of 180 rpm for 24 h.

The hydrolysates prepared from the microwave processing stage were filtered, adjusted to pH 4 with 2 molar NaOH solution, then diluted to a 1 : 2 ratio with a salt solution. For *M. pulcherrima*, the salt solution was composed of the following: KH_2PO_4 (2 g L^{-1}), $MgSO_4 \cdot 7H_2O$ (0.376 g L^{-1}), $MgCl_2 \cdot 6H_2O$ (2.16 g L^{-1}), $ZnSO_4 \cdot 7H_2O$ (0.04 g L^{-1}), $(NH_4)_2SO_4$ (0.126 g L^{-1}), NH_4Cl (0.708 g L^{-1}), yeast extract (2.00 g L^{-1}), $CaCl_2 \cdot 2H_2O$ (0.300 g L^{-1}) and tartaric acid (30.0 g L^{-1}). The pH was adjusted to 4 before diluting. For *C. curvatus*, the concentrated salt solution was composed of the following: KH_2PO_4 (14.0 g L^{-1}), Na_2HPO_4 (5.00 g L^{-1}), $MgSO_4 \cdot 7H_2O$ (3.00 g L^{-1}), $FeCl_3 \cdot 6H_2O$ (0.300 g L^{-1}), $ZnSO_4 \cdot 7H_2O$ (0.040 g L^{-1}), $MnSO_4 \cdot 1H_2O$ (0.120 g L^{-1}), $CaCl_2 \cdot 2H_2O$ (0.300 g L^{-1}), $(NH_4)_2SO_4$ (1.00 g L^{-1}) and yeast extract (2.00 g L^{-1}). The pH was adjusted to 6.5 before diluting.

1 mL cultures were inoculated with 25 μL of inoculum in 24-well plates and incubated at 180 rpm at 20 °C and 25 °C for *M. pulcherrima* and *C. curvatus*, respectively, for 168 h. Throughout the fermentation process the growth of the cultures was estimated by analysis of the absorbance at 600 nm and the biomass was recovered by centrifugation in 1.5 Eppendorf tubes (10 000 rpm, 5 min). The supernatant and the pellets were stored separately at −20 °C prior to further analysis.

For the characterization of the fatty acid profile, an adapted literature method was used,[26] whereby the biomass pellets were suspended in methanol/H_2SO_4 1% v/v and heated at 90 °C for 3 h in sealed pressure tubes. The resulting fatty acid methyl esters were extracted with hexane, washed with water and analysed using GC-MS.

The DP1 and DP2 saccharides were analysed using an Agilent 1260 Infinity HPLC system (USA) equipped with an Agilent Hi-Plex H (300 × 7.7 mm, 8 μm particle size) column. For levoglucosan, glucose, fructose, xylose, cellobiose, rhamnose and the organic acids (lactic, formic and acetic acids), an isocratic (no gradient) mobile phase of 0.005 M H_2S0_4, a flow-rate of 0.4 mL min^{-1}, a column temperature of 60 °C, a refractive index detector at 55 °C, a total run time of 35 minutes, and an injection volume of 5 μL were used. For inhibitor analysis (furfural, levoglucosenone and 5-HMF), the following parameters were used: an ACE C18 column (250 × 4.6 mm, 5 μm particle size), an isocratic (no gradient) mobile phase of acetonitrile : water (25/75), a flow-rate of 0.8 mL min^{-1}, a column temperature of 30 °C, a 220 nm DAD detector, a total run time of 22 minutes, and an injection volume of 5 μL. For both experiments, five samples of mixed standard chemicals (0.5, 0.75, 1.0, 1.5 and 2.0 mg mL^{-1}) were used for the calibration.

The samples also had their oligosaccharide profiles analysed using a Dionex ICS-5000 HPLC system (Thermo Scientific, USA) with a Dionex CarboPac SA10 column (4 × 250 mm) and guard column with a 1.5 mL min^{-1} flow rate using pulsed amperometric detection. The separation was carried out at 25 °C for 10 min using gradient elution with 1 mM sodium hydroxide (NaOH) at 1.5 mL min^{-1} and a 10 μL injection volume. The samples were appropriately diluted in Milli-Q water and filtered through a 0.22 μm syringe filter prior to analysis.

Results and discussion

Analysis of the acidolysis residue

Fig. 1 shows the influence of temperature on the microwave lignin isolation (Fig. 1a–c). Clearly, for MSP and WE the residual mass decreased sharply from 170 °C to 180 °C. For WE, the residual mass kept decreasing until 190 °C and remained at 20% at higher holding temperatures. In contrast to woody biomass, WS had a relatively low residual mass even at 170 °C. The C/H results show similar trends. For MSP and WE, the C/H content was relatively stable when a holding temperature of ≥190 °C was applied, while for WS, the carbon content stabilised at 180 °C but the hydrogen content kept decreasing with increasing temperature. For all three feedstocks, the residue colour kept getting darker when the holding temperature was increased. It is important to note that after 200 °C treatment, the samples became totally black, suggesting that these temperatures may be causing the destruction of the lignin, and may therefore be too high to be of value. Microwave-assisted acidolyses were also carried out with different types of acid and solution concentrations (Fig. 1d). It was found that the concentration had a significant influence on the acidolysis process. When concentrated H_2SO_4

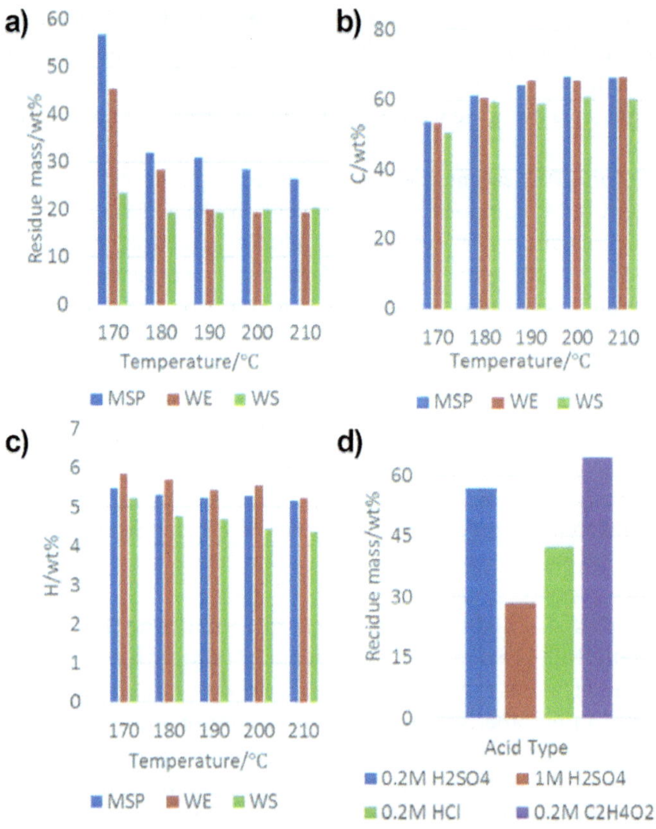

Fig. 1 Influence of temperature preparation on (a) residue yield, (b) carbon and (c) hydrogen content; (d) nature of the acid influence on the residue mass yield. Conditions: holding time of 10 min, dynamic mode, 0.2 M acid solutions.

(1 mol L^{-1}) was used, the residual mass was much less than when diluted H_2SO_4 was used, thus showing that a higher concentration solution could accelerate separation even at relatively low temperatures. The mass of residue formed by acetic acid (0.2 mol L^{-1}) was only slightly higher than that formed by sulphuric acid. This suggested there was the potential to perform acidolysis with acetic acid. Acetic acid can be produced from biomass, which is a more benign and potentially more attractive option than sulphuric acid. Among the three residues treated with a diluted acid solution (0.2 mol L^{-1}), the one treated with hydrochloric acid produced the lowest residual mass, thus indicating that hydrochloric acid can cause a higher extent of solubilisation of biomass in acidolysis.

By using the method shown in ESI Fig. 1,† the purities and yields are provided in Table 1. In terms of the purity and yield, MSP was shown to be a very suitable feedstock for microwave-assisted acidolysis at 190 °C (10 min holding time, 0.2 M sulphuric acid), however, for WE and WS the protocol still requires improvement. Compared with hardwood and herbaceous biomass, it was reported that softwood has the lowest content of acid-soluble lignin, being only 0.2–0.5% of the total lignin.[25] For MSP, 82 wt% of the acid-insoluble lignin could be retained in the residue and, after the removal of soluble matter, the purity of the lignin residue could be as high as 93 wt%. Compared to MSP, the yield of WE was much lower, being at 30%, but the purity of 87% was still relatively high. Wu and Argyropoulos[27] produced MWL by a 14 day milling process with black spruce (softwood) and poplar (hardwood). The yields were 28.5 wt%/28.7 wt% and the purities were 88.3 wt%/84.7 wt%. Compared with these data, microwave lignin isolation could compete with the MWL protocol in terms of both the yield and purity, especially when the feedstock was softwood where both a high yield and purity were achieved. More importantly, microwave lignin isolation only took 10 min, which was far more efficient than the MWL method.

The purity of WS was 63%, which was lower than that of MSP and WE. However, the low purity was caused not only by sugar contamination, but also by ash content. The ash content (extractive-free basis) of 190 °C WS residue was as high as 25 wt% while the ash contents of the MSP and WE residues were both <3 wt%. The results given in ESI Table 1† show that the WS had an ash content of 14 wt%. This indicates that the ash content of the feedstock could greatly affect the quality of the residual lignin. Most herbaceous energy crops contain high amounts of ash,[28,29] thus suggesting that an ash-wash is necessary for effective lignin isolation from herbaceous biomass. Fig. 2a–d show the mass and DTG curves of three feedstocks and their isolated residues at 170/190 °C. The TG curves of the three feedstocks and the treated residues shared some similar trends. For all of the feedstocks, the maximum DTG peaks occurred in the temperature range of 330–360 °C. According to the literature,[30,31] the strong peaks are attributed to the pyrolysis of cellulose. There were also well-pronounced shoulders at 280–330 °C, which are related to the pyrolysis of hemicellulose;[30,31] after microwave treatment, the residues were more recalcitrant to pyrolysis. After acidolysis, polysaccharides were removed and lignin was retained in the residues. The 190 °C residues had a higher final mass than those produced at 170 °C, thus showing that a higher temperature benefited acidolysis. The 190 °C WS residue had the highest final residual mass and the lowest maximum mass loss rate, which was perhaps caused by the much higher ash content in straw (ESI Table 1†). For the 170 °C residues, the peaks at 330–360 °C were weakened significantly, but the

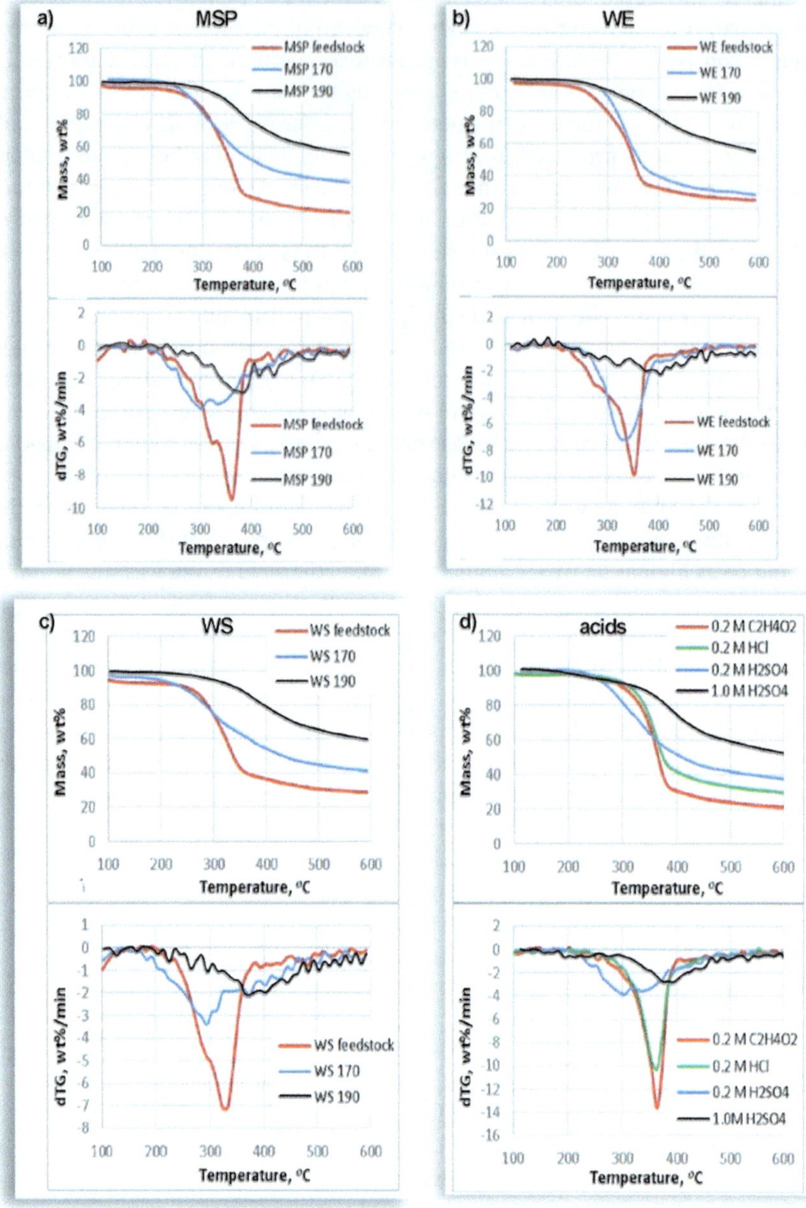

Fig. 2 Comparison of TG and dTG curves for (a) MSP, (b) WE and (c) WS; (d) influence of the acid nature on the properties of MSP. Conditions for (a)–(c): holding time of 10 min, dynamic mode, 300 W, treatment temperatures of 170/190 °C. Conditions for (d): treatment temperature of 170 °C.

peaks between 280–330 °C were still sharp, thus indicating that microwave-acidolysis at 170 °C could remove cellulose to a certain extent, but was not able to solubilize hemicellulose effectively. This was most obvious for the 170 °C WE residue, showing that it had the most polysaccharide among the three 170 °C residues; for the three 190 °C residues, the TG and DTG curves were very similar. The DTG curves only showed weak peaks at around 370–410 °C, which were caused by the pyrolysis of lignin. This demonstrated that microwave acidolysis at 190 °C could solubilize the saccharide component and lignin was predominant in the residues. This provided evidence that a relatively thorough separation of the lignin and polysaccharides could be achieved by acidolysis at 190 °C using microwave assisted heating.

Fig. 2c and d shows the TG curves of four 170 °C residues processed by different acid solutions. The peak of the DTG curve of the 1 M H_2SO_4 residue was at 393.86 °C, thus showing that it contained highly pure lignin. Therefore, its final residual mass of 49.85% was the highest. This illustrated that the lignin could be efficiently isolated at a relatively low temperature with concentrated sulphuric acid. In Fig. 2d, the DTG curves of the HCl and acetic acid residues showed sharp peaks at 350 °C, while the curve of the H_2SO_4 residue only had two broad peaks at 294.23 °C and 333.99 °C. Furthermore, in Fig. 2c, the final residual mass of the H_2SO_4 residue was shown to be lower than that of the HCl and acetic acid residues. The higher thermal stability of the H_2SO_4 residue suggested that it contained more lignin. This indicated that sulphuric acid was more effective for lignin isolation. Although dilute HCl and acetic acid could dissolve similar amounts of biomass to diluted H_2SO_4 (as shown in Fig. 1), the fact that their dTG curves are somewhat similar to that of the original feedstock suggests that selective dissolution has not been achieved. This demonstrates that further optimisation is needed for the use of HCl or acetic acid in microwave acidolysis.

Fig. 3a shows the FT-IR spectra of MSP and its residual lignin at different temperatures. For the three biomasses, as the holding temperature rose, the bands at 1601 cm^{-1}, 1508 cm^{-1}, 1451 cm^{-1} and 1424 cm^{-1} (assigned as aromatic skeleton bands[32–34]) intensified. These strong peaks suggested high aromatic content in the residues after treatment. Similarly, the peak at 1208–1215 cm^{-1} which was ascribed to the OH deformation of phenol was strengthened as the processing temperature rose, thus indicating that more phenolic units were present in the higher temperature residues. The peak at 1030 cm^{-1}, assigned to the primary alcohol of polysugars,[34] weakened as the holding temperature increased, thus suggesting the better removal of polysaccharide at higher temperatures. At temperatures higher than 190 °C, the weakening of this peak became less obvious, thus showing that the residues were relatively stable after processing by microwave acidolysis at temperatures of 190 °C or higher.

Besides the general trends discussed above, there were also some minor differences among the three lignin residues (Fig. 2b), such as at 1250–1270 cm^{-1} (C–O stretching of the guaiacyl unit[33,34]) and at 1100–1125 cm^{-1} (C–H in-plane deformation of the syringyl-unit[33,34]). This is caused by the variety of lignin, because softwood lignin contains only guaiacol units, while hardwood and herbaceous lignin contain both guaiacyl and syringyl units in different proportions. Based on this, lignin from softwood, hardwood and herbaceous biomass could be used for the production of different chemicals after further activation. For example, softwood lignin could produce more vanillin, while if the target product is acetosyringone or phenol, lignin from hardwood and straw would be a better raw material.

Fig. 3c shows the spectra of residues when different acid types and concentrations were applied at 170 °C. The spectrum of the residue treated by

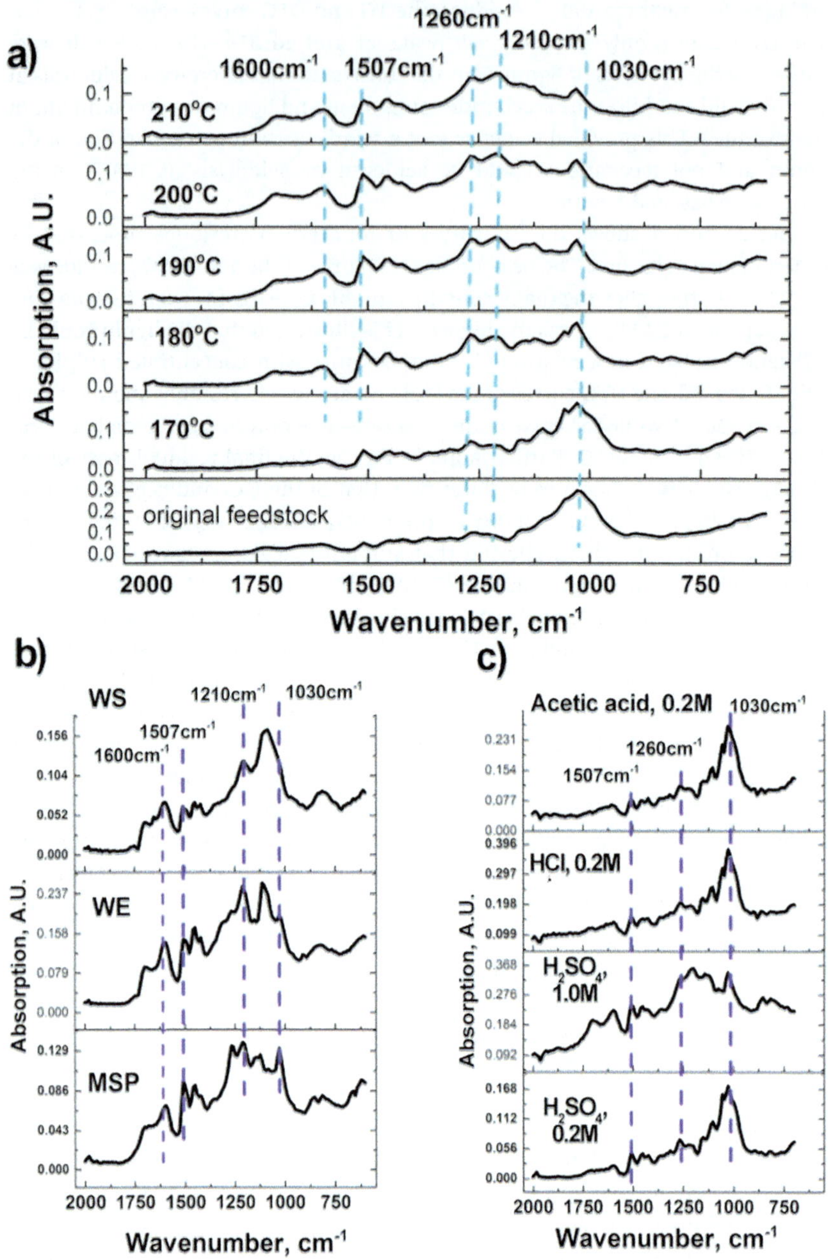

Fig. 3 FTIR analysis of feedstocks and the processed residues. (a) Temperature influence on the spectra of the MW treated MSP residue. The conditions are: holding time of 10 min, dynamic mode, 300 W, MSP, 0.2 M H_2SO_4 solution; (b) influence of feedstock nature on the spectra of the MW residues obtained at 190 °C. The conditions are: holding time of 10 min, dynamic mode, 300 W, MSP, 0.2 M H_2SO_4 solution; (c) influence of acid nature on the spectra of the MW residues obtained at 170 °C. The conditions are: holding time of 10 min, dynamic mode, 300 W, MSP.

concentrated sulphuric acid showed notable differences compared with the spectra of the residues treated by dilute acid. The sugar peak (1030 cm^{-1}) was much weaker compared to that of the other three spectra. On the other hand, the aromatic peaks (1601 cm^{-1}, 1508 cm^{-1}, 1451 cm^{-1}, 1424 cm^{-1} and 1213 cm^{-1}) were very sharp. These signals showed that the processed residue was highly pure lignin, which proved that concentrated acid could accelerate the separation between the lignin and sugar at a relatively low temperature. This result is consistent with the TG analysis, which showed that lignin was dominant in the residue processed by 1 M sulphuric acid.

Solid State NMR (SSNMR) spectra are shown in Fig. 4. Obviously, after microwave acidolysis, the peaks between 109–162 ppm were much stronger. According to Jingdong Mao *et al.*,[35] the peaks in the range between 60–108 ppm can be ascribed to aliphatic carbon mainly from carbohydrate (a portion of the side-chain of lignin also contributes to these peaks) while the peaks between 109–162 ppm are aromatic. Thus the spectra clearly show the existence of lignin as a major component of the material.[35,36] This demonstrated that lignin was well isolated from the biomass, especially at a holding temperature of 190 °C. Comparing the SSNMR spectra of the 190 °C residues with those of the 170 °C residues, the aromatic peaks were seen to be stronger, while the carbohydrate peaks were weaker, thus proving that the isolation of lignin by microwave-assisted acidolysis is affected greatly by the holding temperature. At high temperature (190 °C), microwave acidolysis could dissolve most cellulose and hemicellulose, while at a relatively low temperature (170 °C) the treated residues still contained severe sugar contamination. The peak at 55 ppm was attributed to the methoxy groups of lignin.[35,37] The change of this peak at 55 ppm proved that lignin was well isolated at a holding temperature of 190 °C.

The SSNMR spectra of the three types of biomass also showed some differences besides the general trend observed above. One of the most obvious differences was the peaks located in the range from 50–15 ppm that were attributed to aliphatic carbon.[35,38] Compared with those of the MSP residues, these peaks in the spectra of the WE and WS residues were much stronger. As shown in Table 1, the MSP residue was the purest among the three isolated residues. It had the least sugar contamination, therefore, this peak was very weak in the MSP residue spectra. A difference was also found for the peak at ~105 ppm for the three 190 °C residue spectra. The 190 °C residues of hardwood and straw had well-pronounced shoulders at 105 ppm which was not the case for softwood. This peak was assigned as the C2 and C6 carbon in the syringyl unit.[38] Because softwood lignin only has G units, this peak was weak in the spectrum of the 190 °C MSP residue.

Fig. 5 shows the contents of the pyrolytic compounds (phenolic) of MSP, WE and WS and their 190 °C residual lignin obtained by py-GC/MS analysis. After microwave acidolysis, the content of the phenolic compounds increased significantly as indicated in Fig. 5. For the feedstocks, compounds from polysaccharides, such as furfural and mannose, were dominant. On the other hand, for the residue lignins, phenolic compounds were the main pyrolytic products. As expected, the three types of residual lignin produced pyrolytic products with different distributions, due to softwood, hardwood and herbaceous lignin having different monomers. Softwood was mainly composed of G-units, hardwood of G-units and S-units, and herbaceous lignin of all three units (G/S/H-units). Thus, the 190 °C MSP residual lignin had more vanillin (G-unit pyrolytic product), while the

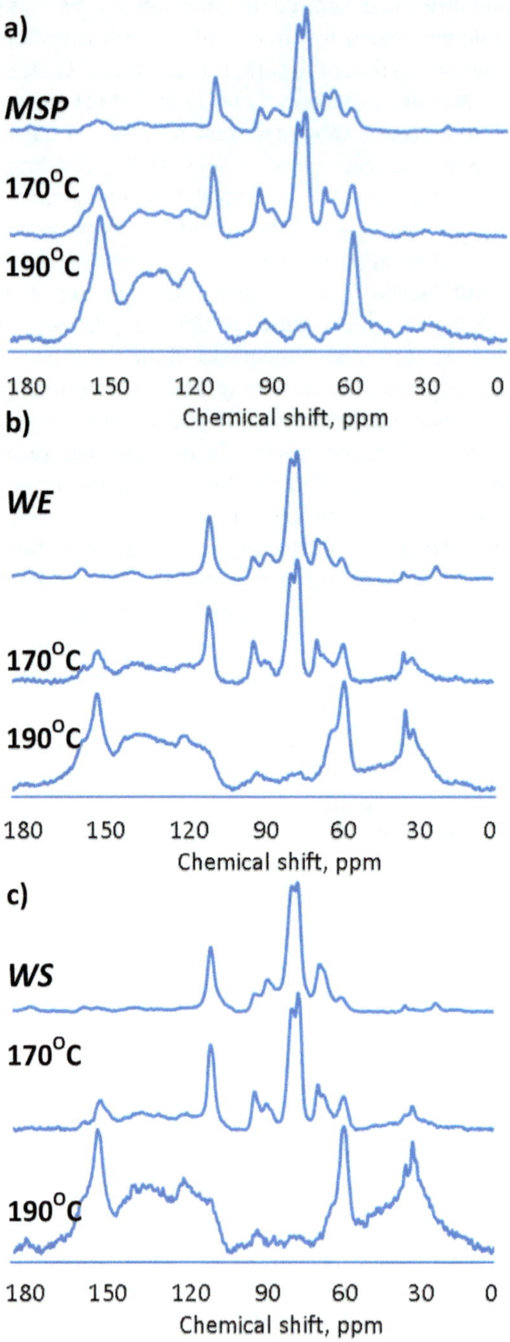

Fig. 4 Solid State NMR spectra of three feedstocks and their processed residues. The conditions are: holding time of 10 min, dynamic mode, 300 W, MSP, 0.2 M H_2SO_4 solution. (a) Spectra of MSP and its processed residue; (b) spectra of WE and its processed residue; (c) spectra of WS and its processed residue.

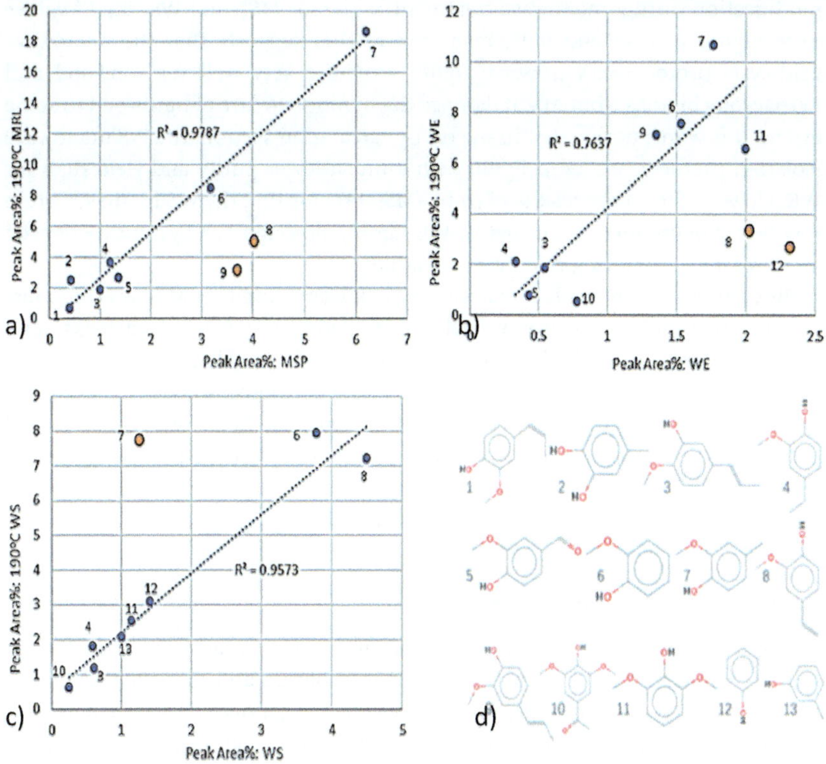

Fig. 5 Py-GC/MS analysis. Comparisons of the GC peak areas of the phenolic compounds of the feedstock and the 190 °C residue. Conditions: holding time of 10 min, dynamic mode, 300 W, MSP, 0.2 M H_2SO_4 solution, 190 °C. (a) MSP samples; (b) WE samples; (c) WS samples; (d) structures of the standard compounds plotted in the figure. (1) *cis*-Isoeugenol; (2) 4-methylcatechol; (3) isochavibetol; (4) 4-ethylguaiacol; (5) vanillin; (6) guaiacol; (7) 2-methoxy-4-methylphenol; (8) 4-vinylguaiacol; (9) *trans*-isoeugenol; (10) acetosyringone; (11) syringol; (12) phenol; (13) *m*-cresol (the compounds represented by red dots are outside of the trend line).

residual lignin of WE and WS had more phenol (H-unit pyrolytic product) and acetosyringone (S-unit pyrolytic product).

In each of Fig. 5a–c, a trend line was drawn to show the linear relationship between the content of phenolic compounds of the feedstock and the residual lignin. The gradient of the line is approximately the mass ratio of the biomass to the residue. Thus, if the phenolic compounds are stable during microwave acidolysis, they should be concentrated in the residues to the extent that the mass is reduced, and thus they should sit on the line; if not, it suggests that structure modification occurred during the treatment. Therefore, the coefficient of determination (R^2) of the trend line can be used as an index to show the extent of lignin structure modification during isolation. Obviously, most phenolic compounds obeyed the trend except for only one or two compounds in each figure, thus showing that the lignin structure was generally well maintained during microwave acidolysis. Comparing the three trend lines, the R^2 value of Fig. 5b was the lowest (0.76), thus indicating that WE lignin experienced relatively more modification during microwave treatment, while for MSP and WS, the R^2 values were higher at 0.97 and 0.95, respectively. This suggests that the microwave acidolysis protocol can preserve lignin structure very well for softwood and herbaceous biomass, but might damage the lignin structure of hardwood to some extent. It is worth noting that in the purity calculation section, the WS microwave isolation protocol produced lignin with a much lower purity and yield than for WE. However, from the results of py-GC/MS, it seems that the WS lignin structure was better maintained than that of WE lignin, although WS lignin experienced more dissolution during microwave treatment.

By combining the residue analysis above, it was found that MSP was the most suitable material for microwave-acidolysis among the three feedstocks, and produced lignin in the highest yield, and with the least ash and sugar contamination. The lignin and polysaccharide of softwood biomass could be separated thoroughly by a 10 min microwave treatment while the lignin structure was well preserved in the residue, thus making softwood the most potentially useful feedstock for a zero-waste microwave biorefinery. Therefore the aqueous phase after acidolysis of MSP was used for fermentation.

Fermentation of the aqueous phase

A proportion of the saccharide feedstock partitions into the aqueous phase; this is a dilute solution that contains some monosaccharide and disaccharide sugars (DP1 and DP2), organic acids, furans, and a small amount of lower molecular weight oligosaccharides (DP3–DP5) and larger oligosaccharides (DP6+). Conventional yeast based processes, such as the production of bioethanol using *S. cerevisiae*, require inhibitor concentrations of below 10 mM and high hexose sugar concentrations.[39] However, several oleaginous yeasts are known to be able to survive in highly inhibitory conditions and can even metabolise some of these types of carbon sources.

For example, *Cryptococcus curvatus* has been demonstrated to have a high threshold for furfural inhibitors,[40] while being able to metabolise acetate and some oligosaccharides.[41] Similarly we recently reported on the yeast *Metschnikowia pulcherrima*, which is able to grow in extremely acidic conditions[42] and has been shown to produce a range of cellulases.[43]

Yeast derived lipids have generated an increasing amount of interest in recent years, and have been demonstrated to be suitable for a range of applications including as a feedstock for biodiesel,[44] advanced biofuels[45] and alternative chemical transformations.[46] However, one of the most sought after applications could be in the replacement of terrestrial crops, such as palm oil, which lead to wide spread deforestation. For this application the product must be highly saturated, with the majority of the rest of the lipid being composed of oleic acid.[47] To determine the suitability of the solubilised fraction for fermentation, *C. curvatus* (Cc) and *M. pulcherrima* (Mp) were cultured on the solubilised fraction produced from the microwave processing (Fig. 6). Both yeasts grew reasonably well on the material depolymerised at 170 °C with 2 g L^{-1} of yeast biomass produced for both species. Under these conditions, 1.5 g L^{-1} of mono- and disaccharides were available for fermentation but less than 0.5 g L^{-1} of acids and furfurals were produced. The high yeast co-efficients strongly suggest that both yeasts are using the larger oligosaccharides in the broth as well. Upon increasing

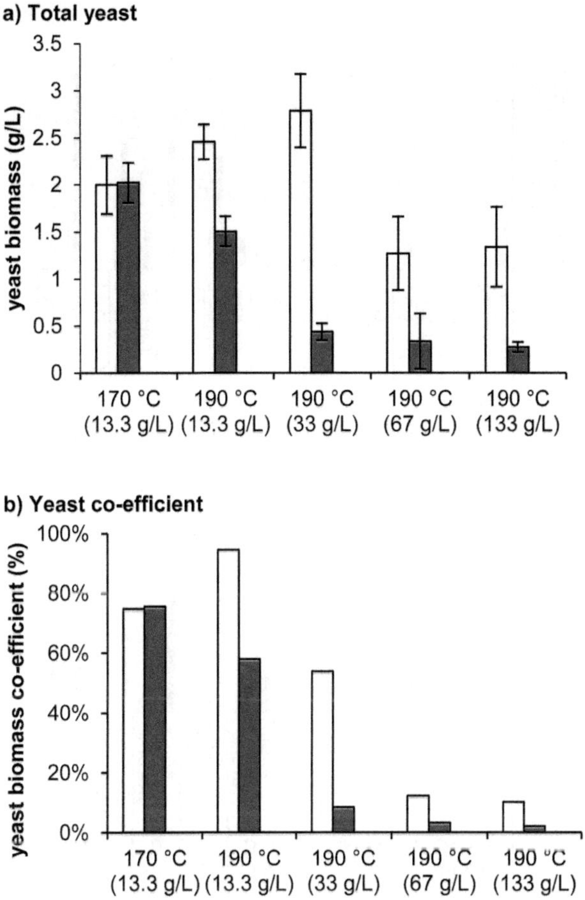

Fig. 6 Yeast biomass produced from the various hydrolysate samples, given as the initial loadings of wood biomass at the temperature it was depolymerised, for *C. curvatus* (white bars, pH 6.5, 180 rpm, 168 h) and *M. pulcherrima* (blue bars, pH 4, 180 rpm, 168 h), where (a) is the total yeast biomass and (b) is the yeast biomass coefficient, based on the proportion of the yeast grown compared to the initial level of saccharides (DP1–DP5), organic acids and furan inhibitors.

the processing temperature to 190 °C, the acid and furfural concentrations were increased at the expense of the monosaccharide concentration. *C. curvatus* was able to metabolise some of this feedstock, although the impact on *M. pulcherrima* was more pronounced. Upon increasing the loading of initial wood biomass to 33 g L^{-1}, the mono- and di-saccharide concentrations were only increased partially, although over 2.5 g L^{-1} of organic acids were now present in the hydrolysate. Presumably as a result, the growth of *M. pulcherrima* was reduced substantially, though *C. curvatus* continued to thrive on this feedstock, producing 2.8 g L^{-1} yeast biomass. A co-efficient of 0.54 on the available acids, furans and saccharides (DP1–DP5) in the media was applied. While higher loadings of biomass promoted higher sugar concentrations, the furfural and acids produced also increased substantially. With initial wood biomass loadings of 67 g L^{-1} and

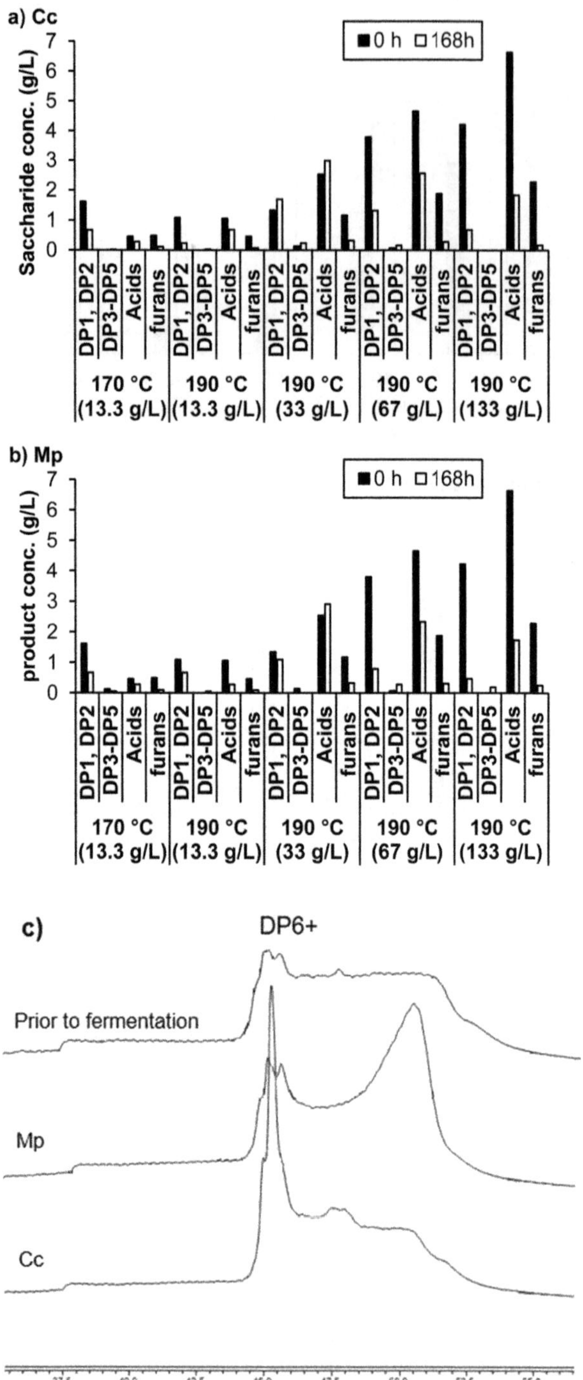

Fig. 7 The quantification of saccharides (DP1–DP5), acids (formic, lactic and acetic) and furans (5-HMF and furfural) before and after fermentation over the range of hydrolysates for (a) *C. curvatus* (pH 6.5, 180 rpm, 168 h) and (b) *M. pulcherrima* (pH 4, 180 rpm, 168); (c) qualitative oligosaccharide chromatograph demonstrating DP6 and larger oligosaccharides for the original hydrolysate (depolymerised at 170 °C, 13.3 g L^{-1} wood biomass) and those left after fermentation with *M. pulcherrima* (pH 4, 180 rpm, 168 h) and *C. curvatus* (pH 6.5, 180 rpm, 168 h).

100 g L^{-1}, little yeast biomass was produced from either yeast. Both yeasts metabolised the majority of the mono- and di-saccharides over the 7 days, though *M. pulcherrima* was more capable of metabolising the small amount of DP3–DP5 oligosaccharides in the broth than *C. curvatus* (Fig. 7a and b). Both yeasts demonstrated some ability to break down some of the larger oligosaccharide feedstock too (Fig. 7c). *M. pulcherrima* appeared to reduce some of the larger oligosaccharides, thus presenting a different oligosaccharide profile of the broth after the fermentation was complete. More strikingly, *C. curvatus* seemed able to convert a proportion of the larger oligosaccharides, presumably producing saccharides in the DP6 and smaller ranges in the process. This ability supports the higher levels of yeast biomass observed with *C. curvatus*, and demonstrates why the yeast could grow so well on the hydrolysed feedstock.

Interestingly, both yeasts seem to have similar mechanisms for dealing with the inhibitory compounds. At a 33 g L^{-1} loading, neither *M. pulcherrima* nor *C. curvatus* converted the acids in the hydrolysate, and the majority present at the start of the fermentation remained in the broth. However, at higher concentrations, where both yeasts grew poorly, the acids were seemingly metabolised. The furfural concentrations in the mixtures were found to be below the detectable

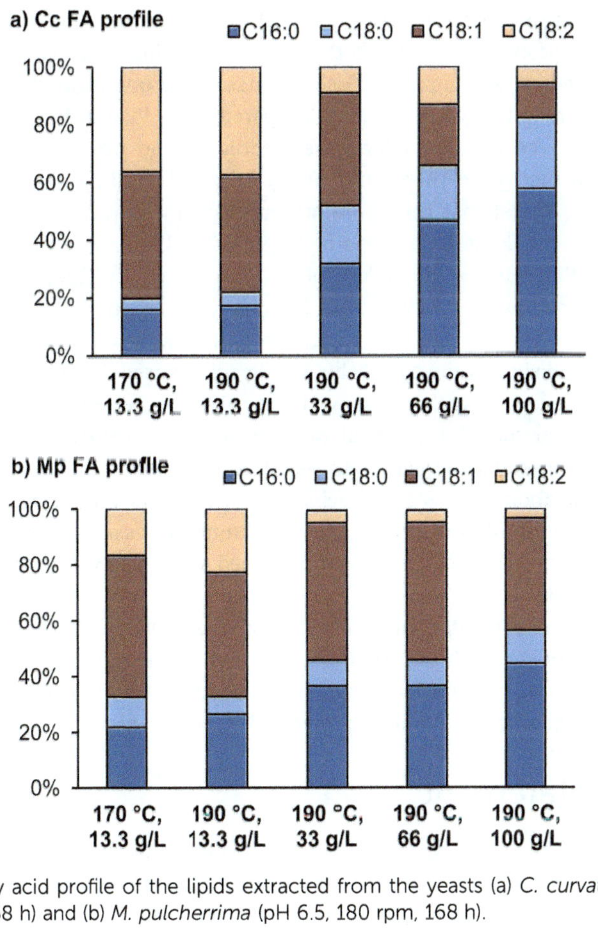

Fig. 8 Fatty acid profile of the lipids extracted from the yeasts (a) *C. curvatus* (pH 6.5, 180 rpm, 168 h) and (b) *M. pulcherrima* (pH 6.5, 180 rpm, 168 h).

limits for both yeasts, under all conditions, while the 5-HMF concentration was also severely reduced. This suggests that the yeasts were converting the furan compounds and producing a less toxic substrate, which is a mechanism commonly seen with these types of yeasts.[48]

Both yeasts are oleaginous, and as such the lipid produced was investigated as a possible palm oil substitute (Fig. 8). Despite both yeasts having been reported to be effective producers of saturated lipid, less than 20% of the cell weight was recovered under any of the conditions analysed. In addition, the lipid profile was found to be highly dependent on the feedstock processing temperature. Under these conditions *C. curvatus* produced a lipid that was more similar to rapeseed oil at low loadings of initial biomass, but the saturation increased substantially upon increasing the initial biomass loading. Similarly, *M. pulcherrima* produced a lipid that was analogous to palm oil at increased loadings of biomass. This suggests that highly stressful fermentation conditions can contribute to increasing saturation. This opens up the possibility of being able to stress these yeasts into producing desirable lipid profiles for industrial production, although more work would be necessary to increase the lipid content up to suitable levels.

Conclusions

A new method of rapid microwave-assisted acidolysis for a potential biomass biorefinery is proposed to achieve the zero-waste utilization of lignocellulosic biomass. Three types of lignocellulosic biomass were investigated for the separation of lignin and polysaccharides. Compared with WE and WS, MSP was the most suitable feedstock for microwave acidolysis, and produced high quality lignin with a high yield (82%) and purity (93%). According to the py-GC/MS results, it was found that the lignin structure could be well preserved during acidolysis, thus making this unmodified lignin suitable for phenolic compound production. Sulphuric acid could isolate lignin with the highest quality. If concentrated sulphuric acid was used in acidolysis, lignin could be isolated at a relatively low temperature in comparison with that required for dilute acid, although it is likely that the use of this strong acid caused extensive degradation of the saccharide feedstock. To convert the inhibitor rich aqueous phase, two yeasts known for their inhibitor tolerance were selected (*Cryptococcus curvatus* and *Metschnikowia pulcherrima*). Both yeasts could grow on the saccharide feedstock and metabolise a range of acids, furans, sugars and oligosaccharides. However, at high inhibitor loadings the yeast biomass was severely reduced. Both yeasts are known to be oleaginous and produced an oil that was akin to palm oil from this feedstock.

Based on these results, microwave-assisted acidolysis offers a potentially powerful tool for lignocellulosic biomass utilization, especially for softwoods. Due to the efficiency and selectivity of microwave heating, 10 min is enough time to produce high quality lignin and a fermentable aqueous phase.

Access statement

All data used in the preparation of this manuscript for the sections funded by the EPSRC grants EP/K014773/1 and EP/N013522/1 are contained within this document, the ESI,† or available upon request from the DOI.

Acknowledgements

This work was supported by the EPSRC (EP/K014773/1) and Industrial Biotechnology Catalyst (Innovate UK, BBSRC, EPSRC) to support the translation, development and commercialisation of innovative Industrial Biotechnology processes (EP/N013522/1). Solid-state NMR spectra were obtained at the EPSRC UK National Solid-state NMR Service at Durham. We are thankful to the Biorenewables Development Centre (Department of Chemistry, University of York) for their help with the sample milling and Py-GCMS experiments. We thank Dr Christopher Ibenegbu for the oligosaccharide analysis and Dr Hannah Briers for the monosaccharide, di-saccharide and inhibitor analysis.

Notes and references

1 G. Altinay and E. Karagol, *Energy Econ.*, 2004, **26**, 985–994.
2 A. B. Awan and Z. A. Khan, *Renewable Sustainable Energy Rev.*, 2014, **33**, 236–253.
3 J. Goldemberg and S. Teixeira Coelho, *Energy Policy*, 2004, **32**, 711–714.
4 P. Kumar, P. Kumar, D. M. Barrett, D. M. Barrett, M. J. Delwiche, M. J. Delwiche, P. Stroeve and P. Stroeve, *Ind. Eng. Chem. Analytical Ed.*, 2009, **48**, 3713–3729.
5 J. J. Bozell and G. R. Petersen, *Green Chem.*, 2010, **12**, 539.
6 F. Cherubini, *Energy Convers. Manag.*, 2010, **51**, 1412–1421.
7 B. Kamm and M. Kamm, *Appl. Microbiol. Biotechnol.*, 2004, **64**, 137–145.
8 P. Kaparaju, M. Serrano, A. B. Thomsen, P. Kongjan and I. Angelidaki, *Bioresour. Technol.*, 2009, **100**, 2562–2568.
9 M. Ni, D. Y. C. Leung, M. K. H. Leung and K. Sumathy, *Fuel Process. Technol.*, 2006, **87**, 461–472.
10 A. Toledano, A. García, I. Mondragon and J. Labidi, *Sep. Purif. Technol.*, 2010, **71**, 38–43.
11 J. J. Bozell, *Clean: Soil, Air, Water*, 2008, **36**, 641–647.
12 D. D. Laskar, M. P. Tucker, X. Chen, G. L. Helms and B. Yang, *Green Chem.*, 2014, **16**, 897.
13 T. Q. Yuan, F. Xu and R. C. Sun, *J. Chem. Technol. Biotechnol.*, 2013, **88**, 346–352.
14 Y. Wan, P. Chen, B. Zhang, C. Yang, Y. Liu, X. Lin and R. Ruan, *J. Anal. Appl. Pyrolysis*, 2009, **86**, 161–167.
15 R. Singh, A. Shukla, S. Tiwari and M. Srivastava, *Renewable Sustainable Energy Rev.*, 2014, **32**, 713–728.
16 Q. Xie, F. C. Borges, Y. Cheng, Y. Wan, Y. Li, X. Lin, Y. Liu, F. Hussain, P. Chen and R. Ruan, *Bioresour. Technol.*, 2014, **156**, 291–296.
17 J. H. Clark, V. Budarin, F. E. I. Deswarte, J. J. E. Hardy, F. M. Kerton, A. J. Hunt, R. Luque, D. J. Macquarrie, K. Milkowski, A. Rodriguez, O. Samuel, S. J. Tavener, R. J. White and A. J. Wilson, *Green Chem.*, 2006, **8**, 853–860.
18 H. Li, Y. Qu, Y. Yang, S. Chang and J. Xu, *Bioresour. Technol.*, 2016, **199**, 34–41.
19 W. Gindl-Altmutter, M. Obersriebnig, S. Veigel and F. Liebner, *ChemSusChem*, 2015, **8**, 87–91.
20 E. Rojo, M. S. Peresin, W. W. Sampson, I. C. Hoeger, J. Vartiainen, J. Laine and O. J. Rojas, *Green Chem.*, 2015, **17**, 1853–1866.

21 V. Budarin, A. B. Ross, P. Biller, R. Riley, J. H. Clark, J. M. Jones, D. J. Gilmour and W. Zimmerman, *Green Chem.*, 2012, 3251–3254.

22 Y. Yuan and D. J. Macquarrie, *Bioresour. Technol.*, 2015, **198**, 819–827.

23 D. Hernández, M. Solana, B. Riaño, M. C. García-González and A. Bertucco, *Bioresour. Technol.*, 2014, **170**, 370–378.

24 Q. Bu, H. Lei, S. Ren, L. Wang, Q. Zhang, J. Tang and R. Ruan, *Bioresour. Technol.*, 2012, **108**, 274–279.

25 T. Tappi, *2002–2003 TAPPI Test Methods*, 2002.

26 W. W. Christie, *Adv. Lipid Methodol.*, 1993, **2**, e111.

27 S. Wu and D. Argyropoulos, *J. Pulp Pap. Sci.*, 2003, **29**, 235–240.

28 H. Kludze, B. Deen and A. Dutta, *Fuel Process. Technol.*, 2013, **109**, 96–102.

29 D. Porbatzki, M. Stemmler and M. Müller, *Biomass Bioenergy*, 2011, **35**, S79–S86.

30 G. Wang, W. Li, B. Li and H. Chen, *Fuel*, 2008, **87**, 552–558.

31 E. Biagini, F. Barontini and L. Tognotti, *Ind. Eng. Chem. Res.*, 2006, **45**, 4486–4493.

32 J.-Y. Chen, Y. Shimizu, M. Takai and J. Hayashi, *Wood Sci. Technol.*, 1995, **29**, 295–306.

33 Y. Huang, L. Wang, Y. Chao, D. S. Nawawi, T. Akiyama, T. Yokoyama and Y. Matsumoto, *J. Wood Chem. Technol.*, 2012, **32**, 294–303.

34 L. M. Kline, D. G. Hayes, A. R. Womac and N. Labbé, *BioResources*, 2010, **5**, 1366–1383.

35 M. Jingdong, K. M. Holtman, J. T. Scott, J. F. Kadla and K. Schmidt-Rohr, *J. Agric. Food Chem.*, 2006, **54**, 9677–9686.

36 G. R. Hatfield, G. E. Maciel, O. Erbatur and G. Erbatur, *Anal. Chem.*, 1987, **59**, 1659–1664.

37 M. Bardet, M. F. Foray and Q. K. Trân, *Anal. Chem.*, 2002, **74**, 4386–4390.

38 A. T. Martínez, G. Almendros, F. J. González-Vila and R. Fründ, *Solid State Nucl. Magn. Reson.*, 1999, **15**, 41–48.

39 C. J. Chuck, F. Santomauro, L. A. Sargeant, F. Whiffin, T. Chantasuban, N. R. A. Ghaffar, J. L. Wagner and R. J. Scott, *Biofuels*, 2014, **5**, 293–311.

40 X. Yu, Y. Zheng, K. M. Dorgan and S. Chen, *Bioresour. Technol.*, 2011, **102**, 6134–6140.

41 Z. Gong, H. Shen, X. Yang, Q. Wang, H. Xie and Z. K. Zhao, *Biotechnol. Biofuels*, 2014, **7**, 158.

42 F. Santomauro, F. M. Whiffin, R. J. Scott and C. J. Chuck, *Biotechnol. Biofuels*, 2014, **7**, 42.

43 M. L. A. Strauss, N. P. Jolly, M. G. Lambrechts and P. Van Rensburg, *J. Appl. Microbiol.*, 2001, **91**, 182–190.

44 L. A. Sargeant, C. J. Chuck, J. Donnelly, C. D. Bannister and R. J. Scott, *Biofuels*, 2014, **5**, 33–43.

45 J. L. Wagner, V. P. Ting and C. J. Chuck, *Fuel*, 2014, **130**, 315–323.

46 R. W. Jenkins, L. A. Sargeant, F. M. Whiffin, F. Santomauro, D. Kaloudis, P. Mozzanega, C. D. Bannister, S. Baena and C. J. Chuck, *ACS Sustainable Chem. Eng.*, 2015, **3**, 1526–1535.

47 F. Whiffin, F. Santomauro and C. J. Chuck, *Biofuels, Bioprod. Biorefining*, 2016, **10**, 316–334.

48 J. R. M. Almeida, T. Modig, A. Petersson, B. Hähn-Hägerdal, G. Lidén and M. F. Gorwa-Grauslund, *J. Chem. Technol. Biotechnol.*, 2007, **82**, 340–349.

Faraday Discussions

DISCUSSIONS

Conversion technologies: general discussion

Dimitris Argyropoulos, Harry Bitter, Agnieszka Brandt-Talbot, Vitaliy Budarin, Claudio Chesi, James Clark, Marta Coma, Claudia Crestini, Bruce Dale, Ines Graca, Jason Hallett, Changwei Hu, Xiaoming Huang, George Huber, Trevor Hughes, Andrew Hunt, Eero Kontturi, Yiping Luo, Mark Mascal, Avtar Matharu, Valentina Matveeva, A Mount, Xianhong Ouyang, Roberto Rinaldi, Gadi Rothenberg, Joseph Samec, Simo Sarkanen, Christoph-Maximilian Seidel, Christian Stevens, Vaishali Thaore, Keith Waldron, Karen Wilson, Fei Xie and Douwe Sjirk Zijlstra

DOI: 10.1039/C7FD90049J

Yiping Luo opened the discussion of the paper by George W. Huber by asking: Did you carry out experiments using different ratios of water and solvent? Why did you choose the ratio 80 : 20? Also, regarding recovery of the lignin, after the experiment at 130 °C, is the structure of the lignin different to that of the original lignin?

George Huber responded: For LGO production, we have done experiments using different ratios of water and solvent as described in our recently published *Green Chemistry* paper.[1] The overall yield of 1,6-hexanediol precursors was highest using a solvent containing primarily THF with 1–2.5 wt% H_2O. The reaction rates for the desired reactions are a lot higher in polar aprotic solvents so that is why we choose this ratio. Also, with these solvents, a much lower concentration of acid is required.

According to the NMR analysis, the isolated lignin is very similar in structure to the original lignin. Please see ref. 2 for a discussion of this.

1 J. He, M. Liu, K. Huang, T. W. Walker, C. T. Maravelias, J. A. Dumesic and G. W. Huber, *Green Chem.*, 2017, **19**, 3642–3653.
2 J. S. Luterbacher, A. Azarpira, A. H. Motagamwala, F. Lu, J. Ralph and J. A. Dumesic, *Energy Environ. Sci.*, 2015, **8**, 2657–2663.

Xiaoming Huang asked: You have claimed that using your process you co-produced a lignin stream which still contains a large amount of the native structure based on NMR analysis – have you tested this lignin using a hydrogenolysis depolymerization approach using transition metal catalysts such as Pd/C? This is a good way to test the reactivity of lignin, which has been demonstrated in a recent *Science* paper published last year by the group of Luterbacher.[1] Also, you used multiple steps to produce 1,5-PDO from furfural. Is it possible to shorten the chemical route by integrating some of the steps using a bifunctional catalyst? For example, for the transformation from furfural to DHP, the hydrogenation and dehydration steps might be achieved in one-pot using an acidic bifunctional metal supported catalyst.

1 L. Shuai, M. T. Amiri, Y. M. Questell-Santiago, F. Héroguel, Y. Li, H. Kim, R. Meilan, C. Chapple, J. Ralph and J. S. Luterbacher, *Science*, 2016, **354**, 329–333

George Huber replied: We have not tested the lignin for hydrogenolysis with Pd/C. We are however using other approaches for lignin conversion.

The dehydration and hydrogenation steps cannot be integrated because the dehydration step occurs in the gas phase without a solvent. A hydration step where water is added would be required. The hydration and hydrogenation steps can be integrated, which we have done. However, we did not see much improvement by integrating the hydration and hydrogenation steps.

Vaishali Thaore enquired: In the synthesis of 1,6-hexanediol, were there any byproducts from the two hydrogenation reactions? What were the reaction conditions? Were the conditions the same for both of the hydrogenation reactions? Or have you shown the intermediate formed and then the product 1,6-hexanediol? It is possible to get two isomers – were two isomers observed? What was the yield? Can it be increased?

George Huber responded: The conversion of LGO and HMF to 1,6-HDO is a 3-step process: (i) simultaneous hydrogenation of LGO to Lgol, and HMF to THFDM, over Ni/SiO_2 at 100 °C; (ii) hydrogenolysis of Lgol to THFDM in the presence of HMF-derived THFDM over 1% Pd/SiAl at 150 °C; (iii) hydrogenolysis of THFDM (and the side-product THP2M5H) to 1,6-HDO over a Pt-based catalyst at 160 °C. Both the hydrogenation of LGO to Lgol and the hydrogenation of HMF to THFDM are quantitative, and no by-products were detected. At low conversion, the partially hydrogenated intermediates Cyrene (from LGO) and furandimethanol (from HMF) may be observed. Lgol can be converted with 58% selectivity to THFDM over a Pd/SiAl catalyst at 150 °C, with parallel side-products THP2M5H (20%) and 2MTHFA (17%). The product yield can be maintained in the presence of the HMF-derived THFDM, allowing LGO and HMF to be upgraded simultaneously.

Yes, two stereoisomers were observed in the hydrogenation of LGO to Lgol and the hydrogenolysis of Lgol to THFDM. In the hydrogenation of Cyrene to Lgol, an excess of *exo*-Lgol over *endo*-Lgol was observed over Pd/C (*exo*/*endo* ~ 2.0) and Pd/Al_2O_3 (*exo*/*endo* ~ 4.5), most likely due to steric effects. The Lgol stereoisomer ratio was found not to depend on the temperature, particle size, or conversion over Pd/Al_2O_3 catalysts, but it does depend on the choice of catalyst support. In the hydrogenolysis of Lgol to THFDM over 1% Pd/SiAl, an excess of *cis*-THFDM over *trans*-THFDM (*cis*/*trans* ~ 2.5) was observed. In contrast, HMF hydrogenation produces predominantly *cis*-THFDM. In Lgol hydrogenolysis, the THFDM *cis*/*trans* ratio is independent of the Lgol *exo*/*endo* ratio, indicating that the mechanism involves passing through an intermediate which erases the stereochemistry of the reactant molecule.

Work is ongoing to understand how to improve the selectivity towards the desired products in these reactions. For more details on LGO hydrogenation, see our recent publication.[1]

1 S. H. Krishna, D. J. McClelland, Q. A. Rashke, J. A. Dumesic and G. W. Huber, Hydrogenation of levoglucosenone to renewable chemicals, *Green Chem.*, 2017, **19**, 1278–1285.

Agnieszka Brandt-Talbot commented: The production of 1,5-pentanediol and 1,6-hexanediol requires a lot of hydrogen (4 and 5 equivalents, respectively). For a scaled-up process, where would this hydrogen come from? Do you think it is okay to have a green–brown hybrid product, meaning a renewable carbon backbone surrounded by non-renewable hydrogen?

George Huber answered: We have modeled the hydrogen coming from natural gas, which is the cheapest source of hydrogen. Industrially, the hydrogen would most likely come from natural gas. The hydrogen can come from any source, but producing hydrogen from renewable resources is not a goal of this project.

Claudia Crestini asked: What was the concentration of the acid used in the biomass fractionation? What was the resilience time?

George Huber responded: I believe this question is about the concentration and resilience time for extraction of cellulose, lignin and hemicellulose from the biomass using GVL/water/H_2SO_4 mixtures. The details of this approach are provided in ref. 1. Quoting from the paper: "In a typical experiment, 6 g of grinded white birch and 24 g of a liquid solution of GVL/water 70/30 by weight and 0.1 M sulfuric acid were added to the reactor and heated at 125 °C for 3 h".

1 D. M. Alonso, S. H. Hakim, S. Zhou, W. Won, O. Hosseinaei, J. Tao, V. Garcia-Negron, A. H. Motagamwala, M. A. Mellmer, K. Huang, C. J. Houtman, N. Labbé, D. P. Harper, C. Maravelias, T. Runge and J. A. Dumesic, *Sci. Adv.*, 2017, **3**(5), e1603301, DOI: 10.1126/sciadv.1603301

Changwei Hu said: Have you tried to recycle your solvent? GVL is quite expensive, so this procedure should also be expensive. Have you considered the action of the solvent?

George Huber answered: We have modeled separation and recycling of the solvent using ASPEN. Details of this analysis are provided in ref. 1. For the production of LGO, we use THF. We have also used other solvents. More details about LGO production can be found in ref. 2.

1 D. M. Alonso, S. H. Hakim, S. Zhou, W. Won, O. Hosseinaei, J. Tao, V. Garcia-Negron, A. H. Motagamwala, M. A. Mellmer, K. Huang, C. J. Houtman, N. Labbé, D. P. Harper, C. Maravelias, T. Runge and J. A. Dumesic, *Sci. Adv.*, 2017, **3**(5), e1603301, DOI: 10.1126/sciadv.1603301
2 J. He, M. Liu, K. Huang, T. W. Walker, C. T. Maravelias, J. A. Dumesic and G. W. Huber, *Green Chem.*, 2017, **19**, 3642–3653.

Joseph Samec asked: For LGO to HMF, have you used helium? Also, how do you prevent humin formation?

George Huber replied: We could have used any inert gas for this reaction. The purpose of the gas was to pressurize the reactor. Humin formation can be decreased by working with polar aprotic solvents, wherein the rates for conversion to the desired products are higher.

Vitaliy Budarin commented: In your system, the dehydration of cellulose to LGO and HMF took place under high pressure (around 70 bar). Under these conditions, water produced during pyrolysis could become solubilised in the THF, affecting both the total yield of LGO and HMF and the LGO/HMF ratio. This could make the process uncontrollable. Did you try to remove water from the system during the cellulose activation?

George Huber responded: The high pressure was used to maintain the solvent in a liquid phase. The amount of water that was produced was around 0.5 wt%. Water is soluble in THF under these conditions and the process is controllable. We have tried to use molecular sieves (3 Å) to remove the water. The LGO selectivity did improve using this adsorbent.

Mark Mascal asked: In the paper you write that your cellulose "was converted into LGO and HMF in 48% yield using THF–water mixtures and a dilute sulfuric acid catalyst". Did the mass balance consist of humic material or could you isolate or identify any other components?

George Huber replied: The mass balance consisted of humeric materials.

Xianhong Ouyang opened the discussion of the paper by Christoph-Maximilian Seidel by asking: Did you check for sugars in the water? After steam explosion, the crude material is wet. Is some of the hemicellulose converted to sugars? Is it possible that the enhancement of the cellulose enzymatic digestibility is caused by the hemicellulose release?

Christoph-Maximilian Seidel responded: Yes, we checked the supernatant for sugars, and some were present since hemicellulose is converted to sugars during the pretreatment. Yes, the removal of hemicellulose from biomass usually enhances its enzymatic digestibility. Nevertheless, the yields from the hemicellulosic sugars were the same for the exploded and non-exploded biomass samples, which means it did not have an effect on our study.

Claudia Crestini requested: It has been reported in the literature that steam explosion induces a decrease in cellulose digestibility due to the formation of inhibitors. Could you comment on the discrepancy between your data and the previous literature reports?

Christoph-Maximilian Seidel answered: Yes, it is true that during steam-explosion or hot water pretreatment inhibitors can be formed. According to the literature, they mainly affect the fermentation and have less of an effect on the glucose yields during enzymatic hydrolysis. We tested this with some samples to see if washing the pretreated biomass can have an effect on the enzymatic digestibility, which was not the case. Compared to Söderström's report,[1] who has already stated that washing of spruce wood has no influence on the hydrolysis, there is no discrepancy with the literature. Nevertheless, in a biorefinery, washing of the pretreated biomass may be necessary, especially when simultaneous saccharification and fermentation is done, where inhibitors have a stronger influence.

1 J. Söderström, M. Galbe and G. Zacchi, *Biotechnol. Prog.*, 2004, **20**, 744–749.

Joseph Samec asked: Did you investigate the degree of crystallinity before and after the steam explosion? It will definitely affect the enzyme digestibility. What about the enzyme stability?

Christoph-Maximilian Seidel replied: We did not investigate the cellulose crystallinity. It would be interesting to see if it is also affected by the explosion pressure and we will definitely have a look at it. The enzyme stability we investigated only indirectly through the different enzyme dosages.

Karen Wilson asked: Do you know what controls the particle size of the exploded cellulose? Is it connected to the size of the crystalline domains?

Christoph-Maximilian Seidel answered: We did not investigate the particle size of the exploded cellulose itself, since we do not know if or how it might be possible to determine the cellulose particle size inside the biomass. We investigated only the macroscopic particle size of the biomass through sieving. The particle size of the pretreated biomass is mainly influenced by the pretreatment severity and the explosion pressure difference.

George Huber commented: Concerning the two different feedstocks, the explosion one looks like it would be very hard to feed into a reactor. How would you feed the steam explosion feedstock into a reactor without it becoming blocked?

Christoph-Maximilian Seidel replied: For continuous feeding into the steam reactor, I would suggest something like a screw feeder.

Keith Waldron asked: Did you mill the materials to similar particle sizes in advance of digestion? You say that the steam exploded material had a smaller particle size than the non-steam exploded material. This might have had an effect on diffusion of the enzymes into the material, and on the digested products coming out of the material.

Christoph-Maximilian Seidel responded: No, we did not mill the biomass to similar particle sizes in advance of the enzymatic hydrolysis. However, it is true that the particle size has a strong influence on the enzymatic hydrolysis, which we showed through wet sieving. If we were to mill the biomass in advance of the hydrolysis, probably full cellulose conversion would be reached and then no effects of the explosion itself could be seen.

Roberto Rinaldi requested: I was wondering if you can comment on the influence of the initial porosity of the lignocellulose on the outcome of the steam explosion process?

Christoph-Maximilian Seidel responded: Since we never investigated the initial porosity of our different biomass samples, I can only refer to the literature, where it is stated that corn stover (68–95%) and spruce (around 75%) have rather high porosities and beech a rather low porosity (around 45 percent).[1] Since the explosion has quite a low influence on the digestibility of corn stover compared to beech or spruce there is no direct correlation.

1 (a) J. Zhao and H. Chen, *Chem. Eng. Sci.*, 2013, **104**, 1036–1044; (b) M. Plötze and P. Niemz, *Eur. J. Wood Wood Prod.*, 2011, **69**, 649–657.

Karen Wilson asked: How do you incorporate waste water treatment into your economic/process model? Do you consider recycling it or treating it fully so it can be returned into the normal stream?

Christoph-Maximilian Seidel answered: The steam explosion pretreatment produces the treated biomass, a pretreatment liquor and steam. Only some components of the pretreatment liquor have to be treated, but since there are also sugars within it, it cannot be completely considered as a waste stream. A biorefinery in Crescentino, Italy that uses steam-explosion pretreatment completely treats/recycles the water, which proves that in commercial scale biorefineries with steam-explosion pretreatment, the water can be treated/recycled completely.[1]

1 http://www.betarenewables.com/en/crescentino/the-project.

Dimitris Argyropoulos opened the discussion of the paper by Joseph S. M. Samec by commenting: Your paper presents a very relevant issue as far as biomass utilization and the variability of feedstocks are concerned. Please keep in mind that there have been numerous COST Actions and relevant round robin studies, with tangible published conclusions. The community, especially newcomers to this area, need be informed of and provided with guidance on the many diligent early efforts of the lignin, cellulose and biomass, paper and pulp communities.

Joseph Samec answered: I agree. Here, editors, reviewers and the scientific community as a whole have an important role in ensuring correct knowledge transfer. If we had reference samples for biomass, maybe these studies would progress the research field. I am aware that even then, research groups will come to different conclusions. Even though the publication of tangible conclusions may seem bad for a community, I personally think that it shows that we are active in a very exciting research area where there is a real discourse. Research communities have probably solved even more difficult problems, such as those involving biochemistry and the genome.

Claudia Crestini commented: It should be taken into consideration that the extreme diversity and variability of biomass feedstocks and biomass derived materials as well as their intrinsic reactivity causes the identification and selection of "standards" to be a challenging task. The scientific community should not be oblivious to the previous efforts aimed at standardising specific analytical protocols using common samples through round robin efforts carried out within several COST Actions. This approach allowed critical improvement and to some extent standardisation of the available analytical techniques within the research groups involved, allowing for meaningful comparisons to be made with significant ramifications in different research areas.

Gadi Rothenberg commented: I completely agree that sustainable sources indeed need reliable standards, but this is not going to be easy to implement unless stakeholders start asking for it. Nor is it going to be simple. As part of the CatchBio project in the Netherlands, for example, we participated in a round robin analysis of lignin, which took many months to complete and even then it was still difficult to agree on what we were actually reporting. Comparison references are probably more feasible, and hopefully can be standardised.

Joseph Samec responded: We are in limbo as the stakeholders have not awakened yet. Personally, I think that governments should step in and bridge this transition as the U.S. Department of Energy does with the National Renewable Energy Laboratory. In Europe, Asia *etc.*, the required institutes should be formed.

Keith Waldron stated: Even when researchers quote a method in a research paper, there is always the possibility that they have used the method incorrectly. Standards could be helpful as a type of quality control between laboratories.

Joseph Samec replied: I totally agree! The comprehensive methods out there are multistep and make use of analytical instruments, so there is at least a chance that the analysis may vary depending on technical skills and instrumentation. Standards would overcome this to some extent.

Simo Sarkanen commented: Reliable analytical methods are indeed of central importance for the effective valorization of lignocellulose. One of the most egregious examples may be found in the notorious lack of reliability in the molecular weight distributions of lignin samples. The problem can only be resolved by using absolute molecular weight determinations (light scattering or ultracentrifuge sedimentation equilibrium measurements, for example) to calibrate empirical distribution profiles. How should the matter be handled system wide?

Joseph Samec responded: Experts in the field, of which you are certainly one, should discuss, develop, evaluate, and agree what methodologies are best suited to evaluate different parts of the feedstock and even products from fractionation/valorization. If we had standardized feedstock samples, researchers could still compare their methodology with each others in terms of the yields and efficiency of valorization. Also, analytical chemists could compare new feedstock samples to standard samples and evaluate/develop novel analytical techniques.

Karen Wilson opened a general discussion based on the papers by George W. Huber, Christoph-Maximilian Seidel and Joseph S. M. Samec by asking: Does anyone have an opinion on which feedstocks steam explosion is most suitable for? Do any of you have any comments on the different pre-treatment methods, and which may be more suitable for certain biomass sources?

Changwei Hu responded: Different kinds of biomass materials can be treated differently – that is what the community is currently doing.

Bruce Dale commented: If industry is going to be persuaded to commercialize biomass conversion processes, adequate comparative data is needed so that companies can choose among the many alternatives. Therefore, researchers doing pretreatment and conversion studies should collaborate to compare yields, rates and titres for different pretreatment and conversion approaches under the same conditions using multiple biomass materials. The group effort led by Dr Charles Wyman about 15 years ago provides a good example of how to do this.[1]

Some of our recent publications with collaborators compare the AFEX, ionic liquid and dilute acid pretreatments from sugar yields through to ethanol yields.[2] The whole system really needs to be compared, since pretreatment is fundamentally about system-wide economics. I really think pretreatment researchers should provide mass balance information at a minimum, as George Huber suggests, and whenever possible we should work with other experts to compare different pretreatments, or use the same pretreatment on different biomass materials, as in Keith W. Waldron's paper (DOI: 10.1039/C7FD00044H).

1 (a) C. E. Wyman, V. Balan, B. E. Dale, R. T. Elander, M. Falls, B. Hames, M. T. Holtzapple, M. R. Ladisch, Y. Y. Lee, N. Mosier, V. R. Pallapolu, J. Shi, S. R. Thomas and R. E. Warner, *Bioresour. Technol.*, 2011, **102**, 11052–11062; (b) C. E. Wyman, B. E. Dale, R. T. Elander, M. Holtzapple, M. R. Ladisch and Y. Y. Lee, *Bioresour. Technol.*, 2005, **96**, 2026–2032.
2 N. Uppugundla, L. da Costa Sousa, S. P. S. Chundawat, X. Yu, B. Simmons, S. Singh, X. Gao, R. Kumar, C. E. Wyman, B. E. Dale and V. Balan, *Biotechnol. Biofuels*, 2014, **7**, 72, DOI: 10.1186/1754-6834-7-72.

Changwei Hu replied: Systematic considerations will be also important.

Jason Hallett commented: I disagree with the idea that there cannot be a universal approach to pretreatment. It is possible to have a generic pretreatment process (ionic liquids are one example) that can be tailored based on feedstock considerations (lignin content, biomass composition, *etc.*) in a manner representative of process control. As long as the solvent/medium is adjustable, this is possible. It is the premise of using acidic ionic liquids.

Bruce Dale replied: If you consider the pretreatment as a stand-alone process, you can perhaps think this way. If instead you think of the pretreatment as embedded in a system where the composition and moisture content of the biomass, the composition of the enzymes used, the amount and composition of the sugars produced by hydrolysis, the need (if any) for wash streams, the recovery of the pretreatment chemicals, the effect of pretreatment on the fermentation process and the waste streams produced are all important, I think one's point of view changes. I do think there could be a "universal approach" for pretreatment, but it must incorporate much, much more than just pretreatment and hydrolysis; it must address all of these other system-level issues. I think that pretreatment design must include consideration of the yield of all the potentially fermentable sugars in the biomass, the concentration of these sugars and the rates of both hydrolysis and fermentation. In other words, it must provide most, if not all, of the data needed for economic analysis and comparison with other pretreatment technologies. At a minimum, there must be a mass balance. If it is to be more than an academic exercise, pretreatment research needs all these data to be useful. These are issues we have tried to summarize in ref. 1, a paper on why and how pretreatment research must improve.

1 B. E. Dale and R. G. Ong, *Biotechnol. Prog.*, 2012, **28**, 893–898.

Bruce Dale also commented: The choice of pretreatment is fundamentally an economic question. For example, we can grind any biomass fine enough so that it will be converted to sugars quantitatively. Or we can add a lot of the enzyme and obtain high yields that way. But we can never afford to do either of those things so those approaches, and many, many other pretreatment approaches, are irrelevant to building a real industry because they never will be used in practice. All pretreatment research should be accompanied by a mass balance and the beginnings, at least, of an economic analysis. (A mass balance is the foundation of an economic analysis.)

Historically, in the petrochemical industry and the petroleum refining industry, the key economic parameters for the refinery are yield, concentration, rate and scale. These four parameters are likely to have the same importance for biorefineries. I hope we, as a bioeconomy community, will pay attention to them and report them for our own systems.[1]

1 B. E. Dale and R. G. Ong, *Biotechnol. Prog.*, 2012, **28**, 893–898.

Karen Wilson asked: There has been a lot of discussion about problems relating to the reproducibility of results from different groups – how do we address this as a community? Should we also be standardizing the product or just the feedstock? How far through the conversion process do we have to get before any work is deemed reliable?

James Clark communicated: Bio-based products will soon be standardised (through the European CEN TC411) and this will include issues such as the sustainability of the feedstock and hopefully of the process. To some extent we will need to live with the natural variability of biomass and perhaps learn to focus more on the key properties and application potential of partially refined mixtures, and not always chase pure single compounds as our traditional chemistry education has taught us.

Dimitris Argyropoulos commented: How can we hope to see products in the market when we cannot be 100% sure that we can deliver those products to a consistent specification? Our community needs to make sure we are acutely aware of such variations and ensure that process and product variability is kept at a minimum.

Joseph Samec replied: Biomass will of course vary and maybe the stakeholders need to have this variation in mind. I think that it would be worthwhile for the scientific community to try to understand what factors are important in these variations. With knowing what factors are important, measures to ease these variations or adjust the technology to buffer variations in the feedstock would be feasible

Avtar Matharu made a general comment: Feedstock-agnostic is a term that has appeared and is often used to imply that a process is feedstock-independent. However, the meaning of agnostic is "a person who believes that nothing is known or can be known of the existence or nature of God. Synonyms: sceptic, doubter, questioner, doubting Thomas, challenger, scoffer, cynic". Thus, the use of feedstock-agnostic is incorrect if we are implying "feedstock-independent". I believe the term 'agnostic' may have been used in error instead of 'agonist' and 'antagonist'. May I recommend the use of the term "feedstock-independent" from now on.

Karen Wilson asked: Should journals be applying stricter guidelines on what they will accept in terms of baselines? What should the bare minimum be that we expect from papers for publication? Should this be, for example, reporting full elemental analysis or the mass balance, or acceptable temperature ranges for the reaction?

James Clark communicated: Perhaps we can follow the guidelines in the new standards (CEN TC411) and require information on the origin and sustainability of the biomass. However, we will also need to remember that there has been a largely unsuccessful campaign over many years to get more "green checking" into the publication process, *e.g.* regarding the toxicity of new compounds.

Dimitris Argyropoulos stated: Some journals have asked us to provide crystallographic data, I ask for this to be replaced with data that describes the biomass.

Harry Bitter commented: On the question of whether standards are needed or not with respect to feedstocks and materials, I can see why this would be required from an application point of view. However, from a fundamental point of view, I am not sure whether this is needed. In a scientific approach, one should always compare his/her results with others or with the state of the art. So when taking a scientific approach, a kind of standard is present inherently.

Christian Stevens remarked: Renewable resources are, almost by definition, of an altered quality every year due to changes in the weather conditions *etc.*

The food industry deals all the time with renewable resources that are always changing in quality, so a lot can be learned from this field. Even if it is impossible to have a good standardized resource, the "Champagne standard" technique could be used. Grapes are of a different quality, however a fairly constant quality of Champagne can be produced each year by mixing batches of grapes from different seasons with a varying quality. This technique is also used to obtain flax fibre batches of a constant quality.

Joseph Samec responded: Having a "pool" of several blends of a certain biomass, *e.g.* birch, could definitely serve as a reference for this particular biomass. This could be a fast way to generate reference samples for the community. It would also be worthwhile to develop an understanding of the individual contributions of the different samples. However, this would take a longer time to attain.

Agnieszka Brandt-Talbot asked: The petroleum industry is able to utilize a wide range of different raw oils, gases and coals (an oil from Saudi Arabia does not have the same composition as a Canadian tar sand) – regarding coping with the variability of lignocellulosic biomass for conversion to fuels and chemicals and standardization of analytical techniques, can our field learn from the fossil fuel industry?

Joseph Samec communicated in reply: We certainly have a lot to learn from the petroleum industry, especially in terms of efficiency. It is correct that crude oil varies a lot. However, one specific refinery cannot take in different feeds in any proportion but rather can blend the incoming feeds to certain specifications, which involve a mix of higher and lower boiling hydrocarbons (related to the suggested Champagne blend for biomass). A lot of the operations in refineries are turn key solutions delivered by a catalyst company that ensures that the specifications are upheld.

At the same time, certain refineries specialise in specific blends such as the Venezuela blend that is very heavy. I think that the Champagne blend idea proposed during this meeting is a great idea for a reference sample and also that different specific reference samples should be provided so that researchers have both samples for the comparison of processes and also for analytical studies.

Dimitris Argyropoulos commented: One means and a possible viable step toward accountability and the standardization of feedstocks could be *via* the use of paper submission portals. For example, the journal *Green Chemistry* requires evidence for the characterization of new compounds, which can include crystallographic data or other results. In a similar manner, journals may want to ask for fundamental biomass characterization data when relevant papers are submitted.

Ines Graca opened the discussion of the paper by V. G. Matveeva by asking: What is the advantage of using an immobilised enzyme instead of 100% enzyme? Do you know what the influence of the support material is *e.g.* the effects of silica *versus* alumina?

Valentina Matveeva responded: The advantage of the immobilized enzyme is that it is reusable. When selectivity is 100%, the yield is 90%. The influence of the support on the enzyme is a small decrease of the active centers.

Mark Mascal asked: How long does the reaction take? What is the scale and loading? Have you done comprehensive analysis of the available catalytic methods? The enzymatic methods have great selectivity, but you can process a lot of material in a short space of time using catalytic oxidation.

Valentina Matveeva replied: In a batch reactor, the reaction proceeds for 60 min. We used an optimal 4 wt% glucose solution. We have done comprehensive analysis of the available catalytic methods. The concentration of glucose used for the enzymatic oxidation is lower in comparison with other catalytic methods, but the selectivity and purity of the product are higher.

Marta Coma queried: A magnetic support is used so that you can reuse the enzyme – how many times can you reuse it? Did you try to reuse it more than 10 times? Is there a cycle number at which the activity decreases?

Valentina Matveeva replied: We reused the magnetic biocatalysts 10 times and the activity was stable.

Karen Wilson asked: What is the absolute activity of your immobilised enzyme? You need to look at the kinetics as well as the product yields. Are there any plans to look at continuous flow reactions or actual measurements of the kinetics of the reaction? How do you compare the activities of your materials on different supports if the enzyme loading is different?

Valentina Matveeva responded: The absolute activity of the immobilised enzyme was measured according to standard methods. We plan to look at continuous flow reactions and we plan to perform a detailed study of the kinetics.

We compared the activities of the biocatalysts on different supports using the same loading of the enzyme.

Karen Wilson asked: What are the challenges on going from batch to flow processes, when using immobilised enzymes?

Valentina Matveeva responded: In this case, it is important to perform the correct calculations and design the flow reactor correctly. The same high selectivity and conversion should be obtained using the flow reactor as in the batch process.

Karen Wilson said: Your method uses a crosslinking approach, how similar is that to the approach used by CLEA Technologies?

Valentina Matveeva replied: The enzymes in the Cross-Linked Enzyme Aggregates (CLEAs) are covalently bound to each other through a cross-linker. In many cases, this increases the stability of the enzyme against pH, temperature, solvents or proteases. Leakage is extremely low to non existent.[1] We use a similar approach.

1 http://www.cleatechnologies.com/products/cross-linked-enzyme-aggregates-cleas/.

Gadi Rothenberg remarked: In your paper, you report reactions using a 4 wt% solution of glucose. This will have to increase notably for any practical application, and concentrated glucose solutions behave very differently to dilute ones (for a comparison, see ref. 1). Do you expect to get substrate inhibition of the enzyme at higher concentrations?

1 Z. W. Srokol and G. Rothenberg, *Top. Catal.*, 2010, **53**, 1258–1263.

Valentina Matveeva replied: When the glucose concentration is more than 10%, inhibition of the enzyme by the substrate occurs. The influence of the glucose concentration on the activity of the catalyst was studied at a concentration of 4 wt% and inhibition was not observed.

A Mount commented: One possible interpretation of Fig. 5 and 6 in your paper is that there are two states of immobilised enzyme, giving rise to two peaks in each of these plots. The first peak in each case (at a lower pH and lower temperature, respectively) appears to give rise to a similar response to that of an enzyme in solution and could therefore correspond to an immobilised enzyme which is in a similar environment to that in solution, whilst the second peak at a higher pH and higher temperature could correspond to a form of immobilised enzyme which is in a different environment to that in solution and therefore gives rise to a different response. Such a two-state system could result in bimodal kinetics – is there any evidence of this? Would optimisation of the surface conversion reaction then require consideration of how to optimise the catalytic activity of one or both of these immobilised forms of enzyme (*e.g.* their reaction kinetics, degree of conversion and/or stability)?

Valentina Matveeva answered: We agree with this comment. We plan to perform a detailed study of the kinetics.

Karen Wilson asked: When you compare different supported enzymes, how do you assess them? Are you measuring activity per gram of supported material? Do you have the same number of enzyme macromolecules loaded on each support?

Valentina Matveeva replied: The absolute activity of the immobilised enzyme was measured according to standard methods.

Mark Mascal opened the general discussion of the paper by Eero Kontturi by asking: Does your treatment with formic acid generate formyl esters at the surface?

Eero Kontturi answered: No – or at least not to a substantial extent. This has already been published by another group who pioneered the use of formic acid as a dispersing agent for cellulose nanocrystals.[1]

1 O. van den Berg, J. R. Capadona and C. Weder, *Biomacromolecules*, 2007, **8**, 1353–1357.

Gadi Rothenberg commented: During your presentation, you asked "why does formic acid work so well?". The reason could be due to the fact that formic acid molecules have basically no organic residue (just a hydrogen atom, and that is very little). The formic acid molecules could adsorb (or, more correctly, envelop) the crystallites of the cellulose and many of them could fit right next to each other. They would interact with the –OH groups on the crystallite surface, forming a close-packed solvation shell.

To examine this, I suggest that you try using acetic acid (glacial). If there is a large difference between the results with the acetic and formic acids, this would indicate that it is perhaps the methyl group that causes the difference.

Additionally, formic acid is known to form dimers in its pristine state, and acetic acid also does this, but the dimers of formic acid would be much more stable (in my opinion). Perhaps the dimers of formic acid can still solubilise the cellulose crystals, but the dimers of acetic acid cannot?

George Huber asked: The Forest Products Laboratory[1] has developed a technology to produce nanocellulose during the last 10 years. They have had difficulty finding customers who want the nanocellulose. How different is your nanocellulose to their technology? Can you compare the two technologies?

1 https://www.fpl.fs.fed.us/research/facilities/nanocellulose_pilot-plant.php.

Eero Kontturi responded: The pressing issue with nanocellulosics, especially cellulose nanocrystals, is: what is the application? Many players have been swift to introduce a production line before an actual application exists. Hence the lack of customers. From an application perspective, however, the price of production is a decisive issue. Lower prices encourage the development of more diverse applications. In this respect, this method has potential to outlive all the present methods which are based on a liquid/solid system. The gas/solid system presented in our paper overcomes many problems, *e.g.* purification, of the state of the art production systems. Also, the yield is superior. In addition, recycling of the gaseous acid is far more effortless than for a liquid acid that is confined within a reaction mixture. All these factors are bound to bring down the price of cellulose nanocrystals, if a production line relying on the gas/solid system is set up.

However, our study is fundamental in nature and I cannot provide any realistic quantitative figures to compare the methods in practice.

Fei Xie asked: Have you investigated the chirality of your cellulose nanocrystal (CNC) samples made using the HCl vapour method?

Eero Kontturi responded: No, we have not looked into that. Basically, there is no reason as to why they should not form chiral nematic liquid crystal phases like the CNCs that are produced with the normal sulphuric acid method in aqueous dispersions.[1] The cellulose crystal is *de facto* chiral and this is the reason for the chiral nematic phase formation. This has been shown to occur also in non-aqueous systems,[2] so even chiral nematic liquid crystal formation in the proof-of-concept dispersion of these CNCs in formic acid should be possible if the concentration becomes higher than the nematic limit. That said, the chirality can depend on a number of factors. In aqueous systems, a CNC rod that is very highly charged can lose its chirality because of long range electrostatic forces that form a substantial charge cloud around the crystal.[3]

1 X. M. Dong, J.- F. Revol and D. G. Gray, *Cellulose*, 1998, **5**, 19–32.
2 S. Elazzouzi-Hafraoui, J.-L. Putaux and L. Heux, *J. Phys. Chem. B*, 2009, **113**, 11069–11075.
3 J. Araki and S. Kuga, *Langmuir*, 2001, **17**, 4493–4496.

Avtar Matharu commented: There is a patent that reports use of morpholine to produce nanocellulose, which would avoid the use of formic acid. Also, we have just published an acid-free hydrothermal microwave-assisted selective scissoring (Hy-MASS concept) of cellulosic residues to yield nanofibrillar cellulose and nanocrystalline cellulose.

Karen Wilson asked: You are using HCl vapour in your process – what is the general safety of this and would industry find this acceptable (if you are looking to scale up the process)?

Eero Kontturi answered: We have recently scaled up the process, in the laboratory, in order to accommodate some tens or hundreds of grams of cellulose in one batch. This was done with a reactor that utilises dry gaseous HCl, instead of the HCl vapour used for our present paper. To my knowledge – although I am not an expert on this – use of dry HCl is not an issue. It is used in the chemical industry in huge quantities and possesses very few environmental concerns if handled correctly.

Karen Wilson asked: What happens to the amorphous cellulose during your process?

Eero Kontturi replied: The fate of the amorphous cellulose was elaborated on within our preceding paper.[1] In brief, the crystallinity of the cellulose increases simultaneously as it degrades. This happens without any mass transfer in a gas/solid system, thus, it is not an artefact of varying mass balances. We have a thermodynamic hypothesis for the reasons behind the crystallization, which is presented in ref. 1.

Regarding amorphous cellulose, in general, most chemists and physicists working with fundamental cellulose science do not like this term. The terms 'disordered cellulose' and 'dislocations in the crystallites' are more favoured. This is because the disordered segments in native cellulose microfibrils are very short indeed, on the order of perhaps 4–5 anhydroglucose units (*ca.* 2 nm).[2] They are completely different to the bulky amorphous domains visible in semicrystalline synthetic polymers.

1 E. Kontturi, A. Meriluoto, P. A. Penttilä, N. Baccile, J.-M. Malho, A. Potthast, T. Rosenau, J. Ruokolainen, R. Serimaa, J. Laine and H. Sixta, *Angew. Chem., Int. Ed.*, 2016, **55**, 14455–14458.

2 Y. Nishiyama, U.-J. Kim, D.-Y. Kim, K. S. Katsumata, R. P. May and P. Langan, *Biomacromolecules*, 2003, **4**, 1013–1017.

Simo Sarkanen opened the general discussion of the paper by J. P. Hallett by asking: Is there any information on the purity of the lignins (in regard to ash and carbohydrate content) that are isolated using the ionoSolv pretreatment?

Jason Hallett responded: The ash content of the lignins is below detection (much less than 1%). Carbohydrates are only apparent at the "partially cooked" stage, and then they are hydrolysed and removed by the full pretreatment.

Joseph Samec asked: Previously, in my work, we analyzed an eutectic solvent extracted lignin. The group that provided the lignin sample claimed they had washed the sample many times. However, we only saw the solvent peak by NMR. Did you have any problems with getting rid of the solvent in your lignin samples?

Jason Hallett replied: It is a bit easier with the ionic liquids used than with deep eutectic solvents. We did not detect any cations (no N-content *via* CHNS analysis) in the lignin after minimal washing. We did detect some S-content, which has been analysed as $CaSO_4$ crystals (it is likely that some Na_2SO_4 and K_2SO_4 are also present).

Christian Stevens remarked: Many people are using FT-IR analysis and are drawing conclusions from very minor peaks. Do you feel that these conclusions are fair and that they can be trusted since biomass is a very complex matrix?

Changwei Hu responded: It is true that the composition and structure of biomass materials are complex, and that the baselines of the FT-IR spectra are strong while the variations are small. For a qualitative purpose, I think these data can give information on the variation in the biomass components and some of the interactions among these components.

Mark Mascal queried: Do you know the degree of polymerization in the residual cellulose? Does it need to be bleached?

Jason Hallett answered: We have not measured this for hardwoods. For grasses (miscanthus), the degree of polymerisation reduces by about 25% during the phase of pretreatment in which the lignin is removed. For softwoods, the effect is less pronounced.

George Huber remarked: How do you propose to recover the solids from the ionic liquids? Have you tracked the different minerals in the biomass to make sure they do not accumulate in the ionic liquid? Also, how do you recover the solvent from the lignin solid?

Jason Hallett responded: The solid recovery is currently done using filtration, although there might be better options (*e.g.* centrifugation).

We do track the minerals (Na, K, Ca, Si and others) and these tend to either crystallise in the cellulose pulp or complex with the lignin, depending on the conditions. Generally, mineral salts are not very soluble in ionic liquids.

The lignin is washed with water, and that water is added back into the liquor. In a continuous process, it would be used for the lignin precipitation.

Douwe Sjirk Zijlstra said: The saccharification yields from the 3rd and 4th recycle of the ionic liquid are increased compared to the first two runs – what is the reason for this increase?

Jason Hallett replied: This is quite an interesting observation, and it is generically true that the recycled acidic ionic liquids offer higher saccharification yields. We attribute this to a slight reduction in the acidity of the recycled solvent, which has been partially neutralised through the cleavage of esters, such as acetate, in the biomass. This leads to a higher glucan content in the pulp (less glucan is hydrolysed), and therefore a higher saccharification yield. The actual enzyme activity is not appreciably higher.

Karen Wilson remarked: You analysed the fractionated biomass for nitrogen species to make sure no decomposition of the ionic liquid had occurred. How accurately can you measure the nitrogen content? How confident can you be in that analysis? Are better methods needed?

Jason Hallett answered: We used standard CHNS analysis, which is only accurate to 0.30%. However, this is enough for our purposes.

In general, yes, a much more refined analysis method would be beneficial, as would some way of distinguishing between different nitrogen sources (cations *vs.* proteins, for example).

Simo Sarkanen opened the discussion of the paper by Vitaliy Budarin: The isolation of 93% pure lignin using microwave-assisted acidolysis of softwood lignocellulose is a truly remarkable accomplishment. How much of an increase in the purity of the lignin can be achieved by extending the reaction time beyond 10 minutes? Is this lignin soluble in any solvent, solvent mixture or solution?

Vitaliy Budarin replied: Extension of the reaction time increases the impact of secondary processes. The obtained lignin is insoluble in water and ethanol.

Eero Kontturi remarked: The work that you do appears to be the inverse of purified cellulose production, almost a form of inverse pulping or holocellulose production. Have you done any microscopy on the remaining lignin? Is it a porous material? One would expect it to be porous in the same way that a delignified fibre is porous.

Vitaliy Budarin replied: Investigation of the textural properties of the obtained lignin should be very important. However, unfortunately, we have not yet done any microscopy or nitrogen adsorption measurement with this material.

Changwei Hu asked: You used only water as the solvent – what is the advantage of the microwave method compared with conventional heating methods?

Vitaliy Budarin responded: At the moment, we do only microwave-assisted hydrolysis. The microwave irradiation very efficiently interacts with water molecules, therefore some advantages of microwave heating (such as the fast heating rate and energy efficiency) are more pronounced in the case of hydrolysis. However, other microwave features, such as specificity, could be less pronounced in the presence of water. So, there is always an optimum water to sample ratio.

Andrew Hunt queried: The solution of sugars obtained from acidolysis was used to do the fermentation – did you see any inhibition of the fermentation due to other molecules generated in the process?

Vitaliy Budarin responded: The fermentation was not affected by the co-products from microwave-assisted hydrolysis

Claudio Chesi said: Some residual acid-soluble lignin may still be present in the saccharide solution after the process. This lignin content may inhibit enzymatic fermentation of the sugars to some extent. Have you assessed the potential inhibiting effect of such a residual lignin content on the fermentation of the saccharides?

Vitaliy Budarin answered: Yes, we have found some phenols in the hydrolysate. However, the concentration of these phenols is low and does not effect the fermentation

Mark Mascal remarked: From an industrial perspective, is it practical and cost-effective to process a hundred tonnes of feedstock a day using microwaves?

Vitaliy Budarin replied: Microwave technology is very developed, especially in the areas of the food industry, wood drying and ceramic production. The microwave power used in these applications is sometimes higher than 1 megawatt. The standard cost of a 100 kW magnetron is around £100 000. Therefore, I believe that microwave-assisted activation of biomass can be cost-effective and could be applicable for megatonne scale processes.

Karen Wilson opened a general discussion based on the papers by V. G. Matveeva, Eero Kontturi, J. P. Hallett and Vitaliy Budarin by stating: It is clear that to advance the uptake of bio-products by industry and make processes more viable, we need a better fundamental understanding of how biomass properties are affected by different processing conditions. To achieve this, we require a more detailed understanding of the reaction mechanisms and the impact of impurities on these mechanisms. Such fundamental work is a big undertaking, so what are the current thoughts as to how we can convince industry or other funding bodies that this is something important that they should invest resources into?

James Clark communicated: Hopefully this will be seen as important very soon. The chemical industry is slowly moving towards bio-resources, but the level of underpinning understanding will need to improve if they are to become major feedstocks. Chemists are beginning to be seen as essential in the bio-resources area, although still not as well represented in terms of numbers or key science inputs as biology. Chemists are seen as valuable for developing downstream conversions to important bio-based chemical products – we now need to show our value in developing a core understanding of biomass and its deconstruction to chemicals, fuels and materials.

Trevor Hughes commented: From my perspective within the oil industry, I would like to add that, in some situations, a multi-component (and potentially cheaper) mixture produced by processing biomass may well be useful for some specific applications. Andrew Hunt presented a very interesting paper (DOI: 10.1039/C7FD00049A) on producing a pure bio-solvent, whose solvency properties were characterised using Hansen solubility parameters. Perhaps a similar methodology could be used to produce multi-component solvents that have useful applications as industrial solvents/cleaners?

Karen Wilson asked: If we want a better understanding, do we need better analytical techniques? Where are the challenges in biomass analysis? Are there any dream measurements you would like to make or techniques that you would like to see developed?

Changwei Hu answered: I think that new analytical instruments or methods should be developed to confront the complexity of the biomass systems.

Karen Wilson queried: Are there any opportunities to use computational methods to understand the mechanisms of biomass conversion? What role does computational modelling have to play?

Changwei Hu responded: Computational methods could be applied to model the structural variation of biomass materials.

Faraday Discussions

PAPER

Feedstock selection for polymer and chemical production: feedstock-specific recalcitrance

Ian P. Wood, Enriqueta Garcia-Gutierrez, Nikolaus Wellner and Keith W. Waldron*

Received 30th January 2017, Accepted 8th February 2017
DOI: 10.1039/c7fd00044h

Plant cell wall materials derived from a range of waste biomass sources have great potential as a source of sustainable alternatives to petrochemicals. Perhaps the most straightforward way of realising this potential would be to hydrolyse the most efficiently fermentable polymers into their constituent sugars and use yeast to ferment these into useful chemicals. However, it also makes sense to pre-extract components which have a greater value in polymeric form. This is particularly true for non-cellulosic polymers, which are rich in poorly-fermentable pentose sugars. Liquid hot water (LHW) pretreatment can be used to extract non-cellulosic carbohydrates in a cost-effective manner, leaving a cellulose-rich substrate which is easier to hydrolyse using commercial cellulases. However, inherent differences in the plant cell wall structure and composition mean that some biomass sources may be more suitable for exploitation than others. Here, we examine eight different feedstocks (two each from hardwood, softwood, cereal straws and dicotyledonous crops), expose them to 26 different LHW pretreatment conditions and hydrolyse the entire pretreated slurry with a commercial cellulase. This enables side-by-side comparisons, in terms of saccharification yield, of the feedstocks. The results clearly demonstrate considerable differences in suitability between the feedstocks, in relation to the quantity of products released and the processes needed to obtain them.

1. Introduction

Plant-based carbohydrates have great potential as feedstocks for biotechnology. Traditionally, easily accessible but potentially edible sources of fermentable sugar, such as sucrose or enzymatically hydrolysed starch, are used for fermentation. However, if the fermentable sugar constituents of lignocellulose could be efficiently released, non-edible forms of waste biomass could also be converted

Quadram Institute Bioscience, Norwich Research Park, Colney, Norwich NR47UA, UK. E-mail: ian.wood@quadram.ac.uk; enriqueta.garcia@quadram.ac.uk; klaus.wellner@quadram.ac.uk; keith.waldron@quadram.ac.uk; Fax: +44 (0)1603 507723; Tel: +44 (0)1603 255385

into usable products.[1] Ideally, a range of biomass sources from various origins could be used. Perhaps the most usable would be primary "waste" residues, such as those produced by forestry and agriculture,[2] which are collected in bulk, have undergone limited degradation, and are therefore more predictable and uniform in composition.

To convert lignocellulose into a form that can be degraded by hydrolytic enzymes, the structure must first be disrupted using a pretreatment (Fig. 1). Many potential pretreatments exist for lignocellulose, including physical, chemical and biological processes.[3] One of the most plausible routes for exploitation is pretreatment using liquid hot water (LHW), which largely relies on the autocatalytic breakdown of non-cellulosic polymers at high temperatures. Liquid hot water has several advantages, the biggest being that the estimated capital cost for a LHW facility is lower compared to chemical pretreatments.[4] Unlike other pretreatments, the equipment needed for LHW pretreatment does not have to withstand the corrosion caused by additional chemicals (acids or alkalis). Fermentable sugars released from these processes can be converted to a range of petro-chemical substitutes using yeast or other microorganisms (Fig. 1).

Currently, the main bottleneck for this system is the hydrolysis of lignocellulose to its constituent sugars. Biomass from lignocellulosic sources is more or less resistant to degradation in its native form – reflecting its main evolutionary purpose, to provide protection from pathogens and the environment. Enzymatic hydrolysis of biomass cellulose is strongly influenced by at least three factors: feedstock composition, pretreatment severity and cellulase composition. Changing any one of these factors is likely to have an impact on the final saccharification yield. Yet, with no standard assay for saccharification potential, it

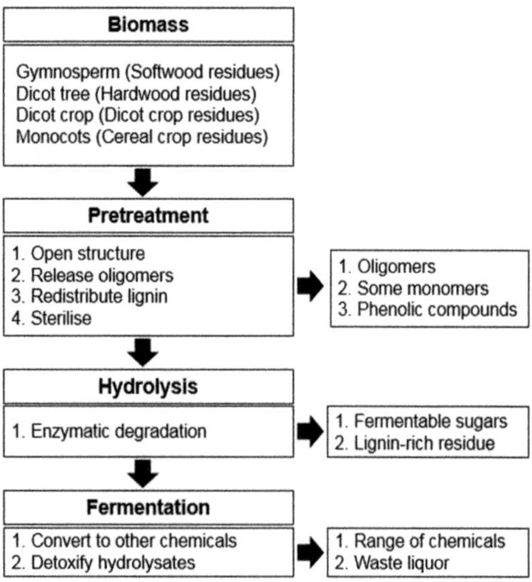

Fig. 1 Simplified process diagram for the conversion of biomass into chemicals and the products generated at each stage. Fundamental differences in the cell wall chemistry of the initial feedstock will influence the yield and nature of products generated during processing.

is difficult to make clear and reliable comparisons between feedstocks or genotypes.[5]

For pretreatment at a low severity, the saccharification yield is limited by the accessibility of the substrate to hydrolytic enzymes, but at higher severities, the yield is determined by the amount of thermochemical degradation of the fermentable products to *e.g.* furfural derivatives.[5] Therefore, assays restricted to single conditions rarely address feedstock-specific responses to pretreatment severity fully. For example, the same yield could be obtained from two feedstocks, which differ in recalcitrance, if the conditions were not intense enough for one but too severe for the other.

Also, biomass is generally considered as a homogeneous resource that is limited by availability.[6] However, biomass composition is not uniform. Most exploitable sources of biomass, which could be used for provision of a sustainable chemical industry in temperate climates, may be derived from: (i) gymnosperm (softwoods), (ii) dicotyledonous plant (herbaceous crop residues such as oilseed rape straw), or (iii) monocotyledonous plant (cereal residues) cell walls. The main chemical constituents of gymnosperms, dicots and monocots differ, with important implications for their use. Gymnosperm cell walls are generally rich in glucomannan and lignin which is comprised of almost exclusively guaiacyl-subunits.[7] Unlike gymnosperms, dicot-derived biomass contains more glucuronoxylan and pectic components instead of glucomannans, and the lignin contains a greater variety of subunits. The majority of agricultural residues are derived from grass species,[6] which differ from both dicot and non-grass monocot cell walls as they are rich in arabinoxylan and mixed-linkage glucan, as opposed to pectic components.[7]

Considerable variation in composition and structure also exists within these groups and even within species. This includes both genetic and environmentally determined variation in growth, including tissue differentiation/various degrees of cell wall thickening. Examples include the varied compositions of wheat straw between accessions, largely caused by variation in the proportions of different tissues.[8] This can make it difficult to make clear comparisons between studies as both the reactor configurations and original material composition are likely to differ.

Some studies have begun to explore side-by-side comparisons of feedstocks, noting differences between hardwood and softwoods,[9] different agricultural residues,[10] different genotypes[11–13] and tissues.[14] However, there are few comprehensive, direct comparisons of feedstock recalcitrance using a consistent range of pretreatment conditions.

Here, we discuss eight feedstocks that differ widely in their cell wall composition – softwood chippings (cedar, pine), hardwood chippings (beech and poplar), monocotyledonous straws (wheat, rice straw) and non-tree dicot residues (oilseed rape straw and hemp fibre). The samples were milled to pass through a 250 μm sieve, pretreated with varying severity (140–210 °C, 2.5–160 min) in liquid hot water (LHW), and the whole-slurries were enzymatically saccharified using excess cellulase. The resulting monomeric glucose (Glc) and reducing sugars (RS) were quantified to obtain a comparative overview of the response of difference feedstocks when processed using LHW at different pretreatment severities.

2. Experimental

2.1. Materials

Dust-extracted oilseed rape and hemp residues were sourced from Hemp Technology Ltd. Suffolk, UK and used as examples of dicotyledonous crop residues. Virgin de-barked, dust-extracted wood chips from two dicotyledonous wood species, poplar (aspen) and beech, were sourced from Prorep, UK. Sawdust from pine (Living World, Canada) and cedar chips (Native by Nature, UK) was used as examples of gymnosperm cell wall material. Rice straw was grown and collected in Vietnam (*Oryza sativa*, cv. KhangDan18)[15] and dust-extracted wheat straw was obtained from Norfolk Straw Products Ltd., UK. These were used as examples of monocot cereal straws.

2.2. Microwave assisted liquid hot water pretreatment

All the feedstock sources were milled to pass through a 250 μm sieve using a Cyclone Mill TWISTER (Retch®, Germany). Air-dried samples of the milled material (750 mg) were weighed into 10–20 mL Biotage vials and a magnetic flea was added. Each sample was mixed thoroughly with water (14.25 mL) and capped immediately before pretreatment. The samples were pretreated at various temperatures and durations (outlined in Table 1) using a Biotage Initiator$^+$ Microwave Synthesiser (Fig. 2 – heating at 2–5 °C s^{-1}, cooling to <100 °C within 1–2 min and <50 °C within 6 min), and frozen at −20 °C immediately after pretreatment (Table 1).

The vials were defrosted and decapped, and a sample of the liquor was removed (1 mL) and buffer with an anti-microbial was added to each vial (0.4 M sodium acetate + 0.04% thiomersal, 6 mL, pH = 5). The whole pretreated slurries were vortexed and enzymatically hydrolysed with Cellic® CTec 2 (188 μL. Approx. 30 FPU g^{-1} biomass). Then the vials were recapped and incubated for 96 h (50 °C, 100 RPM). Samples of the hydrolysates were analysed for their reducing sugar and monomeric Glc concentrations using a multiplex dinitrosalicylic acid (DNS) assay

Table 1 Experimental design. All biomass sources were pretreated using the listed conditions, exposing the material to various pretreatment severities. The pretreatment severity for each of the test conditions entered in the table was calculated following Overend & Chornet (1987): $\log_{10}(R_0) = \text{time} \times \exp((\text{temp} - 100)/14.75)$

	Duration (min)			
Temp (°C)	2.5	10	40	160
Severities calculated for the different durations and temperatures				
140	1.576	2.178	2.780	—
150	1.870	2.472	3.074	—
160	2.165	2.767	3.369	—
170	2.459	3.061	3.663	4.265
180	2.753	3.355	3.958	4.560
190	—	3.650	4.252	4.854
200	—	3.944	4.546	5.148
210	—	4.239	4.841	5.443

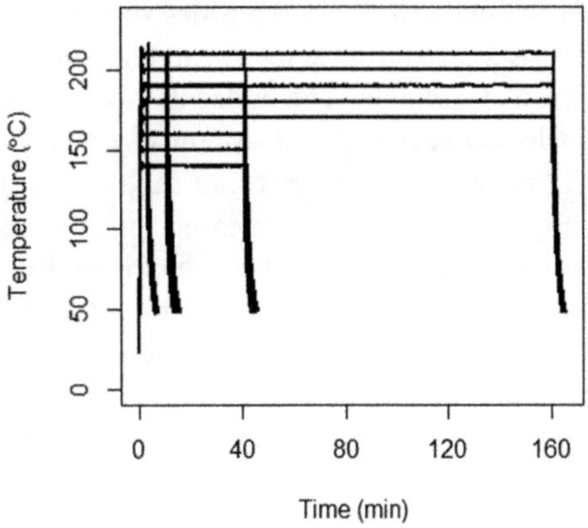

Fig. 2 Pretreatment using a Biotage® Initiator^{+} Microwave Synthesizer, using the hold-times outlined in Table 1 (wheat straw samples shown as an example). The temperature of each vial was very precisely controlled during pretreatment, and heating and cooling were very rapid.

without dilution (Wood *et al.* 2012) and a GOPOD assay after dilution (Megazyme, Ireland) respectively. The amount of Glc and reducing compounds (Glc equivalents) in the cellulase Cellic® CTec 2 was assayed separately and subtracted from the results for all the vials to obtain the sugar content from saccharification of the substrate alone.

2.3. Fourier transform infrared spectroscopy (FT-IR)

The material remaining after pretreatment and hydrolysis was centrifuged, the supernatant discarded, and the pellet washed in ddH_2O. FT-IR ATR (attenuated total reflectance) spectra were obtained for each wet substrate, against ddH_2O as a background, using a dynamic alignment FT-IR spectrophotometer (Bio-Rad FTS 175C, Bio-Rad Laboratories, Cambridge, USA), equipped with a GoldenGate™ single reflection diamond ATR accessory. The spectra were truncated to the 800–1800 cm^{-1} region, the baseline corrected and the area normalised before analysis. Spectral data interpolation was conducted using the 'loess' function (locally weighted scatterplot smoothing) for each wavenumber in Matlab®, to show spectral changes at regular intervals of severity.

3. Results and discussion

3.1. Saccharification yields from whole, unwashed, pretreated slurries

The main aim of pretreatment is to open the structure of the biomass and make the constituent polysaccharides available to cell-wall degrading enzymes in a cost effective manner. The main products of interest are fermentable sugars, particularly Glc, but also other potentially usable sugars, oligomers and carbohydrates.

We obtained eight different feedstocks, exposed them to 26 different pretreatment conditions (temperature and hold-time) and quantified the amount of monomeric Glc released following saccharification using the GOPOD assay, and used a less-specific DNS assay which quantified the total reducing sugars (in Glc equivalents) in the hydrolysates. This allowed us to monitor both the Glc yields and the other solubilised oligosaccharides or fermentable sugars in the media. Reducing groups derived from some sugar degradation products may also contribute to the overall absorbance. The molecular weights of these compounds are not the same as Glc, hence the DNS assay tends to overestimate yields, in Glc equivalents, by *ca.* 17% for pentose sugars and *ca.* 30–47% for furfural/furfural derivatives. Conversely, oligosaccharides released into the pretreatment liquors are likely to be underestimated, as only the reducing end is quantified. In this study, we did not wash the material after pretreatment, therefore pretreatment-soluble oligosaccharides and fermentation inhibitors were retained in the final hydrolysates.

The Glc yields generally increased with severity, following a sigmoidal curve, indicating an increase in the availability of glucan to saccharification with minimal loss of the monomeric Glc to break-down products at severities of $\log(R_0)$ > 5 (Fig. 3). The total reducing sugar (RS) concentrations, as assayed using DNS,

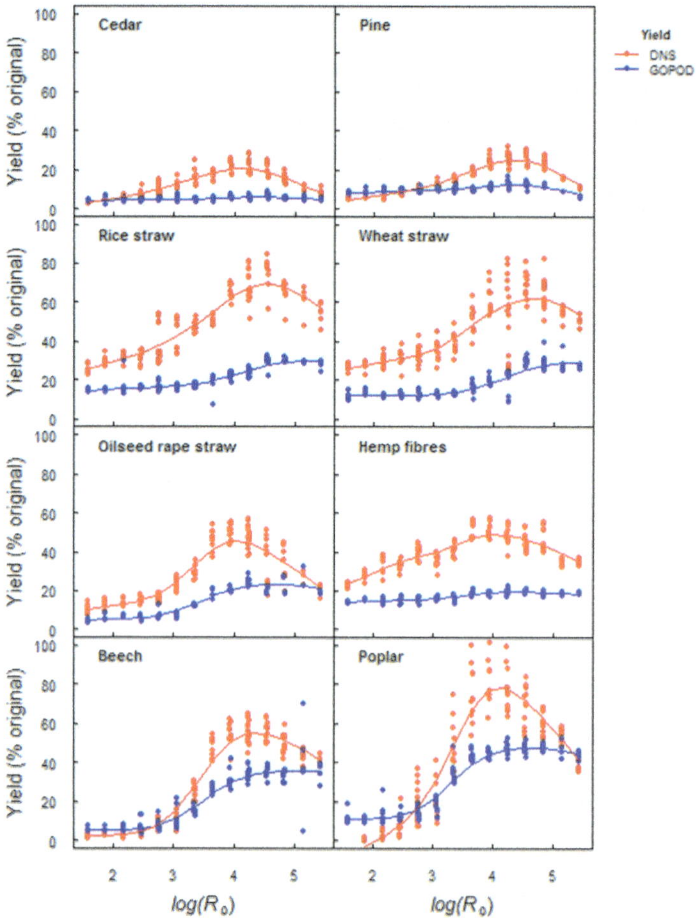

Fig. 3 Monomeric glucose (GOPOD assay) and reducing sugar (DNS assay) yields (mass Glc equivalents) from different feedstocks pretreated at a range of severities (see Table 1) and enzymatically hydrolysed as whole-slurries using the same cellulase dose. Yields expressed as percentage monomeric sugar per weight of original biomass.

increased more steeply with an increased severity up to $\log(R_0) \approx 4$. Unlike the Glc yields, pretreatment at higher severities resulted in a notable decrease in the RS following saccharification. This change in the RS abundance is likely to reflect the intermediate release, and subsequent destruction, of non-cellulosic oligomers and/or furfural into the pretreatment liquors. Recent detailed analysis of oligomers solubilised from OSR straw using LHW at intermediate severities (*circa* $\log[R_0] = 4$) has shown that xylo-oligosaccharides are the most abundant compounds released into the liquors.[16] It is therefore likely that these oligomers are those detected by the DNS assay.

Considering the potential properties of non-cellulosic oligomers, and limitations in their fermentation, they could provide alternative, valuable products. Yeast have a strong preference to ferment hexose sugars rather than the pentoses released from xylan saccharification. Therefore xylan oligomers may be of greater value for other products, such as prebiotics[17] or biodegradable films.[18]

The severity required to obtain the maximum amount of reducing compounds in the hydrolysate was $\log(R_0) \approx 4.23$–4.26 for all the wood types (beech, poplar, cedar and pine), whereas lower severities ($\log(R_0) \approx 3.95$) were needed for the less-woody dicotyledonous plants (OSR & hemp). For both monocot straws (wheat & rice), notably higher severities ($\log(R_0) \approx 4.54$) were required. This is the equivalent of a *ca.* 20 °C increase in the pretreatment temperature when using a 10 min hold time. This was also true for Glc yields, as monocots generally required a higher severity pretreatment (*circa* $\log R_0 \approx 5$) than for the dicotyledonous materials ($\log R_0 \approx 4.3$) to achieve an optimum yield. These differences are most likely linked to the increased thermal stability of the main non-cellulosic compounds, which decompose to oligosaccharides, monomers, and furfural, detected by the DNS assay. Arabinoxylan, found in monocot cell walls, requires higher temperatures for depolymerisation and degradation compared to glucuronoxylan, which is more abundant in dicot species.[19] The comparatively low temperatures needed to release the maximum amount of reducing compounds into the hydrolysate for dicotyledonous plants (OSR and hemp) are likely to reflect the abundance of pectic components and/or common uronic acid and acetyl-substitutions associated with xylan, which are cleaved at comparatively low temperatures and lower the pH of the pretreatment liquors ($\log R_0 \approx 4.3$, compared to ≈ 5). It is likely that more thermally-resistant arabinoxylans and/or mixed linkage glucans limit access of the cellulase to the cellulose at lower pretreatment severities.

The Glc yields also generally conformed to trends associated with each cell wall type. The hardwoods (BCH & POP) responded well to LHW pretreatment, providing the highest overall Glc yields (*circa* 500 g monomeric Glc per kg of sawdust – 4–6-fold increase) at severities greater than $\log(R_0) = 4.25$. Whereas, the cell wall material from softwood (cedar and pine) and hemp residues showed hardly any improvement in Glc yield over the severities used ($\log R_0 < 5.44$). Recent applications of LHW pretreatment to pine sawdust also showed that severities of $\log R_0$ 3.33–4.69 were not sufficient to increase the Glc yield, but could improve the saccharification of poplar wood.[9] Nevertheless, LHW did release some non-Glc sugars/oligomers from the softwoods used in this study – cedar and pine – with the RS yields peaking at $\log R_0 \approx 4.2$ (Fig. 3). A similar optimal severity for xylan release (*ca.* 4.25) was also proposed recently for Douglas fir, when treated using LHW.[20] It is therefore likely that most softwood species will consistently respond to LHW pretreatment in this way.

Results found elsewhere in the literature suggest that saccharification yields might be slightly improved when lower pretreatment durations and higher temperatures are used to achieve the same severity (for example, see ref. 9). In this study, we did not see any consistent associations between the pretreatment duration/time and sugar yields that could not be explained by differences in the combined severity factor (R_0). We suspect that the disparities in yield observed in other studies are a reflection of the heating and cooling times associated with each system. For example, the reactor used by Nitsos *et al.* (2016) was heated at *circa* 7 °C min^{-1} as opposed to >180 °C min^{-1} in this study. In cases where the heating and cooling times are significant, a more reliable estimate of the pretreatment severity might be obtained by logging the reactor temperature and integrating the severity over the course of the pretreatment.

To document the chemical changes that had occurred for each biomass type following pretreatment and enzymatic hydrolysis, FT-IR spectra were obtained from the recalcitrant residues (Fig. 4). Changes in the FT-IR spectra of the recalcitrant residues, in relation to the pretreatment severity, were remarkably similar between substrates from the same group, but there were differences in the overall intensity and in the fingerprint patterns between the groups. The most visible difference was in the negative bands at 1029, 983 and 895 cm^{-1}. Softwood (cedar and pine) samples showed fewer spectral changes, whereas the hardwood (beech and poplar) and monocot (rice and wheat) straw samples showed very sizeable spectral changes. The extent of these changes mirrored directly the varying yield of monomeric glucose, quantified using the GOPOD assay, in the hydrolysates (Fig. 3) as the severity was increased. The size of the positive bands varied accordingly, consistent with cleavage and loss of the carbohydrate components and enrichment of the other resistant components, which remained in the indigestible pellet.

The pattern of the positive bands varied between sample groups, reflecting the different lignocellulosic compositions. The more subtle changes included the removal of bonds with an infrared absorbance at 1731 cm^{-1} for all the substrates, but particularly for the more readily hydrolysable dicotyledonous materials such as beech, poplar and OSR. This change likely indicates severity-dependent hydrolysis of pectic alkyl esters or the acetyl-groups on xylan and lignin at severities of $\log(R_0) < 4$. The release of acetyl-groups into the reaction mixtures is likely to catalyse depolymerisation of these materials, and may be partially responsible for the differences in the saccharification yields observed.[5] The poor response of softwood to LHW pretreatment may also be explained by fewer acetyl- or uronic acid groups, which could lower the pH of the pretreatment liquors, and/or the relative abundance of xylans which are hydrolysed at comparatively low temperatures. In terms of autocatalytic processes, these groups appear to be essential catalysts for biomass pretreatment without additional chemicals.

Interestingly, reducing cell wall acetylation has been suggested as a strategy that might improve the use of lignocellulose in biotechnology. This makes sense when considering untreated biomass, as acetylation generally makes xylans more hydrophobic and therefore more difficult to depolymerise using enzymes.[21] High concentrations of acetic acid in the hydrolysate may also inhibit yeast growth.

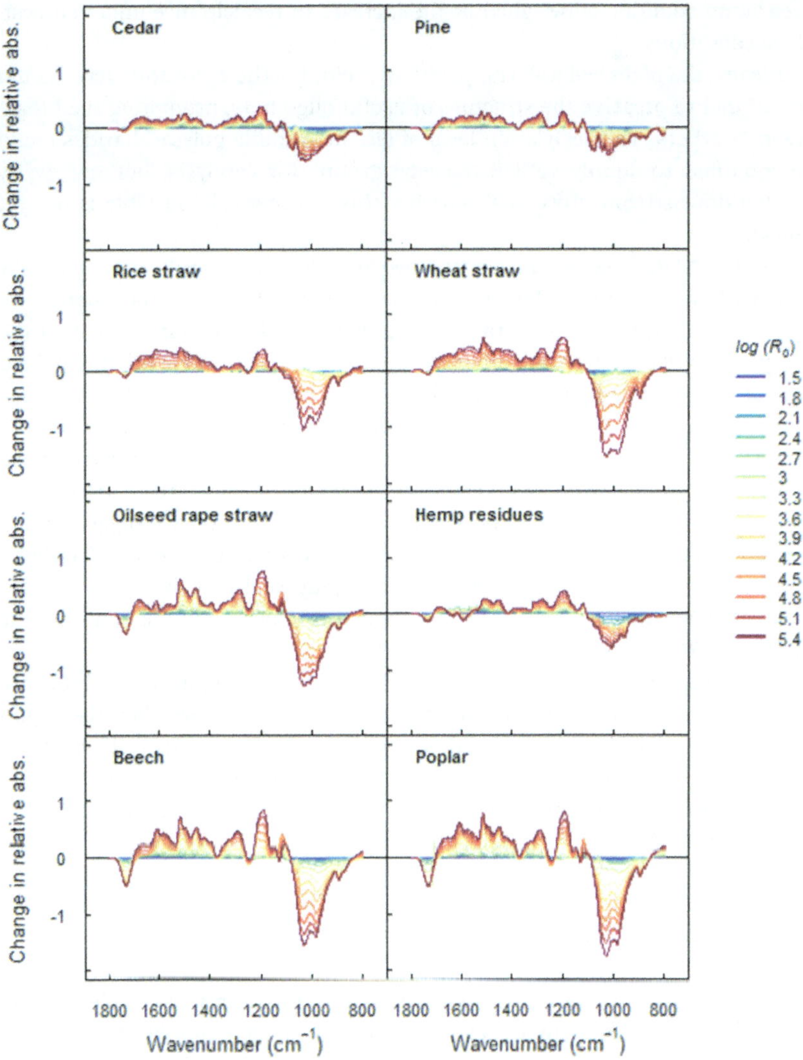

Fig. 4 Changes in the FT-IR spectra obtained from the recalcitrant residues following pretreatment and enzymatic hydrolysis. For clarity, the spectra have been interpolated at regular intervals (log(R_0) 0.3) and expressed relative to the material pretreated at the lowest severity (log(R_0) 1.5). This shows the severity-dependent removal (negative values) and enrichment (positive values) of linkages in the recalcitrant pellet following pretreatment and hydrolysis.

Therefore, taking into consideration downstream processes, acetylation appears to have a negative effect.

However, as shown in this study and others, acetyl-groups are quite readily hydrolysed at low severities and are useful catalysts for depolymerisation. Some form of pretreatment will always be needed to convert a non-edible plant material, with a reasonable fitness (resistance to environment, pests and pathogens), into something that can be easily digested by commercial cellulase cocktails. The perceived benefits to saccharification and fermentation of reducing cell wall acetylation could be outweighed by the increase in recalcitrance under autocatalytic conditions.

Ideally, the plant cell wall composition, including the acetylation level, could be tailored to preserve the structure of useful oligomers, promoting the breakdown of others. The acetylation level of less heat-stable polysaccharides could be modified to supply sufficient acetyl groups for catalysis, but not overly burden downstream processes. Whether this is genetically feasible is not yet known.

In the FT-IR spectra, the absorbance for all other wavelengths generally increased in relation to the severity, owing to the loss of carbohydrates *via* saccharification, but the patterns did not change. It is therefore likely that the composition of the lignin and other components in the insoluble pellet remained similar after pretreatment and hydrolysis. Alternatively, chemical changes made to the lignin could be difficult to resolve from differences in the overall spectra. Almost unperceivable differences in absorbance were observed at wavenumbers typically associated with pseudo-lignin formation – such as 1697, 1611, 1512, 1360, 1300, 1203, 1020, 867 and 800 cm^{-1}.[22] Hu *et al.*, (2012) have suggested that pseudo-lignin production might be suppressed by limiting oxygen availability.[22] In this study, biomass was pretreated in tightly sealed vials with a limited headspace, which may have produced semi-anaerobic conditions, limiting the breakdown of sugars into pseudo-lignin.

Although some re-localisation and modifications of the lignin are likely to have occurred, much higher severities are generally required to alter the lignin composition considerably.[23] Liquid hot water pretreatment may therefore be a good choice for biorefining as it is less destructive to other components, such as lignin, which could be made into valuable products such as antioxidants and antimicrobials.[24]

4. Conclusions and discussion points

(1) Feedstock usability is not just determined by the quantity available, but also fundamental differences in the cell wall composition. For example, residues from forestry are often considered together in terms of usability, despite clear differences in their chemistry and suitability for refining.

(2) LHW pretreatment was able to solubilise oligomers from most of the feedstocks at severities *circa* $\log(R_0) = 4$, with slightly higher severities needed for monocot straws. Considering the potential properties of these oligomers, and limitations in their fermentation, they could provide alternative, valuable products.

(3) Although acetyl-groups can hinder the saccharification of untreated biomass,[21] these groups are important catalysts during autocatalytic pretreatments and are relatively easy to cleave at low pretreatment severities. Clear strategies must be devised which take into consideration the whole process chain.

(4) LHW might be a good method for preserving the structure of lignin for valorisation.

(5) In cases where the heating and cooling times are significant, they should be included to obtain a reliable estimate of the combined pretreatment severity (R_0) of the systems.

(6) More generally, issues in the supply chain also need to be addressed. We need to be pragmatic regarding what alternatives to petro-chemical products can be made from biological sources destined for waste, and make better use of these.

Acknowledgements

The authors receive financial support from the Biotechnology and Biological Sciences Research Council (BBSRC) *via* the Institute Strategic Programme 'Food and Health' (BB/J004545/1) and the Renewable Industrial Products from Rapeseed (RIPR) Programme (BB/L002124/1). This work was also supported by the European Commission's Lifelong Learning 'Leonardo da Vinci Programme'. The authors would also like to thank Henri Tapp for conducting spectral interpolation in Matlab and Andrew Watson for computational assistance.

References

1 M. Arshadi, T. M. Attard, R. M. Lukasik, M. Brncic, A. M. D. Lopes, M. Finell, P. Geladi, L. N. Gerschenson, F. Gogus, M. Herrero, A. J. Hunt, E. Ibanez, B. Kamm, I. Mateos-Aparicio, A. Matias, N. E. Mavroudis, E. Montoneri, A. R. C. Morais, C. Nilsson, E. H. Papaioannou, A. Richel, P. Ruperez, B. Skrbic, M. B. Solarov, J. Svarc-Gajic, K. W. Waldron and F. J. Yuste-Cordoba, *Green Chem.*, 2016, **18**, 6160–6204.

2 R. Slade, R. Saunders, R. Gross and A. Bauen, *Energy from biomass: the size of the global resource*, Imperial College Centre for Energy Policy and Technology and UK Energy Research Centre, London, 2011.

3 D. P. Maurya, A. Singla and S. Negi, *3 Biotech*, 2015, **5**, 597–609.

4 L. Tao, A. Aden, R. T. Elander, V. R. Pallapolu, Y. Y. Lee, R. J. Garlock, V. Balan, B. E. Dale, Y. Kim, N. S. Mosier, M. R. Ladisch, M. Falls, M. T. Holtzapple, R. Sierra, J. Shi, M. A. Ebrik, T. Redmond, B. Yang, C. E. Wyman, B. Hames, S. Thomas and R. E. Warner, *Bioresour. Technol.*, 2011, **102**, 11105–11114.

5 M. Pedersen and A. S. Meyer, *New Biotechnol.*, 2010, **27**, 739–750.

6 S. Kim and B. E. Dale, *Biomass Bioenergy*, 2004, **26**, 361–375.

7 P. Sarkar, E. Bosneaga and M. Auer, *J. Exp. Bot.*, 2009, **60**, 3615–3635.

8 S. R. A. Collins, N. Wellner, I. M. Bordonado, A. L. Harper, C. N. Miller, I. Bancroft and K. W. Waldron, *Biotechnol. Biofuels*, 2014, **7**, 121.

9 C. K. Nitsos, T. Choli-Papadopoulou, K. A. Matis and K. S. Triantafyllidis, *ACS Sustainable Chem. Eng.*, 2016, **4**, 4529–4544.

10 M. Michelin and J. A. Teixeira, *Bioresour. Technol.*, 2016, **216**, 862–869.

11 J. Lindedam, S. B. Andersen, J. DeMartini, S. Bruun, H. Jorgensen, C. Felby, J. Magid, B. Yang and C. E. Wyman, *Biomass Bioenergy*, 2012, **37**, 221–228.

12 F. Matsuda, M. Yamasaki, T. Hasunuma, C. Ogino and A. Kondo, *Biosci., Biotechnol., Biochem.*, 2011, **75**, 1603–1605.

13 I. P. Wood, N. Wellner, A. Elliston, D. R. Wilson, I. Bancroft and K. W. Waldron, *Biotechnol. Biofuels*, 2015, **8**, 1–11.

14 M. J. Zeng, E. Ximenes, M. R. Ladisch, N. S. Mosier, W. Vermerris, C. P. Huang and D. M. Sherman, *Biotechnol. Bioeng.*, 2012, **109**, 390–397.

15 I. P. Wood, H. G. Cao, L. Tran, N. Cook, P. Ryden, D. R. Wilson, G. K. Moates, S. R. A. Collins, A. Elliston and K. W. Waldron, *Biotechnol. Biofuels*, 2016, **9**, 193.

16 Z. W. Wang, M. Q. Zhu, M. F. Li, J. Q. Wang, Q. Wei and R. C. Sun, *Biotechnol. Biofuels*, 2016, **9**, 142.
17 A. K. Samanta, N. Jayapal, C. Jayaram, S. Roy, A. P. Kolte, S. Senani and M. Sridhar, *Bioact. Carbohydr. Diet. Fibre*, 2015, **5**, 62–71.
18 S. Sousa, A. Ramos, D. V. Evtuguin and J. A. F. Gamelas, *Ind. Crops Prod.*, 2016, **94**, 682–689.
19 K. Werner, L. Pommer and M. Brostrom, *J. Anal. Appl. Pyrolysis*, 2014, **110**, 130–137.
20 R. Zhu and V. Yadama, *Biomass Bioenergy*, 2016, **90**, 78–89.
21 M. J. Selig, W. S. Adney, M. E. Himmel and S. R. Decker, *Cellulose*, 2009, **16**, 711–722.
22 F. Hu, S. Jung and A. Ragauskas, *Bioresour. Technol.*, 2012, **117**, 7–12.
23 M. P. Pandey and C. S. Kim, *Chem. Eng. Technol.*, 2011, **34**, 29–41.
24 H. A. Ruiz, R. M. Rodriguez-Jasso, B. D. Fernandes, A. A. Vicente and J. A. Teixeira, *Renewable Sustainable Energy Rev.*, 2013, **21**, 35–51.

Faraday Discussions

PAPER

Effect of methanol in controlling defunctionalization of the propyl side chain of phenolics from catalytic upstream biorefining†

Paola Ferrini, ‡[a] Claudio Chesi,[b] Nicholas Parkin[b] and Roberto Rinaldi *[b]

Received 16th February 2017, Accepted 19th April 2017

DOI: 10.1039/c7fd00069c

In recent years, lignin valorization has gained upward momentum owing to advances in both plant bioengineering and catalytic processing of lignin. In this new horizon, catalysis is now applied to the 'pulping process' itself, creating efficient methods for lignocellulose fractionation or deconstruction (here referred to as Catalytic Upstream Biorefining or 'CUB'). These processes render, together with delignified pulps, lignin streams of low molecular weight (M_w) and low molecular diversity. Recently, we introduced a CUB process based on Early-stage Catalytic Conversion of Lignin (ECCL) through H-transfer reactions catalyzed by RANEY® Ni. This approach renders a lignin stream obtained as a viscous oil, comprising up to 60 wt% monophenolic compounds (M_w < 250 Da). The remaining oil fraction (40 wt%) is mainly composed of lignin oligomers, and as minor products, holocellulose-derived polyols and lignin-derived species of high M_w (0.25–2 kDa). Simultaneously, the process yields a holocellulose pulp with a low content of residual lignin (<5 wt%). Despite the efficiency of aqueous solutions of 2-propanol as a solvent for lignin fragments and an H-donor, there is scant information regarding the CUB process carried out in the presence of primary alcohols, which often inhibit the catalytic activity of RANEY® Ni, as revealed in model compound studies performed at low temperature. Considering the composition of the lignin oils obtained from CUB based on ECCL, the processes commonly render *ortho*-(di) methoxy-4-propylphenol derivatives with a varied degree of defunctionalization of the propyl side chain. In this contribution, we present the role of the alcohol solvent (methanol or 2-propanol) and Ni catalyst (Ni/C or RANEY® Ni) in control over selectivity of phenolic products. The current results indicate that solvent effects on the catalytic processes could hold the key for improving control over the degree of functionalization

[a]*Max-Planck-Institut für Kohlenforschung, 45470 Mülheim (Ruhr), Germany*

[b]*Imperial College London, Department of Chemical Engineering, SW7 2AZ, UK. E-mail: rrinaldi@imperial.ac.uk*

† Electronic supplementary information (ESI) available. See DOI: 10.1039/c7fd00069c

‡ Current address: KU Leuven, Center for Surface Chemistry and Catalysis, Celestijnenlaan 200f, 3001 Leuven, Belgium

of the propyl side-chain in the lignin oil obtained from CUB, offering new avenues for lignin valorization at the extraction step.

Introduction

The deconstruction of lignocellulosic materials generating streams of low molecular weight (M_w) with controlled molecular diversity corresponds to the primary challenge of catalytic upstream biorefining (CUB). In this context, the fractionation of lignocellulose by solvent treatment performed in the presence of a hydrogenation catalyst under hydrogen pressure or employing an H-donor solvent (*e.g.* 2-propanol) has opened up novel avenues for lignin utilization as a raw material to produce fuels, chemicals, and materials.[1–4] In the presence of a hydrogenation catalyst, the lignin fragments released from the lignocellulosic matrix undergo several reductive processes, which decrease their reactivity and propensity to recondensation. As a result of the Early-stage Catalytic Conversion of Lignin (ECCL), the lignin fraction is isolated as a stream of low molecular weight (M_w 100–600 Da) with a high content of individual components (5–50% or even higher for genetically modified plants).[5–10]

Considering the composition of the lignin oils obtained from CUB based on ECCL, there is a common trend among the reported processes. Typically, ECCL renders *ortho*-(di)methoxy-4-propylphenol derivatives.[5–14] Nonetheless, the degree of functionalization of the propyl side chain (R) constitutes the distinguishing feature of the catalytic systems already reported. Among the products, dihydro-*p*-lignols (R: $-CH_2CH_2CH_2OH$) are the primary products of H-transfer CUB process performed in the presence of RANEY® Ni employing 2-propanol as part of the solvent system and as the primary H-donor for the reductive processes[5–7] or Pd/C under H_2 pressures (2 MPa).[11] In turn, dehydro-*p*-lignols (R: $-CH{=}CHCH_2OH$) were reported for processes performed in the presence of Ni/C catalysts under low H_2 pressure (0.5 MPa).[12] 4-Propylguaiacol and 4-propylsyringol (R: $-CH_2CH_2CH_3$) are the major products obtained from CUB conducted in the presence of Ru/C, Zn(II)/Pd/C or Ni/C catalytic systems under moderate H_2 pressures (1–3.4 MPa).[8–10,12,13] Propenyl derivatives (R: $-CH{=}CHCH_3$) of lignols were reported as the main products from CUB processes performed in the presence of Pd/C under no H_2 pressure[14] or Ni/C catalytic system under low H_2 pressure (0.5 MPa).[12]

In a recent study, Luo *et al.* employed a nickel catalyst supported on activated carbon in the production of depolymerized lignin upon fractionation, using methanol as a solvent and reportedly an H-donor/source.[12] The results attracted our attention owing to the high selectivity towards dehydro-*p*-lignols from the depolymerization of birch sawdust lignin. Despite being detrimental for RANEY® Ni's activity and stability, as reported by us and other researchers,[15–17,25] the current pieces of evidence suggest that methanol could act as an H-source at temperatures of 240–300 °C.[18–20] Moreover, the results by Luo *et al.* starkly contrast with those previously reported by Song *et al.*[21] In that report, just three monophenolic products were detected and identified by GC-MS analysis, in the characterization of the lignin fraction extracted from Birch wood. Luo *et al.* showed that, in line with our previous results,[5] the product mixture is more complex.

Herein, we revisit the CUB process performed in the presence of Ni catalysts in methanol as the solvent and hypothetically an H-donor. To shed light on the role of alcohol type in the selectivity of the CUB process in the absence of H_2 pressure, we compare the performance of a Ni/C catalyst with RANEY® Ni in the CUB process performed on birch or poplar (hardwood). The product selectivity and apparent molecular weight distribution of the lignin oils obtained under varied conditions are assessed in detail.

Results and discussion

Table 1 lists the experimental conditions of the fractionation processes presented in this work. As a reference experiment (or blank), lignin was extracted by an organosolv process performed in methanol with no added hydrogenation catalyst. In the CUB experiments catalyzed by Ni/C, the suspension containing extracted birch sawdust, catalyst and solvent (2-propanol or methanol) was reacted at 200 °C for 6 h. In addition, CUB of non-extracted birch sawdust was carried out in MeOH in the presence of RANEY® Ni at a lower temperature (180 °C) and for a shorter period (3 h). Since RANEY® Ni stores a considerable amount of hydrogen in its structure (up to 2 mmol g^{-1}), the spent catalyst was reused in the second run of the ECCL process to ensure that methanol is working as an H-donor under the CUB conditions.

To investigate changes in the M_w distribution, the soluble products obtained from fractionation were analyzed by Gel-Permeation Chromatography (GPC, Fig. 1). The lignin obtained by the organosolv process showed an overall M_w lower than the organosolv lignins obtained from poplar in aqueous 2-propanol.[5] Strikingly, the chromatogram displayed a peak of strong intensity at around 200 Da, suggesting the presence of a fraction of monophenolics. As will be discussed later, 2D GC × GC analysis confirmed the presence of volatile compounds embedded in the lignin solid sample (Table 2). Conversely, the formation of such a fraction of low M_w compounds was not observed for the organosolv lignin from poplar produced in aqueous 2-propanol at the same temperature.[5]

In the CUB process, the presence of a Ni/C or RANEY® Ni catalyst formed a product mixture with GPC signals of increased intensity in the region of $M_w >$ 270 Da. Moreover, the products of the CUB process performed on birch wood in 2-propanol showed a substantial shift in the onset of M_W distribution to below 10 kDa. Clearly, both Ni catalysts can catalyze to a different extent H transfer reductive processes in either solvent. However, more extensive depolymerization was clearly achieved by using 2-propanol as the primary H-donor, owing to the intrinsic reactivity of this alcohol. In fact, 2-propanol is converted on a nickel

Table 1 Experimental conditions for birch wood catalytic upstream biorefining processes

Entry	Catalyst	Catalyst/biomass (wt/wt)	Solvent	T (°C)	Time (h)
1	None	—	MeOH	200	6
2	Ni/C	0.05	MeOH	200	6
3	Ni/C	0.05	2-PrOH	200	6
4	RANEY® Ni (fresh)	0.5	MeOH	180	3
5	RANEY® Ni (spent)	0.5	MeOH	180	3

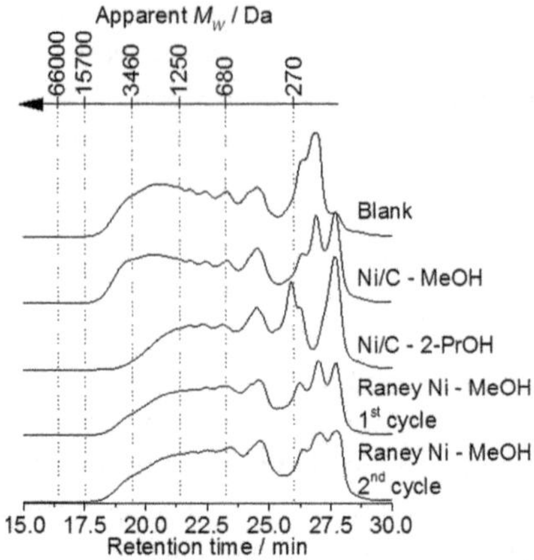

Fig. 1 GPC of lignin products obtained from the organosolv (blank) and CUB processes under the conditions listed in Table 1. Chromatograms were detected by a diode-array detector and normalized by sample concentration.

surface into acetone, which is readily desorbed leaving adsorbed hydrogen atoms on the metal surface.[22] In contrast, as a result of strong (dissociative) chemisorption of primary alcohols on Ni surfaces, methanol and other primary alcohols give rise to products that cannot be liberated easily from the catalyst surface (*e.g.* alkoxide groups).[23,24] In spite of that, the CUB process could proceed without a hydrogen pressure applied to the reaction mixture. This observation suggests that methanol could act as an H-donor under the CUB process conditions. Notably, this was not the case in studies carried out on model molecules at temperatures of 80–120 °C.[25] Under these conditions, the residence time of methanol-derived species chemisorbed on RANEY® Ni seems to be substantially high, fully inhibiting the catalytic hydrogen transfer processes.

Considering that RANEY® Ni contains a significant content of hydrogen stored in its structure, we decided to re-utilize the spent RANEY® Ni from the CUB process in methanol. The GPC results from the lignin oils obtained by using fresh or spent RANEY® Ni showed comparable M_W distribution, indicating that the catalyst can also sustain activity for lignin depolymerization in methanol. Although this result may suggest the utilization of methanol as an H-source, hemicellulose sugars formed by cooking wood in polar solvents could eventually also serve as an alternative H-source *via* reforming processes, as reported by Galkin *et al.*[14] Therefore, there is still a compelling need for a more extensive analysis of the CUB process regarding utilization of different sources of hydrogen present in the liquor.

2D GC × GC-FID/MS analysis probes the volatile species occurring in the lignin oil or streams. Organosolv lignin and catalytic upstream biorefining products from birch wood (Fig. 1, ESI†) exhibited compositions remarkably different from those previously obtained from poplar wood fractionation.[5,7] Compared to the Organosolv lignin obtained in aqueous 2-propanol previously reported,[5] the lignin showed a higher number of monophenolic derivatives. However, through the inspection of 2D GC × GC-FID images, the content of volatile species significantly increased in the presence of a hydrogenation catalyst.

Table 2 Content of selected lignin products occurring in the lignin oil obtained from the CUB process carried out under varied conditions (listed in Table 1), as determined by 2D GC × GC-FID

		Content (wt%)					
			Ni/C		RANEY® Ni in MeOH		RANEY® Ni in 2-PrOH/H_2O
Selected component		Blank[a]	MeOH	2-PrOH	1st cycle	2nd cycle	7 : 3 (v/v)[5]
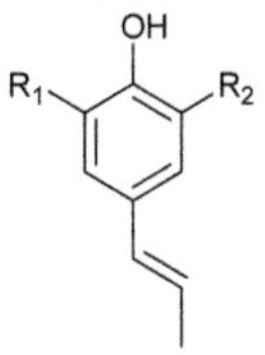	**1**: R_1: H, R_2: OCH_3	0.3	3.0	3.4	2.8	3.3	—
	2: $R_1 = R_2$: OCH_3	3.0	5.1	6.1	7.0	7.7	—
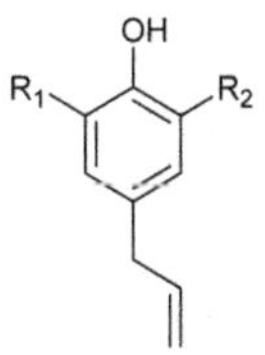	**3**: R_1: H, R_2: OCH_3	—	0.2	0.2	0.1	0.2	—
	4: $R_1 = R_2$: OCH_3	0.3	0.7	0.7	0.3	0.5	—
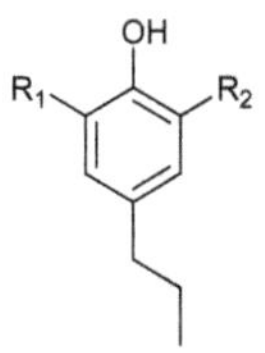	**5**: R_1: H, R_2: OCH_3	—	—	—	0.9	0.8	0.2
	6: $R_1 = R_2$: OCH_3	—	—	—	1.9	2.0	2.3
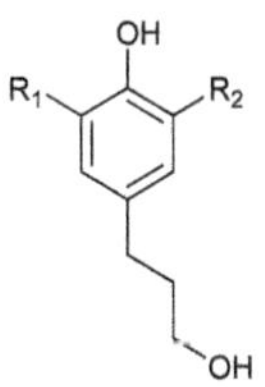	**7**: R_1: H, R_2: OCH_3	—	—	—	—	—	2.7
	8: $R_1 = R_2$: OCH_3	—	—	—	—	—	7.4
Sum		3.6	9.0	10.4	13.0	14.5	12.6

[a] The lignin sample was a solid material.

To analyse the differences between the product mixtures, the most abundant phenolic derivatives were identified and quantified (Table 2).

Table 2 reveals that the main lignin products, obtained from the CUB process performed in methanol, show unsaturated propyl side chains (**1–2**, R: $-CH{=}CHCH_3$ and **3–4**, R: $-CH_2CH{=}CH_2$). Unlike the CUB process carried out in 2-propanol as the solvent and H-donor, which together with phenolic products also renders cyclohexanols (in a total content of 3.8% in the lignin oil; individual contents: 4-ethylcyclohexan-1-ol, 2.9%; 4-propylcyclohexan-1-ol, 0.9%),[5] no cyclohexanol derivatives in the product mixtures obtained from CUB in methanol were detected. This observation, together with the absence of compounds **5–8**, indicates that the lignin stream undergoes less extensive hydrogenation when the process is performed in MeOH rather than in 2-propanol. Notably, even in methanol, RANEY® Ni is capable of reducing the unsaturation on the propyl side chain, resulting in *ca.* 2% yield of 2,6-dimethoxy-4-propylphenol (**5**) in the oil. Conversely, the combination of RANEY® Ni with 2-propanol promoted full saturation of the side chains in the monophenolic products. Interestingly, dihydro-*p*-lignols (**7–8**) were the most abundant species,[5] suggesting a lower activity of RANEY® Ni towards hydrodeoxygenation in comparison with Ni/C.

To provide further insight into the nature of the lignin oil obtained under various CUB conditions, we chose to examine the char formation under pyrolysis conditions. Accordingly, thermogravimetric analysis was performed on the lignin oil samples under an argon atmosphere (Fig. 2a). The pyrolysis of organosolv lignin (from the reference experiment) yielded 35 wt% char at 800 °C, in agreement with the values reported for other lignins.[26,27] In contrast, the products of catalytic upstream biorefining can undergo larger volatilization/decomposition, resulting in less char formation at 800 °C. The products of the CUB process in the presence of RANEY® Ni are more volatile than their counterparts from the CUB with Ni/C. The amount of char formed by the lignin products at 800 °C decreased as follows: blank (35 wt%) > Ni/C–MeOH (32 wt%) > Ni/C–2-PrOH (30 wt %) > RANEY® Ni – MeOH, 2nd run (24 wt%) > RANEY® Ni – MeOH – 1st run (20 wt %). Therefore, these results demonstrate that, even in methanol, RANEY® Ni is capable of reducing part of the most reactive groups present in the lignin fragments, yielding a lignin oil less prone to recondensation under pyrolytic conditions, compared to those obtained with Ni/C catalyst. However, compared to the oil produced from catalytic upstream biorefining in aqueous 2-propanol at 180 °C,[5] which shows only 7 wt% of solid residue at 800 °C (not displayed in Fig. 2), lignin products obtained in methanol showed much higher yields of biochar. This is attributed to the higher molecular weight and the remarkably different chemical composition of the products in these CUB streams (Table 2).

Displayed in Fig. 2b are the first derivatives of TGA, which compare the rates of volatilization of the samples. The products of fractionation in methanol

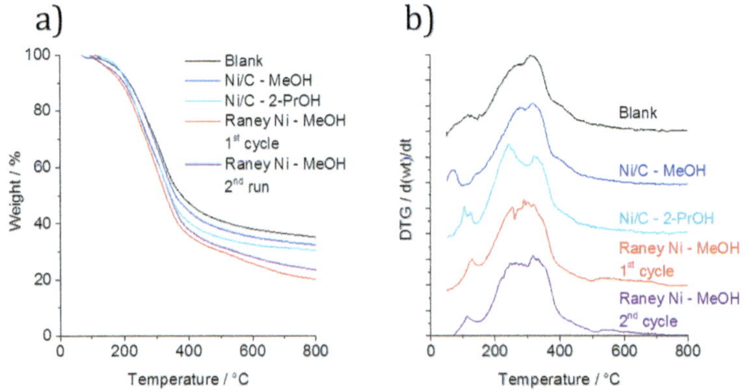

Fig. 2 TGA (a) and DTG (b) of lignin products obtained in the processes listed in Table 1 (analysis performed under Ar flow 60 mL min^{-1}; heating rate: 5 °C min^{-1}; sample size: 1–3 mg).

presented similar DTG profiles regardless of the presence of a catalyst. All products presented a small peak at low temperature (<150 °C), assigned to loss of moisture and dehydration of the product mixture.[26,27] The two main peaks at 280 and 310 °C, which overlapped in oils obtained with RANEY® Ni, were ascribed to the volatilization of the species detected by GC × GC and to the thermal decomposition of lignin compounds.

Interestingly, the products of both RANEY® Ni catalyzed processes in methanol showed the second stage of weight loss at high temperature (500–700 °C), not observed in the DTG profiles of the other products. This lump in the DTG profile, which can be assigned to the thermal decomposition of the phenolic ring (*e.g.* demethoxylation),[27–29] suggests that only the products of RANEY® Ni-catalyzed fractionation are stable enough to undergo this decomposition, while the products of other catalytic systems preferably undergo condensation and char formation at temperatures above 500 °C.

The products of the reaction performed in 2-propanol show a different volatilization profile compared with the products of methanol fractionation. The peak at 100 °C is assigned to the release of moisture and the presence of highly volatile products. An intense peak at 240 °C can be ascribed to the volatilization of the low-M_w species detected by 2D-GC × GC, while at a higher temperature (310 °C) volatilization and decomposition of larger molecules occur. For comparison, the main volatilization peaks are shifted at higher temperatures (280 and 310 °C) in the product mixtures obtained in methanol. This observation may be a consequence of the lower M_w distribution for the lignin oil obtained from CUB in 2-propanol. Indeed, smaller lignin products degrade at temperatures lower than high molecular weight fractions,[28] explaining the shift at a lower temperature of the main degradation peak of Ni/C–2-PrOH product. The major weight loss of the products of CUB in aqueous 2-propanol occurred at *ca.* 210 °C,[5] indicating that employing 2-propanol as the H-donor and RANEY® Ni as the catalyst enhances the production of volatile compounds.

Conclusions

In summary, the study of Ni-based catalytic systems for CUB of hardwoods in methanol revealed an average performance in the catalytic fractionation of lignin. In agreement with the observations of Luo *et al.*,[1] the distribution of lignin products obtained in this work pertain to a complex mixture in all cases, differing from the observation of Song *et al.*,[21] who identified just three monophenols as the main components of the lignin stream. This discrepancy may be ascribed to the analytical technique employed for lignin characterization. Indeed, bidimensional GC × GC-FID/MS offers higher resolution and sensitivity than its monodimensional counterpart,[30] hence revealing the complexity of seemingly simple mixtures according to GC-MS characterization. It is apparent that, in the absence of an external H_2 pressure, reductive processes can also occur when using methanol as a solvent for CUB. However, the inhibition of Ni catalysts due to strong chemisorption of methanol on the Ni surface causes the lignin stream to possess a higher M_W, compared to lignin oils obtained from CUB carried out in (aqueous) 2-propanol.[5] The highest depolymerization was achieved when RANEY® Ni was used in the process. Conversely, depolymerization of birch wood occurring in CUB with Ni/C resulted in the production of significant amounts of dehydro-*p*-lignols, accounting for a lower activity of the catalyst towards hydrogenation and higher activity towards hydrodeoxygenation on RANEY® Ni, which, in turn, promotes a greater extent of saturation on the side chains.

In the broader context, CUB processes render depolymerized lignin oils with a high degree of uniformity in terms of product distribution. CUB therefore starkly contrasts to the solvent-based extraction of lignin (*e.g.* organosolv processes) which obtains lignin as degraded polymers. In the absence of H_2 pressure, the facile control over the defunctionalization of the propyl side chain of phenolics by the choice of solvent adds to the advantages of the CUB process over the conventional solvent extraction of lignin (organosolv processes), increasing the portfolio of lignin platform chemicals obtained by CUB processes.

Experimental section

Materials

RANEY® Ni 2800 slurry, nickel nitrate hexahydrate ($Ni(NO_3)_2 \cdot 6H_2O$), activated carbon (Charcoal Norit® 97876), 2-propanol (99.9%), methanol (99.5%), di-*n*-butyl ether (99%), were purchased from Sigma-Aldrich and used as received. Birch sawdust (Germany) was used without extraction (except where specified).

Ni/C catalyst preparation

Nickel catalyst supported on activated carbon was prepared by incipient wetness impregnation, following the procedure reported elsewhere.[21] Briefly, a nickel solution was prepared by stirring nickel nitrate hexahydrate ($Ni(NO_3)_2 \cdot 6H_2O$, 2.76 g) in distilled water (8 mL) for 45 min. After this, 5 g of activated carbon (charcoal Norit®, total pore volume: 1.08 cm^3 g^{-1}) was added to the solution, and the material was covered and stirred for 24 h. Finally, the catalyst was dried in an oven (120 °C) overnight. Prior to reaction, the catalyst was reduced under a N_2 flow (30 mL min^{-1}), heating to 450 °C for 60 min and maintaining the temperature for 2 h.

Organosolv process

The substrate (17 g birch wood) was suspended in a solvent (140 mL) in a 250 mL autoclave and heated to the desired temperature within 1 h under mechanical stirring. The suspension was processed under autogenous pressure for 3 h. In sequence, the mixture was left to cool down to room temperature. A reddish-brown solution (liquor) was obtained after filtering out the pulp fibers. The pulp was washed several times with methanol portions (20 mL) and dried under a vacuum. Organosolv lignin (containing hemicellulose impurities) was isolated from the liquor by removing the solvent mixture at 60 °C under vacuum using a rotoevaporator. Organosolv lignin was obtained as a reddish-brown solid residue.

Catalytic upstream biorefining process with RANEY® Ni

The substrate (14.5 g birch sawdust, lignin content: 25.7% ± 0.9%), fresh or spent RANEY® Ni (10 g wet) and solvent (140 mL) were placed in a 250 mL autoclave and heated to the desired temperature within 1 h under mechanical stirring. The reaction proceeded under autogenous pressure for 3 h. In sequence, the mixture was left to cool down to room temperature. The liquor was separated from the solids (RANEY® Ni plus pulp) by filtration through a glass fiber filter (GF6, ∅ 90 mm, Whatman). The wet solids were immediately poured into a round-bottomed flask (250 mL) containing methanol. Under overhead mechanical stirring, the solids were resuspended in methanol. RANEY® Ni was attracted to the flask bottom by a magnet externally placed on the bottom of the round-bottomed flask. This procedure was repeated until the catalyst was completely removed from the pulp fibers (5–6 times). The pulp fibers remained in suspension and were recovered by filtration. The spent catalyst was washed and stored in methanol. In the recycling experiment, the recovered catalyst was directly placed into the autoclave vessel for a further cycle. The liquor and the filtrates (from the catalyst separation procedure) were combined. The lignin oil was isolated by removing the solvent at 60 °C under vacuum with a rotoevaporator.

Catalytic upstream biorefining process with Ni/C

For the fractionation process using Ni/C as a catalyst, birch sawdust was ground, sieved (<500 μm) and extracted prior to fractionation. Birch powder (7 g) was placed in a Soxhlet reactor and extracted with an ethanol : benzene (1 : 2 v/v, 500 mL) mixture at 90 °C for 12 h. After this, the solid was washed four times with ethanol and dried in an oven (60 °C) overnight. The extracted birch wood (1 g) was then placed in a 50 mL autoclave, together with the catalyst (Ni/C, 0.05 g, except in the blank experiment) and the solvent (20 mL), and was heated to 200 °C. After 6 h at the reaction temperature, the reactor was quenched to room temperature. The suspension was filtered, and the solid was washed several times with the solvent used (methanol or 2-propanol). The catalyst was separated using the same procedure employed for RANEY® Ni separation. The pulp and the liquor were dried at 60 °C overnight, and the lignin products were analyzed.

Gel-permeation chromatography (GPC)

To analyze the apparent molecular weight distribution of lignin products, the samples (10 to 30 mg) were dissolved in THF (2 mL) and filtered (nylon filters, 0.45 μm) prior to injection. The GPC analyses were performed at 60 °C on a Perkin-Elmer HPLC 200 equipped with four columns (2 × TSKgel Super HZ1000, TSKgel Super HZ2000 and TSKgel Super HZ3000, 4.6 mm × ID 15.0 cm, Tosoh Bioscience), using inhibitor-free THF as eluent (0.4 mL min^{-1}, Aldrich). For detection, a diode-array detector (DAD) recording the signal at 216 nm was used. The detector response was normalized by the sample weight. The apparent molecular weight is given relative to polystyrene standards (200 to 60 000 Da, Aldrich), and thus is only for a relative assessment of the overall changes in the apparent molecular weight distributions.

2D GC × GC-FID/MS analysis

The filtered samples (nylon filters, 0.45 μm) were analyzed using 2D GC × GC-MS (1st column: Rxi-1ms 30 m, 0.25 mm ID, d_f 0.25 μm; 2nd column: BPX50, 1 m, 0.15 mm ID, d_f 0.15 μm) in a GC-MS 2010 Plus (Shimadzu) equipped with a ZX1 thermal modulation system (Zoex). The injector temperature was 300 °C. The temperature program started with an isothermal step at 40 °C for 5 min. Next, the temperature was increased from 40 to 300 °C by 5.2 °C min^{-1}. The program finished with an isothermal step at 300 °C for 5 min. The modulation applied to the comprehensive GC × GC analysis was a hot jet pulse (400 ms) every 9000 ms. The 2D chromatograms were processed with GC Image software (Zoex). The products were identified by a search of the MS spectrum with the MS libraries NIST 08, NIST 08s, and Wiley 9. In some cases, the structure was proposed by the analysis of the EI-fragmentation pattern and by comparison of retention times with other samples, or by comparison with reported bio-oil compounds spectra.[31]

Thermogravimetric analysis (TGA)

The weight loss profile of the indicated samples was measured with a Mettler Toledo TGA/DSC 1 Star System. Typically, 1–3 mg of sample was heated from 25 to 800 °C at 5 °C min^{-1} under an argon flow (60 mL min^{-1}).

Acknowledgements

This work was performed as part of the Cluster of Excellence "Tailor-Made Fuels from Biomass." The authors are grateful to the Max-Planck-Institut für Kohlenforschung for the use of research facilities.

References

1 R. Rinaldi, R. Jastrzebski, M. T. Clough, J. Ralph, M. Kennema, P. C. A. Bruijnincx and B. M. Weckhuysen, *Angew. Chem., Int. Ed.*, 2016, **55**, 8164–8215.

2 R. E. Key and J. J. Bozell, *ACS Sustainable Chem. Eng.*, 2016, **4**, 5123–5135.

3 C. Li, X. Zhao, A. Wang, G. W. Huber and T. Zhang, *Chem. Rev.*, 2015, **115**, 11559–11624.

4 O. Y. Abdelaziz, D. P. Brink, J. Prothmann, K. Ravi, M. Sun, J. Garcia-Hidalgo, M. Sandahl, C. P. Hulteberg, C. Turner, G. Liden and M. F. Gorwa-Grauslund, *Biotechnol. Adv.*, 2016, **34**, 1318–1346.

5 P. Ferrini and R. Rinaldi, *Angew. Chem., Int. Ed.*, 2014, **53**, 8634–8639.

6 P. Ferrini, C. A. Rezende and R. Rinaldi, *ChemSusChem*, 2016, **9**, 3171–3180.

7 C. Chesi, I. B. D. De Castro, M. T. Clough, P. Ferrini and R. Rinaldi, *ChemCatChem*, 2016, **8**, 2079–2088.

8 S. Van den Bosch, W. Schutyser, S.-F. Koelewijn, T. Renders, C. M. Courtin and B. Sels, *Chem. Commun.*, 2015, **51**, 13158–13161.

9 T. Parsell, S. Yohe, J. Degenstein, T. Jarrell, I. Klein, E. Gencer, B. Hewetson, M. Hurt, J. I. Kim, H. Choudhari, B. Saha, R. Meilan, N. Mosier, F. Ribeiro, W. N. Delgass, C. Chapple, H. I. Kenttämaa, R. Agrawal and M. H. Abu-Omar, *Green Chem.*, 2015, **17**, 1492–1499.

10 S. Van den Bosch, W. Schutyser, R. Vanholme, T. Driessen, S.-F. Koelewijn, T. Renders, B. De Meester, W. J. J. Huijgen, W. Dehaen, C. M. Courtin, B. Lagrain, W. Boerjan and B. Sels, *Energy Environ. Sci.*, 2015, **8**, 1748–1763.

11 T. Renders, S. Van den Bosch, T. Vangeel, T. Ennaert, S.-F. Koelewijn, G. Van den Bossche, C. M. Courtin, W. Schutyser and B. Sels, *ACS Sustainable Chem. Eng.*, 2016, **4**, 6894–6904.

12 H. Luo, I. M. Klein, Y. Jiang, H. Zhu, B. Liu, H. I. Kenttamaa and M. H. Abu-Omar, *ACS Sustainable Chem. Eng.*, 2016, **4**, 2316–2322.

13 E. M. Anderson, R. Katahira, M. Reed, M. G. Resch, E. M. Karp, G. T. Beckham and Y. R. Leshkov, *ACS Sustainable Chem. Eng.*, 2016, **4**, 6940–6950.
14 M. V. Galkin, A. T. Smit, E. Subbotina, K. A. Artemenko, J. Bergquist, W. J. J. Huijgen and J. S. M. Samec, *ChemSusChem*, 2016, **9**, 1–9.
15 X. Wang and R. Rinaldi, *ChemSusChem*, 2012, **5**, 1455–1466.
16 U. K. Singh, S. W. Krska and Y. Sun, *Org. Process Res. Dev.*, 2006, **10**, 1153–1156.
17 Y. Jiang, Z. Li, X. Tang, Y. Sun, X. Zeng, S. Liu and L. Lin, *Energy Fuels*, 2015, **29**, 1662–1668.
18 Y. Xiang, X. Li, C. Lu, L. Ma, J. Yuan and F. Feng, *Ind. Eng. Chem. Res.*, 2011, **50**, 3139–3144.
19 G. S. Macala, T. D. Matson, C. L. Johnson, R. S. Lewis, A. V. Iretskii and P. C. Ford, *ChemSusChem*, 2009, **2**, 215–217.
20 K. Barta, T. D. Matson, M. L. Fettig, S. L. Scott, A. V. Iretskii and P. C. Ford, *Green Chem.*, 2010, **12**, 1640–1647.
21 Q. Song, F. Wang, J. Cai, Y. Wang, J. Zhang, W. Yu and J. Xu, *Energy Environ. Sci.*, 2013, **6**, 994–1007.
22 L. J. Shorthouse, A. J. Roberts and R. Raval, *Surf. Sci.*, 2001, **480**, 37–46.
23 G.-C. Wang, Y.-H. Zhou, Y. Morikawa, J. Nakamura, Z.-S. Cai and X.-Z. Zhao, *J. Phys. Chem. B*, 2005, **109**, 12431–12442.
24 Y.-H. Zhou, P.-H. Lv and G.-C. Wang, *J. Mol. Catal. A: Chem.*, 2006, **258**, 203–215.
25 X. Wang and R. Rinaldi, *Energy Environ. Sci.*, 2012, **5**, 8244.
26 E. Jakab, O. Faix and F. Till, *J. Anal. Appl. Pyrolysis*, 1997, **40–41**, 171–186.
27 S. Sahoo, M. Ö. Seydibeyoğlu, A. K. Mohanty and M. Misra, *Biomass Bioenergy*, 2011, **35**, 4230–4237.
28 R. Sun, J. Tomkinson and G. Lloyd Jones, *Polym. Degrad. Stab.*, 2000, **68**, 111–119.
29 A. Tejado, C. Peña, J. Labidi, J. M. Echeverria and I. Mondragon, *Bioresour. Technol.*, 2007, **98**, 1655–1663.
30 L. Mondello, P. Q. Tranchida, P. Dugo and G. Dugo, *Mass Spectrom. Rev.*, 2008, **27**, 101–124.
31 J. Ralph and R. D. Hatfield, *J. Agric. Food Chem.*, 1991, **39**, 1426–1437.

Faraday Discussions

PAPER

An integrated biorefinery concept for conversion of sugar beet pulp into value-added chemicals and pharmaceutical intermediates

Max Cárdenas-Fernández,[a] Maria Bawn,[a] Charlotte Hamley-Bennett,[b] Penumathsa K. V. Bharat,[c] Fabiana Subrizi,[d] Nurashikin Suhaili,[a] David P. Ward,[a] Sarah Bourdin,[a] Paul A. Dalby,[a] Helen C. Hailes,[d] Peter Hewitson,[e] Svetlana Ignatova,[e] Cleo Kontoravdi,[c] David J. Leak,[b] Nilay Shah,[c] Tom D. Sheppard,[d] John M. Ward[a] and Gary J. Lye*[a]

Received 7th March 2017, Accepted 30th March 2017

DOI: 10.1039/c7fd00094d

Over 8 million tonnes of sugar beet are grown annually in the UK. Sugar beet pulp (SBP) is the main by-product of sugar beet processing which is currently dried and sold as a low value animal feed. SBP is a rich source of carbohydrates, mainly in the form of cellulose and pectin, including D-glucose (Glu), L-arabinose (Ara) and D-galacturonic acid (GalAc). This work describes the technical feasibility of an integrated biorefinery concept for the fractionation of SBP and conversion of these monosaccharides into value-added products. SBP fractionation is initially carried out by steam explosion under mild conditions to yield soluble pectin and insoluble cellulose fractions. The cellulose is readily hydrolysed by cellulases to release Glu that can then be fermented by a commercial yeast strain to produce bioethanol at a high yield. The pectin fraction can be either fully hydrolysed, using physico-chemical methods, or selectively hydrolysed, using cloned arabinases and galacturonases, to yield Ara-rich and GalAc-rich streams. These monomers can be separated using either Centrifugal Partition Chromatography (CPC) or ultrafiltration into streams suitable for subsequent enzymatic upgrading. Building on our previous experience with transketolase (TK) and transaminase (TAm) enzymes, the conversion of Ara and GalAc into higher value products was explored. In particular the conversion of Ara into L-*gluco*-heptulose (GluHep), that has potential therapeutic applications in hypoglycaemia and cancer, using a mutant TK is described.

[a]*Department of Biochemical Engineering, University College London, Gordon Street, London WC1H 0AH, UK. E-mail: g.lye@ucl.ac.uk; Tel: +44 (0)2076799585*

[b]*Department of Biology and Biochemistry, University of Bath, Claverton Down, Bath, BA2 7AY, UK*

[c]*Department of Chemical Engineering, Imperial College, London, SW7 2AZ, UK*

[d]*Department of Chemistry, University College London, Christopher Ingold Laboratories, Gordon Street, London, WC1H 0AJ, UK*

[e]*Advanced Bioprocessing Centre, Department of Mechanical, Aerospace and Civil Engineering, Brunel University London, Uxbridge, UB8 3PH, UK*

Preliminary studies with TAm also suggest GluHep can be selectively aminated to the corresponding chiral aminopolyol. The current work is addressing the upgrading of the remaining SBP monomer, GalAc, and the modelling of the biorefinery concept to enable economic and Life Cycle Analysis (LCA).

1. Introduction

Sugar beet is a major crop across much of northern Europe. 8 million tonnes are produced annually in the UK by 3500 farmers across East Anglia and the East Midlands. After harvesting the sugar beet it is transported to one of British Sugar's four advanced manufacturing plants at Bury St Edmunds, Cantley, Newark or Wissington. The Wissington plant typically produces 400k tonnes of sugar every year. Co-products include about 350k tonnes of sugar beet pulp (SBP), 15k tonnes of vinasse and up to 64k tonnes of bioethanol every year.[1]

A schematic overview of the Wissington biorefinery is shown in Fig. 1. This provides the context for the current work. In addition to producing sugar (sucrose) from the harvested sugar beet, the plant also produces a range of other value-added products including bioethanol, animal feeds (*i.e.* SBP and vinasse) and renewable energy. The Wissington plant also features extensive energy integration such that heat from the sucrose evaporators is recovered to warm an 18 hectare glasshouse for British Sugar's horticulture business, and anaerobic digestion is used to generate methane, from other waste streams, that is burnt for energy generation.

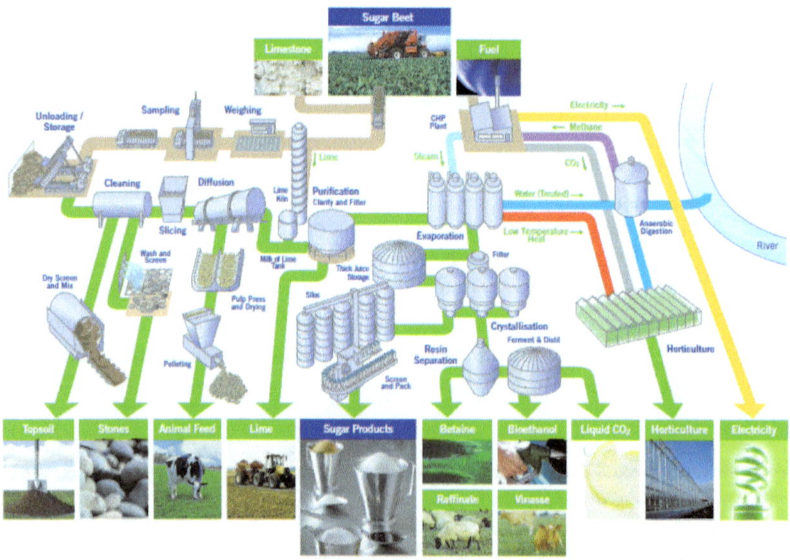

Fig. 1 British Sugar Wissington biorefinery (Norfolk, UK), showing the main unit operations for sugar beet processing, to produce sucrose and a range of by-products including animal feed (from sugar beet pulp), bioethanol, vinasse, *etc.* (Reproduced with permission from British Sugar).

Sugar beet pulp is the plant material left after sucrose extraction. It is found in the aqueous by-product stream from the solid–liquid diffuser where the sucrose is extracted. This stream is released at around 70 °C and is currently dried to approximately 88% w/w dry solids (requiring considerable energy input) before being sold as an animal feed (a relatively low value by-product). The focus of this work is on valorisation of the aqueous SBP stream after it leaves the diffuser; the aim is to create a range of value-added products from the SBP while simultaneously reducing the energy costs and greenhouse gas emissions associated with the SBP drying process.

The conceptual design of a biorefinery process that could be integrated within the Wissington facility must seek to utilise all the main components present in the SBP. As shown in Section 2.1, SBP is largely composed of two biopolymers; cellulose and pectin. Unlike citrus pectin, sugar beet pectin has poor gelling properties due to its high degree of esterification hence its applications are limited.[2] The major monosaccharides present in SBP are D-glucose (Glu), L-arabinose (Ara) and D-galacturonic acid (GalAc). Glu is readily fermented by a range of microorganisms and hence provides a route to a range of fermented products. Ara has been shown to be useful for the production of biopolymers after esterification. It can also be reduced to arabinitol, highlighted as one of the 12 top value-added chemicals from biomass, which can be used as a building block for unsaturated polyester resins.[3] Moreover, chemical methods have been developed for converting Ara into potentially useful starting materials for medicinal chemistry.[4] Oxidised sugars, such as GalAc, have also been highlighted as important building blocks for the production of hyperbranched polyesters and plasticisers.[3]

A particular focus of our previous work has been the enzymatic, and chemo-enzymatic, synthesis of chiral aminoalcohols. These are an important class of pharmaceutically-relevant compounds, and their motif is present in antibiotics, anti-viral glycosidase inhibitors as well as sphingolipids.[5] Enzymes such as transketolase (TK), which catalyses asymmetric carbon–carbon bond formation,[6] and transaminase (TAm), which catalyses selective amine group transfer, have been particularly useful in synthesising a range of such products.[7–10] Chiral aminoalcohols therefore represent an interesting class of higher value products that could be accessed from renewable feedstocks such as SBP. The challenge however is to identify variants of the TK and TAm enzymes that are able to convert SBP monosaccharides in biorefinery process streams with adequate productivity.

2. Results and discussion

2.1 Feedstream compositions

2.1.1 Sugar beet pulp composition. Valorisation of each of the main components present in the SBP stream is the primary goal of this conceptual biorefinery process design. SBP is a rich source of carbohydrates that consists mainly of cellulose (a polymer of D-glucose) and pectin (a co-polymer of various hexose and pentose sugars) as well as other components such as lignin (<3% w/w) and protein (11% w/w). Sugar beet pectin is formed of a linear backbone, from chains of polymeric D-galacturonic acid with intermittent blocks of alternating L-rhamnose (Rha) and D-galacturonic acid residues, and two different types of side chains. These are linear galactans (polymeric D-galactose) and highly branched arabinans (polymeric L-arabinose).[11,12]

The carbohydrate composition of the washed and pressed SBP used in this work was determined as a fraction of the total dry weight after full acid hydrolysis as shown in Fig. 2. These values are similar to those previously reported for other sources of SBP.[12,13] Ara and GalAc, both from pectin, along with Glu, from cellulose, are the main components accounting for 85% of the total monosaccharides. The utilisation of these three monosaccharides is therefore a focus of the subsequent SBP fractionation and bioconversion studies.

2.1.2 Vinasse composition. The cost of enzymes that can be used for biomass solubilisation (Section 2.3.2) or biocatalytic synthesis (Section 2.4) usually has a significant impact on the overall process economics.[14] Within an integrated biorefinery context it is therefore of interest to look for further by-product streams that can be used to support enzyme production by microbial fermentation. This will help reduce the cost of process enzymes if produced on site.

Within the Wissington biorefinery (Fig. 1), vinasse is the by-product of the existing yeast-based bioethanol fermentation. It is the spent fermentation supernatant that is released from the bottom of the first ethanol distillation column. Consequently it will have been concentrated and the thermal treatment promotes side reactions, *i.e.* Maillard reactions between yeast proteins and reducing sugars, as well as precipitation or degradation of heat labile components.

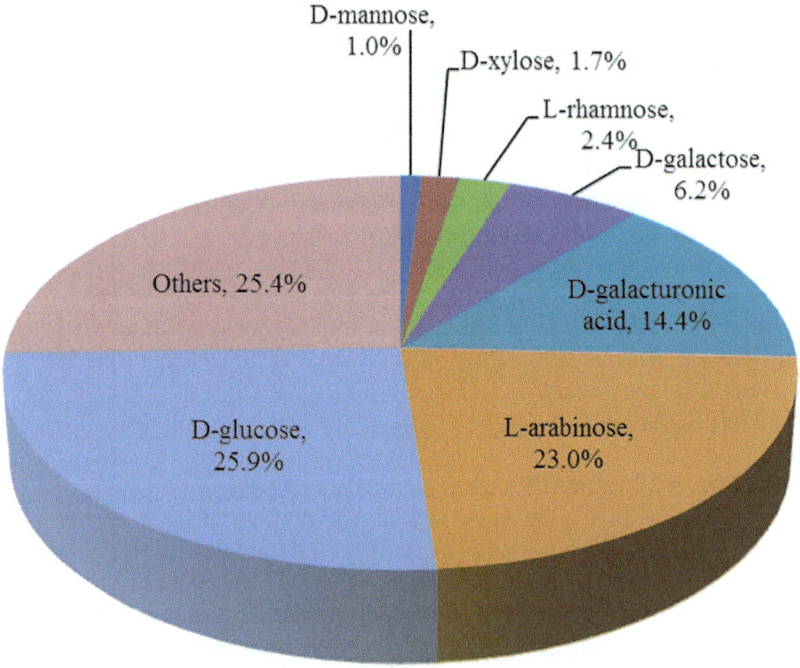

Fig. 2 Composition of the sugar beet pulp used in this work (obtained from the British Sugar biorefinery, Wissington, UK). Monosaccharide concentrations determined using Ion Chromatography (Dionex ICS 5000+, equipped with an electrochemical detector and fitted with a 4 × 250 mm analytical CarboPac PA1 column) after full acid hydrolysis with H_2SO_4 2.5% (v/v) at 121 °C for 1 hour.

The vinasse stream used here appears as a dark liquid containing fine suspended solids and has a pH of approximately 5. It has high levels of glycerol, corresponding to about 35% of the total vinasse (on a dry mass basis), at a concentration of 188 g L^{-1}. As described in Table 1, the vinasse also contains several monosaccharides, such as D-fructose and Gal, as well as low concentrations of sugar alcohols including mannitol, xylitol and dulcitol. The total protein in the vinasse was quantified at 5 g L^{-1}, and the presence of acetate and polyphenols was also determined, which is in agreement with previous publications.[15,16] The high concentration of glycerol, a good carbon and energy source for microbial fermentation, makes the vinasse an interesting fermentation feedstock for enzyme production (Section 2.2) provided simple pre-treatment methods can be established.

2.2 Biorefinery enzyme production (L-arabinofuranosidase)

Enzymes are widely used as industrial biocatalysts due to their high specificity and selectivity. Enzymatic processes are considered environmentally friendly as enzymes are biodegradable and tend to produce less waste.[17] In a biorefinery context, enzymes can be used to breakdown polymers released from complex feedstocks, the classic example being depolymerisation of cellulose into D-glucose monomers by cellulases, or for synthetic purposes, *e.g.* the production of chiral aminoalcohols as described in Section 1.

A range of effective cellulases are already commercially available and we have previously described the cloning and production of TK and TAm enzymes used in chiral aminoalcohol synthesis.[8,10] L-Arabinofuranosidases (AF, EC 3.2.1.55) are exo-type glycosidases that catalyse the successive removal of L-arabinose residues from the non-reducing termini of α-1,2-, α-1,3-, α-1,5- and α-4,6-linked arabinofuranosyl residues.[18] In order to release Ara monomers from the arabinan side chains of SBP pectin we have identified and cloned a thermostable AF from *Geobacillus thermoglucosidasius*. This is a homodimeric enzyme with a MW of 130 kDa that is able to hydrolyse natural arabinans from different sources, such as sugar beet pectin, yielding monomeric Ara and an intact D-galacturonic acid backbone (GABB).

Table 1 Concentration of main components present in the vinasse stream. Sugar and sugar alcohols were determined using ion chromatography, glycerol and acetate using HPLC-RI and polyphenols using spectrophotometry with galic acid as reference. The sugar beet vinasse used in this work was provided by British Sugar (Wissington, UK)

Component	Concentration (g L^{-1})
Glycerol	187.6 ± 2.8
Mannitol	8.0 ± 0.03
Galactose	5.5 ± 0.02
Xylitol	5.4 ± 0.09
Fructose	2.0 ± 0.03
Dulcitol	2.0 ± 0.02
Acetate	1.7 ± 0.02
Total protein	5.1 ± 0.08
Polyphenols	6.0 ± 0.02

As described in Section 2.1.2, vinasse is an interesting feedstock for enzyme production by microbial fermentation due to its high glycerol concentration and the presence of other sugars. Here we have examined the use of vinasse for the production of AF and shown that it can be used as an effective carbon and energy source with minimal pre-treatment. An *E. coli* BL21 (DE3) strain expressing the recombinant AF, was grown in batch culture using a DasBox bioreactor equipped with four 250 mL fermentation vessels (DASbox® Mini Bioreactor System). The culture media contained 6-fold diluted vinasse (to reach an initial glycerol concentration of 30 g L^{-1}) supplemented with 10 g L^{-1} of yeast extract with the pH adjusted to pH 7 using 0.5 M NaOH.

Fig. 3 shows typical cell growth and glycerol consumption profiles as well as the volumetric AF activity (one unit is defined as the amount of AF able to hydrolyse 1 μmole of *p*-nitrophenyl-α-L-arabinofuranoside to *p*-nitrophenol and L-arabinose per minute at pH 7 and 37 °C). The *E. coli* grew rapidly and a maximum biomass concentration of 19.3 g_{dcw} L^{-1} was achieved while glycerol was completely consumed by around 30 h. A high level of AF expression was found with 18.8 kU $g_{dcw}{}^{-1}$ reached after 27 h of fermentation. These results demonstrate the feasibility of using diluted vinasse for the on-site production of processing enzymes using existing feedstocks within an integrated biorefinery. The

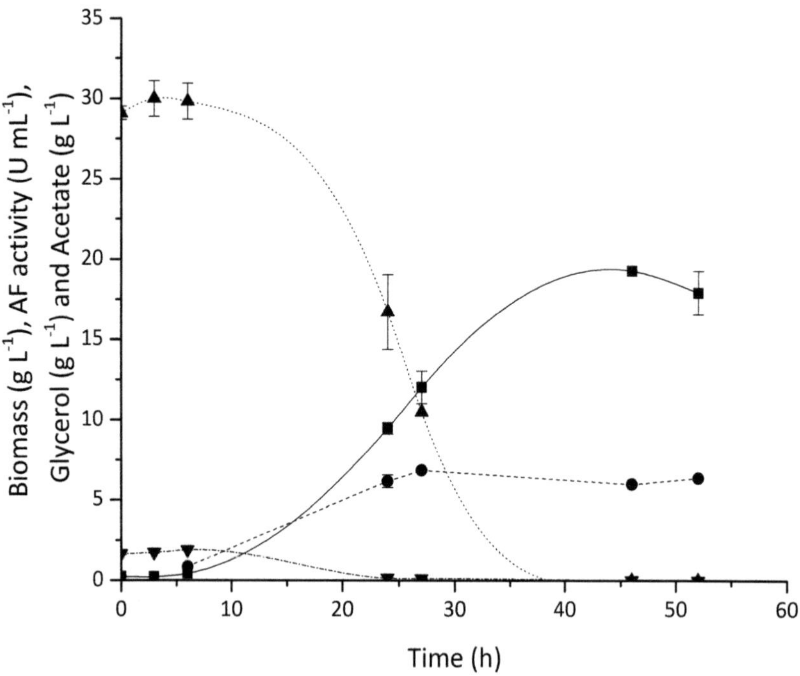

Fig. 3 Production of L-arabinofuranosidase (AF) by *E. coli* fermentation on 6-fold diluted sugar beet vinasse at pH 7: (■) biomass, (▲) glycerol, (●) AF volumetric activity and (▼) acetate. The Dissolved Oxygen Tension (DOT) was maintained at 30% of air saturation by varying the stirrer speed between 200 to 1000 rpm and supplying air at a flow rate of 1 vvm. The pH was controlled at pH 7, at 37 °C with an initial cell concentration of 0.1 g_{dcw} L^{-1} and final working volume of 200 mL. Error bars represent one standard deviation about the mean ($n = 2$).

utilisation of this inexpensive feedstock should have a positive impact on the overall process economics. While the results presented here focus on vinasse utilisation for AF production, the same fermentation process can also be used for TK and TAm production used in the enzymatic upgrading of Ara and GalAc.

2.3 Sugar beet pulp pre-treatment options

2.3.1 Fractionation of SBP by steam explosion. As described in Section 1, the SBP stream exits from the aqueous diffuser in which sucrose is extracted from the sliced sugar beet (Fig. 1). Structurally, the pectin in SBP exists between cellulose microfibrils that are cross-linked by low concentrations of lignin.[12] At this point in the process a mild thermochemical pre-treatment is necessary to 'open up' the SBP structure and release the separate biological polymers. In this work, studies were carried out on dried and frozen SBP although in practice the techniques described here would be applied directly to the wet pulp.

Steam explosion (SE) is a hot water pre-treatment process in which biomass is treated with high-pressure saturated steam. The pressure is then swiftly released, making the material undergo explosive decompression, releasing all water-soluble biomass components into the media.[19] For carbohydrate-rich biomass with low lignin content, like SBP, this can be a relatively mild fractionation method. The mild conditions minimise the formation of sugar degradation by-products, such as furfurals, that can act as inhibitors to subsequent biological processes.[19,20]

SBP fractionation was carried out by SE yielding soluble pectin and insoluble cellulose fractions. SE operating conditions were optimised using a response surface methodology to maximise the release of pectin sugars while minimising cellulose degradation and furfural formation.[21] The two independent variables of time (1 to 30 min) and pressure (4 to 8 Bar) were studied. The yields of polymeric Ara and GalAc from pectin, and Glu from cellulose released into the soluble fraction were measured as experimental responses. The response surface generated (Fig. 4) shows the percentage of Ara solubilised as a function of pressure and time. For certain samples the amount of Glu found in the same fraction is also indicated. The results generally showed high levels of Ara release with low levels of Glu release under the same conditions.

The statistical models generated for both Ara and Glu release were next used to find optimum conditions with the constraints of maximising the yield of Ara (>70%) and minimising the yield of Glu (<6%) in the soluble fraction. Based on the model predictions, the optimum conditions were found to be at a pressure of 5.3 Bar for 24.4 min yielding a theoretical release of 83.2 and 4.8% of Ara and Glu in the soluble fraction, respectively. Verification experiments were carried out under conditions that were slightly modified for an experimental ease (5 Bar and 24 min). The measured levels of Ara and Glu in the soluble fraction were within 5–10% of those predicted by the model. Further analysis showed that the presence of by-products, such as furfurals (*e.g.* HMF) and organic acids (*e.g.* acetic acid), in the soluble fraction were negligible under the optimised conditions.

Overall, these results show that SE is an excellent process for the fractionation of SBP, that allows formation of a soluble pectin fraction and a relatively pure, insoluble cellulose fraction that can be readily used for bioethanol production (Section 2.5).[21]

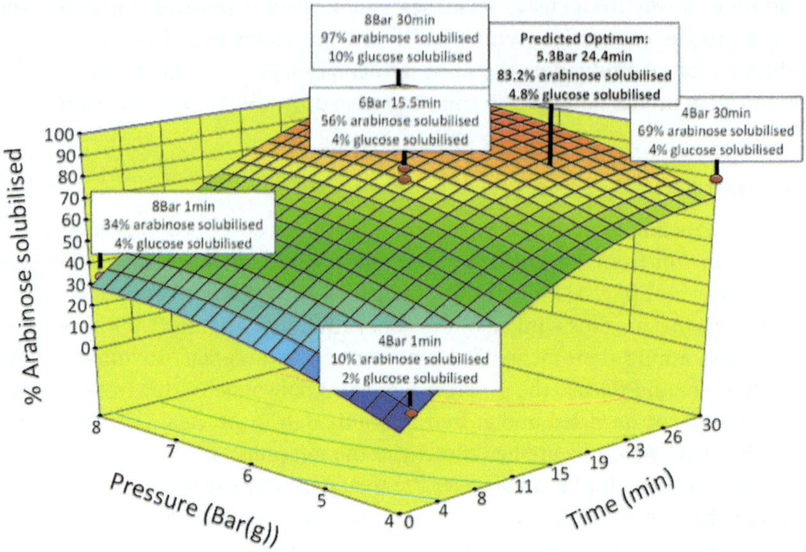

Fig. 4 Surface response diagram showing percentage of L-arabinose solubilised as a function of pressure and time during steam explosion (SE). The red data points represent actual experimental design points. 11 SE pretreatments over various times and operating pressures were performed using a central composite design. SE experiments were carried out in a stirred Parr pressure reactor (1 L capacity, Boston Instruments) with 50 g of whole and washed SBP at the specified pressure and time with agitation fixed at 150 rpm.

2.3.2 Enzymatic hydrolysis of SBP and separation of L-arabinose by ultrafiltration. Following SE, the sugar beet pectin is recovered in the soluble fraction as a largely intact polymer. This requires further hydrolysis to release the main pectin monosaccharides: Ara and GalAc (Fig. 2). The initial focus was on Ara since it is the most abundant. Selective depolymerisation of the arabinan side chains of the pectin using L-arabinofuranosidase (AF) was attempted. This would yield soluble monomeric Ara leaving the polymeric GalAc backbone (GABB) intact.

His-tagged AF was produced by *E. coli* fermentation using vinasse as a carbon and energy source (Section 2.2) and then purified with Ni-NTA agarose beads. The soluble sugar beet pectin following SE (Section 2.3.1) was used as a substrate. After incubation for 8 h (100 mg of pure AF per mL of reaction in 0.1 M bis–Tris buffer at pH 7 and at 60 °C), 87% of the polymeric Ara was released as monomer. No other sugar monomers were detected in solution confirming AF selectivity.

Separation of the Ara monomers from the polymeric GABB was then explored. Membrane filtration is a well established unit operation in the food and (bio) pharmaceutical industries.[22,23] In tangential flow filtration (TFF) processes, the feed flow is directed parallel to the membrane surface creating shear and preventing membrane fouling by retained high MW species. Ultrafiltration (UF) uses membranes with a molecular weight cut-off (MWCO) between 10^3 to 10^5 Da (pore size 1–20 nm) and can be used to separate molecules with at least one order of magnitude difference in size.[23] In this case the GABB would be retained while the Ara monomers (MW = 150 Da) will pass through the membrane.

The performance of two different UF membrane configurations was studied; a flat sheet membrane (FSM, Minimate TFF Capsule, Pall®) and a hollow fiber membrane (HFM, MidiKros, SpectrumLab®) with a total membrane area of 50 and 115 cm,[2] respectively and both with a MWCO of 1 kDa. Both UF processes were carried out using AF-hydrolysed pectin at a feed flow rate of 40 mL min^{-1} and at a constant transmembrane pressure (TMP) of 20 and 40 psi for HFM and FSM respectively. As shown in Table 2, Ara recovery in the permeate was similarly high for both types of membrane (~88%) and the Ara transmission was constant as a function of time. Membrane rejection of GABB was equally high in both cases (>93%). The slightly higher rejection for the HFM is probably related to the different membrane materials (modified polyethersulfone and omega polyethersulfone for HFM and FSM, respectively).

From a process design perspective the HFM configuration appears most favourable since the overall filtration process was 1.5-fold faster and achieved a higher concentration factor for GABB in the retentate while operating at a lower pressure. Initial permeate fluxes (permeate flow rate per membrane area) were 12 and 21 L m^{-2} h^{-1} for HFM and FSM, respectively. However, by the end of the process, the HFM flux was reduced by only 39% while for the FSM it decreased by more than 60% implying higher levels of membrane fouling in the flat sheet configuration.

Overall, the enzyme-membrane process showed high levels of Ara release from the GABB. Monomeric Ara was recovered in equally high yield in the UF permeate in a form suitable for subsequent upgrading (Section 2.4). The GABB retained in the HFM process was concentrated over 10-fold. Current work seeks to identify endo- and exo-polygalacturonases for monomerisation of GalAc residues in the GABB in order to enable complete enzymatic fractionation of the sugar beet pectin.

2.3.3 Acid hydrolysis of SBP and separation of monosaccharides by centrifugal partition chromatography. An alternative to selective enzymatic fractionation (Section 2.3.2) is the complete acid hydrolysis of the sugar beet pectin followed by simultaneous separation of all the pectin monosaccharides *via* the application of a higher resolution separation technique. Here we have investigated the application of Centrifugal Partition Chromatography (CPC). This is a liquid–liquid chromatography technique that partitions solutes between two immiscible liquid phases. One liquid phase is held stationary and retained in the

Table 2 Ultrafiltration process performance for the recovery of monomeric Ara and rejection of GABB following enzymatic hydrolysis of arabinan side chains of SBP. Table shows a comparison of processes using a Hollow Fiber Membrane (HFM) and a Flat Sheet Membrane (FSM). Both processes were carried out at a constant TMP of 20 and 40 psi for HFM and FSM respectively and feed flow rate = 40 mL min^{-1} for both processes

	HFM	FSM
Ara recovery (%)	89.5	86.7
GABB rejection (%)	98.3	92.5
Concentration factor	10.3	7.6
Process time (h)	2	3
Volume (mL)	244	235
Residual flux (%)	60.8	41.4

column *via* centrifugal forces, while the mobile phase is pumped past it through a series of interconnected chambers where the two phases mix and solute partitioning occurs.[24] The lack of a solid stationary phase allows crude process streams to be handled with little pre-treatment which is important in a biorefinery context.

Following steam explosion, the solubilised pectin fraction was fully hydrolysed by adding 2.5% (v/v) sulphuric acid and heating to 121 °C for 1 hour in an autoclave before adjustment to pH 6 with NaOH. This crude hydrolysed material had a total dissolved solids content of approximately 100 g L^{-1} and a total sugar concentration around 20 g L^{-1} as well as containing unknown coloured impurities giving the material a black colour.

This crude hydrolysed SBP mixture was then separated by CPC. An ethanol : aqueous ammonium sulfate (300 g L^{-1}) (0.8 : 1.8 v/v) two-phase system was established that could isolate the monosaccharides as well as remove coloured impurities in the crude sample during phase system formation.[25,26] The lower phase (LP) was utilised as the stationary phase and the upper phase (UP) as the mobile phase with a 950 mL preparative CPC column (Kromaton FCPC-A, Rousselet Robatel Kromaton) operated in ascending mode. The crude sample was prepared in the stationary phase. Intermittent extrusion of the liquids in the CPC column is possible by pumping just the LP (stationary phase).[26]

Fig. 5, shows the separation of 152 mL of crude hydrolysed SBP pectin. Coloured impurities were fully eluted before the monosaccharides emerge in the order Rha, Ara, Gal and GalAc. Extrusion after 72 minutes ensured the rapid elution of GalAc from the column; this can be seen from the sharp GalAc peak at

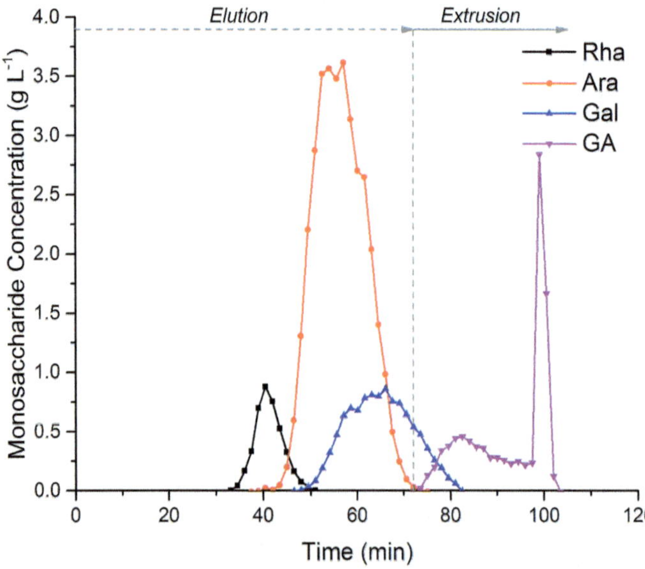

Fig. 5 Centrifugal Partition Chromatography (CPC) of crude hydrolysed sugar beet pectin to separate the main monosaccharides: (Rha) L-rhamnose, (Ara) L-arabinose, (Gal) D-galactose and (GA) D-galacturonic acid. Experiments were performed using a 250 mL semi-preparative CPC column (Kromaton), using an ethanol : ammonium sulphate (300 g L^{-1}) (0.8 : 1.8 v/v) phase system at 1000 rpm in ascending mode, with a mobile phase flow rate of 8 mL min^{-1}. Stationary phase retention was 50%.

100 min. Extrusion reduced the total run time, increased the GalAc fraction concentration and regenerated the column with a fresh stationary phase for subsequent injections. During biorefinery processing three main fractions could be collected from the CPC. Rha could be collected in a fraction between 33-45 min with 93% recovery and 92% purity. Ara could be collected between 46–66 min with 97% recovery and 84% purity due to the co-elution of Gal. Finally, GalAc was isolated between 78-100 min with over 95% recovery and 96% purity.

CPC appears to be a promising separation technique for biorefinery applications. In this case it allowed the processing of crude, hydrolysed SBP facilitating impurity removal and monosaccharide fractionation. Specifically, it was possible to isolate the two main monosaccharides present in SBP, *i.e.* Ara and GalAc, with high yields and adequate purity. The monosaccharide throughput achieved, operating in elution–extrusion mode, was 1.9 g h^{-1} L^{-1} of column volume.

2.4 Enzymatic upgrading of L-arabinose catalysed by transketolase

Ara is the main monosaccharide present in sugar beet pectin (Fig. 2) hence it was the initial focus for catalytic upgrading. Within our biorefinery concept it can be recovered by both enzymatic fractionation (Section 2.3.2) and acid hydrolysis followed by CPC (Section 2.3.3). Ara can be the starting material for production of arabitol, one of the top commodity biochemicals from biomass,[3] however our focus is on enzymatic upgrading to higher added value compounds.

TK is a thiamine diphosphate (ThDP) dependent enzyme that has been used in stereospecific carbon–carbon bond formation.[27] The ability of TK to perform the reaction irreversibly with the loss of CO_2, when using β-hydroxypyruvate (HPA) as the ketol donor, makes it an attractive biocatalyst in industrial synthesis.[6] Utilisation of Ara as an aldehyde acceptor with TK, as shown in Fig. 6 would generate the ketoheptose L-*gluco*-heptulose (GluHep), which has applications in cancer and hypoglycaemia.[28] Natural aldoses accepted by wild-type (WT) TK are phosphorylated, such as D-ribose-5-phosphate and erythrose-4-phosphate, however non-phosphorylated sugars have lower affinities.[29] To achieve a commercially viable conversion at large-scale with the non-phosphorylated substrate Ara, variants of TK with greater specific activity are required.

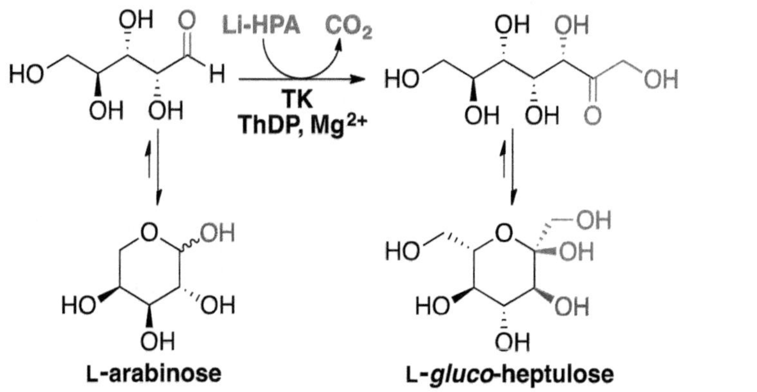

Fig. 6 Transketolase mediated synthesis of L-*gluco*-heptulose from L-arabinose using lithium hydroxypyruvate (Li-HPA) as the ketol donor.

Screening of existing TK libraries previously prepared by saturation mutagenesis of conserved residues in the TK active site identified a range of *E. coli* TK mutants that would readily accept Ara.[30] Docking calculations using Ara as a substrate were also performed which highlighted three potentially important residues that could interact with Ara (R358, H461, and R520). The three libraries were screened using a colorimetric plate-based assay developed for high throughput applications with sugar substrates. All *E. coli* TK enzyme variants identified in the screen were then expressed and activities towards Ara compared to the WT transketolase. As shown in Fig. 7, the greatest improvements in specific activities identified during screening were a 2.1-fold increase for R520Y and a 1.5-fold increase for R520P. Small scale conversions of Ara to GluHep were subsequently performed for all 6 variants. The highest conversion yields were obtained for R520Y and H461Y as these had the highest specific activity and the highest expression levels.

These two high-performing TK mutants were then used for the efficient stereoselective, one-step synthesis of GluHep from pure Ara at preparative scale. The reaction was performed for 24 h and ^{1}H NMR spectroscopic analysis showed a 1 : 1 ratio between the remaining Ara and the GluHep formed. A facile isolation procedure was developed whereby the mixture was filtered through Amberlite® IR120 to remove Tris buffer and some Ara to give GluHep: further purification using column chromatography gave isolated yields of 45% (H461Y) or 40% (R520Y).

Product characterisation confirmed that the TK-catalysed reaction showed exquisite diastereoselectivity leading to a single D-*threo* ketose where only the more stable α-anomer was generated; this was consistent with the

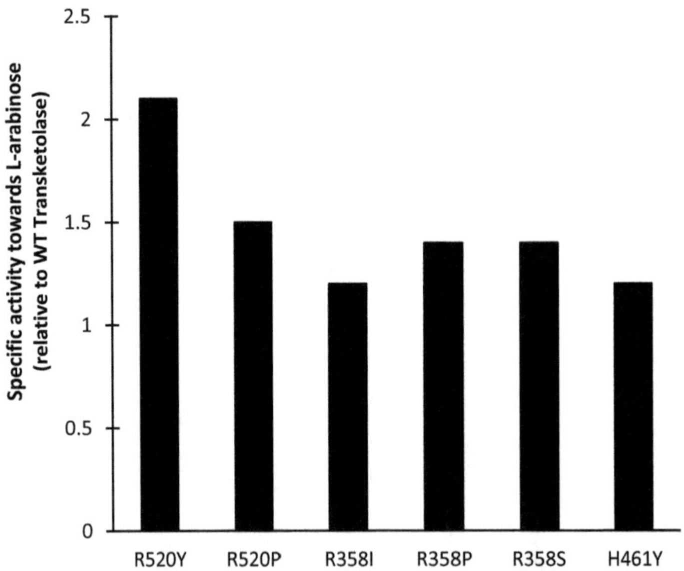

Fig. 7 Specific activities (μmol min^{-1} mg^{-1}) of *E. coli* transketolase mutants toward L-arabinose, relative to the wild-type (WT) enzyme. Experiments performed using clarified lysates with Li-HPA (33 mM), Ara (33 mM), ThDP (1.7 mM) and $MgCl_2$ (6 mM) in Tris–HCl (50 mM) at pH 7.0. The specific activity of the WT-TK was 0.021 μmol min^{-1} mg^{-1}.

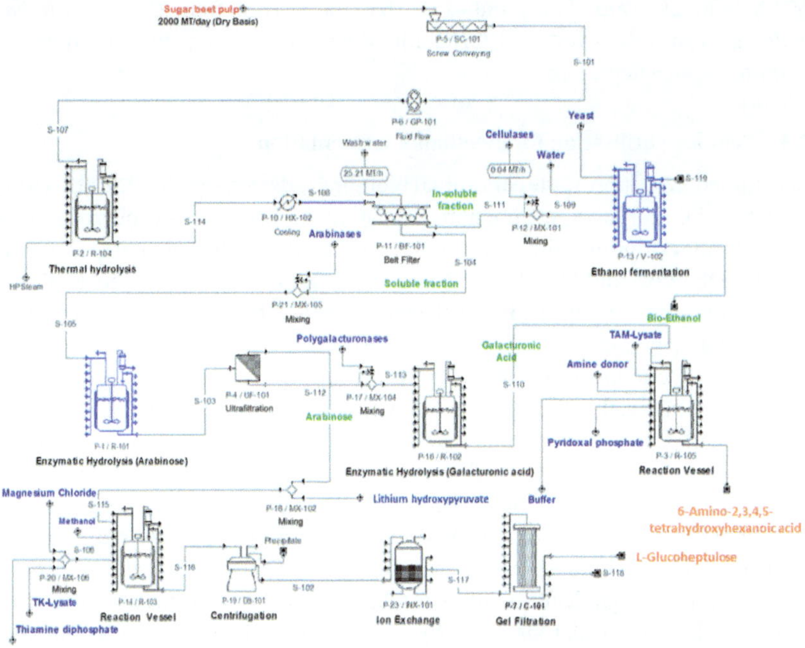

Fig. 8 Indicative process flowsheet of the conceptual biorefinery processes showing: steam explosion (thermal hydrolysis) of the wet SBP; cellulose utilisation for bioethanol fermentation; enzyme-membrane fractionation of sugar beet pectin; bioconversion of D-galacturonic acid (transaminase) and L-arabinose (transketolase). Basis: 2000 metric tonnes of SBP per day.

stereoselectivity observed in other TK reactions with aldoses.[29] The absolute configuration of GluHep was confirmed by optical rotation and comparison of the data to its enantiomer in the literature.[31]

For application in a biorefinery context, it is important to show that this TK catalysed conversion can also be carried out using real process streams. This was demonstrated using a crude Ara stream following ultrafiltration (Section 2.3.2). The ultrafiltration permeate, containing 59 mM of Ara at pH 6.0, was diluted with Tris buffer (50 mM to 33 mM to match the pure Ara reaction conditions) and the pH adjusted to 7 prior to the addition of TK and cofactors. The reaction was stirred at room temperature and after 24 h GluHep was produced in 38% isolated yield, just slightly below the yields obtained with pure Ara using both R520Y and H461Y mutants. Likewise, conversions were also carried out on a mixture mimicking the composition of the Ara-rich fraction from CPC (Section 2.3.3). The mixture contained Ara (26 mM) and some of the co-eluting Gal (6 mM). After a 24 h reaction, TK mutant H461Y gave a 46% conversion yield to GluHep. It also displayed a high selectivity with only 2% of Gal consumed by the end of the bioconversion.

These studies demonstrate that the TK variants identified are able to upgrade Ara in biorefinery process streams into a higher value ketoheptose, GluHep. The commercial value of this compound is currently being explored. Preliminary studies have also shown further upgrading to our target aminoalcohols. Screening of an in-house TAm library indicated that the ω-TAm from *Chromobacterium violaceum* can further convert GluHep into the corresponding chiral aminopolyol 2-amino-L-*gluco*-heptulose.

2.5 Cellulose utilisation for bioethanol fermentation

As outlined in Section 1, the conceptual biorefinery design requires the utilisation of all the main carbohydrate fractions present in the SBP. After steam explosion of the SBP under optimised conditions (Section 2.3.1) a relatively pure, insoluble cellulose fraction could be obtained. Since there is already a bioethanol facility at the Wissington site (Fig. 1), it was decided to direct this glucose-rich stream to bioethanol formation.

The solid fraction recovered after SE was therefore subjected to hydrolysis using a commercial cellulase mix: 0.5 mg of cellulase per g of solid at 50 °C (Cellulase 3L-C013L, Biocatalysts Ltd, Cardiff, UK). After incubation for 24 h, the reaction was stopped by centrifugation and the supernatant recovered for monosaccharide analysis. The concentration of soluble Glu was 19 g L^{-1} which indicated that virtually all of the SBP cellulose had been monomerised. This also confirmed that the mild SE conditions established in Section 2.3.1 did not lead to the formation of significant cellulase inhibitors. Very low concentrations of other sugars such as Ara, Gal, GalAc and Rha, coming from the remaining pectin after SE, were also detected.

The unpurified supernatant from the cellulase hydrolysis was subsequently used as a fermentation medium for the growth of a commercial *Saccharomyces cerevisiae* strain (Alcotec 24 Turbo Yeast, Hambleton Bard Ltd, Chesterfield, UK). This was incubated at 30 °C for 24 h in shake flasks. At the end of the fermentation process, almost all of the Glu was consumed producing 0.48 g of ethanol per g of Glu. For an unoptimised fermentation process this value is very close to the theoretical ethanol/glucose yield of 0.51 g g^{-1} confirming the feasibility of this approach. Analysis of the spent fermentation medium showed that the concentrations of the other monosaccharides present in the cellulose hydrolysis stream remained unaltered over the 24 h incubation. These could be directed to anaerobic digestion (Fig. 1) for energy generation. Likewise, the CO_2 formed during fermentation can be captured and pumped into the glasshouse to support British Sugar's horticulture business and it can also be captured and liquefied for commercial sale (Fig. 1). This would have the added benefit of reducing overall emissions and hence the environmental impact.

2.6 Process economic modelling and life cycle analysis

Having demonstrated SBP fractionation, separation and conversion, it is necessary to establish a whole process flowsheet underpinned by material and energy balances for each of the unit operations. These are needed for estimating the capital and operating costs of the integrated biorefinery. Models of the standard unit operations are well known and we have previously developed mechanistic models of the TK[32] and TAm[33] reactions needed to simulate bioconversion kinetics. Economic and Life Cycle Analysis (LCA) will ultimately inform the commercial feasibility and environmental sustainability of the biorefinery concept. These analyses will also establish process competitiveness against benchmark products: bioethanol production will be benchmarked against gasoline, biopolymers against polylactic acid, and speciality chemicals against succinate and ibuprofen (representing low- and higher-value products, respectively).

Fig. 8 shows an indicative process flowsheet for a facility processing 2000 metric tonnes of SBP (on a dry basis) per day and operating seasonally. The flowsheet depicts the main operations of SBP processing, cellulose conversion to bioethanol, Ara release and bioconversion to GluHep and fractionation of the GalAc backbone followed by conversion to 6-amino-2,3,4,5-tetrahydroxyhexanoic acid. Preliminary simulation results obtained using SuperPro Designer® (Intelligence, Inc., N.J., U.S.A.) identified a costly inventory of lithium hydroxypyruvate, which acts as the ketol donor in the synthesis of GluHep from Ara (Section 2.4) and accounts for a significant percentage of raw materials costs. Finding alternative, cheaper donors would therefore be economically favourable. Once finalised, the flowsheet will allow us to identify rate-limiting steps that require further research and development and to study the financial trade-off between equipment scale and operational efficiency.

3. Conclusions and future work

This work has demonstrated the technical feasibility of an integrated biorefinery concept in which wet sugar beet pulp can be fractionated by steam explosion into pectin and cellulose-rich streams. The three main monosaccharides present in the two biopolymers (Ara, Glu and GalAc) can then be solubilised, using either enzymatic or chemical hydrolysis, and separated into crude fractions suitable for conversion into higher value products.

It has been shown that the Ara-rich fraction can be enzymatically upgraded by TK into GluHep, which has potential therapeutic applications.[28] Likewise, we have demonstrated that the Glu-rich fraction can be fermented by a Yeast strain to produce bioethanol with high yield (in a facility that already exists at the British Sugar Wissington site). Current work is examining the stereoselective amination of the remaining GalAc fraction using transaminases including the *C. violaceum* ω-TAm. Initial studies have suggested that this enzyme can convert GalAc into 6-amino-2,3,4,5 tetrahydroxyhexanoic acid which can be used for the synthesis of biopolymers such as polyhydroxypolyamides[34] and as a precursor for potent antiviral polyhydroxyazepanes.[35] Three of the enzymes used in this study (AF, TK and TAm) have been expressed in *E. coli* which can yield high specific enzyme activities when grown on diluted sugar beet vinasse.

Future studies will address some of the major engineering challenges associated with the integrated SBP biorefinery concept. These are summarised below.

• Increasing the yield and productivity of the Ara and GalAc bioconversions by investigating substrate/product inhibition and fed-batch operation.

• Evaluating the economic and environmental benefits of the proposed biorefinery based on whole process modelling and LCA alongside market analysis of the synthetic products.

• Implementing next generation unit operations, specifically intensified and flow reactor technologies, to further increase space-time yields.

• Exploring the potential of synthetic biology approaches to engineer host cells capable of simultaneous pectin hydrolysis and Ara/GalAc upgrading or cellulose hydrolysis and ethanol fermentation.

Acknowledgements

The authors would like to thank the UK Engineering and Physical Sciences Research Council (EPSRC) for financial support of this work (EP/K014897/1) as part of their Sustainable Chemical Feedstocks call. Input and advice from the project Industrial Advisory Board is also acknowledged. DPW would also like to thank EPSRC and NS the Ministry of Higher Education Malaysia for the award of PhD studentships.

References

1 British Sugar, Personal Communication, 2017.
2 G. Mesbahi, J. Jamalian and A. Farahnaky, *Food Hydrocolloids*, 2005, **19**, 731–738.
3 T. Werpy and G. Petersen, *Results of Screening for Potential Candidates from Sugars and Synthesis Gas, Office of Scientific and Technical Information (OSTI)*, 2004, vol. I.
4 R. W. Foster, C. J. Tame, D.-K. Bučar, H. C. Hailes and T. D. Sheppard, *Chem.–Eur. J.*, 2015, **21**, 15947–15950.
5 M. E. B. Smith, B. H. Chen, E. G. Hibbert, U. Kaulmann, K. Smithies, J. L. Galman, F. Baganz, P. A. Dalby, H. C. Hailes, G. J. Lye, J. M. Ward, J. M. Woodley and M. Micheletti, *Org. Process Res. Dev.*, 2010, **14**, 99–107.
6 J. Shaeri, R. Wohlgemuth and J. M. Woodley, *Org. Process Res. Dev.*, 2006, **10**, 605–610.
7 D. Koszelewski, K. Tauber, K. Faber and W. Kroutil, *Trends Biotechnol.*, 2010, **28**, 324–332.
8 M. F. Villegas-Torres, R. J. Martinez-Torres, A. Cázares-Körner, H. Hailes, F. Baganz and J. Ward, *Enzyme Microb. Technol.*, 2015, **81**, 23–30.
9 C. U. Ingram, M. Bommer, M. E. B. Smith, P. A. Dalby, J. M. Ward, H. C. Hailes and G. J. Lye, *Biotechnol. Bioeng.*, 2007, **96**, 559–569.
10 L. Rios-Solis, M. Halim, A. Cázares, P. Morris, J. M. Ward, H. C. Hailes, P. A. Dalby, F. Baganz and G. J. Lye, *Biocatal. Biotransform.*, 2011, **29**, 192–203.
11 T. Sakamoto and T. Sakai, *Phytochemistry*, 1995, **39**, 821–823.
12 V. Micard, C. M. G. C. Renard and J. F. Thibault, *Enzyme Microb. Technol.*, 1996, **19**, 162–170.
13 A. G. M. Leijdekkers, J. P. M. Bink, S. Geutjes, H. A. Schols and H. Gruppen, *Bioresour. Technol.*, 2013, **128**, 518–525.
14 P. Tufvesson, J. Lima-Ramos, M. Nordblad and J. M. Woodley, *Org. Process Res. Dev.*, 2011, **15**, 266–274.
15 J. M. Salgado, E. M. Carballo, B. Max and J. M. Domínguez, *Bioresour. Technol.*, 2010, **101**, 2379–2388.
16 A. Pramanik, A. Mitra, M. Arumugam, A. Bhattacharyya, S. Sadhukhan, A. Ray, S. Haldar, U. K. Mukhopadhyay and J. Mukherjee, *Folia Microbiol.*, 2012, **57**, 71–79.
17 J. D. Rozzell, *Bioorg. Med. Chem.*, 1999, **7**, 2253–2261.
18 B. C. Saha, *Biotechnol. Adv.*, 2000, **18**, 403–423.
19 B. Zhang, A. Shahbazi, L. Wang, O. Diallo and A. Whitmore, *J. Ind. Microbiol. Biotechnol.*, 2011, **38**, 819–824.

20 K. Ziemiński, I. Romanowska, M. Kowalska-Wentel and M. Cyran, *Bioresour. Technol.*, 2014, **166**, 187–193.
21 C. Hamley-Bennett, G. J. Lye and D. J. Leak, *Bioresour. Technol.*, 2016, **209**, 259–264.
22 R. van Reis and A. Zydney, *J. Membr. Sci.*, 2007, **297**, 16–50.
23 S. M. Cramer and M. A. Holstein, *Curr. Opin. Chem. Eng.*, 2011, **1**, 27–37.
24 W. Murayama, T. Kobayashi, Y. Kosuge, H. Yano, Y. Nunogaki and K. Nunogaki, *J. Chromatogr. A*, 1982, **239**, 643–649.
25 D. P. Ward, M. Cárdenas-Fernández, P. Hewitson, S. Ignatova and G. J. Lye, *J. Chromatogr. A*, 2015, **1411**, 84–91.
26 D. P. Ward, P. Hewitson, M. Cárdenas-Fernández, C. Hamley-Bennett, A. Díaz-Rodríguez, N. Douillet, J. P. Adams, D. J. Leak, S. Ignatova and G. J. Lye, *J. Chromatogr. A*, 2017, **1497**, 56–63.
27 H. C. Hailes, D. Rother, M. Müller, R. Westphal, J. M. Ward, J. Pleiss, C. Vogel and M. Pohl, *FEBS J.*, 2013, **280**, 6374–6394.
28 M. Board, A. Colquhoun and E. A. Newsholme, *Cancer Res.*, 1995, **55**, 3278–3285.
29 F. Charmantray, V. Hélaine, B. Legeret and L. Hecquet, *J. Mol. Catal. B: Enzym.*, 2009, **57**, 6–9.
30 F. Subrizi, M. Cardenas-Fernandez, G. J. Lye, J. M. Ward, P. A. Dalby, T. D. Sheppard and H. C. Hailes, *Green Chem.*, 2016, **18**, 3158–3165.
31 W. D. Maclay, R. M. Hann and C. S. Hudson, *J. Am. Chem. Soc.*, 1942, **64**, 1606–1609.
32 B. H. Chen, E. G. Hibbert, P. A. Dalby and J. M. Woodley, *AIChE J.*, 2008, **54**, 2155–2163.
33 L. Rios-Solis, P. Morris, C. Grant, A. O. O. Odeleye, H. C. Hailes, J. M. Ward, P. A. Dalby, F. Baganz and G. J. Lye, *Chem. Eng. Sci.*, 2015, **122**, 360–372.
34 D. E. Kiely, L. Chen and T. H. Lin, *J. Polym. Sci., Part A: Polym. Chem.*, 2000, **38**, 594–603.
35 T.-L. Shih, M.-T. Liang, K.-D. Wu and C.-H. Lin, *Carbohydr. Res.*, 2011, **346**, 183–190.

Faraday Discussions

PAPER

Bioelectrochemical conversion of CO_2 to chemicals: CO_2 as a next generation feedstock for electricity-driven bioproduction in batch and continuous modes

Suman Bajracharya,[a] Karolien Vanbroekhoven,[a] Cees J. N. Buisman,[b] David P. B. T. B. Strik[b] and Deepak Pant *[a]

Received 30th January 2017, Accepted 29th March 2017
DOI: 10.1039/c7fd00050b

The recent concept of microbial electrosynthesis (MES) has evolved as an electricity-driven production technology for chemicals from low-value carbon dioxide (CO_2) using micro-organisms as biocatalysts. MES from CO_2 comprises bioelectrochemical reduction of CO_2 to multi-carbon organic compounds using the reducing equivalents produced at the electrically-polarized cathode. The use of CO_2 as a feedstock for chemicals is gaining much attention, since CO_2 is abundantly available and its use is independent of the food supply chain. MES based on CO_2 reduction produces acetate as a primary product. In order to elucidate the performance of the bioelectrochemical CO_2 reduction process using different operation modes (batch *vs.* continuous), an investigation was carried out using a MES system with a flow-through biocathode supplied with 20 : 80 (v/v) or 80 : 20 (v/v) CO_2 : N_2 gas. The highest acetate production rate of 149 mg L^{-1} d^{-1} was observed with a 3.1 V applied cell-voltage under batch mode. While running in continuous mode, high acetate production was achieved with a maximum rate of 100 mg L^{-1} d^{-1}. In the continuous mode, the acetate production was not sustained over long-term operation, likely due to insufficient microbial biocatalyst retention within the biocathode compartment (*i.e.* suspended micro-organisms were washed out of the system). Restarting batch mode operations resulted in a renewed production of acetate. This showed an apparent domination of suspended biocatalysts over the attached (biofilm forming) biocatalysts. Long term CO_2 reduction at the biocathode resulted in the accumulation of acetate, and more reduced compounds like ethanol and butyrate were also formed. Improvements in the production rate and different biomass retention strategies (*e.g.* selecting for biofilm forming micro-organisms) should be investigated to enable continuous biochemical

[a]*Separation and Conversion Technology, Flemish Institute for Technological Research (VITO), Boeretang 200, Mol 2400, Belgium. E-mail: deepak.pant@vito.be; pantonline@gmail.com*

[b]*Sub-department of Environmental Technology, Wageningen University, Bornse Weilanden 9, 6708WG Wageningen, The Netherlands*

production from CO_2 using MES. Certainly, other process optimizations will be required to establish MES as an innovative sustainable technology for manufacturing biochemicals from CO_2 as a next generation feedstock.

Introduction

Due to rapid economic growth and increasing energy consumption, the global carbon dioxide (CO_2) emissions increased from approximately 14 gigaton (in 1971) to over 38.2 gigaton in 2014.[1] The CO_2 level in the atmosphere increased from 315 ppm in 1957 to 390.5 ppm in 2011,[2] with the highest value of 403.94 ppm recently recorded in 2016.[3] The rising CO_2 levels contribute to global warming and thus impart a significant threat on ecosystem functioning and the earth's climate. CO_2 capture and fixation techniques have been proposed to mitigate climate change and environmental degradation, which can assist in stabilizing atmospheric CO_2 levels and also utilize CO_2 as a sustainable feedstock for chemical and fuel production.[4] The use of CO_2 as a feedstock for chemicals is gaining much attraction lately, since CO_2 is abundantly available and its use is completely independent of the food supply chain. In current perspectives, the bioproduction of short chain fatty acids (SCFA, mainly acetate, propionate and butyrate) and alcohols from CO_2 is an attractive strategy for the multiple purposes of waste remediation, resource recovery and CO_2 sequestration/utilization.[5] The utilization of CO_2 as a feedstock for the synthesis of fuels and chemicals offers independence from fossil feedstocks in chemical industries. Using CO_2 as a feedstock for the production of bulk chemicals other than urea can potentially circumvent almost 20 million tons of CO_2 emissions per year.[6] Furthermore, when CO_2 can be utilized and converted directly from the emission source within the same location, the costs and energy needed in CO_2 transport and storage can be avoided.

Lately, a large diversity of bioprocesses have been integrated with electrical interactions at solid electrodes to develop bioreactors called bioelectrochemical systems (BESs).[7] Microbial electrosynthesis (MES) is an innovative approach in which microorganisms are employed as biocatalysts on either the cathode or anode to achieve electricity-driven synthesis of chemicals and fuels. CO_2 fixation in MES comprises bioelectrochemical reduction of CO_2 to multi-carbon organic compounds using the reducing equivalents produced at the electrically-polarized cathode.[8] This has several advantages over the use of land-based organic feedstocks of biomass origin. Among the advantages of utilizing CO_2 for MES are its availability from the atmosphere as well as waste gases, no dependency on arable land requirements, avoidance of the environmental degradation associated with intensive agriculture and low toxicity to microbial flora.[8–10] Hence, capturing CO_2 for bioproduction is the key concept for removal of CO_2 from the atmosphere and for the reduction of carbon footprint, which would ultimately benefit the whole environment. However, there are disadvantages associated with the use of CO_2 for bioproduction, including the intensive energy requirements for cell activation, difficulties in autotrophic growth and CO_2 fixation (*e.g.*, energy intensive Wood–Ljungdahl pathways), and nutrient requirements for biocatalyst growth.[10] Moreover, the requirement of a large number of electrons for organic compound synthesis is a major challenge when using CO_2 as a starting material for the MES system.

MES based on CO_2 reduction produces acetate as a primary product when methanogenesis is avoided, employing a mixed culture of wastewater origin as a biocatalyst. Various researchers have demonstrated an electricity-driven CO_2 reduction process for the bioproduction of SCFA, mainly acetate and butyrate, and alcohols.[8,11–14] CO_2 reduction in MES has been studied using selectively enriched mixed cultures with methane-inhibiting chemicals and high acetate production rates from batch operations have been reported.[15–17] However, studies on continuous flow operation for CO_2 reduction in MES with a stable mixed culture biocathode are limited. Continuous mode operations for CO_2 reduction using mixed culture biocatalysts were reported only by Batlle-Vilanova *et al.*[18] and Jourdin *et al.*[17] Due to the prevalence of suspended microbes in CO_2 reduction,[13] obtaining a robust biocathode is still a challenge for operating MES in continuous flow mode. In recent studies, the enrichment of homoacetogens from the anaerobic sludge provided a sustained CO_2 reduction to acetate over a very long span of operation.[14,17] The robustness of the autotrophic behavior of the biocathode should also be compared for different operation modes. In order to elucidate the long-lasting stability of a biocathode in the bioelectrochemical CO_2 reduction process, an investigation was carried out using an MES system with a flow-through biocathode, while operating it in sequential batch and then in continuous mode. In addition, control of the cell voltage in the MES reactor was studied instead of fixing the cathode potential so as to simplify the reactor operation.

Materials and methods

Preparing the inoculum: homoacetogenic activity enrichment from wastewater sludge

The anaerobic sludge from the anaerobic digester, as mentioned in a report by Mohanakrishna *et al.*,[19] was collected. A four stage selective enrichment methodology was adopted for enrichment of the homoacetogens in the inoculum culture, as described earlier by Bajracharya *et al.*[11] The culture after four transfers was used as an inoculum for bioelectrochemical CO_2 reduction in the cathode chamber of the reactor.

MES reactor setup

For the MES experiments, a double chambered microbial electrosynthesis reactor was assembled using three circular cylindrical PVDF rings, each of which forms a 16 mL compartment. The first compartment contained an anode, which was a circular platinum sheet (10 cm^2) laser-welded to a titanium (Ti) plate current collector. A Fumasep FKB CEM PEEK® reinforced membrane was placed between the first and second compartments, separating the anode and cathode compartments. A composite carbon felt cathode with two circular carbon felts (projected surface area (PSA): 10 cm^2) was placed between the second and third compartments. A stainless steel mesh collector was sandwiched between the two carbon felts. The second and third compartments together created 32 mL of cathode compartments. A schematic representation of the reactor is shown in Fig. 1. In this cathode arrangement, the catholyte was forced to flow through the cathode from the third compartment to the second compartment. The forced

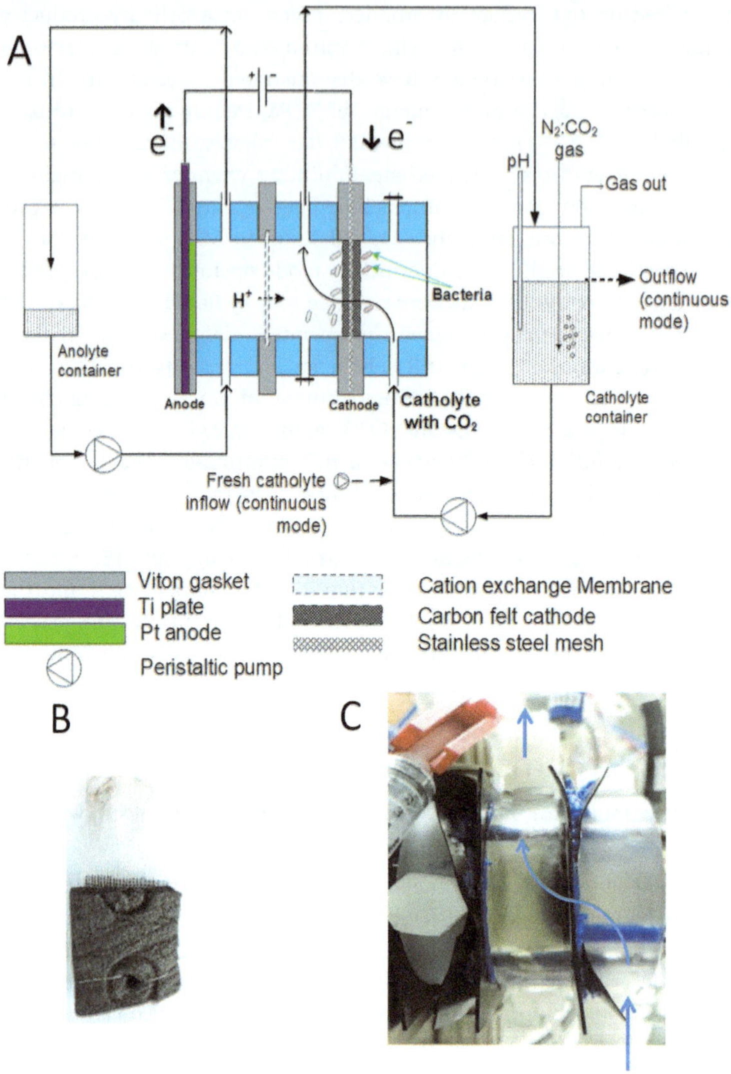

Fig. 1 (A) MES reactor set-up with a flow-through double carbon felt cathode. (B) Double carbon felt cathode with a stainless steel mesh current collector. (C) Photograph of the MES reactor showing the flow of catholyte through the cathode.

flow-through cathode was intended to collect/attach the suspended biocatalyst on the cathode.

Before placing them in the reactor, the composite carbon felt cathode and the membrane were boiled at 70 °C for 1 hour in phosphate buffer solution (PBS). Polyvinyl chloride (PVC) tubes (Ø 6/8) connected each compartment to a separate electrolyte bottle for recirculation. A peristaltic pump (Watson Marlow 323U/D) was used for the recirculation. The recirculation rate was 180 mL min^{-1}.

For the electrochemical measurements, the anode and cathodes were connected to a Potentiostat (VMP3, Biologic Science Instruments, France) in a two-electrode setup: the cathode as the working electrode and the anode as the counter/reference electrode. The potential difference between the cathode and anode, known as the cell voltage, was controlled through the potentiostat. In order to measure the cathode potential, an Ag/AgCl (3 M KCl, +205 mV *vs.* SHE) reference electrode (REF321, Radiometer-Analytical) was placed near the cathode so that the cathode potential could be measured using a multimeter when required.

Electrolytes

The medium used for the start-up phase in the study was a phosphate buffer solution (PBS) composed of (g L^{-1}): 0.33 KH_2PO_4, 0.45 K_2HPO_4, 1 NH_4Cl, 0.1 KCl, 0.8 NaCl, 0.2 $MgSO_4 \cdot 7H_2O$ and 1 yeast extract, which was supplemented with a 20 mL L^{-1} vitamin solution (DSMZ 141), 20 mL L^{-1} trace element solution (DSMZ 141) and 4 g L^{-1} sodium bicarbonate prior to the operation. The media was prepared anaerobically by purging with N_2 gas. During the MES experiment, the yeast extract was removed from the medium. The media without yeast extract, henceforward, is called mineral media. The anolyte was a mineral medium of pH 3–4 maintained by adding 1 M HCl. The pH of the catholyte was adjusted to 7 at the start of the experiment and the reactor temperature was maintained at 30 °C by heating the catholyte in the recirculation bottle on a thermocouple controlled hot plate. The pH of the catholyte was not controlled but due to the flushing with a $CO_2 : N_2$ gas mixture the pH was maintained as slightly acidic. The catholyte in the recirculation bottle was continuously stirred with a magnetic stirrer at 500 rpm.

Start-up and operation of the reactor

After setting up the MES reactor with the flow-through cathode, 500 mL of the catholyte in the glass bottle was recirculated through the flow-through cathode at 180 mL min^{-1}. The reactor was inoculated with an enriched bacterial culture from wastewater sludge (10% v/v), as mentioned previously.[14] In the cathode chamber, at first, bacterial growth was carried out using 2 mM fructose addition (for quick growth) and then the cathode potential was controlled at −0.8 V *vs.* Ag/AgCl using chronoamperometry (CA) and the potentiostat.

Batch operations

After 33 days of the start-up phase, the catholyte was replaced with 550 mL of fresh mineral medium, leaving around 50 mL of the previous electrolyte as an inoculum. MES operation with continuous $CO_2 : N_2$ (20 : 80) gas bubbling in the catholyte was carried out by polarizing the cathode at −1.1 V *vs.* Ag/AgCl until day 49 and then a cell voltage of 2.2 to 3.1 V was applied instead of controlling the cathode potential. The reduction current was recorded every 10 minutes. Water oxidation at the anode served as the electron and proton source. Every new batch was started with the replacement of the previous catholyte with fresh mineral medium. Two batches, termed Batch 1 and Batch 2, of the MES operation were carried out with (20 : 80) $CO_2 : N_2$ gas mixture bubbling and then a continuous catholyte feed operation was carried out. Starting from Batch 2, the catholyte volume was increased to 1 L. An overview of the operational conditions for each batch and the continuous mode operations are provided in Table 1. The batch

Table 1 An overview of the operational conditions during the sequential batch and continuous operations

Operation mode	Starting day of operation	CO_2 source	Cell voltage (V)
Batch 1	Day 33	50 mM $NaHCO_3$ + 20% CO_2 gas	3.1
Batch 2	Day 164	50 mM $NaHCO_3$ + 20% CO_2 gas	3.1
Continuous 1	Day 216	50 mM $NaHCO_3$ + 80% CO_2 gas	3.3
Batch 3	Day 261	50 mM $NaHCO_3$ + 80% CO_2 gas	3.3 & 3.4
Continuous 2	Day 508	50 mM $NaHCO_3$ + 80% CO_2 gas	3.4 & 3.6

operation of Batch 3 was again repeated starting from day 261 after the continuous operation.

Continuous operations

After the operation of Batch 2, the MES operation was continued starting from day 164 under a continuous mode with constant catholyte inflow and outflow rates – 10 mL h^{-1} of fresh catholyte feed and the same rate of outflow – maintained using the recirculation bottle to ensure a constant catholyte volume in the reactor. The total recirculating catholyte volume remained at 1 L.

$$\text{Hydraulic retention time (HRT)} = \frac{\text{catholyte volume}}{\text{inflow rate}} = 100 \text{ h}$$

Starting from this continuous mode of operation, the liquid medium fed into the reactor was continuously bubbled with $CO_2 : N_2$ (20 : 80) whereas the catholyte in the recirculating bottle was bubbled with $CO_2 : N_2$ (80 : 20) gas mixture. The continuous mode operation for the CO_2 reduction was carried out in two phases, namely Continuous 1 and Continuous 2. The cell voltage applied was between 3.3 and 3.6 V during the continuous mode operation.

Sampling and chemical analyses

About 3–5 mL of the catholyte was sampled every 2–3 days for analysis of the short-chain volatile fatty acids (VFAs) and ethanol. To maintain a constant catholyte volume, the sample volume was replaced using the same volume of the mineral medium described earlier in the electrolyte section. The pH of the sample was measured and whenever it had increased above 7.5, 1 M HCl was added to lower it. The CO_2 sources in the reactor remained as the 4 g L^{-1} $NaHCO_3$ and the gaseous CO_2 from gas bubbling. Bicarbonate was intermittently added during the replacement of the sampling volume by adding new mineral medium whenever samples were taken.

For the analysis of formic acid, acetic acid, propionic acid, butyric acid and ethanol, cell-free supernatants were analyzed using an Agilent 1200 series HPLC with an Agilent Hi-Plex H column and an Agilent 1260 infinity refractive index detector. The column was eluted at 60 °C with 0.01 M sulfuric acid, as the mobile phase, with a low rate of 1 mL min^{-1}. The detector temperature was 55 °C.

Calculations

Production rate. The possible products of CO_2 reduction using MES were hydrogen, acetate, propionate, butyrate, methane and ethanol. Based on work by Eerten-Jansen *et al.*,[20] the production amount of each product at each sampling moment t was calculated:

$$\text{Production}_{i,t} = \text{accumulation}_{i,t} + \text{out}_{i,t} - \text{in}_{i,t}$$

For a batch mode, there is no outflow and inflow. Thus the equation becomes

$$\text{Production}_{i,t} = \text{accumulation}_{i,t}$$

For a batch operation, the production (g) of any compound i at time t_2 is given by

$$P_{i,t_2} = V_{\text{cat}}(C_{i,t_2} - C_{i,t_1})$$

and for a continuous mode, the production (g) is given by

$$p_{i,t_2} = V_{\text{cat}}(C_{i,t_2} - C_{i,t_1}) + Q\frac{(C_{i,t_2} + C_{i,t_1})}{2}\Delta t - QC_{\text{in},i}\Delta t$$

where the subscript i refers to the product (that is, all the organics which were quantified), V_{cat} is the catholyte volume (L), C_i is the concentration of product (g L^{-1}), Q is the influent and effluent flow rate (L d^{-1}), Δt is the time between the sampling times t_1 and t_2 (d) and $C_{\text{in},i}$ is the concentration of the product i in the influent (g L^{-1}).

The volumetric production rate (g L^{-1} d^{-1}) was calculated for the batch mode as

$$P_{i,t} = \frac{p_{i,t}}{\Delta t}$$

and for the continuous mode as

$$P_{i,t} = \frac{p_{i,t}}{\Delta t\ V_{\text{cat}}}$$

Electron (e^-) equivalents. The maximum quantity of electrons available in the MES reactor, provided by the electrical current, can be compared to the moles of electrons recovered in the product *e.g.* acetate. One mole of acetic acid ($C_2H_5O_2$, molecular weight 60 g mol^{-1}) is equivalent to eight moles of electrons.

The production in terms of the number of moles of product is given by $N_{i,t} = p_{i,t}/M_i$ and the number of moles of e^- equivalents in the product i is given by $n_{i,t} = N_{i,t} \times f_{e,i}$, where $N_{i,t}$ is the number of moles of the product 'i' analysed at time t, and $f_{e,i}$ represents the molar conversion factor (*e.g.* 8 electron equivalents per mol for acetate).

The electron equivalents provided by the electrical energy input in MES is given by

$$n_{e,t} = \frac{\int_0^t I\mathrm{d}t}{F}$$

where F is the Faraday constant (96 485 C mol^{-1} of electron) and I is the current.

Coulombic efficiency (CE). Coulombic efficiency, also referred to as current efficiency(η) represents the fraction of electrons recovered through generating the products compared to the electrons supplied by the current, and was calculated in % as.

$$\eta = \frac{n_{i,t}}{n_{e,t}} \times 100$$

This represents the efficiency of capturing the electrons from the electrical current to generate the product. In this study, mainly acetate production was considered for the coulombic efficiency calculations. Coulombic efficiencies for every measurement of the organic compounds were calculated based on the product concentrations accumulated from the beginning of the MES operation as well as those intermittently calculated based on the difference in the product concentrations between two consecutive measurements.

Results

Cathode potentials maintained by controlling the cell voltage

During cell voltage controlled operation of the MES, the corresponding cathode potentials *versus* (*vs.*) the Ag/AgCl reference electrode were measured for a number of applied cell voltages (2 to 3.6 V). It was observed that to reach a hydrogen evolution onset potential of −0.61 *vs.* Ag/AgCl, a cell voltage of >2.2 V was required in the MES reactor. Still, a higher cell voltage was required to overcome the electrochemical losses. Based on the measurements of the cathode potential, an equation for a straight line fitting the data-points was obtained from a plot of the cathode potentials and the cell voltages, as shown in Fig. 2.

The slope of the plotted line in Fig. 2 is 0.26, which means that to lower the cathode potential by 0.1 V *vs.* Ag/AgCl, the cell voltage had to be increased by 0.4 V. The plot in Fig. 2 shows that a cell voltage >3 V corresponds to a cathode potential of <−0.9 V *vs.* Ag/AgCl. Thermodynamically, the redox potential for CO_2 reduction to acetate is −0.48 V *vs.* Ag/AgCl under the standard biological condition of pH 7, but a much lower cathode potential had to be applied to compensate for the

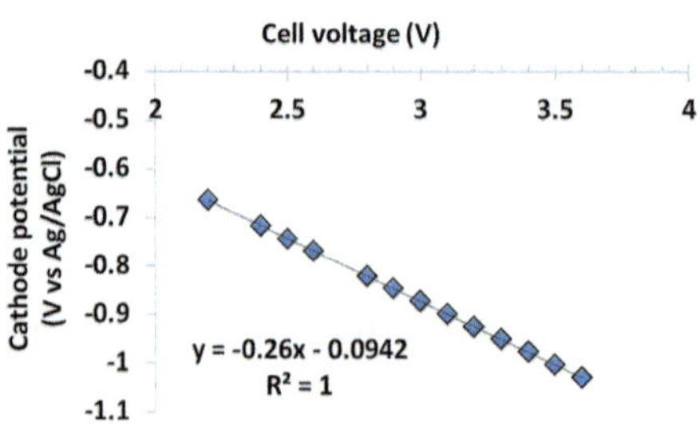

Fig. 2 Plot of the cathode potentials corresponding to the applied cell voltage.

energy losses due to resistance. Cathode potentials of <−0.9 V *vs.* Ag/AgCl have been reported for CO_2 reduction with mixed culture biocatalysts in previous MES studies.[14,16,21]

Acetate production from CO_2 reduction with an applied cell voltage of >3 V

In order to adjust the redox potential for the bioelectrochemical CO_2 reduction, a number of CA techniques with a stepwise increment in the cell voltage from 2.2 to 3.1 V were performed and the concentrations of acetate in the catholyte were monitored during the Batch 1 operation. Multiple mechanisms of electron transfer are possible for the bioelectrochemical CO_2 reduction at the MES biocathode of mixed culture origin, also including the interactions of extracellular enzymes.[22,23] The stepwise shifting of the cell voltage likely allows all possible electron transfer mechanisms responsible for CO_2 reduction at the biocathode to occur. When the applied cell voltage in the MES reactor was between 2.2 and 3 V, the production of acetate was not observed. Instead, the acetate previously present in the catholyte was degraded (acetate level declined to an undetected level from day 52 as shown in Fig. 3). When the applied cell voltage was lower than 3 V, the corresponding cathode potential remained above −0.9 V *vs.* Ag/AgCl which appeared to be insufficient to provide the reducing power for CO_2 reduction at the cathode.[13,14]

The current densities were <5 A m^{-2} unless the applied voltage was higher than 3 V. Only after applying a 3.1 V cell voltage did the current densities also increase, reaching 6–10 A m^{-2}, and the production of acetate started (Fig. 3). From day 68 the concentration of acetate started to rise steadily, during which period the applied voltage was 3.1 V. The highest acetate concentration measured was 2.167 g L^{-1} on day 115, which corresponds to 46.1% current recovery as acetate. The average acetate production rate for the batch operation remained at 44 mg L^{-1} d^{-1}. Interestingly, from day 100, production of butyrate was also observed, which reached a highest concentration of 20 mg L^{-1}. Ethanol was also detected

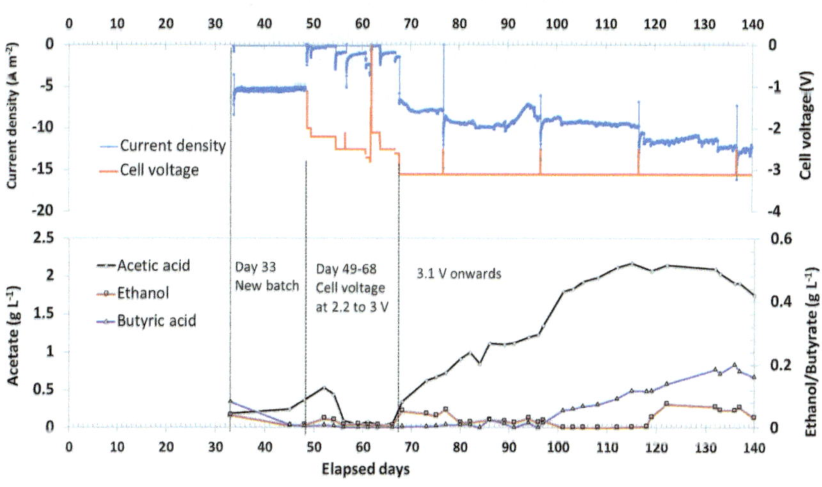

Fig. 3 Production of acetate from CO_2 reduction during the Batch 1 operation of the MES by controlling the cell voltage at 3.1 V.

randomly in the catholyte samples, reaching a concentration of <200 mg L^{-1}, but the trend of accumulation was not consistent.

Sequential batch and continuous modes of CO_2 reduction in MES

Bioproduction in the MES reactor was carried out in a sequence of batch and continuous feed modes (namely, Batch 2—Continuous 1—Batch 3—Continuous 2) in order to examine the bioelectrochemical CO_2 reduction process by varying the operation mode. In Batch 2, the MES was operated after replacing the catholyte with 950 mL of new catholyte, retaining 50 mL of the catholyte from the previous batch. With a 3.1 V applied cell voltage, a slow rise in the acetate concentration was observed and it reached 0.74 g L^{-1} within 42 days of operation in Batch 2 (Fig. 4) with a constant (20 : 80) $CO_2 : N_2$ gas mixture sparging. No methane was detected through the headspace gas analyses carried out during the experiment, which excluded the possibility of CO_2 reduction to methane. The rate of acetate production in Batch 2 reached a maximum of 56.6 mg L^{-1} d^{-1} with a CE of 29.3%.

In order to vary the operation mode, Batch 2 operation was shifted to a continuous feed mode (Continuous 1 in Fig. 4) by maintaining a relatively long (100 h) hydraulic retention time starting from day 216. At the start of the continuous mode, the acetate concentration declined rapidly due to removal of the catholyte with the outflow and dilution with the inflow. In the batch modes, the acetate produced at the cathode was accumulating in the catholyte, as there was no outflow and inflow. Thus, the acetate concentration steadily increased in the batch mode whereas, in continuous feed mode, a steady rise in the acetate concentration was not observed since the accumulation rates were balanced by continuous outflow and inflow. But for comparison purposes, the acetate produced during the continuous modes was cumulatively projected as shown in Fig. 4, which can provide the expected concentration that would be accumulated in the catholyte in a similar way as for the batch modes.

The production rates were negative due to the higher washing out rates at the beginning of the continuous run. In order to increase the production rate of the acetate, the cell voltage was raised to 3.3 V. The current densities were slightly increased (12–15 A m^{-2}) and then the production of acetate was recovered at the maximum rate of 100 mg L^{-1} d^{-1}. This production rate was in a similar range to the rates reported by Batlle-Vilanova *et al.*[18] for continuous mode (~20 h HRT) operation of a CO_2 reduction carried out at −0.8 V *vs.* Ag/AgCl. During the 33 days of operation in continuous mode, a total of 1.5 g of acetate was produced which corresponds to a CE of 50%. The CE and production rates for acetate production were higher for the Continuous 1 operation than for Batch 1.

After operation of the MES in continuous mode for around 45 days, the catholyte became clearer visually than in the batch operations, which was likely due to the wash-out of suspended biomass. The concentration of acetate and the production rates declined after 45 days of continuous operation. Even though a significant amount of suspended bacterial biomass was washed out (visible by eye), the acetate production was sustained at least at a rate of 28 mg L^{-1} d^{-1} during the continuous operation. Thus, it can be inferred that the bacterial biomass retained in the reactor or attached to the cathode could sustain the acetate production from CO_2 reduction regardless of the operation mode.

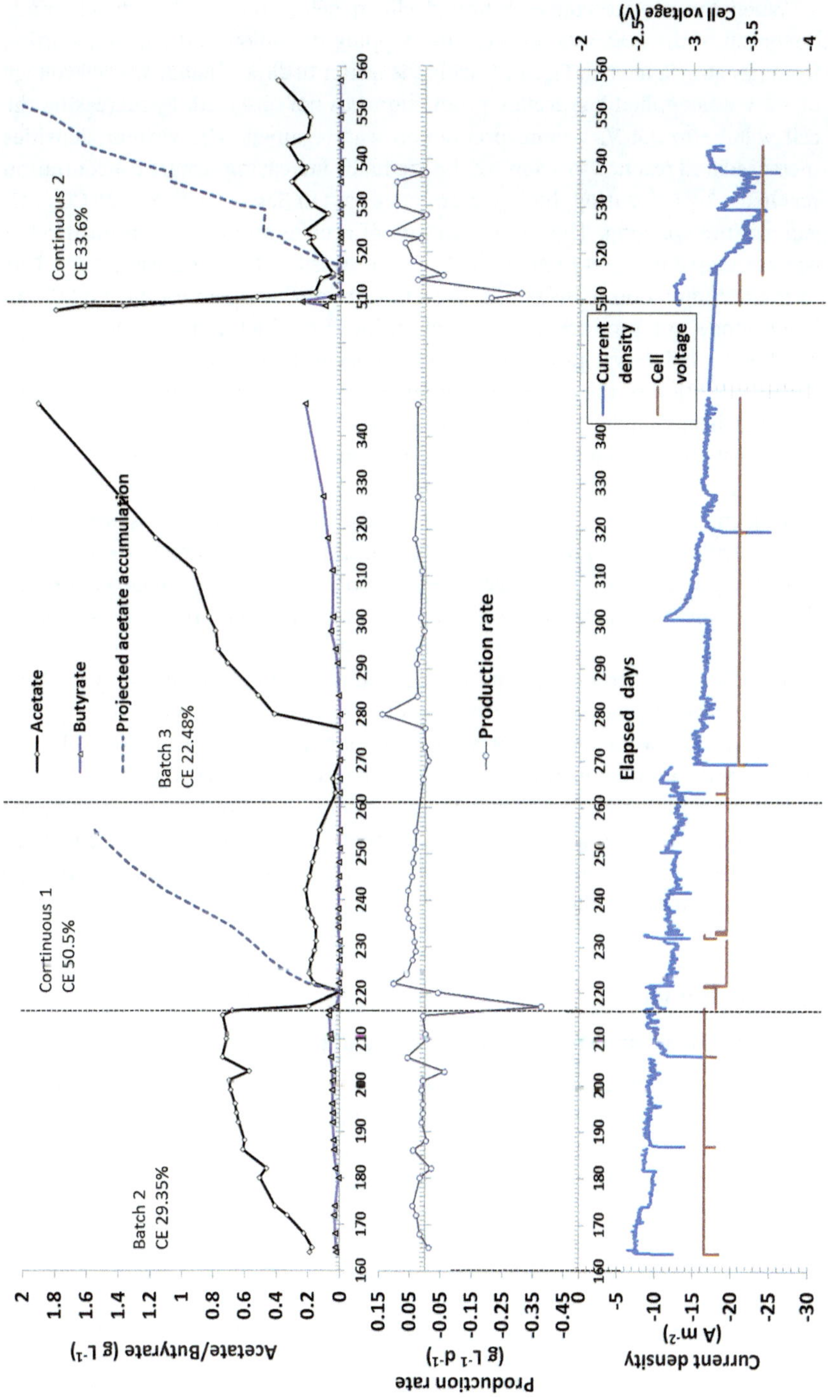

Fig. 4 Production profiles and electrical inputs for sequential operation of the batch and continuous modes (namely, Batch 2—Continuous 1—Batch 3—Continuous 2) of the microbial electrosynthesis for CO_2 reduction.

When the acetate concentration declined below 100 mg L^{-1}, batch mode operation of the MES was repeated by stopping the inflow and outflow, starting from day 261 (Batch 3 in Fig. 4). During the initial 10 days of Batch 3, a cell voltage of 3.3 V was applied but acetate production was not observed. By increasing the cell voltage to 3.4 V, acetate production was regained. The current densities increased and reached >15 A m^{-2}. The gradually increasing acetate concentration reached ~2 g L^{-1} during the 65 days of operation in Batch 3 with 80 : 20 CO_2 : N_2 gas mixture sparging. The maximum rate of acetate production within Batch 3 was calculated to be ~139 mg L^{-1} d^{-1} and for this batch, the CE was 22.5%. Due to the accumulation of acetate in the catholyte during operation of the MES in batch mode, the production of a more reduced product, namely butyrate, also resulted though it was produced in a low concentration (>200 mg L^{-1}).

On repetition of the continuous mode after Batch 3 (Continuous 2 from day 508 as indicated in Fig. 4), the concentration of acetate again declined at the start due to fast washing out and dilution. Upon increasing the cell voltage to 3.6 V, the acetate production rates recovered reaching a maximum of 108 mg L^{-1} d^{-1}. During operation at a 3.6 V cell voltage, a total of 2.56 g of acetate was produced in 50 days with a CE of 33.6%. The highest concentration of acetate in the catholyte was recorded as 0.4 g L^{-1}. An overview of acetate production from CO_2 reduction during sequential batch and continuous operation of the MES is presented in Table 2.

It was apparent from sequential batch and continuous operations of the MES that the acetate production became limited by the applied cell voltage in the reactor when it was operated for a long duration. But it was also noticeable that the CE for the production process declined when a higher cell voltage was applied, indicating a higher loss of energy while overcoming the resistance. Overall, the biocathode supported acetate production from CO_2 reduction in both operation modes, which demonstrates the robustness of the biocathode which adapted to various operation conditions.

Discussion

Sequential batch and continuous operations improved the MES process

A gradual increase in the affinity towards the CO_2 reducing activity of the biocathode can be observed with respect to the changes in cell voltage during these sequential batch and continuous modes of operation. The maximum acetate production rate and the highest concentration attained for Batch 3 were increased

Table 2 An overview of acetate production during the sequential batch and continuous operations

Operation mode	Projected cathode potential (V *vs.* Ag/AgCl)	Acetate production rate [maximum] (mg L^{-1} d^{-1})	CE %	Highest acetate accumulation (g L^{-1})
Batch 2	−0.90	13.8 [56.6]	29.3	0.74
Continuous 1	−0.95	39 [100]	50	1.5
Batch 3	−0.98	31 [139]	22.5	2
Continuous 2	−1.03	51 [108]	33.6	2.56

considerably than the rates and concentrations for Batch 2 and Continuous 1 phase (Table 2). The maximum acetate production rate during the repetition of the continuous mode (the Continuous 2 phase after Batch 3) operated at a 3.6 V cell voltage was similar to the rate for the first continuous operation. However, the CO_2 reduction process in Continuous 2 was more pronounced, and the acetate concentrations were higher and slowly increased to reach almost 0.5 g L^{-1} with the same outflow/inflow rate of the catholyte.

The CEs for acetate production of the continuous operations were higher than the CEs calculated for the batch operations. But the CO_2 reduction process in batch mode also produced additional organic products like ethanol and butyrate due to further reduction of the acetate and those products were not included in the CE calculation. Due to the slower rate for accumulation of acetate using a continuous mode, the production of other reduced compounds was apparently excluded. In the continuous mode of operation, acetate was the only reduced product and the highest CE of 50% was obtained for Continuous 1 phase (Table 2). The higher CE of acetate production for the continuous operations could most likely be due to the continual replenishment of micronutrients during the continuous operations. The replenishment of micronutrients can additionally support the biocatalyst in the electron mediation process during CO_2 reduction. Which justifies the higher CEs for continuous operation. A micronutrient supplement could be beneficial for activating the metabolism of CO_2 reduction in microbes. Blanchet *et al.*[24] have indicated that under nutrient limited conditions without yeast extract, no bioproduction was obtained using a *Sporomusa ovata* culture with $H_2 : CO_2$.

Apparently due to the higher applied cell voltage, the production of acetate in Continuous 1 reached higher rates than in the preceding Batch 2 mode. But most likely this could also be contributed by the continual replenishment of the catholyte, which provided essential trace elements and vitamins for the bacterial metabolism. The acetate production rate during Continuous 1 was higher than the rate for Batch 2 and also the expected acetate accumulation for Continuous 1 (1.5 g L^{-1}) was two times higher than the titer (0.74 g L^{-1}) obtained for Batch 2. Likewise, the average acetate production rate (51.4 mg L^{-1} d^{-1}) and expected concentration (2.5 g L^{-1}) of acetate when accumulated in Continuous 2 were also higher than the average production rate (31 mg L^{-1} d^{-1}) and titer (2 g L^{-1}) of Batch 3. Thus, continual improvement in the MES of acetate was observed with sequential batch and continuous operation. To some extent, the removal of the suspended biocatalyst along with the outflow of the catholyte in the continuous mode also created a selection pressure for the biocatalyst to adapt towards the CO_2 reduction process. Thus, the production rates of Batch 3 increased when the MES process was reinstated following the wash-out of a certain portion of the biocatalyst during the Continuous 1 operation.

A high cell voltage was required to continue CO_2 reduction during long-term operation

In this study, the applied cell voltage was between 3.1 and 3.6 V, which corresponded to an estimated cathode potential between −0.9 and −1 V *vs.* Ag/AgCl (Table 2). This range of cathode potential for CO_2 reduction in MES has been frequently reported in relation to a reduction mediated by hydrogen evolution.[13,15]

In the electrochemical reactor context, due to the control of the cell voltage instead of the electrode potential, the cathode potential was subjected to change during the MES with a small change in the catholyte pH or other parameters. Change of the operation mode also induces variation in the process parameters. Gradual changes in the electrochemical resistance of the MES reactor can be expected due to changes in the bacterial density (biofilm-forming or suspended) and fouling of the membrane, and also due to the accumulating products. Previously, Bajracharya *et al.*[14] mentioned the requirement for a dynamic cathode potential to supply the reducing equivalents required to maintain continual acetate production during long-term operation. In this context, a few MES studies reporting CO_2 reduction used current controlled operation which supplied a constant amount of reducing equivalents at the cathode regardless of the changes in other process parameters.[21,25]

It was observed that the cell voltage had to be increased in order to achieve considerable production rates in our experiment. Changes in the process parameters like the pH, concentrations, membrane & electrode resistance, and other electrochemically active species possibly involved could be the reason. In addition, under the cell voltage controlled conditions, the variation in the anode potential could also be a reason for the requirement of an increment in the cell voltage during long-term operation. The interpolated cathode potentials for all the applied cell voltages in this study were more negative than the onset potential of hydrogen evolution. Most likely, hydrogen evolution mediated the bio-electrochemical CO_2 reduction in this MES study. Whenever the operation modes were changed, the hydrogen availability in the reactor could have become limited, so a higher cell voltage had to be applied to provide hydrogen. Evidently, for the few examples of headspace gas analysis carried out at the beginning of the Continuous 2 phase, no hydrogen was detected, even though the cell voltage was 3.4 V. Thus, a higher cell voltage was applied to create more reduced cathode conditions that allow a higher hydrogen production. A noteworthy development in this MES study is that the application of the electrical energy for the MES was simplified through the cell voltage controlled operation, which could be a benefit in streamlining the electrotrochemical configuration for up-scaled and long-term operation.

MES with CO_2 reduction continued with retention of the biocatalyst

Variations in the operational mode did not hamper the CO_2 reduction process as biocatalyst retention within the cathode was maintained. In a continuous mode, the catholyte outflow washed out the suspended microorganisms, but still, the microbes retained in the reactor or attached to the electrode continued the reduction process. During the first continuous mode operation, due to the washing out of bacterial biomass, the catholyte turned clear after 30–40 days and very slow acetate production rates <5 mg L^{-1} d^{-1} were sustained. The production was stopped when the production rates were slower than the outflow. By switching the operation to batch mode, the bacterial biomass was retained and after a few days, the acetate concentration started to accumulate in Batch 3. The recurrence of the acetate production in Batch 3 indicates a dominance of the suspended bacteria over the biofilm forming bacteria in the CO_2 reducing biocathode. Thus, the bioelectrochemical CO_2 reduction was sustained during both operation modes.

Regarding the loss of biomass in the continuous mode, it could be suggested that both the biofilm and suspended (*i.e.* planktonic) bacteria participate in the CO_2 reduction process but that a domination of the suspended biocatalyst can be expected. When planktonic bacteria were lost, the production went down when the reactor was operated for a longer duration in continuous mode. The solution could be to find ways to create selection pressure for biofilm-forming microorganisms instead of planktonic bacteria. Use of a flow-through cathode in the experiments was also intended to allow attachment of the biocatalyst on the carbon felt cathode. Recently it has been proposed that increasing the sodium chloride stress induces biofilm formation by *Clostridium ljungdahlii* on glass, graphite and glassy carbon electrode materials.[26] Thus, significant microbial retention in the reactor is required for a continual CO_2 reduction process.

Microbial electrosynthesis in the scenario of 'CO_2 as a next generation feedstock'

Conventionally, organic substrates that originate from sugars are used as an organic feedstock for the bioproduction of chemicals and fuels as these compounds are already partially reduced and hence are associated with a lower reduction power demand for the bioproduction process.[27] Nevertheless, refining the substrate to its pure form, to be used as a starting material in bioproduction, could be expensive. Additionally, using sugar based substrates for bioproduction creates undesirable interference with their other utilities. On the other hand, waste organics that have a low utility value such as glycerol, fatty acids, and more importantly, in the context of the recent developments in MES, CO_2 have become promising substitutes for the highly expensive organics.

As an example for the utilization of CO_2 as substrate in bioproduction, a microbial electrosynthesis based on CO_2 reduction has been demonstrated with production of acetate at rates of ~150 mg L^{-1} d^{-1} and a CE of 50%. The production rates and efficiencies were still lower than the rates and CEs reported for other long-term MES studies,[14–16] indicating that the process and the reactor configuration were not fully optimized. However, the feasibility of the CO_2 reduction based MES process and the robustness of the CO_2 reducing biocathode towards batch and continuous operation modes were clearly shown, although a multitude of disturbances occurred during the process. Additionally, a re-activation of the biocathode was also demonstrated during the operational transition from continuous to batch. In a very recent development, it was shown that production of acetate can be achieved at high rates up to 66 kg m^{-3} d^{-1} with 100% specificity using CO_2 reduction based MES.[17]

Given such developments in CO_2 reduction based MES, further development of energy-efficient and selective biocatalysts and optimization of the MES process can be achieved. Integration of the CO_2 reduction in MES with other bio-electrochemical processes (electrofermentation) will add commercial value to the MES by producing value-added bulk and speciality chemicals,[28–30] which will attract scale up of the bioproduction. Through these scenarios, CO_2 can been prospected as a promising next generation feedstock for bioproduction using a microbial electrosynthesis model.

Conclusions

The continual production of acetate from CO_2 during long-term operation of these sequential batch and continuous modes for MES exhibited the robustness of the enriched biocathode for CO_2 reduction during MES. When the cathode potentials could be maintained to supply reducing equivalents or hydrogen in the reactor, the CO_2 reduction process continued. Acetate production rates of 100 mg L^{-1} d^{-1} and coulombic efficiencies of ~50% were attainable using a continuous mode whereas, in batch mode, a 139 mg L^{-1} d^{-1} acetate production rate was obtained in this study. The acetate production rates from CO_2 reduction using MES are still low and likely not high enough to obtain economic gain from the MES process, in comparison to conventional biochemical production routes (like the fermentation of sugars). Nevertheless, the development of a robust mixed culture biocathode has been shown to be feasible for producing acetate and other value-added compounds using batch and continuous modes of operation. Further process optimization with such robust biocathodes could establish MES as an innovative sustainable technology for the manufacture of renewable chemicals from CO_2 as a next generation feedstock.

Acknowledgements

Suman Bajracharya was supported by a PhD grant from VITO's strategic research funds (project no. 1310225). The authors would like to acknowledge Tim van den Eynde for conducting a part of the experiment.

References

1 IEA, *Energy and Climate Change*, International Energy Agency (IEA), 2015.
2 P. Ciais, C. Sabine, G. Bala, L. Bopp, V. Brovkin, J. Canadell, A. Chhabra, R. DeFries, J. Galloway, M. Heimann, C. Jones, C. Le Quéré, R. B. Myneni, S. Piao and P. Thornton, in *Climate Change 2013-The Physical Science Basis. Contribution of Working Group I to the Fifth Assessment Report of the Intergovernmental Panel on Climate Change*, ed. T. F. Stocker, D. Qin, G.-K. Plattner, M. Tignor, S. K. Allen, J. Boschung, A. Nauels, Y. Xia, V. Bex and P. M. Midgley, Cambridge University Press, Cambridge, United Kingdom and New York, NY, USA, 2013, pp. 465–570.
3 E. Dlugokencky and P. Tans, *NOAA/ESRL*, 2016.
4 M. Aresta and A. Dibenedetto, *Dalt. Trans.*, 2007, 2975–2992.
5 F. M. Liew, M. Köpke and S. D. Simpson, in *Liquid, Gaseous and Solid Biofuels - Conversion Techniques*, INTECH, 2013, pp. 125–173.
6 A. Otto, T. Grube and D. Stolten, *Energy Environ. Sci.*, 2015, **8**, 3283–3297.
7 S. Bajracharya, M. Sharma, G. Mohanakrishna, X. Dominguez Benneton, D. P. B. T. B. Strik, P. M. Sarma and D. Pant, *Renewable Energy*, 2016, **98**, 153–170.
8 K. P. Nevin, T. L. Woodard, A. E. Franks, Z. M. Summers and D. R. Lovley, *MBio*, 2010, **1**, e00103.
9 D. R. Lovley, *Environ. Microbiol. Rep.*, 2011, **3**, 27–35.
10 J. Desloover, J. B. A. Arends, T. Hennebel and K. Rabaey, *Biochem. Soc. Trans.*, 2012, **40**, 1233–1238.

11 K. Rabaey and R. A. Rozendal, *Nat. Rev. Microbiol.*, 2010, **8**, 706–716.
12 C. W. Marshall, D. E. Ross, E. B. Fichot, R. S. Norman and H. D. May, *Appl. Environ. Microbiol.*, 2012, **78**, 8412.
13 S. Bajracharya, A. Ter Heijne, X. Dominguez, D. P. B. T. B. Strik, K. Vanbroekhoven, C. J. N. Buisman and D. Pant, *Bioresour. Technol.*, 2015, **15**, 14–24.
14 S. Bajracharya, R. Yuliasni, K. Vanbroekhoven, C. J. N. Buisman, D. P. B. T. B. Strik and D. Pant, *Bioelectrochemistry*, 2017, **113**, 26–34.
15 C. W. Marshall, D. E. Ross, E. B. Fichot, R. S. Norman and H. D. May, *Environ. Sci. Technol.*, 2013, **47**, 6023–6029.
16 L. Jourdin, T. Grieger, J. Monetti, V. Flexer, S. Freguia, Y. Lu, J. Chen, M. Romano, G. G. Wallace and J. Keller, *Environ. Sci. Technol.*, 2015, **49**, 13566–13574.
17 L. Jourdin, S. Freguia, V. Flexer and J. Keller, *Environ. Sci. Technol.*, 2016, **50**, 1982–1989.
18 P. Batlle-Vilanova, S. Puig, R. Gonzalez-Olmos, M. D. Balaguer and J. Colprim, *J. Chem. Technol. Biotechnol.*, 2015, **91**, 921–927.
19 G. Mohanakrishna, J. S. Seelam, K. Vanbroekhoven and D. Pant, *Faraday Discuss.*, 2015, **183**, 445–462.
20 M. C. A. A. Van Eerten-Jansen, A. Ter Heijne, T. I. M. Grootscholten, K. J. J. Steinbusch, T. H. J. A. Sleutels, H. V. M. Hamelers and C. J. N. Buisman, *ACS Sustain. Chem. Eng.*, 2013, **1**, 513–518.
21 S. A. Patil, J. B. A. Arends, I. Vanwonterghem, J. van Meerbergen, K. Guo, G. W. Tyson and K. Rabaey, *Environ. Sci. Technol.*, 2015, **49**, 8833–8843.
22 F. Kracke and J. O. Krömer, *BMC Bioinformatics*, 2014, **15**, 410.
23 J. Deutzmann, M. Sahin and A. Spormann, *MBio*, 2015, **6**, 1–8.
24 E. M. Blanchet, F. Duquenne, Y. Rafrafi, L. Etcheverry, B. Erable and A. Bergel, *Energy Environ. Sci.*, 2015, **8**, 3731–3744.
25 S. Gildemyn, K. Verbeeck, R. Slabbinck, S. J. Andersen, A. Prévoteau and K. Rabaey, *Environ. Sci. Technol. Lett.*, 2015, **2**, 325–328.
26 J. Philips, K. Rabaey, D. R. Lovley and M. Vargas, *PLoS One*, 2017, **12**, e0170406.
27 S. Roy, A. Schievano and D. Pant, *Bioresour. Technol.*, 2016, **213**, 129–139.
28 R. Ganigué, S. Puig, P. Batlle-Vilanova, M. D. Balaguer and J. Colprim, *Chem. Commun.*, 2015, **51**, 3235–3238.
29 S. M. T. Raes, L. Jourdin, C. J. N. Buisman and D. P. B. T. B. Strik, *ChemElectroChem*, 2016, **4**, 386–395.
30 S. Venkata Mohan, J. A. Modestra, K. Amulya, S. K. Butti and G. Velvizhi, *Trends Biotechnol.*, 2016, **34**, 506–519.

Faraday Discussions

PAPER

Polysaccharide-derived mesoporous materials (Starbon®) for sustainable separation of complex mixtures†

Vânia G. Zuin, [*ab] Vitaliy L. Budarin,[*b] Mario De bruyn,[b] Peter S. Shuttleworth, [c] Andrew J. Hunt, [b] Camille Pluciennik,[b] Aleksandra Borisova,[b] Jennifer Dodson,[b] Helen L. Parker[b] and James H. Clark[b]

Received 5th February 2017, Accepted 7th March 2017

DOI: 10.1039/c7fd00056a

The recovery and separation of high value and low volume extractives are a considerable challenge for the commercial realisation of zero-waste biorefineries. Using solid-phase extractions (SPE) based on sustainable sorbents is a promising method to enable efficient, green and selective separation of these complex extractive mixtures. Mesoporous carbonaceous solids derived from renewable polysaccharides are ideal stationary phases due to their tuneable functionality and surface structure. In this study, the structure–separation relationships of thirteen polysaccharide-derived mesoporous materials and two modified types as sorbents for ten naturally-occurring bioactive phenolic compounds were investigated. For the first time, a comprehensive statistical analysis of the key molecular and surface properties influencing the recovery of these species was carried out. The obtained results show the possibility of developing tailored materials for purification, separation or extraction, depending on the molecular composition of the analyte. The wide versatility and application span of these polysaccharide-derived mesoporous materials offer new sustainable and inexpensive alternatives to traditional silica-based stationary phases.

Introduction

There is an ever increasing demand for natural compounds as supplements for a healthy lifestyle, with polyphenols found in fruit, seeds, roots and agro-

[a]*Department of Chemistry, Federal University of Sao Carlos, Rod. Washington Luís, km 235, Sao Carlos, SP, Sao Paulo, Brazil, 13.565-905. E-mail: vaniaz@ufscar.br*

[b]*Green Chemistry Centre of Excellence, Department of Chemistry, University of York, Heslington, York, YO10 5DD, UK. E-mail: vania.zuin@york.ac.uk; vitaliy.budarin@york.ac.uk*

[c]*Department of Physics, Elastomers & Energy Applications, CSIC c/Juan de la Cierva, 3, Madrid, 28006, Spain*

† Electronic supplementary information (ESI) available: Preparation of polysaccharide-derived mesoporous materials with by-products of microwave assisted hydrothermal process of orange peel. Recovery of the selected phenolic compounds by SPE using C18, PGC and sustainable mesoporous carbonaceous materials as sorbents. See DOI: 10.1039/c7fd00056a

industrial residues of particular interest.[1–5] The global market for these compounds is expected to grow annually at a rate of 8.7%, reaching 24 992 tonnes or $1.03 billion by 2020.[6] Amongst the polyphenols, flavonoids and coumarins appear to play an important role in human health, having beneficial properties to prevent human diseases, such as cardiovascular problems, gastric or duodenal ulcers, and cancer.[7] They can also be used as active compounds in phytomedicines, nutraceuticals and food.[8,9]

Efficient, green and selective extractions and purifications of these high value and low-volume extractives are a considerable challenge for the development of economically-viable biorefineries.[10–12]

In the literature, many analytical separation procedures based on liquid extraction, supercritical or subcritical fluids, microwaves, ultrasound and solid phase extraction (SPE) have been described.[13] Among these, SPE is one of the most important as it is straightforward, easy to use, already commonly applied in polyphenol profiling and does not require any chemical alteration of the compounds.[14] The enrichment efficiency and selectivity are directly related to the nature of sorbents with specific active sorption sites, *i.e.* characteristic structures and physicochemical properties. Numerous types of sorbents have been developed and used for SPE and correlated techniques, such as non-polar alkyl bonded silicas (*e.g.* C18, C8), polar-functionalised bonded silicas (*e.g.* LC–CN, LC–NH_2, LC–diol), polar media (LC–Si, LC–Florisil, ENVI–Florisil and LC–alumina), and carbonaceous media (ENVI–Carb, Porous Graphitic Carbon (PGC)), among others. However, the scale-up and industrial application of SPE have been limited.[15] Reasons for this include the costs of the most common commercial SPE phases available and the instability of silica-bound functional groups, combined with the presence of residual silanol groups on the silica surface that limit the pH and mobile phase that can be used. In addition, other reasons include the unsustainable energy-intensive pyrolysis conditions and harsh chemicals required for the PGC template and the formed microporosity removal, and the inability to model the analyte–sorbent relationships that are the basis of any efficient and controllable extraction and separation processes.[16–19] Furthermore, despite demonstrating high selectivity, affinity and extraction capacity, the mesoporous synthetic polymers or functionalised silica are also limited in terms of variable surface functionality and structure control.[20,21]

A systematic investigation of the performance of a varied class of sustainable materials towards a range of analytes, representative enough to predict the best sorbent for a specific application, is lacking. More recently, novel renewable carbonaceous mesoporous materials with tuneable surface functionality and structure have been reported.[22] These materials are derived from abundant and inexpensive residual polysaccharides, such as starch, alginic acid or pectin, some known as Starbon®.[23] Their production exploits the natural tendency of polysaccharide gels to form nano-channelled biopolymer structures. The nature of the polysaccharides determines the morphology of these materials and the temperature, the surface functionality.[24,25] These characteristics help develop a varied spectrum of sustainable materials which have the versatility needed for a more complete investigation. To date, the successful utilisation of Starbon®, *e.g.* to remove contaminants from wastewater,[26,27] the retention of metals[28] and LC–MS separation of standard carbohydrates, has been described.[29] However, further comprehensive studies related to the design and use of polysaccharide-derived materials as stationary media phases for sample preparation, preparative or even downstream purposes have not been reported to date. Herein, this work examines how the texture and functionality of these low-cost and sustainable porous carbonaceous materials influence the recovery of selected bioactive compounds by SPE. Statistical analysis highlights the key factors, related to both analytes and sorbents, affecting the recovery of all analytes. This therefore creates opportunities for a targeted modification of the sorbents towards a particular separation challenge in the scope of biorefinery and circular economy for the exploitation of complex mixtures containing high-value compounds.[30]

Materials and methods

Chemicals

All ten flavonoid and coumarin standards, formic acid (FAc), HPLC-grade methanol, ethanol and acetonitrile (ACN) were purchased from Sigma-Aldrich (Gillingham, Dorset, UK). All other reagents were of analytical grade. Ultrapure water was obtained from a Milli-Q water purification system (Millipore, Bedford, MA, USA). Standard stock solutions of analytes (1 mg mL^{-1}) were prepared in methanol and diluted to the required concentration. All solutions were stored at 4 °C prior to use.

Preparation of the polysaccharide-derived materials

As received, the pectin was homogenised in *t*-butanol solution to give a total polysaccharide concentration of 4.3 wt%. The pectin solution was left at 4 °C for 24 h. Starch was rapidly gelled in *t*-butanol solution (9.1 wt%) for 10 min at 140 °C and 200 W in a CEM SP Discover microwave and was left at 4 °C for 48 h to retrograde. Alginic acid solution (4.8%) was gelled in a *t*-butanol solution at 90 °C for 2.5 h, cooled and left at 4 °C for 24 h. The polysaccharide solutions were frozen in round-bottomed flasks in liquid nitrogen and freeze dried for at least 24 h in an SP-Scientific Sentry 2.0 freeze drier. The drying conditions were as follows: vacuum set to 100 mTorr and condenser temperature set to −103.9 °C. The dry aerogels were carbonised to the required temperature at 1 °C min^{-1} under an inert atmosphere to produce the desired Starbon® material.[24,31,32]

Characterisation

Scanning electron microscopy (SEM) measurements were performed using a JEOL JSM-6490LV. Samples were mounted on alumina plates and coated with a 7 nm layer of Au/Pd using a high resolution sputter SC-7640 coating device prior to analysis. Transmission electron microscopy (TEM) images were recorded using a Tecnai 12 BioTwin at 120 kV. Samples were suspended in ethanol and deposited onto carbon grids (200 mesh) *via* solvent evaporation.[23,32] Porosimetry analyses were performed using a Micromeritics ASAP 2010. Porous polysaccharide samples were degassed at 60 to 120 °C under vacuum ($P < 10^{-2}$ Pa) for at least 3 h prior to analysis. Data processing was performed using ASAP 2010 version 5.02 and Origin 7.5 software.[33,34]

Solid phase extraction (SPE) studies

The SPE recoveries of the ten selected phenolic compounds were analysed using 0.1 g of each sorbent under study, based on the maximum and minimum values obtained in each case (*i.e.* the amount measured as a percentage of the total analyte originally added to the system or matrix). Standard solutions containing all analytes in methanol at the usual concentration level of the real samples (μg mL^{-1}) were used to conduct the recovery experiments. Methanol (10 mL) was used as the eluting solvent. The tests were conducted in triplicate (R%) providing information on repeatability, selectivity and the effects of possible interferences eluted from the solid phases, as well as some operational aspects related to the handling of the polysaccharide based materials (*e.g.* stability, density, robustness, *etc.*) (see ESI† for additional information).[35] Furthermore, the recovery efficiency of Starbon® was compared with most other common commercial sorbents used to extract analytes, such as C18 and PGC SPE cartridges (0.1 g; 1 mL), obtained from Dionex Corporation (Sunnyvale, CA, USA) and Thermo Fisher Scientific Inc. (Waltham, MA USA).

Chromatographic measurements were carried out using a Hewlett Packard Agilent Series 1100 HPLC-UV equipped with a quaternary pump, autosampler, UV detector (wavelength 280 nm) and degasser. A Kinetex phenyl–hexyl column (5 μm, 250 × 4.6 mm I.D.) was used with the mobile phase ACN (0.1% FAc): H_2O (0.1% FAc) from 19 : 81 to 50 : 50, and a flow rate of 1 mL min^{-1}. Data processing was carried out using Agilent Clarity ChemStation Software.

Linear regression analysis

To model the impacts of the material surface properties and the analyte structures, linear and partial least squares (PLS) regression analyses based on 150 experimental points were conducted using Origin 9.1 software.

Results and discussion

Mesoporous carbonaceous materials as sorbents: recovery studies of bioactive phenolic compounds

Twelve mesoporous sorbents produced from starch [Starbon (S)], alginic acid [Algibon (A)] and pectin [Pecbon (PcB)] at 300, 450, 600 and 800 °C were characterised by SEM, TEM and porosimetry (Fig. 1).[26,36] PcB0, the pectin-derived sorbent prior to carbonisation, was also tested due to the lower than expected phenolic recovery using the PcB prepared at 300 °C (PcB300).[37] The materials are denoted using the first letter(s) of the starting material and the final carbonisation temperature (*e.g.* S300 indicates expanded starch carbonised to 300 °C). Significant differences in the morphology of the materials can be observed in the scanning electron microscopy (SEM) images of the samples carbonised at 800 °C (Fig. 1). The differences are even clearer in the transmission electron microscopy (TEM) images (Fig. 1D–E). S800 has a more amorphous, less clearly defined structure, whilst A800 has a more heterogeneous fine fibrous structure and PcB800 contains nanostructured platelets or spheres.

N_2 adsorption confirmed the mesoporous nature of the materials. An increase in micropore formation and surface area for all three polysaccharide sorbents are observed with the rising carbonisation temperature (Fig. 1), as previously observed for the starch-derived Starbons®.[36] In separation science, the generation of pore systems comprising both meso- and macroporous domains improves the mass transfer efficiencies whilst microporosity generally increases adsorption.[38]

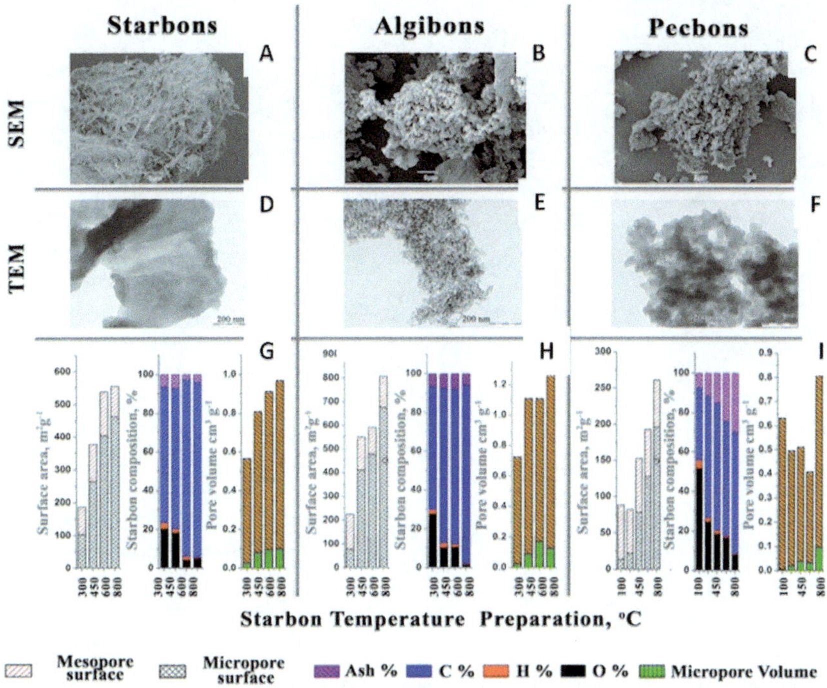

Fig. 1 SEM and TEM images of the polysaccharide-derived materials: (A, D) S800, (B, E) A800 and (C, F) PcB800. N_2 sorption data, ash content and elemental composition for the (G) S-series, (H) A-series and (I) PcB-series.

The CHO and ash contents also vary with the carbonisation temperature and the oxygen content progressively decreases when the temperature is increased. Above 550 °C, the materials have previously been shown to convert into aromatic 2D graphite-like systems.[24,26,30]

The thirteen polysaccharide-derived materials were tested as the stationary phase in SPE cartridges for the adsorption and desorption of ten bioactive phenolic compounds (Fig. 2).

The analytes were selected on the basis of both their widespread presence in a variety of fruits, seeds and vegetables, including biomass waste, and their diverse chemical functionality. The porous carbonaceous materials display varying desorption efficiencies for the sorbed analytes dependent on the starting material, carbonisation temperature and nature of the phenolic compounds. Ordering of the carbonaceous sorbents based on recovery efficiency of the phenolic compounds demonstrates a strong correlation with the material's temperature of preparation (Fig. 3).

Higher levels of recovery for all of the analytes are obtained for materials carbonised at more reduced temperatures, with the S300 and A300 presenting mean recoveries of 80 $\pm$ 4% and 78 $\pm$ 3%, respectively, which are considered adequate for the compounds studied according to analytical validation protocols.[39] Due to the lower recovery levels from PcB300 compared to the Starbons® prepared at the same temperature, the non-pyrolysed material, PcB0, was also studied. Similar or better recovery levels for PcB0 (89 $\pm$ 5%) compared to those

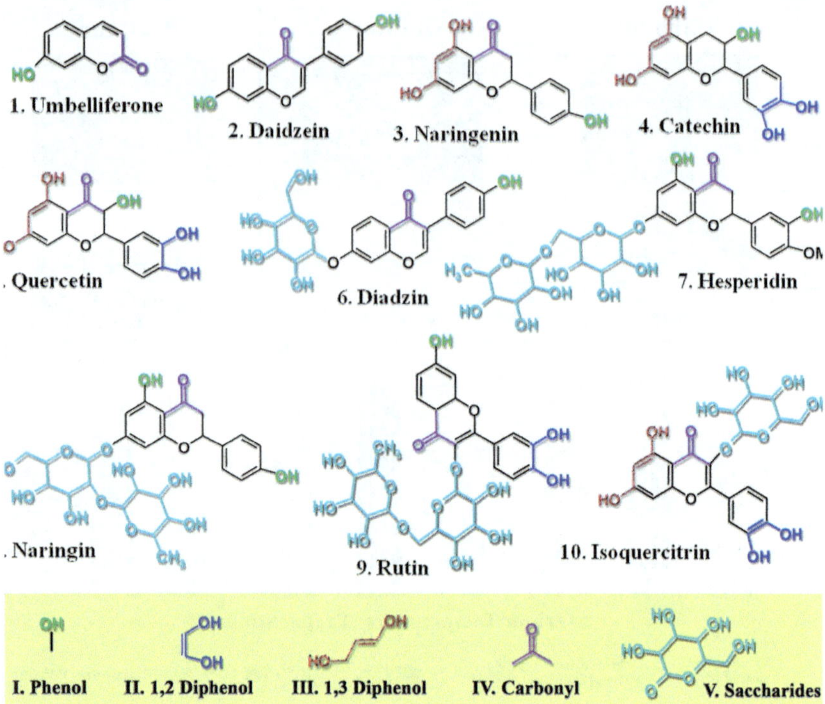

Fig. 2 Chemical structures and representative functional groups of the selected flavonoids and coumarin.

shown for S300 and A300 were obtained. However, the non-carbonised material can be mechanically unstable in more extreme separation conditions than those used here. It is well-known that the sorbents should offer enough mechanical strength to support the pressures applied to the column for the packing and process separation. The stationary phase should not swell or shrink in different mobile phases, pH, temperatures or pressures in order to ensure the uniformity and robustness of the column.[19]

According to the data presented in Fig. 1 for the materials obtained using the same starting polysaccharide-type for each Starbon® series and a correlated investigation,[40] the lower recoveries from PcB300 compared to S300 and A300 can be due to the higher ash content of PcB, compared to the S- and A-series, which showed higher basicity in comparison to other Starbons®. As a result, the pH_{pzc}, that is the pH at which the surface of the materials has a net neutral surface charge, for S300 is 4.3 and for A300 it is 5.4, while PcB300 has a pH_{pzc} of 9.9.

Surface charge is a key parameter in the chemical and physical interactions, caused by the presence of polar or ionic groups, inorganic impurities and vacant bond valences, which play an important role in the surface phenomena. For instance, in aqueous systems, the material's surface interacts with hydrogen and hydroxyl ions, which, in turn, influences the pH of the surrounding environment and the surface charge. This consequentially determines the material application, as in the case of separation processes. Therefore, a low pH_{pzc} signifies an acidic surface, whereas a high pH_{pzc} shows a basic character of the material. In solutions

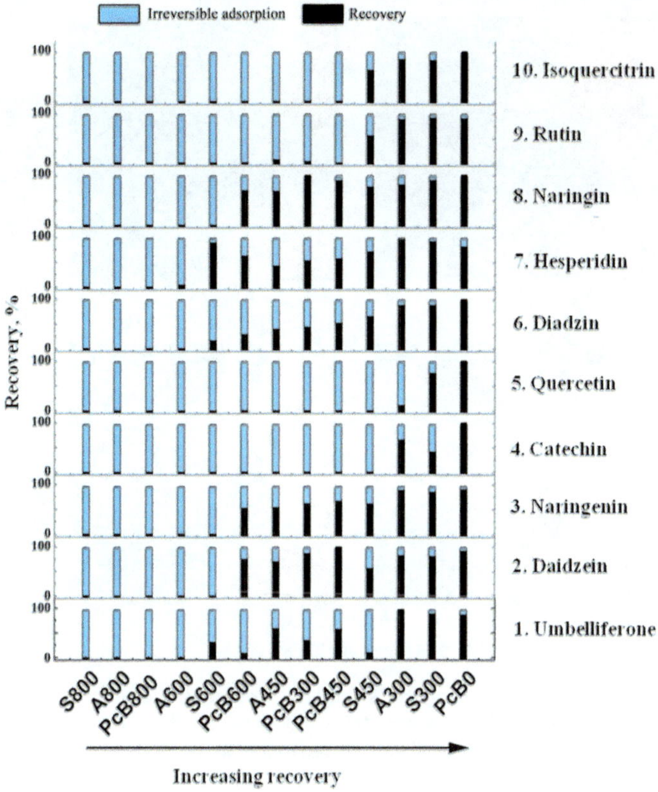

Fig. 3 Recovery results of the bioactive phenolic compounds (10 μg mL^{-1}) obtained using the 13 Starbon® SPE (0.1 g cartridges) and 10 mL of methanol as the eluent.

with a pH above the pH_{pzc}, the net surface charge is negative, whereas the net surface charge is positive below the pH_{pzc}. Moreover, a low pH_{pzc} is linked to hydrophilic surfaces and a high pH_{pzc} to hydrophobic surfaces.[40]

Overall, the sustainable mesoporous materials can be divided into the following classes; (a) those providing excellent recovery of all phenolics, making them useful for extraction and concentration purposes (*i.e.* absorbs and then easily desorbs the phenolic); (b) those that retain the phenolics well but do not facilitate easy recovery making them ideal for retention; (c) those that show good recovery for some phenolic types allowing for selective recovery strategies (see labels at the bottom of Fig. 3).

Based on the initial finding that the carbonisation temperature influences the adsorption–desorption properties of the chosen analytes, a more detailed analysis was carried out to study how the carbonaceous material properties influence the recovery efficiency. This highlighted that a key determining factor was the micropore to mesopore surface ratio, which represents the degree of surface micro-heterogeneity and is defined here as the surface roughness of the material (SR) (Fig. 4A–C).

Generally, microporosity is known to increase the adsorption and retention of small molecules due to size constriction and strong surface interactions.[33]

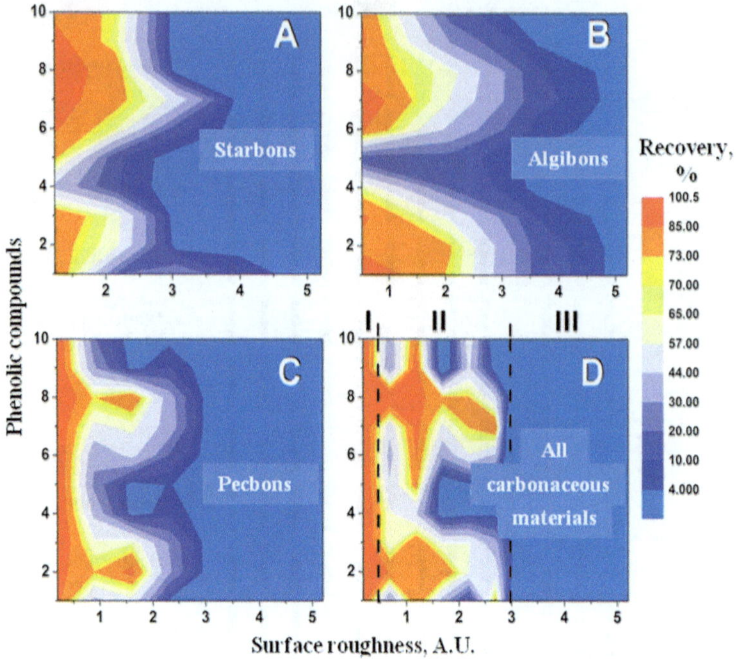

Fig. 4 Recovery efficiencies of the selected compounds correlated to the surface roughness (SR) in: (A) starch-, (B) alginic acid- and (C) pectin-derived materials. The overall experimental recovery results for all synthesised materials (D) (see Fig. 2 for the compound numbers).

Pyrolysis has a strong effect on the textural properties of Starbon® and the materials generated are more microporous at higher temperatures, which are detrimental to Starbon® as these pores can result in irreversible binding of analytes. The ability to adsorb–desorb depends on the analyte's capacity to interact with the stationary phase, by which the larger the number of interactions, the higher the resolution and the better the separation of the compounds of interest. As a consequence, a fundamental property required for stationary phase separation media is its accessibility to as high a proportion of the surface area as possible in order to maximise the number of sites of interaction, without causing undesirable irreversible retention.[41]

Interestingly, the observed recovery results are not proportional to the absolute quantity of micropores but rather relate to the SR index. Conceptually, materials with a high SR index can be viewed as strongly jagged materials, while those with a small SR number are relatively smooth. The link between the SR index and the recovery efficiency also suggests a balance between the influence of micro- and mesoporosity, and thus, between strong adsorption sites and rapid mass transport efficiencies. Across all carbonaceous materials, the three regions of sorption behaviour highlighted earlier can be observed distinctly (Fig. 4D): an SR smaller than 0.5 indicates the preferred application of the materials as a concentrating agent (I); an SR larger than 3 best fits purification purposes (III); and finally, an intermediate SR (0.5–3) offers materials tailored to selective separation (II). As can be inferred from Fig. 5, the SR index correlates well to the recovery of phenolics

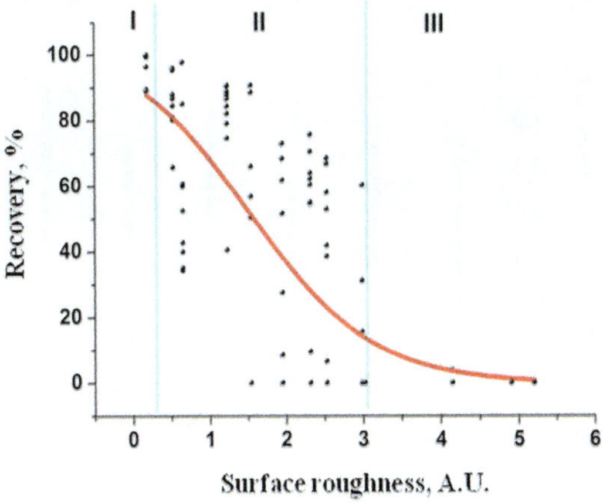

Fig. 5 Recovery values of the phenolic compounds correlated to the SR of all carbonaceous materials studied. (I): concentrating (low temperature prepared materials); (II): selective sorption; and (III): purification (high temperature prepared materials).

for regions (I) and (III), with those materials having a low SR value most favourable to analyte recovery.

Thus, in the intermediate region (II) the SR index alone is not capable of explaining the observed recoveries across all sorbents and analytes. For this reason, the influence of a range of other parameters was evaluated using a statistical approach. This was based on a 3rd order polynomial function of the SR index that satisfied boundary conditions, recovery (0) = 100%, recovery (5) = 0%, as can be seen in the proposed eqn (1).

$$\text{Recovery} = 100 + 200\text{sr} - 100\text{sr}^2$$

$$-a_3(\text{sr} - 2\text{sr}^2 + \text{sr}^3) \quad (1)$$

where a_3 is a function of the analyte and sorbent properties and sr is the surface roughness index (micropore to mesopore surface ratio, SR).

Initially, twelve parameters related to the properties of the phenolic analytes and the carbonaceous materials were systematically taken into account: volume, molecular mass, dipole moment, chemical structure, polarisability, number of specific functional groups within the analyte (Fig. 2); and sorbent C, O, H and ash content, BET surface area and micro- and mesoporous surface area. The functional groups were defined as representative entities frequently found in flavonoids and coumarins, highlighted in Fig. 2 as (1) phenolic; (2) 1,2-diol; (3) 1,3-diol; (4) carbonyl; (5) attached saccharides.

After a number of iterations, six independent variables were found to influence the analyte recovery alongside the surface roughness: (1) number of 1,2-diphenol groups, (2) molecular volume of the analytes, (3) O content, (4) H content, (5) ash content, and (6) the BET surface area (Fig. 6A). All of these variables show a complex correlation with the analyte recovery. Together, these factors are

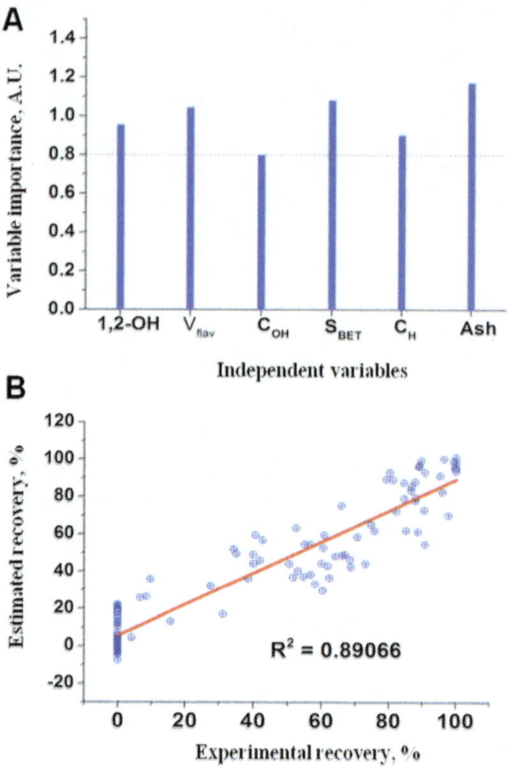

Fig. 6 (A) Comparative importance of the variables related to the recovery of the analytes [(1,2-OH) 1,2-diol groups; (V_{flav}) analyte volume; (S_{BET}) surface area; (C_{OH}) O content; (C_H) H content] and (B) the correlation between the experimental and estimated recoveries established by the proposed 3rd order polynomial function.

compatible considering the experimental results and the modelling data for phenolic compounds recovery (Fig. 6B).

For the sorbent–analyte systems studied, the volume of the phenolic compounds played a vital role as analytes with a larger volume desorbed less efficiently. The overall BET surface area of the materials was also found to be a key factor in the analyte recovery. Both of these factors can be explained by the impact of increased surface interactions between the sorbent and analyte.

Another important parameter is the presence of 1,2-diphenol groups on the flavonoids and coumarin compounds. Increasing numbers of such functional moieties, or even 1,3-diphenol groups, demonstrated an overall reduced recovery pattern (naringenin, hesperidin, diadzin, daidzein and umbelliferone). Other factors responsible for decreasing the analyte recovery include ash, O and H contents of the sorbent. For instance, as commented previously, the pectin-derived materials consistently show higher levels of ash. This suggests that a chelating mechanism between 1,2-diphenol and inorganic species or oxygen-containing functional groups in the carbonaceous materials strongly increases the retention of these analytes. The structure of the other selected analytes indicates that the primary retention mechanisms for such compounds with lower temperature prepared materials (300–600 °C) is mainly due to reversible polar interactions with the groups at the surface of the polysaccharide-derived materials (hydroxyl, C–O, C–O and C=C), such as hydrogen bonds, dipole–dipole and π–π interactions.

As a stationary phase, Starbons® presenting the better recoveries seem to differ from the flat, hexagonally arranged carbon atom sheets PGC, *i.e.* another carbonaceous material studied (see ESI† for additional information), for which the interaction mechanism is mostly based on the polar retention effect on graphitic stacking, affected by the analyte polarity and planarity which depends on the hydrophobic eluent–analyte repulsions and the interaction of analyte polarizable or polarized functional groups with the delocalised π-electrons of the graphitic planar media. In this case, the orientation of the analyte molecule at the surface has an impact on its retention, in which the greater the molecular surface area in contact with the graphitic surface, the greater the retention of that compound, probably explaining how the analytes that orientate their polar functional groups close to the graphitic surface are more strongly or permanently retained by the PGC stationary phase. For Starbons®, the fullerene curved-like surface obtained at higher carbonisation temperatures can induce specific affinities for planar and nonplanar molecules, facilitating adsorption.[41,42]

These results indicate potential routes to improve the recovery of different analytes, either *via* chemical modification of the analytes (such as chelation prior to separation) or reduction of the micropore or even ash content of the solid phase materials. The validity of this latter approach is shown by the reduction of the micropores in A800 using by-products as blocking agents obtained from low-temperature microwave hydrothermal processing of orange waste (mixture of saccharides).[43] The modified blocked A800 was used with or without the carbonisation process at 300 °C prior to using it in the SPE studies. This approach resulted in new materials with a smaller micropore volume (from 0.247 $cm^3\ g^{-1}$ to 0.001 and 0.003 $cm^3\ g^{-1}$ for A800, A800-blocked, and A800-blocked + carbonised, respectively) and reduced surface roughness (0.02 A.U.), significantly enhancing the recoveries of 6 analytes, from 10% to 88.6% (catechin, isoquercitrin, hesperidin, umbelliferone and naringenin) (see ESI† for additional information). This opens up an innovative route to be further explored, aimed at modifying the morphological and chemical structure of such sustainable carbonaceous sorbents for specific and efficient sustainable separations.

Conclusions

To the best of our knowledge, for the first time, an extensively combined statistical treatment of several properties of mesoporous polysaccharide-derived materials relevant to their interaction with bioactive analytes has been carried out. The batch-to-batch variation of the key chemical and textural (surface area and pore volume) properties of a series of prepared Starbon® materials was studied, and their performance as a stationary phase media analysed. This led to the production and selection of a set of sustainable sorbents that can fit a wide variety of applications, enabling purification, separation or concentration of the phenolic compounds, depending on the necessary requirements. In general, the sorbents providing excellent recoveries for all studied phenolics and ideal for extraction were those carbonised at lower temperatures with lower ash contents, such as S300 and A300. The materials obtained at higher carbonisation temperatures retaining more strongly or permanently the analytes were found to be excellent for purification purposes, *i.e.* S800, A800 and PcB800. In these cases, the fullerene curved-like surfaces can induce specific affinities for both planar and nonplanar molecules, promoting adsorption. Materials prepared at intermediate temperatures were interesting, presenting good recoveries for some of the studied phenolics, allowing for more selective separation, and demonstrating the key role of the sorbent–analyte systems that were investigated. So, the textural and functional characteristics of both materials and the analyte were shown to influence the recovery. The key factors included: micro- to mesoporous surface area ratio (described conceptually as surface roughness (SR)), the presence of 1,2-diol moieties on the analyte surface and molecular volume, as well as the O, H and ash contents of the carbonaceous materials.

Additionally, a new protocol to redesign the carbonaceous sorbents with a high SR by using saccharides as micropore blocking agents was evaluated, enabling vastly improved recoveries. Compared to analogous carbonaceous sorbents, the advantages of Starbon® materials are lower preparation temperatures, reduced processing steps and, importantly, improved environmental footprints. Overall, these sustainable sorbents open up new technological perspectives in a flourishing field of separation science based on tuneable morphological and chemical properties, encompassing green efficiency, selectivity, simplicity and flexibility. This versatility for the intelligent and sustainable separation of a wide range of compounds and high-value chemicals from agro-industrial waste, for instance, is determinant to promote circular, integrated and bio-based industrial chemistry contexts.

Acknowledgements

This work was supported by the São Paulo Research Foundation (FAPESP 2013/12052-5 and 2014/50.827-1), and the Coordination for the Improvement of Higher Education Personnel (CAPES 002032/2014-07). The researchers also acknowledge funding from the EPSRC and FAPESP through the CONFAP-RCUK Newton Fund Researcher Links Programme (EP/M028763/1). P. S. Shuttleworth gratefully acknowledges the Consejo Superior de Investigaciones Científicas for an international collaboration grant (i-LINK0636) and the Spanish Ministry of Economy and Competitivity (MINECO) for the concession of a Ramón y Cajal fellowship (RYC-2014-16759) and a proyecto de I+D+I para jóvenes investigadores (MAT2014-59674-JIN). We acknowledge the contribution of Joseph A. Houghton.

References

1 L. A. Pfaltzgraff, M. de Bruyn, E. C. Cooper, V. Budarin and J. H. Clark, *Green Chem.*, 2013, **15**, 307–314.

2 B. Perlatti, M. R. Forim and V. G. Zuin, *Chem. Biol. Technol. Agric.*, 2014, **1**, 1–9.

3 R. A. D. Arancon, C. S. K. Lin, K. M. Chan, T. H. Kwan and R. Luque, *Energy Sci. Eng.*, 2013, **1**, 53–71.

4 M. R. Lauro, L. Crasci, C. Carbone, R. P. Aquino, A. M. Panico and G. Puglisi, *LWT–Food Sci. Technol.*, 2015, **62**, 169–176.

5 L. Alexandru, A. Binello, S. Mantegna, L. Boffa, F. Chemat and G. Cravotto, *C. R. Chim.*, 2014, **17**, 212–217.

6 Grand View Research, Inc., http://www.grandviewresearch.com/press-release/global-polyphenols-market.

7 K. R. Martin and C. L. Appel, *Nutr. Diet. Suppl.*, 2010, **2**, 1–12.

8 I. Parveen, T. Wilson, M. D. Threadgill, J. Luyten, R. E. Roberts, P. R. H. Robson, I. S. Donnison, B. Hauck and A. L. Winters, *Phytochemistry*, 2014, **105**, 186–196.

9 J. Gonzalez-Gallego, S. Sanchez-Campos and M. J. Tunon, *Nutr. Hosp.*, 2007, **22**, 287–293.

10 T. Inoue, S. Tsubaki, K. Ogawa, K. Onishi and J. Azuma, *Food Chem.*, 2010, **123**, 542–547.

11 C. Cheigh, E. Chung and M. Chung, *J. Food Eng.*, 2012, **110**, 472–477.

12 V. G. Zuin, *Pure Appl. Chem.*, 2016, **88**, 29–36.

13 A. Scoma, C. Pintucci, L. Bertin, P. Carlozzi and F. Fava, *Chem. Eng. J.*, 2012, **198–199**, 103–109; A. K. Savage, J. P. M. van Duynhoven, G. Tucker and C. A. Daykin, *Magn. Reson. Chem.*, 2011, **49**, S27–S36.

14 Q. Ren, H. Xing, Z. Bao, B. Su, Q. Yang and Z. Zhang, *Chin. J. Chem. Eng.*, 2013, **21**, 937–952.

15 L. Chebil, C. Chipot, F. Archambault, C. Humeau, J. M. Engasser, M. Ghoul and F. Dehez, *J. Phys. Chem. B*, 2010, **114**, 12308–12313.

16 P. Pollet, C. A. Eckert and C. L. Liotta, *Chem. Sci.*, 2011, **2**, 609–614.

17 S. Parveen, S. Qamar and A. Seidel-Morgenstern, *Chem. Eng. Sci.*, 2015, **122**, 64–77.

18 X. Sang, L. Peng, J. Zhang, B. Han, Z. Xue, C. Liu and G. Yang, *Chem. Commun.*, 2014, **50**, 8128–8130.

19 A. S. Marriott, C. Antonio and J. Thomas-Oates, *Application of carbonaceous materials in separation science*, 2015, pp. 103–126, DOI: 10.1039/9781782622277-00103.

20 J. Haginaka, *J. Sep. Sci.*, 2009, **32**, 1548–1565.

21 M. Tian, J. Qiao and K. H. Row, *Anal. Lett.*, 2013, **46**, 1331–1341.

22 V. Budarin, J. H. Clark, J. J. E. Hardy, R. Luque, K. Milkowski, S. J. Tavener and A. J. Wilson, *Angew. Chem.*, 2006, **118**, 3866–3870; V. Budarin, J. H. Clark, R. Luque, D. J. Macquarrie, K. Milkowski and R. J. White, Mesoporous carbonaceous materials, preparation and use thereof, World Intellectual Property Organisation, patent WO2007104798, 2007.

23 M. M. Titirici, R. J. White, N. Brun, V. L. Budarin, D. S. Su, F. Monte, J. H. Clark and M. J. MacLachlang, *Chem. Soc. Rev.*, 2015, **44**, 250–290.

24 R. J. White, V. Budarin, R. Luque, J. H. Clark and D. J. Macquarrie, *Chem. Soc. Rev.*, 2009, **38**, 3401–3418.

25 R. J. White, V. L. Budarin and P. S. Shuttleworth, Other Approaches and the Commercialisation of Sustainable Carbonaceous Material Technology, *Porous Carbon Materials from Sustainable Precursors*, 2015, pp. 377–406, DOI: 10.1039/9781782622277-00377.

26 H. L. Parker, A. J. Hunt, V. Budarin, P. S. Shuttleworth, K. L. Miller and J. H. Clark, *RSC Adv.*, 2012, **2**, 8992–8997.

27 M. A. Tony, H. L. Parker and J. H. Clark, *J. Environ. Sci. Health, Part A*, 2016, **51**, 974–979.

28 A. Muñoz García, A. J. Hunt, V. L. Budarin, H. L. Parker, P. S. Shuttleworth, G. J. Ellis and J. H. Clark, *Green Chem.*, 2015, **17**, 2146–2149.
29 R. J. White, C. Antonio, V. Budarin, E. Bergstrom, J. Thomas-Oates and J. H. Clark, *Adv. Funct. Mater.*, 2010, **20**, 1834–1841.
30 V. G. Zuin, *Current Opinion in Green and Sustainable Chemistry*, 2016, **2**, 40–44.
31 A. Borisova, M. de Bruyn, V. L. Budarin, P. S. Shuttleworth, J. R. Dodson, M. L. Segatto and J. H Clark, *Macromol. Rapid Commun.*, 2015, **36**, 774–779.
32 P. S. Shuttleworth, V. Budarin, R. J. White, V. M. Gun'ko, R. Luque and J. H. Clark, *Chem.–Eur. J.*, 2013, **19**, 9351–9357.
33 J. R. Dodson, V. L. Budarin, A. J. Hunt, P. S. Shuttleworth and J. H. Clark, *J. Mater. Chem. A*, 2013, **1**, 5203–5207; R. J. White, P. S. Shuttleworth, V. L. Budarin, M. D. Bruyn, A. Fischer and J. H. Clark, *ChemSusChem*, 2016, **9**, 280–288.
34 V. L. Budarin, P. S. Shuttleworth, R. J. White and J. H. Clark, *Green Chem.*, 2015, **32**, 53–81; J. Weber, *Porous Carbon Materials from sustainable precursors*, 2015, pp. 277–310, DOI: 10.1039/9781782622277-00277p; S. Shuttleworth, N. Baccile, R. J. White, E. Nectoux and V. L. Budarin, *Porous Carbon materials from sustainable precursors*, 2015, pp. 311–354, DOI: 10.1039/9781782622277-00311.
35 B. Magnusson and U. Örnemark, *Eurachem Guide: the fitness for purpose of analytical methods: a laboratory guide to method validation and related topics*, ISBN 978-91-87461-59-0, 2014.
36 V. L. Budarin, J. H. Clark, R. Luque and R. White, *Mater. Matters*, 2009, **4**, 19–22; G. Durá, V. L. Budarin, J. A. Castro-Osma, P. S. Shuttleworth, S. C. Z. Quek, J. H. Clark and M. North, *Angew. Chem., Int. Ed.*, 2016, **55**, 9173–9177.
37 R. White, V. L. Budarin and J. H. Clark, *Chem.–Eur. J.*, 2010, **16**, 1326–1335.
38 R. J. White, N. Brun, V. L. Budarin, J. H. Clark and M. M. Titirici, *ChemSusChem*, 2014, **7**, 670–689; H. L. Parker, V. L. Budarin, J. H. Clark and A. J. Hunt, *ACS Sustainable Chem. Eng.*, 2013, **1**, 1311–1318.
39 L. Huber, *Validation and Qualification in Analytical Laboratories, Interpharm*, Informa Healthcare, New York, 2007.
40 J. H. Clark, V. L. Budarin, S. W. Breeden and D. J. Macquarrie. *Starbons®: a new family of bio-based mesoporous materials derived from polysaccharides. Technical Proceedings of the 2012 NSTI Nanotechnology Conference and Expo, NSTI-Nanotech*, 2012, pp. 493–494; A. Borisova, Advances in characterisation, preparation and application of polysaccharide-derived mesoporous carbons for environmental remediation, PhD thesis, University of York, 2015.
41 A. S. Marriott, E. Bergström, A. J. Hunt, J. Thomas-Oates and J. H. Clark, *RSC Adv.*, 2014, **4**(1), 222–228.
42 M. M. Titirici, *Sustainable Carbon Materials from Hydrothermal Processes*, West Sussex, Wiley, 2013.
43 A. M. Balu, V. Budarin, P. S. Shuttleworth, L. A. Pfaltzgraff, K. Waldron, R. Luque and J. H. Clark, *ChemSusChem*, 2012, **5**, 1694–1697; J. H. Clark, L. A. Pfaltzgraff, V. Budarin and J. Sherwood, *Pure Appl. Chem.*, 2013, **85**, 1625–1631.

Faraday Discussions

PAPER

Use of near infrared spectroscopy for the rapid low-cost analysis of waste papers and cardboards

D. J. M. Hayes,*[a] M. H. B. Hayes[b] and J. J. Leahy[b]

Received 25th February 2017, Accepted 2nd March 2017
DOI: 10.1039/c7fd00081b

Analytical data and quantitative near infrared (NIR) spectroscopy models for various lignocellulosic components (including Klason lignin and the constituent sugars glucose, xylose, mannose, arabinose, galactose, and rhamnose), ash, and ethanol-soluble extractives were obtained for 53 samples of paper and cardboard. These samples were mostly the type of materials typically found in domestic wastes (*e.g.* newspapers, printing paper, glossy papers, food packaging). A number of the samples (48) were obtained by separating a sample, after milling, into two particle size fractions. It was found that the fractions containing the smaller particles typically had higher ash and Klason lignin contents and lower glucose and xylose contents than the larger particle size fractions. Nevertheless, all of the sample types had attractive total sugars contents (>50%), indicating that these could be suitable feedstocks for the production of biofuels and chemicals in hydrolysis-based biorefining technologies. NIR models of a high predictive accuracy (R^2 of >0.9 for the independent validation set) were obtained for total sugars, glucose, xylose, Klason lignin, and ash, with values for the Root Mean Square Error of Prediction (RMSEP) of 2.36%, 2.64%, 0.56%, 1.98%, and 4.87%, respectively. Good NIR models (R^2 of >0.8) were also obtained for mannose, arabinose, and galactose. These results suggest that NIR spectroscopy is a suitable method for the rapid, low-cost, analysis of the major lignocellulosic components of waste paper/cardboard samples.

1 Introduction

Lignocellulosic biomass is that which is primarily composed of the polymers cellulose, hemicellulose, and lignin.[1] There has been interest in recent years in processing lignocellulosic feedstocks in biorefining technologies to produce biofuels, such as ethanol, and chemicals.[2] Biorefineries using the hydrolysis platform (*e.g.* enzymatic and acid hydrolysis) obtain saleable products from the

[a]*Celignis Limited, Tierney Building, University of Limerick, Limerick, Ireland; Web: www.celignis.com. E-mail: dan@celignis.com; Tel: +353 85 749 7372; +353 61 518 440*

[b]*Carbolea Biomass Research Group, Department of Chemical and Environmental Sciences, University of Limerick, Ireland; Web: www.carbolea.ul.ie*

sugars liberated from cellulose and hemicellulose, with the lignin fraction and the majority of the ash forming the process residual solids. Waste papers can be a particularly attractive feedstock for hydrolysis technologies since the cellulose contents can often be high and the lignin contents relatively low, particularly in comparison with other waste resources (*e.g.* composts and garden wastes).[3] Cardboards can also have good carbohydrate contents, but typically have higher lignin contents than many papers.[4]

The compositions of papers/cardboards depend on the types of woods from which they are made. Hardwoods and softwoods both contain cellulose as their main constituent; however, their hemicelluloses can differ greatly in content and composition. In softwoods, galactoglucomannan is the principal hemicellulose polysaccharide and constitutes around 20% of the dry weight.[5] The glucose to mannose ratio is about 1 : 3, whereas the ratio of galactose to glucose can vary from 1 : 1 to 1 : 10.[6] Arabinoxylans are also present in softwoods but in lower quantities.[7] In hardwoods, the xylans are the principal hemicellulose, but the type of xylan found is typically a glucuronoxylan rather than an arabinoxylan. The concentration of this xylan in hardwoods varies between 15 and 30% by weight.[8] In a few species, for example in some birches, the xylan content can reach as high as 35%.[6] Paper made from hardwoods tends to be smoother (and therefore easier to write or print on) than paper from softwoods, and softwood paper also tends to be weaker. The fibres from hardwoods and softwoods can be blended into a single paper type, with the proportional quantities varying according to preferences for strength, whiteness, roughness, *etc.* Furthermore, the type and severity of pulping used will determine the amount of lignin that is retained in the final product. Papers/cardboards can also include fillers, such as $CaCO_3$, that can greatly increase the ash content of the material.

As a result of all these variables, the chemistry of papers/cardboards, and hence their potential yields when processed in biorefineries for biofuel production, can differ substantially. The standard method of chemical analysis for the important lignocellulosic constituents (see Section 2.2) can be laborious and costly, putting limitations on the number of samples that can be evaluated. The combination of near infrared spectroscopy (NIR, 700–2500 nm) and chemometric techniques (*e.g.* partial least squares regression[9]) has allowed for numerous models for the analysis of forage and grains to be developed based on sample spectra.[10] The senior author has also developed[11] NIR models to quantitatively predict the lignocellulosic composition of Miscanthus samples, including those that are still wet and of a heterogeneous particle size. The use of NIR spectroscopy can reduce analysis times from weeks to minutes with similarly dramatic reductions in costs. However, no NIR models for the lignocellulosic constituents of paper and cardboard samples could be found in the literature. Therefore, this paper examines whether NIR spectroscopy could be used for the rapid quantitative analysis of a wide variety of papers and cardboards, with a particular focus on the types most often found in domestic wastes.

2 Materials/methods

2.1 Sample collection and preparation

A report[12] that characterised, by mass, the various waste fractions in Irish household and non-household (commercial/industrial) sectors was used to select a total of 24 different types of waste paper and cardboard samples (*e.g.* printing paper, till receipts, cereal boxes, *etc.*, see Table 2). Representative samples of each of these types were then collected and processed for chemical and NIR analyses. Sample preparation involved reducing the particle size of the samples using a FOSS Cyclotec 193 Sample Mill. The milled product was then placed in a Retsch AS 200 Digit sieve shaker to allow for the separation of the following particle sizes: $x > 850$ μm; $180 < x < 850$ μm ("DS" fraction); $x < 180$ μm ("DF" fraction). Any material with a particle size greater than 850 μm was re-milled and re-sieved until all of the sample had been collected as either DS or DF with each of these samples kept separate for subsequent NIR and chemical analysis. In addition to these 48 samples (a DS and DF fraction of each of the 24 samples), another 5 samples (one newspaper sample and four samples of unbleached Kraft pulp obtained from a paper mill in Austria) were milled but not separated between DS and DF fractions, providing a total of 53 samples for analysis.

2.2 Wet-chemical analysis

The reader is referred to an earlier publication[11] for a detailed description of the wet-chemical analytical protocol. In brief, the samples were extracted with 95% ethanol using a Dionex Accelerated Solvent Extractor (ASE 200) prior to a two-stage acid hydrolysis to hydrolyse the structural polysaccharides. The liberated monosaccharides were determined using ion chromatography. The acid insoluble residue was dried, weighed, and then ashed to find the Klason lignin (KL) content. All analyses were carried out in duplicate.

2.3 NIR analysis and model development

The absorbance spectra of the samples, over the 400–2500 nm spectral region, were collected, using a FOSS XDS monochromator, as described previously.[11] Two spectra per sample were collected, and then averaged. The spectra were then imported as NSAS files into The Unscrambler 10.1 (Camo Software AS) for subsequent treatment and model development. In order to ensure that the validation set would be truly independent from the calibration set, both the DS and DF fractions of any given sample were assigned to one set only. For the 24 samples that were split into DS/DF fractions, 7 were randomly selected for inclusion in the validation set (*i.e.* 14 spectra, 7 of DS and 7 of DF) with the remainder included in the calibration set. Of the 5 remaining samples (which had only one particle size fraction), one sample was randomly selected for inclusion in the validation set with the rest assigned to the calibration set. Hence, there were 38 samples in the calibration set and 15 in the validation set.

Partial least squares regression using one *Y* variable (*i.e.* PLS1) was used for the development of the NIR models. These models were developed on the raw spectra and on spectra treated with various techniques including: multiplicative scatter correction, extended multiplicative scatter correction, standard normal variate (SNV), standard normal variate and detrend (SNVDT), and Savitzky–Golay (SG) derivatives (1st to 4th order). Leave-one-out cross-validation was used as the cross-validation method for all models. The Haaland and Thomas[13] criterion was used to select the number of PLS factors in the models, as described previously,[11] with a maximum limit of 8 factors. Models were also developed (and tested using cross-validation rather than an independent validation set) for: DS samples (24), DF samples (24), and all samples (53).

A number of statistics (see Workman[14] and ASTM International[15] for descriptions and formulae) are used to compare the regressions (model-predicted *versus* reference analysis composition). Of these statistics, the most important are those for the independent validation set. The values for the coefficient of determination for this set (R_{pred}^2), the Root Mean Square Error of Prediction (RMSEP), and the Range Error Ratio (RER_{pred}) are particularly important. The RER is equal to the range in constituent values divided by the Standard Error of Prediction (SEP). AACC Method 39-00.01 (ref. 16) provides quality thresholds for model performance based on the RER values: RER $\geq$ 4 – the calibration is acceptable for sample screening; RER $\geq$ 10 – acceptable for quality control; RER $\geq$ 15 – good for quantification.

3 Results and discussion

3.1 Chemical composition

Table 1 presents descriptive statistics for all the samples analysed and for the DS and DF fractions. It includes values for the Standard Error of Laboratory (SEL), determined as described in a previous paper,[11] and SELR (the SEL as a percentage of the mean analyte concentration). For the 24 samples that were separated into DS and DF fractions, the compositional data for each fraction are presented in Table 2 and the weighted averages (considering the relative mass proportions of DS and DF) are presented in Table 3 along with the compositional data for the other 5 samples.

Table 1 shows that there are wide ranges in compositional values for many constituents, for example the total sugars content ranges from 27.1 to 95.0%. Importantly, it is the DF fraction that has the greater range for all but one constituent (mannose). Fig. 1 presents histograms for the value obtained when the DF value is subtracted from the DS value, for each sample, for 9 constituents. It shows that the total sugars and glucose contents are significantly greater in the DS samples than in their corresponding DF samples and that the ash and Klason lignin (KL) contents tend to be much greater in the DF samples. These significant differences are important since they show that dividing between DS and DF samples provides materials of varying compositions that will be of value in NIR model development.

Importantly, Table 3 shows that all of the samples have a total sugars content greater than 50% and so could potentially provide attractive yields of biofuels/chemicals when processed in biorefineries. Table 3 also shows that glucose is the main mass constituent of all samples. This is expected given that cellulose is the major component of papers and cardboards. For most samples, xylose is the next most abundant sugar, although in some cases (*e.g.* in newspaper samples) mannose is present in greater amounts. The other sugars (arabinose, galactose, rhamnose) are typically present in much lower quantities. The KL content varies significantly between samples with the highest quantities in the phone directory and newspaper samples, followed by cardboard samples, with the lowest amounts in tissue papers and kitchen rolls. The ash content also varies greatly and is typically greater in cardboard samples, although the highest ash content was present in the glossy paper sample (which comprised a number of advertising

Table 1 Descriptive statistics of the concentration values for the samples. SELR is the standard error of laboratory (SEL) expressed as a percentage of the average (AV.); SD = standard deviation; KURT = kurtosis

Variety	#	AV.	SD	MAX	MIN	RANGE	KURT	SKEW	SEL	SELR
All samples										
Total sugars	53	74.49	15.52	95.01	27.12	67.88	0.82	−0.91	0.84	1.13
Glucose	53	59.35	14.63	80.29	19.63	60.66	−0.18	−0.61	0.63	1.06
Xylose	53	9.85	3.31	15.53	3.26	12.27	−1.13	0.11	0.19	1.91
Arabinose	53	0.43	0.29	1.19	0.03	1.16	−0.09	0.72	0.02	5.61
Galactose	53	0.58	0.56	2.18	0.05	2.13	1.70	1.53	0.03	4.98
Rhamnose	53	0.11	0.07	0.29	0.02	0.27	−0.33	0.46	0.02	19.71
Mannose	53	4.16	2.74	11.29	0.09	11.21	0.25	0.87	0.10	2.48
Klason lignin	53	9.04	7.77	26.50	0.05	26.45	−0.36	0.78	0.41	4.56
Ash	53	11.09	11.25	52.31	0.25	52.06	4.08	1.84	0.48	4.33
Extractives	53	1.63	0.97	4.91	0.00	4.91	1.43	0.81	0.21	12.88
DS samples										
Total sugars	24	79.36	9.89	95.01	60.40	34.61	−0.74	−0.14	0.76	0.96
Glucose	24	63.86	10.31	80.29	42.62	37.66	−0.58	−0.27	0.56	0.88
Xylose	24	10.13	2.70	14.78	5.36	9.41	−1.02	0.16	0.23	2.32
Arabinose	24	0.41	0.22	0.82	0.07	0.75	−0.80	0.35	0.01	3.32
Galactose	24	0.49	0.40	1.37	0.10	1.28	0.10	1.11	0.02	4.27
Rhamnose	24	0.11	0.06	0.25	0.03	0.22	−0.16	0.66	0.02	15.53
Mannose	24	4.36	2.60	10.17	0.93	9.23	0.36	0.85	0.11	2.44
Klason lignin	24	7.60	6.84	22.44	0.05	22.39	−0.22	0.79	0.50	6.55
Ash	24	7.51	4.96	19.57	0.25	19.31	0.47	0.56	0.48	6.38
Extractives	24	1.47	0.72	2.76	0.00	2.76	−0.53	0.20	0.22	14.80
DF samples										
Total sugars	24	67.65	18.29	94.17	27.12	67.05	−0.23	−0.36	0.97	1.44
Glucose	24	52.99	16.35	78.87	19.63	59.24	−0.60	−0.11	0.74	1.39
Xylose	24	9.82	3.89	15.53	3.26	12.27	−1.40	0.01	0.15	1.48
Arabinose	24	0.41	0.32	1.19	0.05	1.14	0.32	1.04	0.03	8.18
Galactose	24	0.65	0.66	2.18	0.05	2.13	0.90	1.37	0.04	5.80
Rhamnose	24	0.14	0.06	0.29	0.03	0.26	0.12	0.42	0.03	21.51
Mannose	24	3.62	2.56	9.57	0.53	9.04	1.02	1.20	0.11	3.04
Klason lignin	24	10.51	8.24	26.23	0.83	25.40	−0.77	0.51	0.34	3.26
Ash	24	16.75	13.88	52.31	0.46	51.86	1.10	1.10	0.53	3.14
Extractives	24	1.91	1.15	4.91	0.10	4.81	0.87	0.65	0.22	11.46

flyers/pamphlets with abundant colour printing), with that sample having an ash content of 51.9% for the DF fraction.

Table 4 presents correlation coefficients between the 10 constituents for the 53 samples. As would be expected, the total sugars content is highly positively correlated with the glucose content, and it is also negatively correlated with the KL and ash contents. Interestingly, while the major sugars glucose and xylose are also negatively correlated with KL, mannose and the minor sugars are positively correlated with KL, with a correlation coefficient of 0.94 between galactose and KL. The minor sugars also tend to be positively correlated between each other but negatively correlated with glucose and xylose.

Table 2 Compositions of the DS (180 μm < x < 850 μm) and DF (x < 180 μm) fractions of the 24 samples that were fractionated after milling

	Total sugars		Glucose		Xylose		Mannose		Arabinose		Galactose		Rhamnose		Klason lignin		Ash		Extractives	
Sample	DS	DF	DS	DF	DS	DF	DS	DF	DS	DF	DS	DF	DS	DF	DS	DF	DS	DF	DS	DF
Fast food wrapping (card)	62.92	27.12	48.23	19.63	7.44	3.26	5.93	3.14	0.47	0.36	0.75	0.66	0.09	0.08	14.79	16.86	17.21	47.65	2.34	4.91
Brown paper bag	83.91	73.53	68.75	58.00	9.09	9.57	4.91	4.61	0.58	0.55	0.53	0.63	0.06	0.16	7.85	11.11	4.64	9.51	2.08	2.40
White paper bag	92.18	87.88	75.41	69.76	11.78	14.53	4.37	3.03	0.41	0.27	0.18	0.16	0.03	0.12	1.59	1.93	2.63	4.91	0.80	1.57
Mail envelopes	82.40	74.52	66.17	58.29	13.56	13.83	2.24	1.87	0.21	0.16	0.16	0.21	0.05	0.17	2.01	3.39	8.93	19.48	1.89	2.42
Glossy paper (flyers)	70.10	37.79	57.53	30.72	8.19	4.66	3.72	1.87	0.27	0.17	0.32	0.32	0.06	0.05	6.29	9.93	19.57	52.31	0.98	2.17
Food cardboard packaging (pizza)	78.33	53.07	62.03	40.66	9.62	7.08	5.34	3.88	0.61	0.54	0.66	0.76	0.06	0.14	10.97	14.72	8.87	25.85	1.40	1.93
Breakfast cereal cardboard boxes	72.00	55.86	55.69	40.79	8.98	7.19	5.62	5.52	0.67	0.80	0.96	1.34	0.09	0.21	12.90	17.20	10.01	23.59	1.08	1.69
Birthday cards	79.59	68.09	62.30	50.71	14.78	14.38	1.94	2.24	0.21	0.25	0.27	0.41	0.09	0.12	5.87	9.46	9.63	17.14	1.12	1.18
Till receipts	78.99	61.86	65.99	50.13	11.53	10.71	1.19	0.76	0.15	0.07	0.10	0.10	0.04	0.09	4.69	8.81	10.60	20.81	2.62	3.71
Printing paper	83.34	71.19	69.16	57.96	12.98	12.45	0.97	0.60	0.07	0.05	0.13	0.05	0.04	0.08	1.00	2.81	11.43	22.01	0.62	1.37

Table 2 (*Contd.*)

	Total sugars		Glucose		Xylose		Mannose		Arabinose		Galactose		Rhamnose		Klason lignin		Ash		Extractives	
Sample	DS	DF	DS	DF	DS	DF	DS	DF	DS	DF	DS	DF	DS	DF	DS	DF	DS	DF	DS	DF
Egg cartons	67.02	37.73	50.75	26.28	7.36	4.53	7.11	4.76	0.61	0.64	0.96	1.23	0.22	0.20	15.86	20.15	13.41	36.31	2.03	3.64
Yellow pages (directory)	74.10	61.32	56.07	43.09	7.34	6.42	8.60	8.60	0.76	0.99	1.16	2.00	0.17	0.23	15.65	25.35	5.40	5.88	2.07	2.85
Phone directory	66.67	60.39	48.04	41.52	6.02	5.77	10.17	9.50	0.82	1.19	1.37	2.12	0.25	0.29	22.16	26.23	5.15	6.32	0.90	1.71
Cardboard box (beer bottles)	80.41	66.09	65.27	52.33	10.57	8.24	3.67	4.32	0.37	0.44	0.40	0.65	0.14	0.12	6.98	12.12	7.28	17.33	1.47	1.36
Kitchen roll	92.74	94.07	78.12	78.87	9.84	11.20	4.08	3.39	0.36	0.30	0.23	0.20	0.11	0.11	0.05	0.83	0.25	0.55	0.94	1.06
Toilet paper	93.70	94.17	77.68	76.58	13.66	15.53	1.89	1.63	0.18	0.15	0.17	0.16	0.13	0.12	0.42	1.58	0.57	0.46	1.11	0.71
Cardboard box (washing powder)	72.09	53.84	58.41	41.51	8.07	6.95	4.51	4.01	0.45	0.47	0.50	0.74	0.14	0.16	10.72	14.36	9.20	24.74	1.76	2.72
Corrugated cardboard	78.29	68.57	64.03	54.73	9.26	8.93	3.99	3.45	0.43	0.54	0.44	0.72	0.14	0.20	7.59	11.07	6.73	17.06	1.20	1.68
Newspaper (financial times)	60.40	61.49	42.62	43.05	5.36	5.51	10.02	9.57	0.82	1.01	1.37	2.18	0.20	0.17	22.44	25.04	5.86	9.02	2.76	2.29
Cardboard file dividers	87.83	85.34	74.09	70.26	8.99	12.62	4.05	2.07	0.41	0.16	0.18	0.14	0.12	0.09	0.86	1.82	4.89	10.89	1.30	0.22

Table 2 (*Contd.*)

Sample	Total sugars		Glucose		Xylose		Mannose		Arabinose		Galactose		Rhamnose		Klason lignin		Ash		Extractives	
	DS	DF	DS	DF	DS	DF	DS	DF	DS	DF	DS	DF	DS	DF	DS	DF	DS	DF	DS	DF
Cardboard rolls (toilet/kitchen)	77.87	65.09	63.62	51.22	8.87	8.65	4.40	4.12	0.44	0.39	0.41	0.53	0.14	0.18	8.34	12.17	8.48	18.47	2.69	2.29
Paper tissue (kleenex)	95.01	93.27	80.29	77.76	13.44	14.66	0.93	0.53	0.09	0.06	0.14	0.12	0.12	0.13	0.12	0.96	0.26	0.57	0.58	0.15
Laboratory paper	91.88	89.34	76.13	73.10	12.21	13.74	3.04	2.08	0.27	0.18	0.19	0.20	0.03	0.04	2.75	2.76	0.96	1.70	0.00	0.10
Postcards	82.76	81.92	66.24	64.94	14.22	15.27	1.99	1.46	0.17	0.13	0.11	0.09	0.03	0.03	0.39	1.60	8.32	9.49	1.47	1.63

Table 3 Weighted averages (considering the relative mass proportion of the DS and DF fractions) for the 24 samples that were fractionated according to their particle size and for the other 5 samples that were analysed with no fractionation[a]

	DS (% mass)	Total sugars	Glucose	Xylose	Mannose	Arabinose	Galactose	Rhamnose	Klason lignin	Ash	Extractives
Fast food wrapping (cardboard)*	73.1	53.29	40.53	6.31	5.18	0.44	0.73	0.09	15.34	25.40	3.03
Brown paper bag	70.0	80.80	65.53	9.23	4.82	0.57	0.56	0.09	8.83	6.10	2.17
White paper bag	82.9	91.44	74.45	12.25	4.14	0.39	0.18	0.04	1.65	3.02	0.93
Mail envelopes*	67.1	79.81	63.58	13.65	2.12	0.20	0.18	0.09	2.47	12.40	2.06
Glossy paper (flyers)	52.9	54.88	44.90	6.53	2.85	0.22	0.32	0.06	8.00	34.99	1.54
Food cardboard packaging	63.2	69.03	54.16	8.69	4.81	0.58	0.70	0.09	12.35	15.12	1.59
Breakfast cereal cardboard boxes	68.1	66.35	50.94	8.41	5.59	0.71	1.08	0.13	14.27	14.34	1.28
Birthday cards	70.3	76.17	58.85	14.66	2.03	0.22	0.31	0.09	6.93	11.86	1.14
Till receipts	93.8	77.93	65.01	11.48	1.16	0.14	0.10	0.04	4.95	11.23	2.69
Printing paper	73.2	80.08	66.15	12.84	0.87	0.06	0.11	0.05	1.49	14.26	0.83
Egg cartons	91.5	64.53	48.67	7.13	6.91	0.61	0.98	0.22	16.22	15.35	2.16
Yellow pages (directory)	92.0	73.08	55.04	7.27	8.60	0.78	1.23	0.17	16.43	5.44	2.13
Phone directory*	74.3	65.05	46.36	5.95	9.99	0.92	1.57	0.26	23.21	5.45	1.11
Cardboard box (beer bottles)	33.4	70.88	56.65	9.02	4.10	0.42	0.56	0.13	10.40	13.97	1.40
Kitchen roll*	48.3	93.43	78.51	10.54	3.72	0.33	0.22	0.11	0.46	0.41	1.01
Toilet paper	44.5	93.96	77.07	14.70	1.74	0.16	0.17	0.12	1.07	0.51	0.89
Cardboard box (washing powder)*	41.5	61.41	48.52	7.41	4.22	0.46	0.64	0.15	12.85	18.29	2.32
Corrugated cardboard*	29.4	71.43	57.47	9.03	3.61	0.50	0.64	0.18	10.04	14.02	1.54
Newspaper (financial times)	78.3	60.63	42.71	5.39	9.92	0.86	1.55	0.19	23.01	6.55	2.66
Cardboard file dividers	35.9	86.23	71.64	11.31	2.78	0.25	0.15	0.10	1.48	8.74	0.61
Cardboard rolls (toilet/kitchen)	31.0	69.05	55.06	8.72	4.21	0.41	0.49	0.17	10.99	15.38	2.41
Paper tissue (kleenex)	64.3	94.39	79.39	13.87	0.79	0.08	0.13	0.12	0.42	0.37	0.43
Laboratory paper*	73.8	91.21	75.34	12.61	2.79	0.25	0.19	0.03	2.75	1.15	0.03
Postcards	82.4	82.61	66.01	14.40	1.90	0.16	0.11	0.03	0.61	8.52	1.50
Newspaper (irish independent)	N/A	65.51	44.91	6.07	11.29	1.10	1.99	0.16	26.50	1.01	2.40

Table 3 (*Contd.*)

	DS (% mass)	Total sugars	Glucose	Xylose	Mannose	Arabinose	Galactose	Rhamnose	Klason lignin	Ash	Extractives
Kraft pulp #1	N/A	89.01	74.55	7.51	5.78	0.69	0.45	0.02	5.75	0.46	0.09
Kraft pulp #2	N/A	89.09	74.50	7.50	5.96	0.68	0.42	0.02	5.17	0.60	0.84
Kraft pulp #3*	N/A	88.12	73.51	7.48	5.92	0.71	0.49	0.02	5.89	0.75	0.75
Kraft pulp #4	N/A	88.36	73.68	14.46	0.09	0.03	0.07	0.03	1.31	2.47	1.42

[a] * = sample in independent validation set, all other samples were in the calibration set for NIR model development.

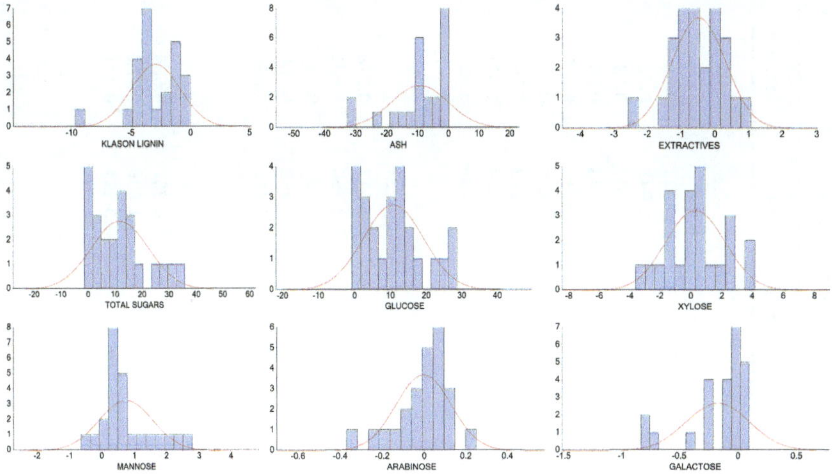

Fig. 1 Histograms, for 9 constituents, for the value of the DS content minus the DF content for each of the 24 samples that were fractionated. A negative value means that the DF content was greater than the DS content for that sample.

Table 4 Correlation table for selected components of the 53 samples. Absolute values greater than 0.5 are highlighted. EXTR = extractives, KL = Klason lignin, ARA = arabinose, GAL = galactose, RHA = rhamnose, GLU = glucose, XYL = xylose, MAN = mannose, TOT = total sugars

	ASH	KL	ARA	GAL	RHA	GLU	XYL	MAN	TOT
EXTR	**0.607**	**0.548**	0.229	0.366	0.338	**−0.747**	**−0.507**	0.204	**−0.757**
ASH		0.291	−0.101	0.062	0.096	**−0.777**	−0.419	−0.163	**−0.849**
KL			**0.833**	**0.942**	**0.692**	**−0.813**	**−0.790**	**0.814**	**−0.739**
ARA				**0.913**	**0.571**	−0.451	**−0.749**	**0.951**	−0.363
GAL					**0.701**	**−0.636**	**−0.705**	**0.878**	**−0.539**
RHA						**−0.509**	−0.430	**0.556**	−0.433
GLU							**0.709**	−0.410	**0.988**
XYL								**−0.724**	**0.712**
MAN									−0.313

3.2 NIR quantitative models

Fig. 2(a) presents the raw absorbance spectra for the 53 samples over the wavelength region 400–2500 nm. It can be seen that there are significant differences

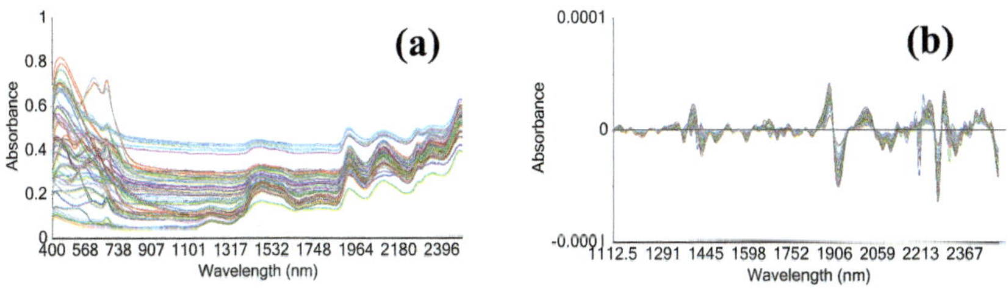

Fig. 2 Absorbance spectra for the 53 samples: (a) raw spectra over the wavelength region 400–2500 nm, (b) second-derivative spectra over the region 1112.5–2487 nm.

Table 5 NIR model statistics for the calibration/validation sets and for datasets comprising either DS, DF, or all samples in the calibration set

	Total sugars	Glucose	Xylose	Mannose	Arabinose	Galactose	Rhamnose	Klason lignin	Ash	Extractives
Calibration (38 samples) and independent validation (15 samples)										
PLS factors	4	3	6	6	6	6	3	6	6	2
R_{calib}^2	0.945	0.954	0.958	0.938	0.943	0.923	0.708	0.972	0.923	0.686
RMSEC	3.375	2.918	0.681	0.704	0.070	0.156	0.033	1.283	2.948	0.489
R_{CV}^2	0.894	0.921	0.892	0.847	0.880	0.846	0.546	0.946	0.801	0.539
RMSECV	4.706	3.863	1.090	1.124	0.103	0.226	0.042	1.778	4.744	0.592
R_{pred}^2	0.984	0.976	0.975	0.884	0.886	0.871	0.574	0.937	0.915	0.650
RMSEP	2.363	2.640	0.561	0.848	0.099	0.194	0.051	1.982	4.872	0.717
SEP	2.433	2.726	0.534	0.864	0.097	0.192	0.052	2.039	4.842	0.741
RPD_{pred}	7.445	6.166	6.012	2.870	2.802	2.779	1.532	3.966	2.629	1.631
RER_{pred}	27.516	21.732	19.804	9.605	10.554	10.179	5.227	16.450	9.789	6.626
DS samples (24)										
PLS factors	2	2	4	5	5	3	3	4	2	1
R_{calib}^2	0.932	0.937	0.944	0.957	0.934	0.935	0.832	0.974	0.677	0.278
RMSEC	2.595	2.589	0.613	0.499	0.056	0.099	0.025	1.083	2.760	0.601
R_{CV}^2	0.884	0.897	0.888	0.885	0.802	0.907	0.705	0.937	0.497	0.141
RMSECV	3.294	3.233	0.884	0.867	0.099	0.121	0.033	1.689	3.462	0.666
RER_{CV}	10.286	11.406	10.434	10.423	7.399	10.367	6.495	12.987	5.465	4.059
DF samples (24)										
PLS factors	5	4	6	4	4	4	3	5	7	1
R_{calib}^2	0.982	0.979	0.985	0.868	0.898	0.918	0.714	0.972	0.982	0.555
RMSEC	2.299	2.259	0.479	0.895	0.100	0.185	0.034	1.351	1.809	0.752
R_{CV}^2	0.947	0.956	0.937	0.751	0.809	0.841	0.500	0.910	0.851	0.431

Table 5 (*Contd.*)

	Total sugars	Glucose	Xylose	Mannose	Arabinose	Galactose	Rhamnose	Klason lignin	Ash	Extractives
RMSECV	4.188	3.383	0.962	1.257	0.139	0.258	0.045	2.444	5.696	0.856
RER_{CV}	15.753	17.172	12.521	7.037	3.048	8.081	5.763	10.200	9.010	5.503
All samples in calibration set (53)										
PLS factors	3	4	5	8	5	7	5	7	6	6
R_{calib}^2	0.954	0.964	0.950	0.959	0.925	0.936	0.783	0.983	0.942	0.854
RMSEC	3.245	2.695	0.734	0.539	0.078	0.136	0.032	1.006	2.675	0.369
R_{CV}^2	0.932	0.952	0.915	0.908	0.884	0.890	0.648	0.964	0.862	0.691
RMSECV	4.004	3.190	0.955	0.825	0.098	0.186	0.040	1.469	4.201	0.543
RER_{CV}	16.806	18.834	12.721	13.465	11.678	11.362	6.835	17.834	12.281	8.959

between the samples, particularly in the visible region (400–700 nm). Furthermore, there was a considerable baseline shift between samples, with the ash content samples typically having higher absorbance values and a lower degree of absorbance variability over the full wavelength region. It was found that superior predictive models could be obtained if the spectra were transformed by a second

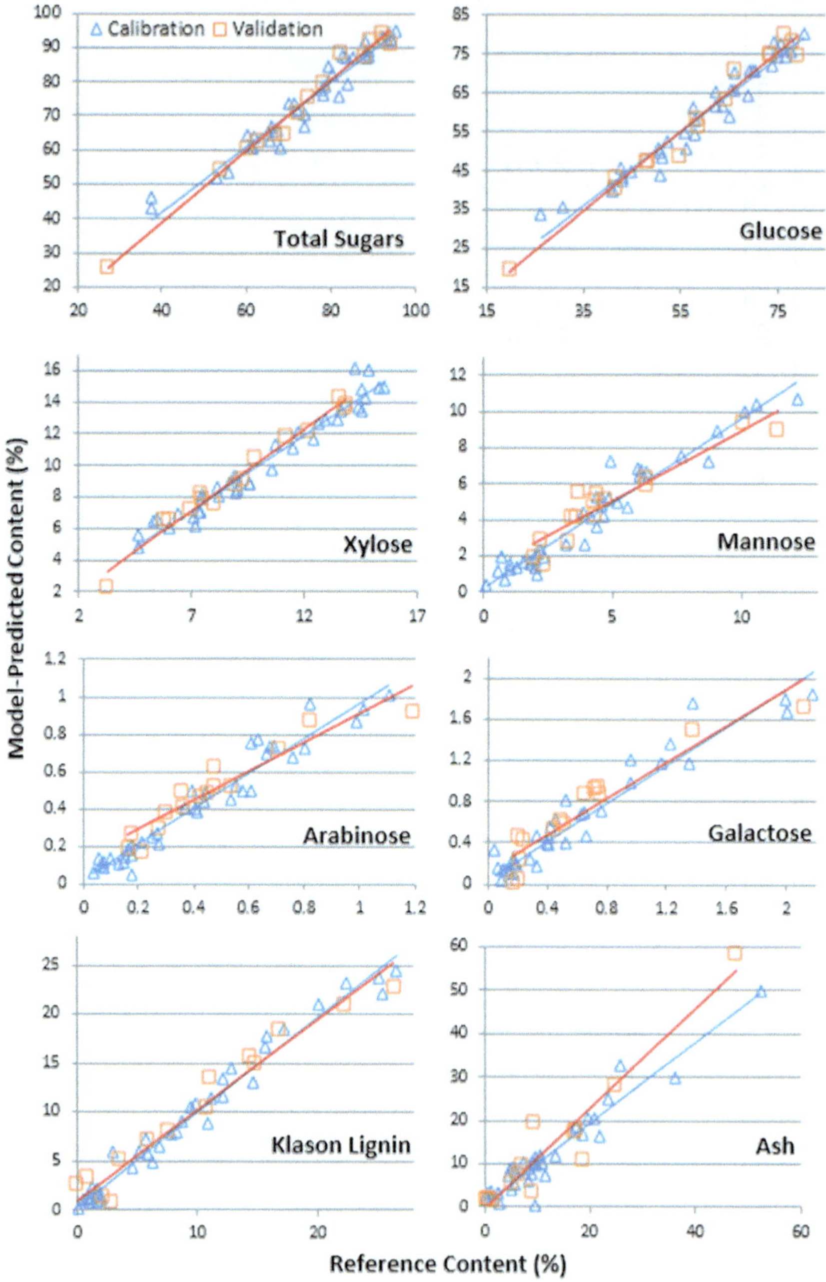

Fig. 3 Model-predicted *versus* reference values for 8 constituents for the calibration (38 samples) and validation (15 samples) sets. All data are expressed on a % dry matter basis.

derivative (Savitzky–Golay, with a second-order smoothing polynomial over a window of 25 nm) and only the region between 1112.5 and 2487 nm was used for model development.

Table 5 presents the NIR model regression statistics for the calibration (38 samples) and validation (15 samples) sets as well as for models developed using either: all DS samples, all DF samples, or all samples in the calibration set. Regression plots for 8 constituents are also presented, for the calibration (38) and validation (15) sets, in Fig. 3.

Some important observations are listed below:

• The range error ratios for the independent validation set (RER_{pred}) are greater than 15, indicating a model suitable for quantification, for total sugars, glucose, xylose, and Klason lignin. Hence, NIR spectroscopy is a good means of quantitative analysis for all major lignocellulosic constituents.

• In some cases (*e.g.* total sugars, glucose, xylose), the R^2 value for the independent validation set (${R_{pred}}^2$) is greater than the values for the calibration set (${R_{calib}}^2$) and cross-validation (${R_{CV}}^2$). This is because the random selection of samples for the validation set chose a sample (fast food cardboard wrapping) that had a DF fraction with by far the lowest values for the total sugars, glucose, and xylose contents (Table 2), as shown by the outlying validation set sample in Fig. 3 for these constituents.

• The predictive accuracy for the ash content (an ${R_{pred}}^2$ of 0.915 and an RMSEP of 4.87%) is good considering that ash does not absorb NIR radiation. The distribution of samples across the compositional range for ash, Fig. 3, is far from ideal with much fewer samples having higher (>30%) than lower ash contents (<20%). The model could potentially be further improved by the inclusion of additional samples to the calibration set to allow for a more even distribution (more similar to the distribution seen for the KL content, for example).

• Rhamnose and ethanol-soluble extractives are poorly predicted with ${R_{pred}}^2$ values of 0.57 and 0.65, respectively. However, this is not of great importance for the evaluation of samples for their suitability for utilisation in biorefining processes.

• Arabinose and galactose, despite being minor sugars in most samples, can be predicted with reasonable accuracy (having RER_{pred} values of 10.55 and 10.18, respectively), indicating that the models are acceptable for quality control.

• The RER_{pred} value for the mannose content is just below 10 (at 9.61), indicating that the model is only suitable for sample screening. However, this may be a result of the random sample selection for the validation set selecting samples with a much lower compositional range than the calibration set (as shown in Fig. 3). For the model developed with all samples in the calibration set (the "All Samples" model in Table 5), the RER_{CV} value (13.47) is significantly greater than the RER_{pred} value and suggests that the model is acceptable for quality control purposes.

• There are a number of trends noticed when comparing the models developed on the DS and DF samples only:

◦ The RER_{CV} values tend to be lower than the RER_{pred} values in the main models, indicating that the increased number of samples in the calibration sets of the main models are of benefit for improved accuracies in prediction.

◦ In all cases, the RMSECV values for the DF models are greater than those for the corresponding DS models.

◦ Despite these greater standard error values, for those constituents (*e.g.* total sugars, glucose, xylose) where there is a greater range in composition for the DF samples compared with the DS samples (Table 1), the R_{CV}^2 values for the DF models tend to be higher than for the corresponding DS models. That resulted in greater RER_{CV} values in those instances.

◦ For the major mass constituents (total sugars, glucose, xylose, Klason lignin, ash), the DS models have a lower number of PLS factors than the DF models. This is to be expected given that the DF samples have higher ash contents and may require more latent variables to properly account for the influence that this constituent can have on spectra.

4 Conclusions

The results show that the lignocellulosic composition of paper and cardboard samples, of the types typically seen in municipal wastes, can vary substantially. However, all samples contained significant quantities of carbohydrates (glucose in particular) to suggest that they could be suitable feedstocks for processing in biorefineries. There are large differences, for many samples types, in the chemical composition of the different particle size fractions obtained after these samples had been milled. In particular, the ash and lignin contents tend to be greater whilst the quantities of the major sugars (glucose, xylose) tend to be lower for the smaller particle size fractions. The differences between the two particle size fractions show that their separate inclusion in NIR model development is warranted since these differences tended to be significantly greater, for the major mass constituents, than the standard error of prediction.

The NIR models, tested under independent validation, show excellent predictive accuracy for the major lignocellulosic constituents of relevance to hydrolysis biorefining technologies. This suggests that NIR spectroscopy could be used as a primary analytical method for the rapid characterisation of paper/cardboard wastes in the future in order to greatly reduce the time and cost involved in analysis, compared with standard wet-chemical techniques. These models can be further improved by the inclusion of additional samples to the calibration set where these samples can help to improve the distribution of samples across the compositional range for important constituents. For example, selective inclusion of more samples with relatively high ash and relatively low total sugars contents could help to reduce standard errors in prediction and produce more robust models.

Nomenclature

calib	Calibration set
RMSEC	Root mean square error of calibration
CV	Cross-validation
RMSECV	Root mean square error of cross validation
DF	Particle size: $x < 180$ μm
RMSEP	Root mean square error of prediction
DS	Particle size: $180 < x < 850$ μm
RPD	Ratio of standard error of performance to standard deviation

KL	Klason lignin
KURT	Kurtosis
SEL	Standard error of laboratory
NIR(S)	Near infrared (spectroscopy)
SELR	Standard error of laboratory, as a percentage of mean analyte concentration
PLS	Partial least squares
SEP	Standard deviation
pred	Independent validation set
SEP	Standard error of prediction
R^2	Coefficient of determination
SG	Savitzky–Golay derivative
RER	Range error ratio

Acknowledgements

The authors acknowledge financial support from the DIBANET project, which was funded by the European Community's Seventh Framework Programme (FP7/2007–2013) under grant agreement #227248-2.

References

1 D. J. M. Hayes, Biomass composition and its relevance to biorefining, in *The Role of Catalysis for the Sustainable Production of Biofuels and Bio-chemicals*, ed. K. S. Triantafyllidis, A. Lappas and M. Stoker, Elsevier B. V., Amsterdam, The Netherlands, 2013, pp. 27–65.

2 D. J. M. Hayes, Second-generation biofuels: why they are taking so long, *Wiley Interdiscip. Rev.: Energy Environ.*, 2013, **2**(3), 304–334.

3 D. J. Hayes and M. H. B. Hayes, The role that lignocellulosic feedstocks and various biorefining technologies can play in meeting Ireland's biofuel targets, *Biofuels, Bioprod. Biorefin.*, 2009, **3**(5), 500–520.

4 R. Yáñez, J. L. Alonso and J. C. Parajó, Production of hemicellulosic sugars and glucose from residual corrugated cardboard, *Process Biochem.*, 2004, **39**, 1543–1551.

5 J. Lundqvist, A. Jacobs, M. Palm, G. Zacchi, O. Dahlman and H. Stålbrand, Characterization of galactoglucomannan extracted from spruce (*Picea abies*) by heat-fractionation at different conditions, *Carbohydr. Polym.*, 2003, **51**(2), 203–211.

6 P. Hakkila, *Utilisation of Residual Forest Biomass*, Springer-Verlag, Berlin, 1989, p. 568.

7 J. P. Casey, *Pulp and Paper: Chemistry and Chemical Technology*, John Wiley and Sons, New York, NY, 3rd edn, 1980, p. 848.

8 E. Sjostrom, *Wood Chemistry: Fundamentals and Applications*, Academic Press, Inc., London, UK, 1981, p. 223.

9 S. Wold, M. Sjostrom and L. Eriksson, PLS-Regression: a basic tool of chemometrics, *Chemom. Intell. Lab. Syst.*, 2001, **58**, 109–130.

10 J. S. Shenk and M. O. Westerhaus in *Forage analysis by near infrared reflectance spectroscopy*, ed. R. F. Barnes, Forages. II, Iowa State University Press, Ames, IA, 3rd edn, 1994.

11 D. J. M. Hayes, Development of near infrared spectroscopy models for the quantitative prediction of the lignocellulosic components of wet Miscanthus samples, *Bioresour. Technol.*, 2012, **119**, 393–405.
12 O. Gaillot, C. Gibson and J. Hogan, *Programme for Municipal Waste Characterisation Surveys*, RPS MCOS, Dublin, 2005.
13 D. M. Haaland and E. V. Thomas, Partial least-squares methods for spectral analyses. 1. Relation to other quantitative calibration methods and the extraction of qualitative information, *Anal. Chem.*, 1988, **60**(11), 1193–1202.
14 J. J. Workman, NIR Spectroscopy Calibration Basics, in *Handbook of Near Infrared Analysis*, ed. D. A. Burns and E. W. Ciurczak, Marcel Dekker, New York, Second edn, 2001.
15 ASTM International, *E1655 – 05-Standard Practices for Infrared Multivariate Quantitative Analysis, West Conshohocken*, ASTM International, Philadelphia, USA, 2012.
16 AACC, *Near-Infrared Methods: Guidelines for Model Development and Maintenance - AACC Method 39-00, Approved Methods of the American Association of Cereal Chemists*, AACC Press, St. Paul, MN, 1999.

Faraday Discussions

PAPER

Automation of route identification and optimisation based on data-mining and chemical intuition

A. A. Lapkin, *[a] P. K. Heer,[a] P.-M. Jacob,[a] M. Hutchby,[b] W. Cunningham,[b] S. D. Bull[b] and M. G. Davidson[b]

Received 20th February 2017, Accepted 29th March 2017
DOI: 10.1039/c7fd00073a

Data-mining of Reaxys and network analysis of the combined literature and in-house reactions set were used to generate multiple possible reaction routes to convert a bio-waste feedstock, limonene, into a pharmaceutical API, paracetamol. The network analysis of data provides a rich knowledge-base for generation of the initial reaction screening and development programme. Based on the literature and the in-house data, an overall flowsheet for the conversion of limonene to paracetamol was proposed. Each individual reaction–separation step in the sequence was simulated as a combination of the continuous flow and batch steps. The linear model generation methodology allowed us to identify the reaction steps requiring further chemical optimisation. The generated model can be used for global optimisation and generation of environmental and other performance indicators, such as cost indicators. However, the identified further challenge is to automate model generation to evolve optimal multi-step chemical routes and optimal process configurations.

Introduction

The bio-based chemicals supply chain is likely to be a significant contributor to the future chemical and biotech industries, given the drive towards a carbon-neutral technological society. A significant challenge in developing a bio-based economy is the adaptation of possible synthesis routes to the simultaneously developing supply chain. Thus, a bio-based economy that relies on gasification to syn-gas with consecutive Fischer–Tropsch synthesis can potentially fit into the conventional supply chain, whereas the supply chain that exploits more complex functionality of bio-derived molecules will have a very different structure. Several studies were published on the possible structure of such a supply chain, for example the most often cited is the detailed analysis of possible routes to most

[a]*Department of Chemical Engineering and Biotechnology, University of Cambridge, Philippa Fawcett Drive, Cambridge CB3 0AS, UK. E-mail: aal35@cam.ac.uk*
[b]*Department of Chemistry, University of Bath, Bath BA2 7AY, UK*

common intermediates based on, primarily, organic acids obtained by fermentation.[1] Different bio-refinery concepts, either based on lignocellulose or grasses, targeting bulk platform molecules[2–5] or high-value chemicals[6] were proposed. In addition, there is a continuous search for new bio-feedstocks, among examples we can list distillers dried grains with solubles (DDGS),[7] food-processing industry waste, such as unused triglycerides,[8] or microalgae.[9] In all the different options for the bio-based supply chain there will be a need for tailoring the manufacturing processes to regional biomass availability and global demands for products. The challenge is to develop the optimal set of chemical and bio-transformations, with the corresponding separation technologies, that allow production adapted for (i) the local conditions of feedstocks and energy production, and (ii) the combination of the local and global consumption of the products.

The concept put forward in this paper is that advances in automated synthesis planning and automated process design could potentially lead to the adaptive evolutionary development of optimal supply chain configurations, and of the corresponding underpinning technologies. There is an increasing body of literature on the analysis of chemical data using network theory, and its application in synthetic pathways design, with or without the use of retrosynthesis heuristics.[10–15] This approach of combining data-mining with heuristics and using network representation of chemical knowledge for automating the analysis, may hold significant promise for process chemists in the design and evaluation of the possible routes to target molecules. Thus, data-mining could, potentially, be used to construct a range of synthetic routes and assemble the data required for the evaluation of numerical values for multiple pre-determined criteria of fitness of the routes, as some of us have recently shown.[16] One current limitation of this approach is the lack of process data in the database records for many organic chemical reactions. Another is the computational power required for analysis of large networks: it is reasonable to analyse pathways of up to 5 linear steps, leading to the initial datasets in the high 100s of thousands of possible routes. However, in the case of highly complex molecules, such as new small molecule active pharmaceutical intermediates (APIs), or large molecule APIs, where linear routes may reach up to 20 steps, a different approach, primarily based on heuristics, is likely to be more realistic.[17]

In order to be useful for developing optimal solutions, any route generation methodology that is chosen will need to be evaluated and optimised. For example, a mass-based metric could be developed to maximise mass-flow through a network of reactions.[18] An alternative is to minimise the material waste for a series of transformations. This could be done using an algorithmic approach based on evaluation of E-factor, the ratio of the mass of generated waste to the mass of useful product.[19] If combined with additional metrics, such as evaluation of the overall energy efficiency, or specific environmental impact of the complete route, then this would lead to a more balanced optimisation approach using a multi-criteria decision making process, for example using outranking algorithms.[16]

What is much harder to do is to link route generation with automated generation of process models and to optimise a complete system that includes multiple synthetic steps and process options. Some ideas on how this could be implemented are discussed in the literature on process optimisation under uncertainty, for example by Engell.[20] Similarly, the idea of linking data-mining with expert systems in combination with evolutionary algorithms has been proposed by some of us earlier.[21]

In this paper we present our initial approach to automating the assembly of possible reaction pathways for the situation where a series of synthetic steps are used to convert a known bio-based feedstock into a commercially valuable product. The selected case study is the multi-step synthesis of paracetamol from limonene, with the process simulated for the hypothetical route shown in Scheme 1, the details of which are discussed in detail in the paper.

Results and discussion

Reaction pathway assembly

The search for possible literature pathways was limited to a maximum of five steps and was done by mining the published reaction data from Reaxys (http://www.reaxys.com). To assemble the data, the approach published previously[16] was further developed by making use of chemical structural information in guiding the route selection.

Firstly, a network was obtained by downloading all reactions from Reaxys involving limonene as a substrate. The products from these reactions were then extracted and used as new search queries. This procedure was repeated iteratively until data covering the first five reaction steps in Scheme 1 was obtained. In this way the data generated covers only the forward direction. Then, the set of species that was previously used as reactants to query Reaxys was used to obtain a new dataset containing all reactions in which they were designated as products. This was carried out using a python2.7 script and the Reaxys application programming interface (API). The raw data was sanitised by removing all reactions that missed product or reactant, *i.e.* those that were only incomplete, "half" reactions. Similarly, if the same reaction was obtained multiple times using different reaction conditions, it was registered only once.

The raw data was converted into a graph using the graph-tool library in python2.7 (ref. 22) by assigning each species contained in the result set as a node to the graph. Subsequently, every time a species acted as a reagent in a reaction,

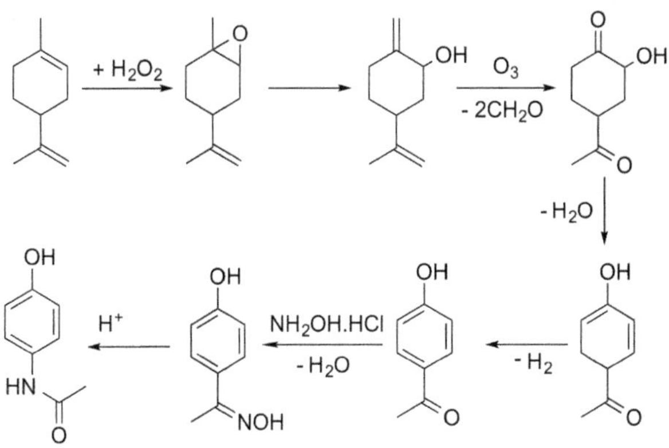

Scheme 1 A hypothetical route from limonene to paracetamol used in process simulation.

an edge was drawn connecting it to the products of that reaction, and each time it acted as a product, edges were drawn connecting it to the reactants of that reaction. This produced a graph of 13 118 083 vertices and 37 954 177 edges. Under the all-to-all wiring scheme, a reaction will have several edges under which it appears, one for each combination of reactant and products. Each "reaction edge" involves one product and one reactant; the remaining species/edges can be identified through looking-up their reaction IDs. Employing such an all-to-all wiring scheme made it necessary to limit the possible intermediates for the reaction steps. Thus, a depth-first search algorithm was run on the graph to find all paths of a length of four steps or less connecting limonene to paracetamol from which around 200 of the most common intermediates were extracted. These intermediates were analysed to determine if they were undesirably small fragments of limonene, or if they contained undesirable elements, for example heavy metals. If either of these were found to be the case for the species in question, it was prohibited from being used as an intermediate in the later route search.

Thus, the restricted chemical space was used to find all routes connecting limonene to paracetamol, again using a depth-first search algorithm implemented in graph-tool, within a maximum path length of five reaction steps. This returned a hit set of 69 793 potential routes that showed connectivity in Reaxys involving 3029 different chemical species. Reaxys does not currently record the stoichiometry of reagents, and yields are missing from around half of all reactions, making it impossible to track mass fluxes across a reaction path. This means that, at this stage, the algorithm would consider a path maintaining a fragment of limonene and introducing almost the entire functionality in the final step to be a valid route. This, obviously, runs counter to the idea of generating paracetamol from sustainably sourced limonene, and was undesirable. The issue could be largely counter-acted by implementing atom mapping in the search algorithm, which enabled the algorithm to disregard such steps automatically. This is, potentially, computationally expensive if carried out for the entire network and, hence, was deemed impractical at this stage. Instead, it was more efficient to obtain all possible paths and to then algorithmically post-process a much smaller set.

The 69 793 potential routes were further analysed using Pybel,[23] a Python wrapper of the Open Babel libraries.[24] The aim was to use computationally cheap, well-implemented criteria for analysis. Because stoichiometry was unavailable and could not be calculated accurately and cheaply, the decision was made to focus on the structural data of the reactants and products of a route. Thus, the relative difference in mass and number of carbon atoms of the reactant and product across an edge were analysed, as well as the number of ring structures present. Limonene is a C_{10} hydrocarbon, containing one ring and weighing 136.2 g mol^{-1}. Paracetamol weighs 151.2 g mol^{-1}, similarly has one ring structure and is a C_8 compound. Thus, any large deviation of an intermediate from these properties would most likely lead to an inefficient route. Therefore, any reaction that resulted in a product that weighed less than 75% of the reactant, or that weighed more than 125% of the reactant were rejected. Similarly, if the change in the number of carbon atoms between a product and a reactant in any reaction was greater than two, that reaction was also rejected. Finally, if either a product or a reactant contained more, or less, than one ring structure, the corresponding step was rejected.

This procedure reduced the size of the hit set by a factor of 65, down to 1068 possible routes. The number of unique chemical species in the network also reduced down from an initial 3029 down to 47. In order to efficiently reduce the number of results to be considered for further analysis, the set of the most frequently occurring reaction steps was manually analysed and steps deemed undesirable, or those too old or obscure to have the papers easily accessible, were further excluded. This reduced the count to 69. The resulting network of routes is shown in Fig. 1, where the possible routes can be traced from limonene at the top of the figure *via* the possible intermediates along the paths indicated by the arrows to paracetamol at the bottom. The network could be further trimmed applying chemical insights, that are currently not automated.

The quality of the records associated to a given entry in Reaxys has a great impact on the feasible network. A potentially interesting reaction step, converting limonene to cumol, was screened out to obtain the results shown in Fig. 1 due to the fact that very little associated data is available in the original paper from 1919, making it not amenable to any automated evaluation. Subsequently, this reaction was manually added back into the reaction network. The changes this caused to the network topography can be observed in Fig. 2: the number of routes immediately increases to 458 and the number of species contained in the network increases to 132. Analysing further the two networks, shown in Fig. 1 and 2, we observe that if the key reaction of direct conversion of limonene to cumol is not available, many routes then go *via* *p*-cymene, for which there are 500 fewer reactions in Reaxys. This shows how crucial the definition of a good set of screening parameters is, while manual analysis is still required due to insufficient data being recorded in Reaxys, see discussion in ref. 16. However, this analysis also highlighted the potential power of this data-mining approach: the 1919 paper had little follow up, most likely because of poor selectivity and little detail in the original paper. However, if this reaction is indeed feasible, or could be further optimised using tools of modern catalysis, it could potentially become a useful part of the network as it opens up many new possible routes. This is highlighted by the number of thicker arrows in Fig. 2, indicating the increased frequency of the appearance of certain species and routes in this network.

Similar to manually adding in old reactions, we can also (automatically) add in-house or proprietary data, not available in the licenced version of Reaxys. In our case, three reaction routes from limonene to paracetamol were developed in house and this set of expert data was added to the network. Largely these new reactions were not present in the previous network, shown in Fig. 1, and several of the intermediate compounds were also new, raising the total count of compounds generated to 3033. The depth-first search algorithm was then re-run and the results further analysed.

Under the all-to-all wiring scheme, the three reaction routes consisted of 23 edges, 10 of which already existed in the network, and contained two new compounds not previously in the network. Adding these 13 edges led to a noticeable increase of 41 additional routes to a total of 69 834 routes involving 3033 unique species. After screening it was found that 78 of these routes, comprising 54 unique species, were promising. The obtained network is shown in Fig. 3, visualising the resulting change in the reaction pathways topology, compared to the network shown in Fig. 1.

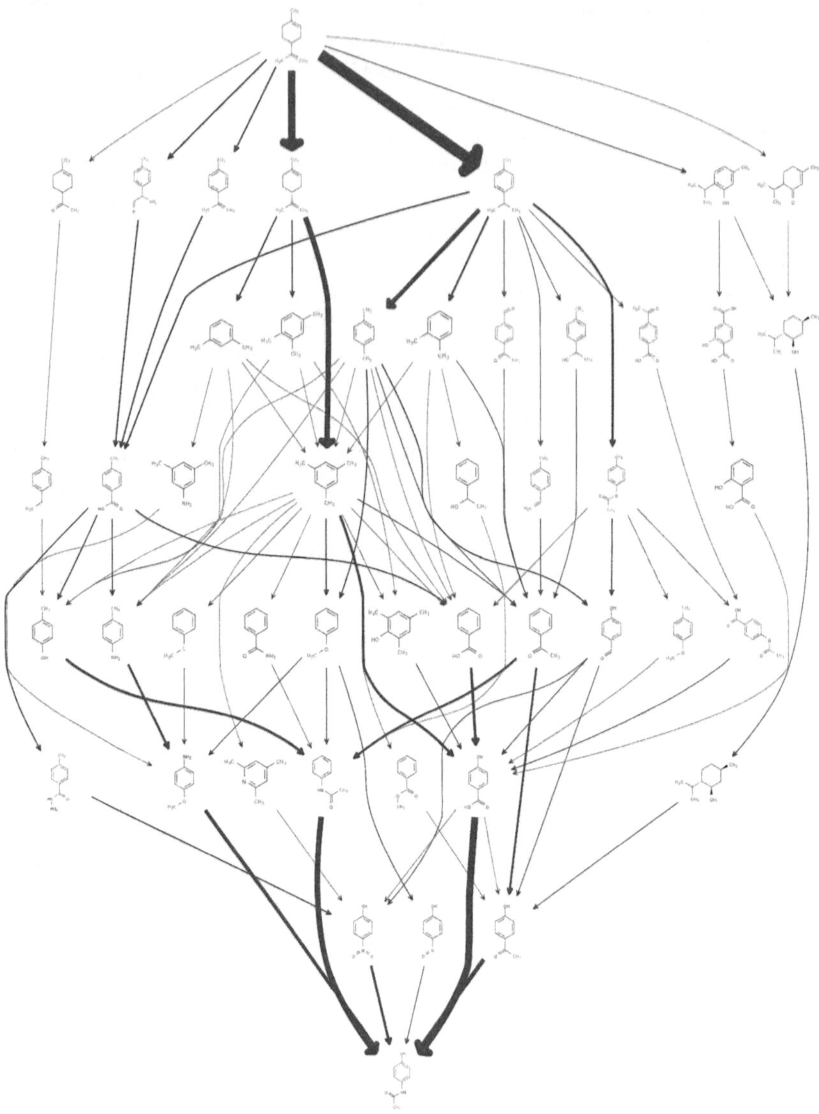

Fig. 1 A network of possible synthetic routes from limonene to paracetamol. The thickness of an edge corresponds to the number of times a route has been reported. The plot is for illustrative purposes and the nodes are positioned by an automated graph algorithm, thus their relative positioning does not necessarily contain any chemical insights.

The total number of routes that seem promising after adding the additional, proprietary, data is thus 9 more than without the proprietary knowledge. It may at first seem counter-intuitive that the addition of three routes should lead to an increase greater than three in the number of routes. This is caused by the fact that the new reactions interlink with the existing network and, by connecting previously unconnected nodes, are able to open up new routes which would otherwise not be there. Also, clearly, several of the new reactions survive the screening stage, illustrating that the screening methodology does not reject promising reactions.

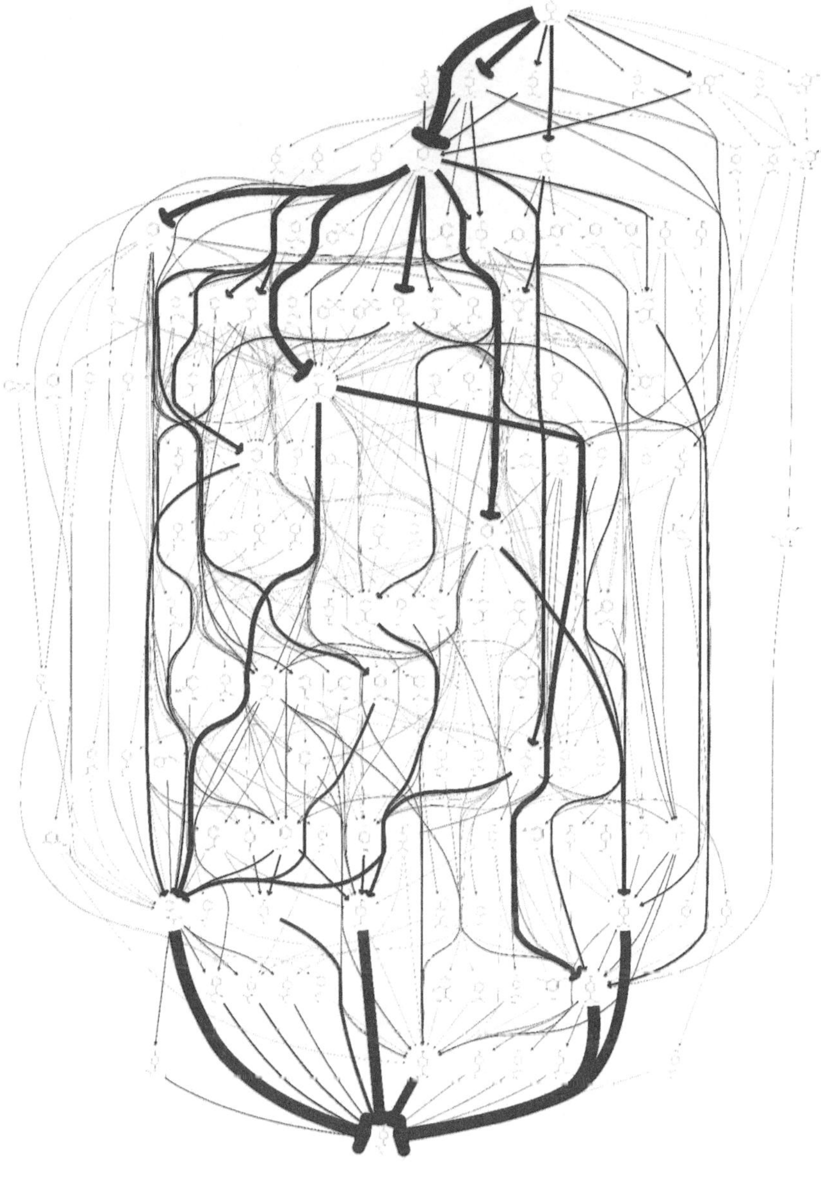

Fig. 2 A network of possible synthetic routes converting limonene into paracetamol. In this scenario, a promising but rarely recorded reaction has not been screened out to illustrate the effect a single, key reaction can have on the network.

The next step in our framework of analysis of alternative routes requires numerical data on the different criteria of efficiency of the routes. Earlier we have developed the approach to route evaluation based on multi-criteria decision making, where criteria of mass and energy efficiency are considered along with route reliability and the selected criteria of environmental impact.[16] However, this approach is missing the crucial step of identification of problematic steps, for example reactions that lead to complex separation problems due to the formation of mixtures of products, or low conversion, or problems due to poor catalyst

Fig. 3 A network of possible synthetic routes from limonene to paracetamol, including the proprietary data, but excluding the 1919 speculative paper.

stability, *etc.* Identification of these types of problems requires much more in-depth analysis of process data, which are not always available in the literature. Thus, a combination of in-house and literature data is potentially desirable to assemble conceptual processes for the most promising routes that can then be used for process simulation to identify the limitations of the proposed processes.

Process simulation

For illustration of the process simulation, we chose one of the hypothetical routes from the network shown in Fig. 3, for which we could use a collection of in-house and literature data, thus combining the results of data-mining and the proprietary data. The chosen route is shown in Scheme 1. For each reaction, a process model was developed using either in-house or literature data. The overall flowsheet for the process is shown in Fig. 4 as an ASPEN flowsheet. Each reaction was simulated individually using gPROMS and we explain the details of each reaction below.

The first step of the process involves epoxidation of limonene to limonene epoxide using a phase transfer catalyst (PTC). Epoxidation is a two-phase reaction wherein limonene reacts with the activated PTC to give limonene epoxide along with small amounts of side products (limonene di-epoxide and diol). The PTC is activated by hydrogen peroxide in the aqueous phase, before transferring to the

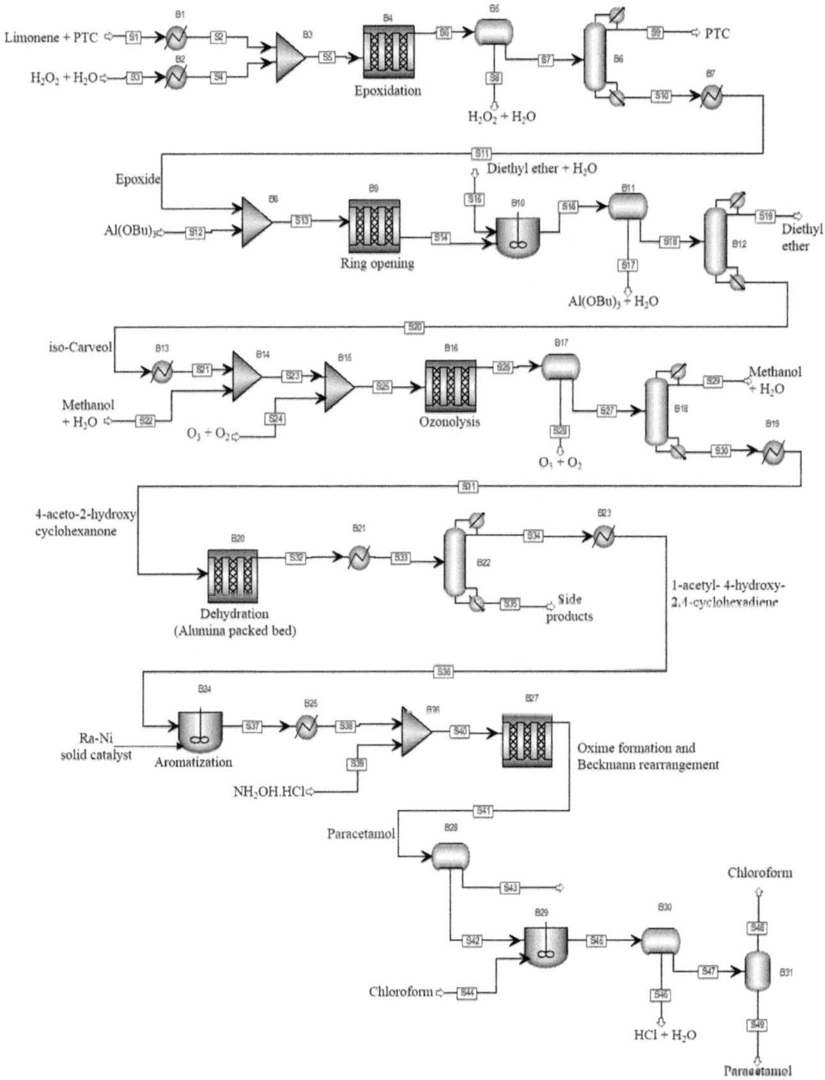

Fig. 4 The proposed process flowsheet detailing all reaction and separation steps. The ASPEN flowsheet is shown for illustration only. All simulations were performed using gPROMS.

organic phase where it catalyses the epoxidation reaction of limonene. The mechanism of this reaction is relatively well-known,[25] albeit its practical implementation, especially in flow, remains a challenge.

The model was formulated as a continuous flow two-phase reaction in a microreactor and was validated against in-house experimental data. The two phases are separated using a decanter and the organic phase is processed further for removal of the catalyst. The separation of the PTC could not be modelled due to a lack of properties of the PTC. However, it was determined from COSMO-RS calculations that vacuum distillation could effectively separate the PTC from the rest of the organics, which was also demonstrated experimentally. In the current process model, therefore, complete separation of the PTC is assumed to occur by vacuum distillation at 333 K and 1 mbar.

In the simulation, the epoxidation reaction achieves 100% conversion at 333 K with a residence time of 1.4 h. However, the yield of limonene epoxide was found to be about 64% under these conditions, with the yield of this step under batch conditions being as high as 95%. Therefore, this step as a continuous flow reaction required further optimisation. The residence time was reduced so as to avoid formation of side products, which resulted in a product yield of 88% under flow conditions. Complete separation of the catalyst was assumed. However, purification of the product was not necessary, since an impure product could potentially be carried through several intermediate reaction steps. The epoxide stream was then used as a substrate for the subsequent ring opening step.

The epoxide is ring-opened in the presence of 10 mol% of aluminium tert-butoxide, $Al(OBu)_3$, as catalyst, to give iso-carveol and by-products. The ring opening of limonene epoxide was modelled as a continuous flow reaction in a microreactor. The kinetics of the system were determined by fitting the in-house batch experimental data. Ring opening with $Al(OBu)_3$ reached a conversion of 100% at 373 K and a residence time of 2.4 h. However, a much shorter residence time of 10 min was sufficient to achieve full conversion in this reaction. Ring opening leads to the formation of carveol and carvone side products, thus, leading to an iso-carveol yield of 59%. The catalyst was separated by solvent extraction using a mixture of diethyl ether and water. The catalyst transfers to the aqueous phase while the organic products remain with the diethyl ether phase, which is later removed by distillation.

Ozonation of iso-carveol is then carried out to give 4-aceto-2-hydroxy cyclohexanone in the presence of aqueous methanol. This was a new reaction, so far not reported in Reaxys. The reaction was performed at −78 °C in methanol, reaching a maximum NMR spectroscopy yield of 60%. The reaction is somewhat problematic and requires considerable further development of the conditions. For this reason, we used the kinetic values of beta pinene ozonolysis to develop the process model; the model parameters were determined by fitting the in-house experimental data, obtained at 298 K.

The ozonolysis was modelled as a gas–liquid Taylor flow in a microreactor. The model accounts for mass transfer from gas to liquid and for the chemical kinetics. The mass transfer coefficient was calculated based on a literature correlation.[26] The microreactor was followed by a gas/liquid separator to remove the gas phase. The organic phase was then distilled to separate out the solvents used in the reaction. In the model, ozonolysis was carried out at a gas-to-liquid ratio of 60, with the organic phase being diluted with aqueous methanol before exposure to ozone. The gas phase was a mixture of O_3 and O_2 (3% O_3). The model of ozonolysis gave 100% conversion at 298 K and a residence time of 1 min, calculated based on the two-phase superficial velocity, with no side product formation being observed. The gas and liquid phases were separated and methanol distilled off, with the organic phase then passed to the dehydration reactor.

Dehydration of 4-aceto-2-hydroxy cyclohexanone at 328 K in the presence of phosphoric acid affords 1-acetyl-4-hydroxy-2,4-cyclohexadiene.[27] This procedure does not give good results. A similar dehydration of the α-hydroxy group is reported to proceed as a heterogeneous gas-phase reaction over activated Al_2O_3 at 310–320 °C.[28] The dehydration process model was based on this literature report. The reaction kinetics were determined by fitting the experimental data reported in the literature.[28] Conversion of about 82% could be reached with a 1-acetyl-4-hydroxy-2,4-cyclohexadiene yield of 69%. The side products formed are separated by distillation. However, further purification of the product was not considered in the current process model.

1-Acetyl-4-hydroxy-2,4-cyclohexadiene was then dehydrogenated to yield 4-hydroxy acetophenone over a Raney Ni catalyst at 553 K. The aromatization reaction is carried out using a solid catalyst and modelled using a continuous stirred tank reactor (CSTR) with suspended catalyst, with continuous removal of hydrogen formed as a side product. The kinetics of aromatization were reported in the literature.[29] A conversion of about 84% was achieved with no side products formed. The product could then be taken directly for oxime formation.

4-Hydroxy acetophenone can be converted to paracetamol in two steps, firstly forming oxime and then converting oxime to the final product.[30] In the current process, the two-steps were carried out using a solventless self-catalysed one-pot method with hydroxylamine hydrochloride.[31] This step is shown in Scheme 1 as a two-step process for clarity of the reaction. Both reactions were modelled to occur in series in a microreactor. The kinetics of the reactions were determined using in-house batch reaction data. 99% conversion of 4-hydroxy acetophenone was attained at 383 K with a yield of 70%. The low yield is due to degradation of paracetamol to 4-aminophenol.

The yields of the different steps are shown in Fig. 5. These yields correspond to both, reactions performed in-house, with models validated using the in-house data, as well as reactions for which models were developed based on the literature data. In this sequence of processes, the ring opening reaction, step 2, has the lowest yield. This step requires further experimental optimisation to reduce side products formation.

Overall optimisation of routes and processes: next steps

Following the generation of multiple possible reaction routes, as shown in Fig. 1–3, a specific route was chosen to generate the overall process simulation data. Simulation of the individual reaction steps required the generation of process models, which, in turn, required the collection of the detailed information about the reaction kinetics, and careful examination of the separation tasks. In this specific case, we used a conventional methodology of process simulation: for each reaction–separation step, specific unit operations were chosen, which fixed the

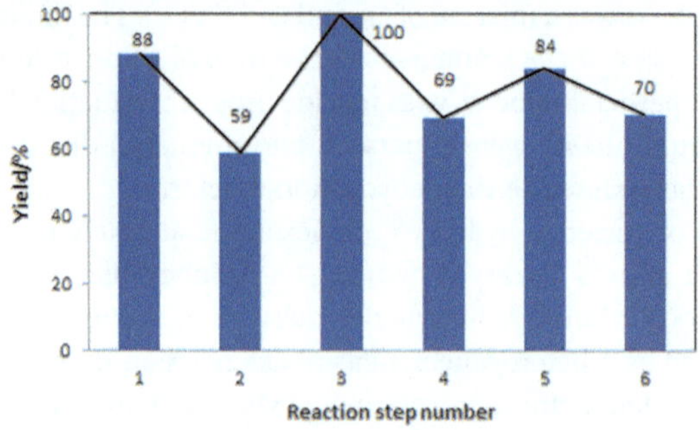

Fig. 5 Yields of the individual reaction steps, corresponding to the flowsheet shown in Fig. 4.

nature of the process model. The linear methodology of assembly of the overall process that was followed allows very few degrees of freedom and results in a process concept, which does not appear to be highly efficient, in terms of the overall yield of the main product. Yet, the process of assembly of the individual models provides insights into the individual steps and their interactions within the overall process. The overall flowsheet is a collection of fairly detailed models, which, of course, allows generation of multiple further numerical indicators, for example process mass intensity (PMI) or E-factor, overall toxicity, and so on. However, this would detract from the more difficult challenge – how to evaluate all feasible routes generated through data-mining?

For this task, the linear process of manual assembly of process models is infeasible, and even the use of a library of pre-formulated models may not be flexible enough. Thus, it would be useful to assemble the overall process model using a less restrictive approach, focusing on reaction kinetics and physical mechanisms that take place in each process, and evaluating multiple feasible assemblies of the phenomena into the overall process.[32] This approach should allow for automating model generation and for evolutionary design of optimal reaction–process configurations, combining network methodology for the identification of reaction pathways with optimisation of conceptual processes. In terms of a purely synthetic chemistry perspective, the network approach, combined with rapid evaluation of reaction efficiencies and overall efficiencies to specific end-points within a network, for example, reactive intermediates of interest, would be a useful methodology, as it provides a tool to compare different synthesis routes, based on the potentially available intermediates within the biorefining paradigm.

Conclusions

We have shown a nascent methodology of assembly and evaluation of multi-step reaction sequences, which combines data-mining and in-house chemical development with conceptual process design. The main limitation of this approach at present is the inability to generate process models automatically. Resolution of this limitation will enable rapid co-evolution of optimal reaction networks and the corresponding processes.

Acknowledgements

This work was funded in part by the EPSRC project "Terpene-based Manufacturing for Sustainable Chemical Feedstocks" EP/K014889. The PhD scholarship of WC is funded by the EPSRC Doctoral Training Centre in Sustainable Chemical Technologies (EP/G03768X/1). We gratefully acknowledge the collaboration with RELX Intellectual Properties SA and their technical support, which enabled us to mine REAXYS. PMJ is grateful to Peterhouse and the Cambridge Trust for PhD scholarships.

References

1 A. Aden, J. Bozell, J. Holladay, J. White and A. Manheim, *Top value added chemicals from biomass*, Pacific Northwest National Laboratory, National Renewable Energy Laboratory, 2004.

2 M. FitzPatrick, P. Champagne, M. F. Cunningham and R. A. Whitney, *Bioresour. Technol.*, 2010, **101**, 8915–8922.

3 X. Pan, C. Arato, N. Gilkes, D. Gregg, W. Mabee, K. Pye, Z. Xiao, X. Zhang and J. Saddler, *Biotechnol. Bioeng.*, 2005, **90**, 473–481.

4 S. Fernando, S. Adhikari, C. Chandrapal and N. Murali, *Energy Fuels*, 2006, **20**, 1727–1737.

5 J. B. van Beilen and Y. Poirier, *Plant J.*, 2008, **54**, 684–701.

6 A. Lapkin, E. Adou, B. N. Mlambo, S. Chemat, J. Suberu, A. E. C. Collis, A. Clark and G. Barker, *C. R. Chim.*, 2014, **17**, 232–241.

7 A. Chatzifragkou, O. Kosik, P. C. Prabhakumari, A. Lovegrove, R. A. Frazier, P. R. Shewry and D. Charalampopoulos, *Process Biochem.*, 2015, **50**, 2194–2207.

8 C. Schotten, D. Plaza, S. Manzini, S. P. Nolan, S. V. Ley, D. L. Browne and A. Lapkin, *ACS Sustainable Chem. Eng.*, 2015, **3**, 1453–1459.

9 H.-W. Yen, I. C. Hu, C.-Y. Chen, S.-H. Ho, D.-J. Lee and J.-S. Chang, *Bioresour. Technol.*, 2013, **135**, 166–174.

10 M. Fialkowski, K. J. M. Bishop, V. A. Chubukov, C. J. Campbell and B. A. Grzybowski, *Angew. Chem., Int. Ed.*, 2005, **44**, 7263–7269.

11 B. A. Grzybowski, K. J. M. Bishop, B. Kowalczyk and C. E. Wilmer, *Nat. Chem.*, 2009, **1**, 31–36.

12 C. M. Gothard, S. Soh, N. A. Gothard, B. Kowalczyk, Y. Wei, B. Baytekin and B. A. Grzybowski, *Angew. Chem.*, 2012, **124**, 8046–8051.

13 M. Kowalik, C. M. Gothard, A. M. Drews, N. A. Gothard, A. Weckiewicz, P. E. Fuller, B. A. Grzybowski and K. J. M. Bishop, *Angew. Chem., Int. Ed.*, 2012, **51**, 7928–7932.

14 S. Szymkuć, E. P. Gajewska, T. Klucznik, K. Molga, P. Dittwald, M. Startek, M. Bajczyk and B. A. Grzybowski, *Angew. Chem., Int. Ed.*, 2016, **55**, 5904–5937.

15 J. N. Wei, D. Duvenaud and A. Aspuru-Guzik, *ACS Cent. Sci.*, 2016, **2**, 725–732.

16 P.-M. Jacob, P. Yamin, C. Perez-Storey, M. Hopgood and A. A. Lapkin, *Green Chem.*, 2017, **19**, 140–152.

17 M. D. Eastgate, M. A. Schmidt and K. R. Fandrick, *Nat. Rev. Mater.*, 2017, **1**, 0016.

18 A. Voll and W. Marquardt, *AIChE J.*, 2012, **58**, 1788–1801.

19 J. Andraos, *Org. Process Res. Dev.*, 2009, **13**, 161–185.

20 J. Steimel and S. Engell, *Comput. Chem. Eng.*, 2015, **81**, 200–217.

21 N. Peremezhney, P.-M. Jacob and A. Lapkin, *Front. Chem.*, 2014, **2**(26), 21–27.

22 T. P. Peixoto, *The graph-tool python library*, Figshare, 2014, DOI: 10.6084/m9.figshare.1164194.

23 N. M. O'Boyle, C. Morley and G. R. Hutchison, *Chem. Cent. J.*, 2008, **2**, 2–5.

24 N. M. O'Boyle, M. Banck, C. James, C. Morley, T. Vandermeersch and G. R. Hutchison, *J. Cheminf.*, 2011, **3**, 33.

25 D. C. Duncan, R. C. Chambers, E. Hecht and C. L. Hill, *J. Am. Chem. Soc.*, 1995, **117**, 681–691.

26 P. Sobieszuk, J. Aubin and R. Pohorecki, *Chem. Eng. Technol.*, 2012, **35**, 1346–1358.

27 Y. K. Kim and J. D. Hatfield, *J. Chem. Eng. Data*, 1985, **30**, 149–153.

28 P. D. Bartlett and G. F. Woods, *J. Am. Chem. Soc.*, 1940, **62**, 2933–2938.

29 Z. Kou, S. Shen, K. Liu, G. Xu, Y. An and C. He, *Int. J. Hydrogen Energy*, 2013, **38**, 11930–11936.

30 F. W. Fitzpatrick and J. D. Gettler, *J. Am. Chem. Soc.*, 1956, **78**, 530–536.

31 E. Rancan, F. Aricò, G. Quartarone, L. Ronchin, P. Tundo and A. Vavasori, *Catal. Commun.*, 2014, **54**, 11–16.

32 H. Freund and K. Sundmacher, *Chem. Eng. Process.*, 2008, **47**, 2051–2060.

Faraday Discussions

DISCUSSIONS

Feedstocks and analysis: general discussion

Leigh Aldous, Dimitris Argyropoulos, Vitaliy Budarin, Carlos I. Cabrera-Rodríguez, James Clark, Marta Coma, David Constable, Jiajun Fan, Florence Gschwend, Daniel Hayes, Changwei Hu, George Huber, Trevor Hughes, Andrew Hunt, Alexei Lapkin, Gary Lye, Mark Mascal, Avtar Matharu, Xindong Mu, Deepak Pant, Emmanouil Papaioannou, Roberto Rinaldi, Gadi Rothenberg, Joseph Samec, Simo Sarkanen, Andrzej Stankiewicz, Christian Stevens, Magdalena Titirici, Keith Waldron, Kirsty Wilson, Anna Zhenova, Long Zhou and Vânia Zuin

DOI: 10.1039/C7FD90050C

James Clark opened the discussion of the paper by Keith W. Waldron: Regarding the point you made about ethanol, is that because it is flat-lining in terms of growth and demand? New bioethanol plants are being built, are they redundant before they are even complete?

Keith Waldron responded: That is a very good question!

Avtar Matharu asked: We are looking for food applications for some of our biobased products/extractives – what are your thoughts on re-use in food applications compared with non food applications?

Keith Waldron answered: There are rules and regulations concerning the development of foods from new sources of ingredients. Therefore, these will have to be explored and the appropriate authorities would have to be consulted.

Andrew Hunt queried: You used a range of different materials that will have different ash and inorganic contents – how does the ash or inorganic content affect your process and where do these species end up?

Keith Waldron responded: We have not yet assessed the ash content. However, insoluble ash will end up in the recalcitrant residue that remains after saccharification and fermentation.

Roberto Rinaldi said: The saccharification of lignocellulose produces a mixture of pentoses and hexoses. I know that it is a challenge to transform both sugar types in one reactor. May I ask you to tell us a bit more about the type of microorganisms that you are employing in the process in order to metabolise both C5 and C6 sugars?

Keith Waldron replied: We are currently working with yeast species, and have focused on their use of C6 sugars.

David Constable said: In commercial production, succinic acid and similar acids are currently obtained through fermentation. Are you considering in your work how to reduce the overall cost of these kinds of molecules *versus* oil? I agree that separation is key to all these processes – are you, or is someone else, looking at alternative separation processes? How would you reduce the overall cost to match the current cost of obtaining these acids considering that the cost of petroleum is so low?

Keith Waldron responded: Petroleum is low cost. However, it is not a sustainable source. Therefore, exploring the use of industrial biotechnology for producing renewable platform chemicals would seem to be a good investment. As a part of that, the separation of such organic molecules from fermentation broths is a challenge which would benefit from further conversation, research and development.

Florence Gschwend commented: The company Clariant uses a sieve-like sponge to adsorb the ethanol for separation from the fermentation broth, to save energy.[1]

1 https://www.youtube.com/watch?v=niZls2dpHjM.

Emmanouil Papaioannou remarked: The process that uses vacuum and a barrier to separate the ethanol from the fermentation broth is called membrane distillation, and in this case hydrophobic membranes are used to achieve a more efficient separation.

James Clark opened the discussion of the paper by Roberto Rinaldi by asking: Can you have sufficient purity in each of the streams to make the whole process commercially viable?

Roberto Rinaldi replied: This is our challenge at the moment.

Gadi Rothenberg commented: Regarding the issue of separation of the resulting monomer mixture obtained from lignin depolymerisation and hydrogenation – we have carried out extensive studies on lignin valorisation[1] and one of our conclusions was that, where possible, it is more beneficial to use the depolymerisation mixture as a mixture, without separation, as this avoids the high costs of distillation and other separation techniques. Such mixtures would not be suitable for polymerisation applications, but they could be suitable for use in formulations for cosmetics, *e.g.* as biobased surfactants. This has also been looked at by several key players in the cosmetics industry, because there is a positive reaction to lignin-based surfactants, which can be termed as "plant extracts".

1 Z. Strassberger, S. Tanase and G. Rothenberg, *RSC Adv.*, 2014, **4**, 25310–25318.

Roberto Rinaldi responded: We definitely agree with your opinion. We have explored utilisation of depolymerised lignin streams as raw-materials for the production of surfactants in an industrial project with the company Henkel GmbH.[1] Our conclusion was that surfactants with special properties can be obtained from depolymerised lignin streams.

1 (a) C. Kropf, A. Schulz, H. Hellmuth, R. Rinaldi and H. J. Estévez Rivera, *Novel anionic surfactants and washing agents containing said surfactants*, WO2017021112, 2017; (b) C. Kropf, A. Schulz, H. Hellmuth, R. Rinaldi and H. J. Estévez Rivera, *Novel anionic surfactants and detergents comprising them*, WO2017021113, 2017; (c) C. Kropf, N. Plath, D. Bedrunka, R. Rinaldi and H. J. Estévez Rivera, *Novel surfactants with lower CMC, and surfactant systems and detergents containing said surfactants*, WO2014117973, 2014.

Joseph Samec commented: This is a very nice study on the effect of the solvent on the outcome of the reaction and also the importance of analysis. Do you see any effect on the hemicellulose fraction due to your solvent choice? Could hemicellulose be acting as a hydrogen donor when methanol is used as the solvent?

Roberto Rinaldi answered: Hemicellulose sugars play a complex role in the catalytic system. The hemicellulose sugars may indeed serve as hydrogen donors for the reaction.

Simo Sarkanen requested: You have reported the catalytic effect of Ni/C and RANEY® Ni on the relative yields of the reduced monomeric phenols obtained through catalytic upstream biorefining – please can you describe how these yields have been estimated and what assumptions, if any, have been made during the calculations?

Roberto Rinaldi answered: The quantification of the selected compounds was performed using flame ionisation detector responses. In the cases for which authentic samples of the components were not available, the corresponding response factors were estimated using a well-established method.[1]

1 J. T. Scanlon and D. E. Willis, *J. Chromatogr. Sci.*, 1985, **23**, 333–340.

Changwei Hu asked: Can you tell us about the stability of the catalyst? Can you comment on the selectivity? Under hydrogen pressure, how do you avoid hydrogenation of the benzyl ring? Also, as lignin is a copolymer of different monomers, you will obtain a mixture of compounds, the separation of which is very complex – what is your strategy?

Roberto Rinaldi replied: To answer your first question, the catalyst RANEY® Nickel is very stable under the catalytic upstream biorefining conditions. We can reuse the catalyst up to seven times without deterioration of the properties of the lignin stream. To answer your second and third questions, RANEY® Ni is a stable and selective catalyst for the targeted hydrodeoxygenation (HDO) of lignin-derived phenolics. Typically, under process conditions, the catalyst is chemoselective towards the HDO of ketones and aldehydes, rendering $-CH_2-$ and $-CH_3$ groups, respectively. Through optimization of the content of the H-donor component in the lignin-extracting solvent mixture (liquor) or the H_2 pressure, saturation of the phenolic ring can be prevented. Finally, to answer your fourth question regarding separation, the product mixture, containing compounds derived from coumaryl (H), guaiacyl (G) and synapyl (S) units, could be chemically funnelled via demethoxylation, thus enriching the mixture in G- or H-derived compounds. We have demonstrated such an approach within the examples reported in ref. 1.

1 X. Wang and R. Rinaldi, *Energy Environ. Sci.*, 2012, 5, 8244–8260.

Dimitris Argyropoulos queried: Considering the composition of the oils that you have *i.e.* the lignin stream that is obtained as a viscous oil, how is a value of up to 60% by weight of monophenolic compounds possible? From 10 g of wood, how much oil do you get and how much cellulose? How do you account for the 60% of monophenolic compounds outlined in the paper?

Roberto Rinaldi responded: In the paper, the content of monophenolics (up to 60%) is reported relative to the weight of the lignin stream obtained from the fractionation process. To illustrate this, let us consider a raw material that contains 30% lignin, and assume that the lignin removal (*i.e.*, delignification) yield is 90%, which can often be achieved for hardwoods under low-severity conditions (such as 180–200 °C and 3 h of heating). For simplification, let us also assume that the hemicelluloses remain fully retained in the pulp. Hence, beginning the process with 10 g of oven-dried wood would result in 7.3 g of pulp and 2.7 g of a lignin oil stream. Therefore, should the content of a single component in the lignin oil be about 10% (*i.e.* 0.27 g out of the 2.7 g weight yield of lignin oil), the weight yield of this component relative to the wood would be 2.7%. At first sight, this hypothetical yield of a single lignin product may seem to be too low for any industrial application. However, it is not. For comparison, the conversion of cellulose into glucose, followed by fermentation, renders a 11% weight-yield of ethanol relative to the weight of the wood processed in the second-generation biorefinery.

Long Zhou said: I have read that oxidation is best for chemical production and hydrogenation is best for obtaining products for bio-fuel applications. Can you comment on the difference and why you used hydrogenation during your biorefining process? In biorefining, high-value added chemicals are generally more profitable than bio-fuels.

Roberto Rinaldi answered: We employ reductive processes during the fractionation of the biomass to avoid repolymerisation of the lignin fragments, which generates degraded polymers that are not very useful. The products of the catalytic upstream biorefining are functionalised phenols, which could well serve as monomers for novel performance polymers. Therefore, although it holds some truth that hydrogenation is very useful for the production of fuels (which should show a low H/C ratio), selective reduction processes will also very much contribute to new strategies for the valorisation of various biomass fractions. The catalytic upstream biorefining or lignin-first approach constitutes a very good example of rational use of catalytic hydrogenation in the biorefinery.

Mark Mascal opened the general discussion of the paper by Gary J. Lye: For biomass conversion, it seems that the default is to form second generation ethanol. However, you could valorize your cellulose more effectively through conversion to levulinic acid and its multiple derivatives, or you could use, for example, George Huber's chemistry (DOI: 10.1039/C7FD00036G). Can you comment?

Gary Lye responded: Indeed, we only followed the cellulose to bioethanol route because we were working within the specific context of the Wissington biorefinery, where a bioethanol facility is already in place. One of the collaborators on this project, Prof. David Leak (University of Bath, UK), is indeed looking to engineer pathways to produce alternative platform chemicals from the cellulose fraction.

Carlos I. Cabrera-Rodríguez commented: A common problem with downstream research is that researchers work on the first recovery step and then stop there. Normally, the further downstream steps are not analyzed. I was wondering, in your case, what is the difference in concentration before and after the centrifugal partition chromatography step? Do you have to concentrate the monosaccharides further before carrying out the next reaction step?

Gary Lye replied: This is a good question. Maintaining high solute concentrations throughout the whole downstream process is critical. In our case, one of the criteria for optimising the initial steam explosion step was to minimise the amount of steam required and hence minimise dilution. With our sugar beet pulp, the total sugar concentration in the aqueous stream following steam explosion was around 20 g L^{-1}. Any chromatographic separation will result in dilution, but in our case we can achieve centrifugal partition chromatography fractions that contain concentrations of the target monosaccharide, L-arabinose, of up to about 5 g L^{-1} (so a 1 : 4 dilution). This is generally better than conventional solid-phase chromatographic separations. Another key point is that the L-arabinose is eluted in a phase system that is compatible with the subsequent enzymatic conversion.

Leigh Aldous said: From your overall biorefinery image, I noticed that limestone appears to be added into the process, as a primary ingredient. Can you comment on the relative scale of the limestone used and it's application in the biorefinery process?

Gary Lye replied: The limestone is added to adjust the pH of the aqueous liquor during the solid-liquid extraction process and (I believe) to help facilitate crystallisation of the sucrose. I am not sure exactly how much limestone is used at the Wissington facility, but I think that the ratio of limestone to sugar beet is small.

Roberto Rinaldi commented: I was wondering if the cellulose fraction has been characterised (in terms of the degree of polymerisation and crystallinity)? People tend to think that purified cellulose is a cheap material, but it is actually a high value product – with a cost around £4000 per tonne. Could you please comment on possible applications of the cellulosic material obtained from your process?

Gary Lye responded: This was not the main focus of our study. All we have shown to date is that the cellulose fraction can be more or less quantitatively recovered following steam explosion, and that this stream can be treated with commercial cellulases to release monomeric glucose, which is then readily fermented by yeast to form ethanol. We followed this route due to an available bioethanol facility at the Wissington site.

I agree that it would be interesting to look at the quality of the cellulose fraction and hence other, higher value, applications.

David Constable asked: Can you comment on the process economics? Also, how much enzyme engineering did you have to do to achieve the desired conversions?

Gary Lye answered: Investigation of the process economics (and Life Cycle Analysis) is work that is currently underway. This is being led by our collaborators at Imperial College London, UK. As shown in the paper, in Fig. 8, we have a model of the preliminary flow sheet in place. All I can say at this stage is that the process as it stands is unlikely to be economic, but the model will highlight where we need to focus on to improve this. A key difficulty is in estimating the market value of the novel products we are producing.

Regarding the enzyme engineering, we already had libraries in place containing the main transketolase and transaminase enzymes that we have investigated for valorisation of the sugar beet pulp monosaccharides. These libraries tend to be single mutant libraries focused on specific points in the enzymes to increase the substrate range. Using some high throughput screens, we were able to identify mutants of both enzymes that could accept both L-arabinose and D-galacturonic acid as substrates. Linking this to the first part of your question, the next challenge in terms of enzyme engineering will be to improve the rate and yield of the conversion in order to start to make the processes economically viable.

Leigh Aldous asked: You mentioned an interest in fermenting the vinasse, but haven't said much about your investigation – can you tell us more? Inhibitors are typically a problem with these methods. Also, there is typically a high potassium content that can be exploited.

Gary Lye responded: Our studies on fermentation of the vinasse were conducted as part of an associated PhD project that will shortly be published.[1] The glycerol concentration in the sugar beet vinasse is around 180 g L^{-1} and the concentration of polyphenols (the usual inhibitor class) is around 6 g L^{-1}. In the work reported in the thesis, we examined a range of methods to successfully remove the polyphenols and looked at supplementation of the vinasse with particular nutrients. The overall finding was that the best strategy simply involved a 6-fold dilution of the vinasse prior to fermentation and supplementation with yeast extract. This was successful for producing a range of enzymes that we are interested in and worked up to small pilot scale.

1 N. B. Suhaili, *Characterisation of biocatalyst production within an integrated biorefinery context*, PhD Thesis, University College London, 2017

Jiajun Fan queried: During discussion of the analysis of your raw material, you mentioned 25% of "other" – what does this "other" consist of? You have a high amount of galactose and galactonic acid present, but surely there are other components too – do you know what the others are?

Gary Lye answered: As shown in Fig. 2 in our paper, the carbohydrate fraction makes up around 75% of the sugar beet pulp. The proportion of all the different monosaccharides is also shown in this figure. The 25% of other material will include proteins, lignin and inorganic material, as mentioned in the paper.

Gadi Rothenberg opened the discussion of the paper by Deepak Pant by saying: I like your idea very much. I am sometimes asked to advise companies on CO_2. I am often asked by industrial business units whether a particular idea will allow them to get rid of a million tonnes of CO_2. For your method, it seems that contact between the electrode and biofilm is essential. I can see this happening nicely on a small scale but not on a large scale, has this been tried?

Deepak Pant answered: As far as systems based on microbial electrosynthesis (MES) are concerned, the technology is very recent and was first proposed in 2010. Since then, rapid strides have been made in terms of the production rates and efficiencies of these systems. However, these are still limited to the lab-scale only. Microbial conversion of CO_2 is well established though and there are plants based on gas fermentation on an industrial scale. The most recent example that I can recall is from Lanza Tech, who have a 100 000 gallon per year (300 tons) pre-commercial plant located at one of Baosteel's steel mills outside of Shanghai, China and they use off-gases from the steel mill (mainly containing carbon monoxide) as a feedstock to produce ethanol. Currently, Europe's first-ever commercial scale production facility for creating bioethanol from waste gases produced during the steel-making process is being built at the site of Arcelor Mittal in Ghent, Belgium. On the other hand, plants such as the Carbon Recycling International (CRI) George Olah Renewable Methanol Plant in Svartsengi, near Grindavik, Iceland have already existed since 2011, and this plant is based on catalytic conversion of CO_2 into methanol. The CO_2 is captured from flue gas released by a geothermal power plant located next to the CRI facility. The origin of the flue gas is geothermal steam emissions.

MES can be compared to microbial fuel cell (MFC) systems, for which, after an intensive research period of almost 15 years, the first pilot and small-scale commercial systems are only emerging now (*e.g.* Cambrian Innovation, USA). Thus there will be a window of 5–10 years before large scale, industry-ready MES systems are available.

George Huber commented: All chemical plants have rules and laws that must be followed, especially with respect to energy and material balances. The Iceland plant works because they have very cheap renewable electricity. CO_2 is expensive to compress and transport. Can you please discuss the energy and material balances of your process and discuss the net energy output of this technology?

Deepak Pant responded: This is indeed the case for both electroconversion and bioelectroconversion of CO_2. The CRI plant in Iceland based on a catalytic process utilizes less expensive electricity from geothermal sources. Even there, the plant normally utilizes CO_2 from a point source and probably not from the atmosphere, which is quite energy intensive and hence expensive. In the case of microbial electrosynthesis, most of the studies, including ours, have been limited to lab experiments only and a full energy and material balance investigation is yet to be carried out. From the studies carried out so far, the main advantages of this process are the lower energy requirement due to the use of a biocatalyst, which lowers the energy required for CO_2 reduction, and also the generation of hydrogen, which in other cases has to be provided externally. Apart from this, we have recently performed a techno-economic analysis of a CO_2 electroconversion refinery. The details of this analysis can be found in ref. 1.

1 A. ElMekawy, H. M. Hegab, G. Mohanakrishna, A. F. Elbaz, M. Bulut and D. Pant, Technological advances in CO_2 conversion electro-biorefinery: a step toward commercialization, *Bioresour. Technol.*, 2016, **215**, 357–370.

Anna Zhenova remarked: You have mentioned that CO_2 is available from the atmosphere. However, as far as I know, there is no mature technology available for cheaply harvesting CO_2 from the air. Can you tell us more about your anticipated source of CO_2 if your concept was to be implemented in the near future?

Deepak Pant replied: Indeed CO_2 is available from the atmosphere but capturing it from this low concentration (~400 ppm) is tricky. There are technologies available that are being used currently, and two of those which I can recall are from Climeworks (Switzerland) and Global Thermostat (USA). While the latter removes CO_2 from ambient air by utilizing readily available, low-cost process heat,[1] the Climeworks technology is based on amine-based filters that bind the CO_2 in conjunction with the moisture in the air.[2] But I agree that harvesting CO_2 from air cheaply is not easy. The CO_2 ElectroRefinery concept that we are proposing can make use of direct air capture using either of the two technologies mentioned above, but more realistically, it will be located close to a point source of CO_2, principally an industrial site that has a direct flue gas source rich in CO_2. This site could be a steel mill or a bioenergy plant based on biomass. In the case of complex gas mixtures, a pre-treatment might be necessary, which in turn will also add to the cost of the overall process.

1 http://globalthermostat.com/the-gt-solution/.
2 http://www.climeworks.com/our-technology/.

Avtar Matharu asked: Is use of dirty CO_2 a concern? Do you have any data from dirty CO_2, especially that from mills?

Deepak Pant answered: Yes, and the ultimate objective is to use off-gases from industry, hopefully without any pre-treatment. This has not been done yet for microbial electrosynthesis but, in our institute, colleagues are working on autotrophic conversion of CO_2 to polymers such as bioplastics (PHB). They tested off-gases from a bioenergy plant in their fermentors and did not see a drastic drop in the performance compared to when pure and defined mixtures of gases were used. Of course, this will remain an important aspect for this research and probably once the technology is optimized at lab-scale, the next step will be to search for microorganisms that are able to tolerate the dirty gas mixtures from mills.

Carlos I. Cabrera-Rodríguez said: Let's continue to discuss the economics. I noticed that you have mainly discussed the electricity required to run the bio-electrochemical system, however the recovery of the product (acetic acid) is also energy intensive since the concentration is low. This will affect the economics, especially for non-expensive products such as methanol or acetic acid. Have you taken this into consideration?

Deepak Pant responded: With the current development of MES, we are at the stage of realization of the concept using a lab-scale reactor. In this regard, the prime focus of research and development is on the microbial electrosynthesis part, rather than the downstream processing of the product. Acetate production using MES has still to reach an economically attractive amount. However, with an intention to show the full production and recovery of the product, we have employed an ion-exchange concept during the recovery of acetate/acetic acid.[1] Indeed, the recovery process is energy intensive and it is yet another challenge to overcome in order to develop this MES technology. Our focus was to present a technology for separation, so economic analysis was not considered at this nascent stage of development. We fully agree that, at this current stage, scaling up of the process together with the production and separation would not be economically sound.

1 S. Bajracharya, B. van den Burg, K. Vanbroekhoven, H. De Wever, C. J. N. Buisman, D. Pant and D. P. B. T. B. Strik, *Electrochim. Acta*, 2017, **237**, 267–275.

Xindong Mu said: An alternative route would be to use electricity to decompose water to obtain hydrogen and oxygen, and then use the hydrogen to drive the reaction – can you compare your route to this established one? What level of electricity power can you envisage to be economically feasible?

Deepak Pant communicated in reply: The alternative route mentioned is indeed also in practice. The mentioned alternative involves water electrolysis for hydrogen production and then use of the hydrogen in CO_2 fermentation. Process-wise, the alternative route has two separate processes whereas in microbial electrosynthesis (MES), a combination of processes is incorporated into a single step *i.e.* direct reduction of CO_2 with electrical energy input. Normally, the losses involved with two separate processes are higher than with a single-step process. With either direct electron transfer or hydrogen evolution at the cathode, both routes can help to minimize the loss of reducing equivalents in MES. Direct electron uptake by microbes conserves energy, whereas with the hydrogen evolution mediated case, hydrogen produced on-site due to water-splitting at the cathode will become instantly available to the microbial biofilm, thus minimizing mass transfer losses. Hydrogen production by water electrolysis requires expensive catalysts and is energy demanding as well. In the second step, the reduction of CO_2 with H_2 also has limitations due to the requirement of a specific catalyst. Overall, the processes involved in the alternative route mentioned are well explored but still, economically, it needs optimization. In contrast, the MES process has still to be optimized and requires more investigation to be established as a commercially viable process. In this regard, economic feasibility analysis has still to be carried out once the system and process are fully optimized.

In terms of the power input in MES, an energy input of as low as 10–18 kWh per kg of acetate was calculated during one of our MES studies.[1] Energy efficiencies of 50–90% are reported in the literature for MES involving CO_2 reduction.[2,3] Higher production rates and titers reported in the literature for CO_2 reduction using MES support the prospect that MES can still present a competitive scenario for economic feasibility.

1 S. Bajracharya, R. Yuliasni, K. Vanbroekhoven, C. J. N. Buisman, D. P. B. T. B. Strik and D. Pant, Long-term operation of microbial electrosynthesis cell reducing CO_2 to multi-carbon chemicals with a mixed culture avoiding methanogenesis, *Bioelectrochemistry*, 2017, **113**, 26–34.
2 L. Jourdin, S. Freguia, V. Flexer and J. Keller, Bringing high-rate, CO_2-based microbial electrosynthesis closer to practical implementation through improved electrode design and operating conditions, *Environ. Sci. Technol.*, 2016, **50**, 1982–1989.
3 C. W. Marshall, D. E. Ross, E. B. Fichot, R. S. Norman and H. D. May, Long-term operation of microbial electrosynthesis systems improves acetate production by autotrophic microbiomes, *Environ. Sci. Technol.*, 2013, **47**, 6023–6029.

Leigh Aldous asked: Coming back to the topic of thermodynamics, you applied a potential difference of 3.1 V – do you know how much of an overpotential this corresponds to? Also, assuming that you have applied an overpotential, do you know where the major resistive losses occur in your prototype cell? You use a relatively dilute electrolyte, so could expect ohmic losses. You use a membrane, so there could be a potential drop across it. Or are the losses almost entirely due to an activation overpotential for electron transfer?

Deepak Pant responded: Under the standard biological conditions, the thermodynamic potential for CO_2 reduction at the cathode is -0.28 V *vs.* SHE and for the water oxidation at the anode it is 0.8 V *vs.* SHE. This corresponds to requirement of a minimum cell voltage of 1.1 V. At the applied potential difference of 3.1 V, the total cell overvoltage was 2 V, including ohmic loses and the activation overpotential. The overvoltage comprises the total electrochemical losses, including the activation overpotential. The measured cathode potential was 0.7 V *vs.* SHE, which corresponds to $0.7 - 0.28 = 0.42$ V of cathode overpotential. Since platinum coated titanium was used as the anode, we can assume a lower anode overpotential. Thus, the ohmic losses and membrane account for the major resistive losses in the reactor. During the experiment, the cathode potential at higher cell voltages was also measured at various stages. The changes in the cathode overpotential were relatively low when the cell voltage was changed (0.26 V for each volt change in the cell voltage). This means that the losses, other than the activation overpotential, were higher. Due to the increase in electrochemical losses as the operation proceeded (*e.g.* due to bacterial growth or membrane fouling), the cell voltage had to be increased for each new stage of the operation.

Marta Coma said: Acetate needs to be accumulated first to a certain extent before it can be converted to butyrate and/or caproate during chain elongation fermentation. Have you done an experiment where this occurs during microbial electrosynthesis? When the butyrate is increasing, is it then converting as well?

Deepak Pant answered: This is certainly the case. From our experiments and some recent results reported from other labs active in this field,[1–3] it can be said that this sort of chain elongation happens when the acetate accumulation is present, the pH is low and there is a longer retention time. In some cases, we also observed caproic acid (a C6 acid) in our MES system.

1 S. Bajracharya, R. Yuliasni, K. Vanbroekhoven, C. J. N. Buisman, D. P. B. T. B. Strik and D. Pant, Long-term operation of microbial electrosynthesis cell reducing CO2 to multi-carbon chemicals with a mixed culture avoiding methanogenesis, *Bioelectrochemistry*, 2017, **113**, 26–34.
2 R. Ganigué, S. Puig, P. Batlle-Vilanova, M. D. Balaguer, J. Colprim, Microbial electrosynthesis of butyrate from carbon dioxide, *Chem. Commun.*, 2015, **51**, 3235–3238.
3 S. Bajracharya, K. Vanbroekhoven, C. J. N. Buisman, D. Pant, D. P. B. T. B. Strik, Application of gas diffusion biocathode in microbial electrosynthesis from carbon dioxide, *Environ. Sci. Pollut. Res.*, 2016, **23**(22), 22292–22308.

David Constable enquired: Do you have any issues with the formation of biofilms? Are these closed systems and how long are they online?

Deepak Pant answered: Of course, biofilms are continuously formed in a bioelectrochemical system but this is mainly an issue in microbial fuel cells (MFCs), the devices used to treat wastewater and produce electrical energy. Even there, mechanisms have been devised to control the biofilm growth such as a shearing mechanism in which the old biofilm is continuously sloughed off and a desired thickness is maintained. In the case of microbial electrosynthesis systems (MES) for CO_2 conversion, it has been noticed that the electro-reduction takes place both in the biofilm and in the planktonic bacteria which are present in the electrolyte. The systems that we use with gas diffusion electrodes are fed-batch types of systems in which the actual cell is much smaller (an electrode surface area of 10 cm^2 and an electrolyte volume of 30 mL), and which are connected to an external reservoir with a larger volume of 500 mL or 1 L. The electrolyte is continuously recirculated using a peristaltic pump at a slow rate and thus part of the biofilm/biomass is removed through this external reservoir, which is also used as a sampling port. In some cases, we have operated these cells for well over a year.

Roberto Rinaldi commented: The micro-organisms are actually growing as well, so the system is also producing biomass. How much does this process cost in terms of nutrients and everything needed to keep the system alive?

Deepak Pant replied: This is true and indeed part of the cost goes towards the microbial metabolism and biomass growth, both in terms of the electrical current supplied as well as nutrients. According to one estimate, 10% of the energy supplied goes into biomass formation. In other words, 10% of the coulombic efficiency (CE) goes towards maintaining growth and biomass formation in MES. In our system, we use phosphate buffer solution (PBS) as the electrolyte, which is supplemented with a carbon source. During the start-up phase only, fructose is used as the carbon source. During the later stages, bicarbonate or carbon dioxide is the sole carbon source. Thus, in terms of the cost of the nutrients, the cost is not that high when compared to the cost of the final product (acids or alcohols). Besides, we do not use any yeast extract in the electrolyte.

David Constable asked: How are you separating the acetate? What is your experience with the resin? Do you see any kind of degradation in its performance over time? Is the process being run continuously? What about recovery and recycling?

Deepak Pant communicated in response: Normally we let the acetate accumulate up to a certain concentration (11–13 g L^{-1}). Afterwards, we introduce a commercially available anion-exchange resin (AmberliteTM FPA53). Acetate sorptions of 10–20 mg per gram of resin were observed from the catholyte broth. The acetate was removed batch-wise from the electrolyte broth. More details on this approach can be found in our recent paper.[1] Briefly, production and separation of acetate from CO_2 was technically feasible, through integration of the MES with an anion exchange resin. However, this may not be the most economic option and other resins and extraction agents needs to be explored. Also, the concentration of acetate needs to be increased significantly to obtain better yields after separation and extraction.

Over time, the performance of the resin dropped slightly during recycling. The MES process was operated in fed-batch mode but the acetate extraction was done batch-wise. The electrolyte after extraction of the acetate was recycled back into the MES cell.

1 S. Bajracharya, B. van den Burg, K. Vanbroekhoven, H. De Wever, C. J. N. Buisman, D. Pant and D. P. B. T. B. Strik, *Electrochim. Acta*, 2017, **237**, 267–275.

Carlos I. Cabrera-Rodríguez said: I must comment since we are talking about anion exchange resins. I read your previous paper and noticed that you used an inorganic salt to regenerate your anion exchange resin. This is not efficient as you are re-diluting your product in the salt solution and a high amount of salt is consumed. I was wondering, what is your strategy to deal with this, and also with the fact that you will be removing phosphate from your buffer with the anion exchange resin?

Deepak Pant answered: NaOH was used to regenerate the anion exchange resin. Re-dilution had to be performed while desorbing the product but the degree of dilution can be controlled using strong bases, which enable a higher degree of desorption of acetate and also limit the extent of dilution of the product. We do agree that the regeneration of the ion-exchange resin consumed a certain amount of NaOH. For any type of separation, either material or energy has to be invested. A comparative cost benefit analysis, including all possible separation processes, is needed in order to choose the most beneficial one. At the current stage of development, it is not straightforward to claim which separation technique is the most suitable.

Indeed phosphate can be removed during ion-exchange processes, but at the stage of product (acetate) separation, the concentration of the acetate (at least 100 mM) was higher than that of the phosphate (10 mM). Thus, exchange of acetate will be the dominant process, rather than exchange of the phosphate. However, the removal of phosphate cannot be completely avoided in anion exchange processes. Thus, after separation of the acetate, the buffer used for the MES may need to be replenished. In our study, only a part of the previous buffer was reused as the catholyte.

Xindong Mu opened a general discussion based on the papers by Keith W. Waldron, Roberto Rinaldi, Gary J. Lye and Deepak Pant by saying: I have a technology transfer job at CAS-QIBEBT. Chemical companies always ask if we can provide an economic feasibility analysis, including comparison with conventional chemicals from oil, natural gas or coal. Over the last two days, I have been very lucky to see a lot of great work on the production of fine chemicals, food additives, *etc.* Are there any potential candidates for competitive mass chemical production from biomass that could compete with petroleum chemicals directly in the market, now or in the near future?

Deepak Pant responded: In terms of any chemicals from biomass that have a commercial potential, there are actually several. There is a recent report from the NREL on "Chemicals from Biomass: A Market Assessment of Bioproducts with Near-Term Potential",[1] which outlines several such products along with near-term commercialization details as well as long-term forecasts. Regarding my work on CO_2 conversion to chemicals, we did a techno-economic evaluation recently and found that formic acid has the greatest potential to be valorized commercially and that significant advantages are present over the current production routes both in terms of energy input as well as the CO_2 footprint.[2]

1 M. J. Biddy, C. Scarlata, C. Kinchin, *Chemicals from Biomass: A Market Assessment of Bioproducts with Near-Term Potential*, National Renewable Energy Laboratory: Golden, CO, USA, 2016, http://www.nrel.gov/docs/fy16osti/65509.pdf
2 A. ElMekawy, H. M. Hegab, G. Mohanakrishna, A. F. Elbaz, M. Bulut and D. Pant, Technological advances in CO_2 conversion electro-biorefinery: a step toward commercialization, *Bioresour. Technol.*, 2016, **215**, 357–370.

Gary Lye responded: This has not really been the focus of our investigation, but what we have shown is that we can convert the L-arabinose and D-galacturonic acid fractions of the sugar beet pulp into higher value (but low volume) fine chemicals of interest to pharmaceutical companies.

Roberto Rinaldi answered: There are several success stories for competitive mass chemical production from biomass. For instance, the process chain from bioethanol to polyethylene (Braskem, 200 000 tons/year)[1] and polylactic acid (which has the second largest production volume for a bioplastic), to mention just two examples.

1 http://www.braskem.com/site.aspx/Im-greenTM-Polyethylene.

George Huber commented: Someone said that petrochemical feedstocks are simple. This is not true. Petrochemical feedstocks are actually very complex and can contain thousands of compounds. It has taken decades to learn how to refine them more efficiently. We can now process more complex feedstocks (like tar sands or heavy oils) and make a variety of products out of streams that were previously only burned for heat. Kerosene was the first product produced from oil. Over time, we have learned how to extract more value out of every part of oil. The same will be true with biomass. There are certain streams that we can process and products we can make. Over time, we will obtain more and more value out of each stream and process each stream more efficiently.

Trevor Hughes responded: I have worked in the oilfield service industry for 32 years. From this perspective, I would like to make a comment about the complexity and characterisation of crude oils, which range from light oils to heavy oils and tar sands. The paper entitled "Sustainable sources need reliable standards" (DOI: 10.1039/C7FD00046D) initiated an important discussion regarding the need to define and agree on a range of standard biomass materials and standardised analytical protocols with which to determine some key compositional properties. I agree that this is a very important step that needs to be implemented in order to compare various different biomass material processing methodologies. In the oil industry, there are a number of standard methods that are used to characterise any given crude oil stock and, in most publications, the results of these analyses are provided prior to subsequent descriptions of the recovery and/or processing methods. In the case of crude oil, some key parameters that define the oil type are: density, viscosity and so-called "SARA" analysis results (SARA analysis provides a simple four component compositional description of the oil in terms of % saturates (S), % aromatics (A), % resins (R) and % asphaltenes (A)). For biomass, I guess that an equivalent rather "broad brush" description of the composition might be the amount of cellulose, hemi-cellulose, lignin, oils and proteins. Of course, in reality, any given crude oil contains thousands of different chemical structures and deeper analytical methods (*e.g.* NMR and mass spectrometry) have been applied to obtain a full compositional description. Good references for these methods are the book "The Chemistry and Technology of Petroleum" by James Speight[1] and the ACS journal *Energy and Fuels*.

1 J. G. Speight, *The Chemistry and Technology of Petroleum*, Marcel Dekker Inc., New York, 3rd edn, Revised and Expanded, 1998.

Carlos I. Cabrera-Rodríguez said: I agree with George Huber's comment that petrochemical oil is also complex as an organic feedstock. I think that the main differences are in the supply (petrochemicals are present in high quantities in one location) and water content of the feedstock. The processing of biomass involves having the products at low concentrations, in the range of 2–10%. What really intrigues me, is why do we try to run distillation columns with such diluted systems? We should treat bio-based chemical purification differently to its petrochemical counterpart. Would you like to comment on this?

Keith Waldron responded: I am not an expert in petrochemical oil. However, I understand that it is a mixture of many different hydrocarbons with varying lengths and complexities. Separation of the components involves fractional distillation. I presume this is well characterised. Furthermore, I understand that petrochemical oil is derived over long time periods from the chemical breakdown of what was living matter. Therefore, it seems to me that it is unlikely to be as complex as the original biological material.

Regarding the purification of bio-based chemicals, this area requires much thought and consideration. My own research group has worked hard at trying to increase the concentration of ethanol derived from cellulose in order to reduce the potential cost of recovery through distillation. So far, we have achieved a concentration of nearly 12%.[1]

1 A. Elliston, S. R. A. Collins, D. R. Wilson, I. N. Roberts and K. W. Waldron, *Bioresour. Technol.*, 2013, **134**, 117–126.

Deepak Pant answered: We do agree that an optimized method for bio-based chemical purification has to be identified/developed. Running a distillation column with a diluted system is definitely not an optimum option. At this current stage, work on concentrating the product should be done. An innovative method should be studied for this aspect. Overall, the conventional petrochemical based separation processes are definitely not optimal for bio-based chemicals.

Gary Lye replied: I agree. All I would add is that for the separation of liquid biofuels from water, *e.g.* for bioethanol, there is considerable scope to look at alternative, less energy intensive recovery methods. There has previously been a lot of interest in techniques such as pervaproration, but I am not aware of the current status of this technology.

Roberto Rinaldi replied: Non-thermal separation processes for biomass-derived products is a topic of high relevance for the development of a bio-refinery. In my group, we have combined catalysis and non-thermal separation methods into a process yielding interesting results for the production of (alkyl) phenols and hydrocarbons from pyrolysis oil.[1]

1 Z. Cao, J. Engelhardt, M. Dierks, M. T. Clough, G.-H. Wang, E. Heracleous, A. Lappas, R. Rinaldi and F. Schüth, *Angew. Chem. Int. Ed.*, 2017, **56**, 2334–2339.

Changwei Hu opened the discussion of the paper by Vânia G. Zuin by asking: Have you investigated other starting materials and compared the results? Have you determined the surface groups of the Starbon®? What is the relationship between the adsorption properties and the surface groups?

Vânia Zuin replied: I have studied only alginic acid, pectin and starch as starting materials for this process. We have determined the surface area of all of the starting materials and the materials produced, as well as their chemical groups. The relationship between adsorption and the surface groups cannot be understood without considering the properties of the analytes.

Anna Zhenova asked: Has any work been performed on attaching groups that would be good for water purification, *e.g.* beta-cyclodextrin, to Starbon®?

Vânia Zuin responded: To the best of my knowledge, this has not yet been done. It is a good idea. I can discuss this with you at a later date.

Andrew Hunt queried: You have used the Starbon® materials for solid phase extraction (SPE) cartridges – what challenges might be faced if this is taken further and we try to use these materials as stationary phases for chromatography?

Vânia Zuin communicated in reply: That is a very good question. One main aspect to be considered is the robustness.

David Constable asked: Not all starches are the same, so how do you achieve consistent results with each new batch of Starbon® material? Do you obtain the same separation characteristics from batch to batch?

Vânia Zuin responded: Actually all of the Starbon® materials were prepared from the same batch of starting material in each case (starch, alginic acid and pectin) and carbonised at different temperatures. This means that it was possible to guarantee some similarities/parameters for the three groups of materials studied.

Vitaliy Budarin communicated: A biorefinery is a facility for conducting the process of biomass fractionation into fuels, chemicals and materials. During this conference, significant attention has been paid to bio-derived fuels and chemicals. However, the cost of bio-derived materials is usually higher than the cost of the produced fuels and chemicals. Therefore, incorporation of bio-derived materials into a biorefinery process could significantly improve its profitability. Production of Starbon® materials – high value mesoporous materials – is one of the possibilities for increasing the profitability of different biorefineries based on precursors containing polysaccharides. These materials can be produced from starch and alginic acid, and also from hemicellulose, which opens up an opportunity for manufacturing them from lignocellulosic materials

Vânia Zuin communicated in reply: I agree with this statement (based on an integrated biorefinery model).

Mark Mascal commented: Alongside Starbon®, Algibon and Pecbon materials, there are also "Ligbons".[1] These are porous solids generated from the residual lignin in CMF processes. This work was performed in collaboration with the Green Chemistry Centre of Excellence at York, so it is appropriate to mention in this context.

1 V. L. Budarin, J. H. Clark, J. Henschen, T. J. Farmer, D. J. Macquarrie, M. Mascal, G. K. Nagaraja and T. H. M. Petchey, Processed Lignin as a Byproduct of the Generation of 5-(Chloromethyl)furfural from Biomass: A Promising New Mesoporous Material, *ChemSusChem*, 2015, **8**, 4172–4179.

Vânia Zuin responded: Thank you for your comment. I am happy to know that there is a new material of this type based on lignin and would like more information.

Vitaliy Budarin communicated in response to Mark Mascal's comment: Yes, this is correct. The mesoporous lignin mentioned can be produced during a CMF process. However, this process is different from the standard starbonisation procedure, which is based on polysaccharide gelation.

Avtar Matharu opened the discussion of the paper by D. J. M. Hayes by saying: Intelligent online monitoring of the feedstock that then allows control of the process to provide a consistent output is really the way forward. I hope that this is the direction we are all moving in.

How does your system work with heterogeneous feedstocks? For example, how would it cope with a bin bag full of different waste materials (a heterogeneous mixture).

Daniel Hayes communicated in reply: We have a series of different models ranging from models for samples that have been ground down to a homogeneous particle size and are dry, to models for samples that have a heterogeneous particle size and are wet. The models tend to be most accurate for the former, but good predictions are still possible for heterogeneous samples (see ref. 1). When dealing with a mix of different samples, the method can work providing that the mix of samples in the analytical cell is representative. This will in fact be easier once the system is transferred to online monitoring, as the equipment can collect spectra from moving samples and could calculate a representative average.

1 D. J. M. Hayes, Mass and compositional changes, relevant to biorefining, in *Miscanthus x giganteus* plants over the harvest window, *Bioresour. Technol.*, 2013, **142**, 591–602.

Vitaliy Budarin said: We are interested in developing an *in situ* method for sugar yield monitoring during microwave hydrolysis experiments. Is your approach suitable for a liquid system?

Daniel Hayes replied: We have recently purchased a NIR unit for liquids and plan to develop models for work similar to that which you describe (as well as for the analysis of oligomers and sugar degradation products).

Florence Gschwend asked: How long will it be until your machine is available and how much will it cost?

Daniel Hayes replied: We do not sell equipment, we develop the predictive models. The plan is, ultimately, that these models will be licensed to be used with various NIR equipment. However, we do not have a timeline for that at this moment.

Leigh Aldous queried: How do you quantify the ash content? Is it basically assumed that the unobserved content is ash? I believe you mentioned that you could quantify two different types of ash, soluble and insoluble ash – is this an additional service, on top of NIR analysis?

Daniel Hayes communicated in reply: The slide presented referred to acid soluble lignin and acid insoluble lignin (*i.e.* Klason lignin). We have NIR models for ash and we have developed preliminary models for acid-/alcohol-/water-soluble ash. The means by which NIR spectra can be used to predict the ash content of samples are unclear as ash does not absorb NIR radiation, however, it does have indirect effects on other analytes that do absorb NIR radiation. We are able to predict the ash content in samples with over 50% ash (as shown in our paper).

David Constable asked: Where do you see this technology taking hold? Who would be your biggest customers? Does your approach always require you to make heterogenous material homogenous?

Daniel Hayes replied: We hope it will take off when more commercial scale biorefineries become operational. Our biggest customers would be the operators of those biorefineries. If it is used inline within a biorefinery, the material need not be so homogeneous, as instead spectra could be collected over a period of time to make an average that is more representative.

Florence Gschwend communicated: One problem we have when using a NREL protocol (see ref. 1), is that we cannot distinguish between acid insoluble lignin and humins/pseudo-lignin after the pretreatment. Is this something you can do?

1 A. Sluiter, B. Hames, R. Ruiz, C. Scarlata, J. Sluiter, D. Templeton and D. Crocker, *Determination of Structural Carbohydrates and Lignin in Biomass, Laboratory Analytical Procedure*, version 08-03-2012, National Renewable Energy Laboratory, 2012.

Daniel Hayes responded: This is not something we can currently do, but we will investigate this. Thank you for the suggestion. We are working on lignin analysis *via* thioacidolysis and this technique may help differentiate between lignin and humins.

Mark Mascal commented: You have described a decrease in the sugar content and an increase in the ash content when the same sample is ground to smaller particle sizes. Is this an artifact, or are you losing sugar content when you reduce the sample to finer particles? If the ash content is not actually increasing within the sample, this means the same material is providing different analysis results when ground to a different extent.

Daniel Hayes responded: It is an often-observed phenomenon, when a sample is ground and then that ground sample is sieved into several fractions according to particle size, that the smaller particle size fractions have a higher ash content. It is not related to a reduction of the sugar content, rather it is simply related to the tendency of different analytes to be concentrated in the more fibrous *versus* the finer particles after milling. When we take a large particle size fraction and analyse its ash and sugar content directly, and then mill it **all** to a very fine particle size, the composition does not change after this milling. It is only when two fractions are collected that there is a bias in the composition.

Alexei Lapkin queried: As the metabolic profile of a plant changes with growth and age, do you account for this variation of the profile in your calibration? Also, there could be potential to use this method to track plant development during the specific circumstances of farming for medicinal plants, when a specific metabolite, or a combination of metabolites, is sought and it is desirable to harvest at a time when the greatest amount of the required target can be obtained.

Daniel Hayes communicated in reply: To obtain good predictions for a feedstock, it is important that the calibration set contains a number of samples of it that represent various different conditions. For example, we sampled many *Miscanthus* plants of different ages to facilitate this (as detailed in ref. 1). Your idea to track medicinal plant development is possible. This would require us to develop new predictive models (*via* reference analysis) for these metabolites.

1 D. J. M. Hayes, Mass and compositional changes, relevant to biorefining, in *Miscanthus x giganteus* plants over the harvest window, *Bioresour. Technol.*, 2013, **142**, 591–602.

Mark Mascal commented: If you are ball-milling, inorganics could be added during the milling process as they could be shearing off of the grinding media. Could this be the source of your increase in the inorganics at smaller particle sizes?

Daniel Hayes replied: It is possible, but then there would be a consistent bias. We have found that the bias is much greater for some samples (*e.g.* paper samples) than others (*e.g.* wood samples). Also, as mentioned, we have taken a large particle size fraction and analysed its ash and sugar content directly, and then milled it **all** to a very fine particle size, and we found that the composition does not change after this milling. It is only when two separate particle size fractions are collected after one round of milling that there is a bias in the composition between those different fractions.

David Constable queried: Do you see any kind of interference from, for example, fillers that are used? What about glues and glossy papers?

Daniel Hayes communicated in reply: These components tend to increase the ash content of the sample, however, their compositions can still be predicted with good accuracy using our NIR models.

Andrzej Stankiewicz opened the discussion of the paper by A. A. Lapkin by saying: I like your approach that integrates a kind of "chemistry synthesis" with process synthesis in order to arrive at an optimal reaction route and optimal process configuration. What was the approximate computer load and time needed to solve the problem presented in your paper?

Alexei Lapkin responded: For the first step – assembly of the process routes – the longest amount of time was taken up by data download. I envisage that this download process will ultimately be done at the database end, and this time will thus be reduced to almost nothing. The actual assembly of the routes and their trimming is very fast – it takes seconds. Other manipulations of the assembled routes do take time: evaluation of the mass metrics is very quick; evaluation of route's energy efficiency is not automated and does take time; assembly of the models is also not yet fully automated; and optimisation of the multi-step reactions does take a considerable amount of time – in the order of hours.

Changwei Hu asked: Do you have one example of a route obtained using the designed computational model? Does your program have the capacity to suggest new syntheses and guide our research?

Alexei Lapkin answered: If your question is whether we already have examples of completely new routes obtained through our approach, then the answer is not as far as having taken the examples into practice. We are currently evaluating the quality of the suggested routes and whether we can take them forward for process development. To answer your second question, we think it could be possible to predict new transformations and guide research into novel reactions – this is 'work in progress'.

Simo Sarkanen commented: One of the significant problems with the synthetic organic chemistry literature is the reporting of overly optimistic product yields. Are you able to decide when adjustments need to be made to egregious claims?

Alexei Lapkin replied: We have not looked at this at all. False reporting and over-exaggerations are general problems within current science. What would help in the future is to codify experimental protocols and use standardised equipment/methods, as then reproducing experiments would become easier and making false claims would become difficult. I have an optimistic view that this will happen relatively quickly.

Avtar Matharu enquired: You use a set of filters – are the filters that you are applying based around greeness/sustainability or just yields? Where does the intellectual property lie in all of this? All of the reactions can be found using a computer, where is the inventiveness?

Alexei Lapkin responded: In this specific study, we used one filter when trimming down the number of routes to remove those that are obviously erroneous, such as routes in which the starting molecule is completely destroyed and the final molecule is synthesised mainly in the last step. In other studies, we have also applied additional 'filters', or criteria for what would make a good route – involving the energy efficiency, mass efficiency and so on. Intellectual property of course lies with the researcher who has developed new processes and discovered new reactions.

Kirsty Wilson remarked: I am curious about the similarities and differences between the data mining techniques and routes generated through your software and those of Chematica.

Alexei Lapkin replied: I am not entirely sure what is contained within Chematica, as it is a 'black box' system. My understanding, from comments made by other experts in chemoinformatics, is that Chematica is based on retrosynthesis rules. Whether it uses data mining as well, like the Wiley system, I do not know. The Wiley system does combine data mining with retrosynthesis rules, which is a very powerful approach.

Anna Zhenova requested: Can you elaborate on where the data for the process simulation comes from?

Alexei Lapkin answered: In this specific case, we developed each individual model based on experimental kinetics and thermodynamics, as well as on literature data. Currently, we cannot conduct model generation any other way. We are working towards a methodology wherein such models would be generated much more quickly, in a semi-automated manner.

Roberto Rinaldi commented: I found it fascinating to hear about the algorithms that you and your group are developing in order to push the frontier of organic synthesis beyond chemists' creativity. I can envisage that such an elegant approach could very well hold promise for the downstream processing of complex mixtures derived from lignocellulose and, particularly, lignin. Considering the algorithm roadmap presented in Fig. 1 of your paper, if we start from the second row, in which seven molecules are listed, how difficult would it be to transform the present algorithm into another one that could be used to search a common path of chemical transformation(s) in order to funnel a complex mixture into a narrower window of products?

Alexei Lapkin replied: That is a very good suggestion. It is something we have not tried before. So far, we have looked at linking a specific feedstock with a specific product. We would need to re-define the search for a network. However, as long as one has the reaction data, then the construction of a network and interrogation of it can be done fairly easily by writing code that asks specific questions. We will look at trying your suggestion!

Simo Sarkanen asked: Ultimately, will your work restrict the ingenuity of organic chemists, or will their efforts be channeled more productively?

Alexei Lapkin replied: I believe the latter. There is no reason to think that what we are doing will restrict organic chemists. During the work we carried out for this specific publication, we saw an increased enthusiasm from synthetic chemists toward our tools, as they provide them with more information and hints, and this in fact increased creativity. Our work would remove some of mundane tasks and help people to avoid missing existing information and knowledge.

Christian Stevens enquired: Does the system presented also cover enzymatic reactions? If yes, it could be used to describe and predict biochemical pathways! Since your model described the pathway from a terpene to aspirin (also a natural product) – how different is your predicted pathway from the natural one? Could your system beat million years of natural selection?

Alexei Lapkin answered: At present, we use only reactions that are available in Reaxys, and there are some enzymatic reactions in there. However, there are other tools that work with biosynthetic pathways. Ultimately, integration of biochemistry and synthetic chemistry routes will be feasible.

We will need to check the natural route to paracetamol before answering your second question. However, most likely it is quite different. We will definitely make the comparison.

Alexei Lapkin commented during a general discussion based on the papers by Vânia G. Zuin, D. J. M. Hayes and A. A. Lapkin: Regarding pathways towards the future bio-based economy and the shape of the future industry, as the industry is a complex hierarchical system, it cannot easily be designed from first principles. Thus, when we ponder which technology is best and how to structure the entire bio-based industry, these are good questions but they have no definitive answers. We need to explore this. Like in any complex system, its dynamic equilibrium state will be arrived at by evolution, not by design. So there will necessarily be failures and changes in which approach is the favourite, but this is inevitable and we should not be afraid of it.

Avtar Matharu remarked following the presentation by Andrzej Stankiewicz: It is my pleasure to conclude this session and the overall Faraday Discussion by firstly thanking Professor Stankiewicz for his brilliant lecture encapsulating three days of intense discussion and elegantly framing it within the concept of 'if Faraday was here today'. Your lecture will resonate with me and many of the attendees for many years to come.

May I now take this opportunity to thank the Royal Society of Chemistry, the organising committee, Burlington House and all the scribes for making this such a successful event. A follow-up event in 5–7 years would be very much of interest, to see how the subject matter has progressed.

I would now like to invite the session chairs Dr David Constable, Professor Chris Stevens and Professor Magda Titirici to share their concluding remarks.

Christian Stevens provided concluding remarks: The success of this Faraday Discussions meeting has been its capability to attract and bring together a range of experts with very different, however, complementary, skills. The development of biorefineries definitively requires the interaction, the respect and the collaboration of several technological fields in order to move this multidisciplinary field forward. This Faraday Discussions meeting succeeded in these aspects. It has involved open and critical discussion of all the aspects of biorefining, which are, and will be, essential for the future success of the practical use of renewable resources.

However, I am convinced that we do have an even more important task! It will be crucial to inform the broader public, as well as the consumers, about the efforts and the advantages connected to biorefining. We have to inform consumers about, and discuss with them, "food *versus* fuel" issues or "land use change" issues related to the production of materials for non-food use. Too many times, false arguments are being used to block sustainable solutions. If we do not succeed in that, it will result in a situation similar to the negative attitude of Europe towards GMOs. We all know now how detrimental that has been to European industry, just on the grounds of emotional arguments. Therefore, we have to be extremely careful that the intentions for sustainable and better production do not get dumped prematurely because of some initial growth pains of a developing system. Petrochemistry took 150 years to get to an optimized system! The renewable resources industry also deserves time to develop into a sustainable system. Growth pains do exist in many different fields, also in this one. If we are not able to explain this to the main consumers and defend our position with realistic, non-hyped, arguments, we are just wasting our time.

Be an ambassador for realistic renewable and sustainable production systems, we need you more than ever.

David Constable remarked: It has been a great pleasure and honor to be a part of the discussions over the past few days. It has been a very stimulating few days of discussion and debate. We are clearly within the early days of transitioning to sustainable use of renewable and bio-based chemicals and materials to generate the products that society wants and needs. The challenges are great, but the opportunities are also great, and my hope is that many of the students in attendance will be inspired to devote their careers to overcoming the many challenges discussed during this meeting. Overcoming these challenges will require a departure from traditional approaches to chemistry and chemical engineering. It will require the development of new catalytic approaches and reagents, the widespread use of synthetic biology, computational methods that lead to artificial intelligence, and a full exploitation of molecular diversity beyond the use of traditional framework molecules and critical elements. I look forward to the next Faraday Discussion on bio-resources and the progress that will have been made.

Magdalena Titirici concluded: This has been a very interesting meeting. It is impressive to see how interdisciplinary the meeting has been and how many different disciplines are involved in solving the challenging sustainability quest. Very important here, will be matching electricity production, which is needed to solve the problems of sustainability implementation in various industries. This will of course require renewable energy based on available materials.

Overall, this has been a very interesting meeting on addressing all of the mentioned sustainability challenges and providing technical solutions to solve some of these challenges.

I look forward to future developments and the next Faraday Discussions meeting on this important and societal topic.

Avtar Matharu said: Thank you all for attending. I now formally close this Faraday Discussion at 13:45 on Wednesday 21st June 2017.

Faraday Discussions

PAPER

Penrose triangles of the fossil-to-bio-based transition

Andrzej Stankiewicz

Received 16th July 2017, Accepted 25th July 2017
DOI: 10.1039/c7fd00178a

Transition within the chemical industry from fossil to green feedstocks is a complex process characterized by the generation of commercially viable feedstock–process–product triangles. The research in this area encompasses a great diversity of relevant topics. A number of those topics have been addressed within this volume of Faraday Discussions and are summarized in this paper. They are categorized and discussed along with seven general questions arising from the feedstock–process–product triangles. Opportunities are identified that should make more of these triangles technically and economically feasible. The future role of renewable electricity as the primary energy source for the bio-based industry is emphasized.

Introduction

The transition from a fossil-based to a bio-based chemical industry is a global process occurring in the context of sustainable development of our planet. Although the process itself is inevitable and irreversible, the vast majority of researchers working in the field agree that the pace at which it is proceeding is still far from satisfactory. The essence of the fossil-to-bio transition lies in fact in the successful generation of commercially viable feedstock–process–product triangles that offer clear benefits to the stakeholders concerned (farmers, industry, governments, end-users, consumers, *etc.*). Unfortunately, for the time being, many of these triangles still appear to be a kind of "Penrose Triangle"† (Fig. 1), described by Lionel and Roger Penrose[1] as "impossibility in its purest form". What challenges need to be met and what questions need to be answered in order to make more of these triangles viable? The Faraday Discussion "Bio-Resources: Feeding a Sustainable Chemical Industry" has allowed debate of the

Delft University of Technology, Leeghwaterstraat 39, 2628 CB Delft, The Netherlands. E-mail: a.i.stankiewicz@tudelft.nl

† Triangular object made of three straight bars that have a square cross-section, which meet at right angles at the vertices of the triangle they form. Also called an "impossible tribar", since such a combination of properties cannot be realized by any three-dimensional object in ordinary Euclidean space. First created by the Swedish artist Oscar Reutersvärd and described by the British psychiatrist Lionel Penrose and his son, mathematician Sir Roger Penrose.[1]

Fig. 1 Penrose triangle for a fossil-to-bio transition.

latest developments in science and technology that address those questions and challenges and create new opportunities.

From fossil-based to bio-based: important differences

Before more deeply discussing the challenges occurring in the transition from a fossil-based to bio-based industry, it is good to realize some important characteristic features and differences between the two types of industry. These features and differences are briefly described below. Some of them may present additional challenges when it comes to development of a viable bio-based process.

Feedstock diversity and variability

Generally speaking, the variability of bio-feedstocks in terms of chemical composition and physical properties is much greater than in fossil fuels. Although both natural gas and crude oil stocks occur in different grades and compositions, the feedstocks for the bioprocesses are more diverse ranging from crops, through to wood, algae, industrial wastewaters, municipal organic wastes, animal wastes, food wastes, *etc.*[2] Furthermore, within biologically the same feedstock differences in the composition and properties are often seen, depending on the climate and water/soil conditions in which the feedstock has been grown. The diversity of the bio-feedstocks is usually considered a drawback because of the increased flexibility of the processing plant that is required to deal with varying feedstocks. However, it can also be seen as an opportunity to bring a more diverse range of products to the market.

Diversity of processes and operations

Usually, a biorefinery includes more diverse types of processes and operations than an oil refinery. Next to "conventional" chemical and catalytic processes that occur in a liquid or gas phase, biorefineries often include mechanical/chemical pre-processing of solid feedstocks, aerobic and anaerobic fermentations, enzymatic reactions, plus a range of microbial processes.[3] This brings additional challenges related to process plant and site integration.

Transporting feedstocks

Contrary to the transport of gas and oil, which largely occurs *via* pipeline infrastructures, the vast majority of bio-feedstocks need to be transported on land or water from the place of harvest to the processing location. Two alternative transport scenarios (centralized plant *versus* decentralized plant) have been considered in the literature,[4] but for both cases transport remains an important cost.

Feedstock *versus* market – regionalization

Since the long-distance transport of biomass is troublesome and expensive, regionalization of bio-based production will occur. Contrary to the current situation, where many oil refineries are built and operated in regions and countries that do not have oil deposits themselves, with a bio-based economy, countries lean in biomass, such as the Gulf states, will have to import bio-based platform chemicals from elsewhere. Furthermore, since the product portfolio of a biorefinery depends strongly on the type of feedstock it converts,[4,5] and the feedstocks often have local character, the manufacturing of certain types of bio-based chemical products will occur locally.

Susceptibility to the force of nature

Lastly, but definitely not least, natural disasters such as floods, droughts, wildfires or agriculture and forest pests can affect the availability of the biomass in a given region and consequently the prices of the respective bio-based products, both regionally and globally. In the long term, climate change will also influence the dynamics of the bio-based economy.

Questions arising from feedstock–process–product triangles

Numerous questions arise within each feedstock–process–product triangle and these questions need to be adequately addressed. The papers presented in this volume of Faraday Discussions can be categorized according to seven general questions dealing with various elements of the triangle, located either at its vertices or between them.

Question 1 (feedstock): how to improve the analytical tools and develop standardized methods for the evaluation of feedstocks?

The need for improvement of the current analytical tools has been addressed by Galkin *et al.* (DOI: 10.1039/C7FD00046D), who have pointed out the shortcomings of the current analytical techniques used for biomass evaluation in the pulping industry. The authors postulated development of analytical tools for targeting all the wood components, including the generated fractions, as well as standardized methods for evaluating and reporting yields. In another paper by Hayes *et al.* (DOI: 10.1039/C7FD00081B), near-infrared spectroscopy was proposed as a suitable method for rapid, low-cost analysis of the major lignocellulosic components of waste paper/cardboard.

Question 2 (feedstock ↔ product): how to select a suitable feedstock for the desired products?

The importance of this issue was demonstrated in the paper by Wood *et al.* (DOI: 10.1039/C7FD00044H) where eight different feedstocks (hardwood, softwood, cereal straws and dicotyledonous crops) were exposed to microwave-assisted liquid hot water pre-treatment. The paper shows that fundamental differences in the cell wall composition resulted in considerable differences in feedstock suitability with regard to the quantity of released products.

Question 3 (feedstock ↔ process): how to pretreat and process feedstocks?

Selection of a pre-treatment technology is of fundamental importance to the entire process economics. Currently, steam-explosion is the most commonly used method for the pre-treatment of the lignocellulosic biomass. Seidel *et al.* (DOI: 10.1039/C7FD00066A) have pointed out that the explosive decompression at the end of this step could enhance the enzymatic cellulose digestibility of hardwood and herbaceous plants. Weigand *et al.* (DOI: 10.1039/C7FD00059F) proposed the use of protic low-cost ionic liquids in order to increase enzymatic glucose yields from willow biomass. In addition, Ferrini *et al.* (DOI: 10.1039/C7FD00069C) have shown the importance of solvent effects during functionalization of the propyl side-chain in lignin oil obtained from the deconstruction of lignocellulosic materials *via* catalytic upstream biorefining. Microwave-assisted acidolysis of the lignocellulosic biomass presents an effective approach to produce high purity lignin and fermentable chemicals from softwood, as demonstrated by Zhou *et al.* (DOI: 10.1039/C7FD00102A). Importantly, the lignin was isolated largely intact and retained the original structure of the native lignin in the feedstock. Interesting findings with regard to microalgae biomass were presented by Zhou *et al.*(DOI: 10.1039/C7FD00065K). These authors pyrolized raw biomass samples along with samples pre-treated through extraction of the lipids or saccharides. It appeared that fractional pyrolysis of the pretreated microalgae not only increased the bio-oil yield but also improved its quality.

Question 4 (feedstock ↔ process ↔ product): how to develop a new process?

This short question actually hides a multitude of situations, strategies and approaches. In some cases, plant flexibility is the central issue when feedstocks (and sometimes also products) vary. In other cases, the optimum process configuration needs to be found for a narrowly defined feedstock and product. Lapkin *et al.* (DOI: 10.1039/C7FD00073A) have proposed an interesting generic method for automation of route identification and optimisation based on data-mining and network analysis. The method was applied to generate multiple possible reaction routes for converting limonene into paracetamol. In the future, this approach should enable rapid concurrent optimization of the reaction network and the corresponding processes. The paper by Coma *et al.* (DOI: 10.1039/C7FD00070G) focusses on flexibility issues during development of a process for converting a highly variable feedstock (organic waste) into platform chemicals. The key to success is having flexible anaerobic fermentation and hydrothermal processes that can treat complex biomass as a whole to obtain a range of products within an integrated biorefinery concept. Cárdenas-

Fernández *et al.* (DOI: 10.1039/C7FD00094D) have presented a concept for developing an integrated biorefinery to convert sugar beet pulp into chemicals and pharmaceutical intermediates. The process is based on steam explosion (thermal hydrolysis) of wet sugar beet pulp, followed by bioethanol fermentation, enzyme-membrane fractionation of the sugar beet pectin, and bioconversion of D-galacturonic acid (transaminase) and L-arabinose (transketolase). Another integrated process, to synthesize 1,5-pentanediol (PDO) and 1,6-hexanediol (HDO) from lignocellulosic biomass, has been proposed by He *et al.* (DOI: 10.1039/C7FD00036G). The route goes *via* furfural and terahydrofuran-dimethanol, respectively, and techno-economic analysis demonstrated that this approach could produce HDO and PDO at a minimum selling price of $4090 per ton. In addition, Bajracharya *et al.* (DOI: 10.1039/C7FD00050B) have studied bio-electrochemical conversion of carbon dioxide, as a sustainable feedstock, into chemicals using microorganisms as the catalyst. This microbial electrosynthesis-based process produces acetate as the primary product.

Question 5 (process): how to improve the effectiveness of the reactions?

The most common answer to this question is through development of new, better catalysts. Liu *et al.* (DOI: 10.1039/C7FD00041C) have disclosed Pt nanoparticles supported on bamboo shoot-derived porous heteroatom doped carbon materials as highly active catalysts for controlled hydrogenation of furfural in aqueous media. They have shown that the product selectivity could be easily modulated by controlling the carbonization temperature of the porous heteroatom doped carbon support and the reaction conditions (temperature and H_2 pressure). Albert (DOI: 10.1039/C7FD00047B) has investigated optimization of polyoxometalate catalysts for a fractionated oxidation of lignocellulosic biomass to produce formic acid and high-grade cellulose. One of those catalysts, the Lindqvist-type POM $K_5V_3W_3O_{19}$, has been shown to catalyse selective oxidation of only the hemi-cellulose and lignin to formic acid, while the cellulose fraction remained untapped. Finally, Huang *et al.* (DOI: 10.1039/C7FD00039A) have investigated the role of acid co-catalysts during selective production of mono-aromatics from lignocellulose over a Pd/C catalyst and they found that HCl and H_2SO_4 showed superior catalytic performances over H_3PO_4 and CH_3COOH.

Question 6 (process): how to improve the product separation?

New and interesting methods for improved product separation have been disclosed within this current volume of Faraday Discussions. Xia and Matharu (DOI: 10.1039/C7FD00035A) have reported, for the first time in the literature, an acid-free subcritical water extraction of pectin from mango peel. Yields of up to 18.34%, with the degree of esterification exceeding 70%, were reported. Lorenz *et al.* (DOI: 10.1039/C7FD00053G) have suggested hydrolyzing cellulose with HCl vapour, in order to facilitate the isolation of cellulose nano-crystals.

Question 7 (product): how to increase the market share of bio-based products?

The enormous variety of bio-based products results in a multitude of answers to the above question and related strategies. According to Bomtempo *et al.* (DOI: 10.1039/C7FD00052A), new platform chemicals derived from biomass should fulfil several criteria, in that they should: "be an intermediate molecule, have a flexible structure to make a wide range of derivatives possible, be cost competitive at the level of the platform molecule and at the level of the derivatives, be capable of generating scale and scope economies in the value chain, be organized within an innovation ecosystem and have associated well-developed mechanisms of governance". Jin *et al.* (DOI: 10.1039/C7FD00049A) have proposed a 10-step procedure that should be applied when developing new bio-based solvents. The procedure includes various functional, technical and economic criteria that a bio-based solvent needs to fulfil, in order to become a marketable product. Bio-based solvents are an example of bio-based products that have been developed for use in chemical manufacturing. Other examples of such products include catalysts and sorbents. In this context, Golikova *et al.* (DOI: 10.1039/C7FD00042A) have studied a new biocatalyst based on glucose oxidase, while Zuin *et al.* (DOI: 10.1039/C7FD00056A) have presented polysaccharide-derived mesoporous materials (Starbon® materials) as sustainable sorbents for solid-phase extraction of naturally-occurring bioactive phenolic compounds. In the case of bio-based polymeric materials, which are closer to the consumer market, new functionalities or improved properties are usually the deciding factor with respect to their commercial success. An example is the polymeric materials composed only of methylated softwood lignin derivatives reported by Wang *et al.* (DOI: 10.1039/C7FD00083A) that can exhibit better tensile behaviour than polystyrene. As pointed out by these authors, mistaken assumptions from the past about the lignin configuration have hindered the development of these materials for more than 50 years. In another paper, Pérocheau Arnaud *et al.* (DOI: 10.1039/C7FD00057J) have presented novel polyesters with higher glass transition temperatures that are based on branched diols from biomass, while Alberts and Rothenberg (DOI: 10.1039/C7FD00054E) have described the development and commercialization of Plantics-GX: a plant-based biodegradable thermoset plastic.

Michael Faraday and the bio-world

The ground-breaking fundamental works of Michael Faraday on electricity and electromagnetism, originally published as a series of articles in the Philosophical Transactions of the Royal Society of London,[6,7] have largely contributed to the utilization of electricity in technology. The link between electricity and the bio-world is clear, at least hypothetically. The works of Miller and Urey[8,9] as well as some later studies (*e.g.* by the group of Bada[10,11]) point at electricity as the "Mother of All Bio" as an electric spark discharged circa 3.5 billion years ago converted the primordial soup into the first amino acids. In other words: "no electricity = no bio". Back to the 21st century, electricity-based technologies attract more and more interest from researchers working in the area of bio-based processing. Electric fields have been investigated in the context of electro-fermentation (*e.g.* Schievano *et al.*[12] and Chandrasekhar *et al.*[13]). Pulsed electric fields have various bio-related applications including within algae treatment, large-scale biomass (*e.g.* sugar beets) processing or the recovery of valuable products from plants and microorganisms (Golberg *et al.*,[14] Frey *et al.*[15]). Algae are harvested using electromagnets and submicron-sized magnetic particles (*e.g.* Xu *et al.*[16] and Cerff *et al.*[17]). Induction heating has been proposed for fast pyrolysis of different types of biomass, including rice straw, sugarcane bagasse, coconut shells, Napier grass, pinewood sawdust and sewage sludge (*e.g.* Tsai[18,19] and Muley[20]). The heating rates in those reactors are as high as 500 °C min^{-1}, which leads to higher quality bio-oils being obtained and lower carbon deposition. Biomass pyrolysis can also be advantageously performed using microwaves[21,22] (a conventional microwave oven is a Faraday cage!). Microwave heating is also applicable to other bio-based processes, for instance esterifications[23] or plant extraction,[24] while a microwave-induced plasma can be used for gasification of waste biomass to synthesis gas.[25] Finally, another form of electromagnetic radiation – light – is widely used in the cultivation of microalgae.[26,27]

On July 7, 1855, Michael Faraday took a short boat trip on the River Thames. That trip resulted in a famous letter, written on the same day, addressed to the Editor of The Times. In this letter, Faraday expressed his deep concerns about pollution of the river, calling it "a real sewer". Two weeks later Punch magazine illustrated this with a cartoon of the famous scholar giving his card to the "Dirty Fellow" – Father Thames. If Michael Faraday were with us today, he would be happy to see that electricity-based processing methods can deliver promising results in the context of wastewater management and sewage treatment. Some of those methods go beyond only treatment and convert sewage sludge into useful products. Examples include production of pyrolytic liquids from industrial sewage sludge using induction heating,[28] production of bio-fuels *via* microwave-assisted pyrolysis of sewage sludge[29] or plasma gasification of sewage sludge.[30,31]

Concluding remarks

The contributions present in this volume of Faraday Discussions not only address some of the questions and challenges of the fossil-to-bio-based transition but also demonstrate new opportunities which, if properly addressed and developed further, should lead to novel, commercially attractive concepts. Electricity-based

Fig. 2 Future bioprocessing plant: modular and green electricity-driven (plant model designed by LEGO® Ideas member Ymarilego). Image used by permission, ©2017 The LEGO Group; source for the portrait of Michael Faraday: "Michael Faraday" by John Hall Gladstone (3rd edn, Macmillan and Co., London, 1874).

technologies that build on Michael Faraday's scientific legacy are examples of such opportunities. In the long term, electricity is destined to play the key role in process industries, especially if obtained from a fully renewable source, as it is the most widely available and most flexible form of energy. This also holds true for the fossil-to-bio-based transition, where energy consumption is presently one of the most important challenges. Bioprocesses are often highly energy-demanding and the energy cost significantly affects the process economy. A bio-based industry that utilizes fossil fuel energy will never be really "green" and a gradual shift to cheap, renewable electricity as the primary energy source is needed. In the long term, such a shift, along with the introduction of modularity during plant design, should result in a dramatic change in the bio-process economy. Future bio-processing plants will be modular and green electricity-driven (Fig. 2). Such an evolution will further help in converting the "Penrose" triangles of today that are associated with the fossil-to-bio-based transition into viable feedstock–process–product solutions.

References

1 L. S. Penrose and R. Penrose, *Br. J. Psychol.*, 1958, **49**, 31–33.

2 J. S. Golden and R. B. Handfield, *Why Biobased? Opportunities in the Emerging Bioeconomy*, 2014, https://www.biopreferred.gov/files/WhyBiobased.pdf.

3 S. K. Maity, *Renewable Sustainable Energy Rev.*, 2015, **43**, 1427–1445.

4 T. T. H. Nguyen, Y. Kikuchi, M. Noda and M. Hiraoa, *Environ. Prog. Sustainable Energy*, 2016, **35**, 174–182.

5 V. Chambost and P. R. Stuart, *Ind. Biotechnol.*, 2007, **3**, 112–119.

6 M. Faraday, *Experimental Researches in Electricity*, Richard and John Edward Taylor, London, 1839, vol. I.

7 M. Faraday, *Experimental Researches in Electricity*, Richard and John Edward Taylor, London, 1844, vol. II.

8 S. L. Miller, *Science*, 1953, **117**, 528–529.

9 S. L. Miller and H. C. Urey, *Science*, 1959, **130**, 245–251.

10 A. P. Johnson, H. J. Cleaves, J. P. Dworkin, D. P. Glavin, A. Lazcano and J. L. Bada, *Science*, 2008, **322**, 404.

11 E. T. Parker, H. J. Cleaves, J. P. Dworkin, D. P. Glavinc, M. Callahan, A. Aubrey, A. Lazcano and J. L. Bada, *Proc. Natl. Acad. Sci. U. S. A.*, 2011, **108**, 5526–5531.

12 A. Schievano, T. Pepé Sciarria, K. Vanbroekhoven, H. De Wever, S. Puig, S. J. Andersen, K. Rabaey and D. Pant, *Trends Biotechnol.*, 2016, **34**, 866–878.

13 K. Chandrasekhar, K. Amulya and S. Venkata Mohan, *Waste Manag.*, 2015, **45**, 57–65.

14 A. Golberg, M. Sack, J. Teissie, G. Pataro, U. Pliquett, G. Saulis, T. Stefan, D. Miklavcic, E. Vorobiev and W. Frey, *Biotechnol. Biofuels*, 2016, **9**, 94.

15 W. Frey, C. Gusbeth, T. Sakugawa, M. Sack, G. Mueller, J. Sigler, E. Vorobiev, N. Lebovka, I. Àlvarez, J. Raso, L. C. Heller, M. A. Malik, C. Eing and J. Teissie, in *Bioelectrics*, ed. H. Akiyama and R. Heller, Springer Japan, Tokyo, 2017, ch. 6, pp. 389–486.

16 L. Xu, C. Guo, F. Wanga, S. Zheng and C.-Z. Liu, *Bioresour. Technol.*, 2011, **102**, 10047–10051.

17 M. Cerff, M. Morweiser, R. Dillschneider, A. Michel, K. Menzel and C. Posten, *Bioresour. Technol.*, 2012, **118**, 289–295.

18 W. T. Tsai, M. K. Lee and Y. M. Chang, *J. Anal. Appl. Pyrolysis*, 2006, **76**, 230–337.

19 M. K. Lee, W.-T. Tsai, Y.-L. Tsai and S. H. Lin, *J. Anal. Appl. Pyrolysis*, 2010, **88**, 110–116.

20 P. D. Muley, C. Henkel, K. K. Abdollahi and D. Boldor, *Energy Fuels*, 2015, **29**, 7375–7385.

21 Y.-F. Huang, P.-T. Chiueh and S.-L. Lo, *Sustainable Environ. Res.*, 2016, **26**, 103–109.

22 H. Marion Morgan Jr, Q. Bu, J. Liang, Y. Liu, H. Mao, A. Shi, H. Lei and R. Ruan, *Bioresour. Technol.*, 2017, **230**, 112–121.

23 N. Azcan and A. Danisman, *Fuel*, 2008, **87**, 1781–1788.

24 R. B. Mato Chain, J. Monzó-Cabrera and K. Solyom, in *Alternative Energy Sources for Green Chemistry*, ed. G. Stefanidis and A. Stankiewicz, RSC, Cambridge, 2016, ch. 2, pp. 34–63.

25 G. S. J. Sturm, A. Navarrete Muñoz, P. V. Aravind and G. D. Stefanidis, *IEEE Trans. Plasma Sci.*, 2016, **44**, 670–678.

26 M. Glemser, M. Heining, J. Schmidt, A. Becker, D. Garbe, R. Buchholz and T. Brück, *Appl. Microbiol. Biotechnol.*, 2016, **100**, 1077–1088.

27 C.-Y. Chen, K.-L. Yeh, R. Aisyah, D.-J. Lee and J.-S. Chang, *Bioresour. Technol.*, 2011, **102**, 71–81.

28 W.-T. Tsai, J.-H. Chang, K.-J. Hsien and Y.-M. Chang, *Bioresour. Technol.*, 2009, **100**, 406–412.

29 W. Zuo, Y. Tian and N. Ren, *Waste Manag.*, 2011, **31**, 1321–1326.

30 J. Balgaranova, *Waste Manage. Res.*, 2003, **21**, 38–41.

31 A. Mountouris, E. Voutsas and D. Tassios, *Energy Convers. Manage.*, 2008, **49**, 2264–2271.

Poster titles

Sustainable carbon catalysts for the oxygen reduction reaction, **K. Preuss, L. C. Tănase, C. M. Teodorescu and M.-M. Titirici**, *Queen Mary University of London, UK*

From biowaste to fluorescent carbon quantum dots, **N. Papaioannou, A. Sapelkin and M. Titirici**, *Materials Research Institute, Queen Mary University of London, UK*

Non-edible fats as source of threo-9,10-dihydroxystearic acid and related esters, **P. Gallart-Sirvent, G. Villorbina, M. Balcells, M. Torres and R. Canela-Garayoa**, *Lleida University, Spain*

Effect of rare sugars for seed germination, **T. Semachi, M. Nagao, S. Usuki, K. Yamatoya and K. Nakata**, *Tokyo University of Science, Japan*

Design of organic peroxide generation with metal-supported photocatalyst and application to sporicidal performance, **K. Iwasaki, Y. Yamaguchi, K. Yamatoya, C. Terashima, K. Katsumata, N. Suzuki, A. Fujishima, H. Sakai and K. Nakata**, *Tokyo University of Science, Japan*

Sterilization of *Anabaena* by TiO_2 photocatalysis and mechanism analysis, **M. Tano, T. Tomo, C. Terashima, K. Katsumata, N. Suzuki, A. Fujishima, K. Yamatoya and K. Nakata**, *Tokyo University of Science, Japan*

Esters production *via* the carboxylate platform: recovery and esterification of carboxylates using CO_2-expanded alcohols, **C. I. Cabrera-Rodríguez, L. A. M. van der Wielen and A. J. J. Straathof**, *Delft University of Technology, Netherlands*

Graphitic carbon materials with chiral nematic liquid crystalline phases derived from nanocrystalline cellulose, **F. Xie, H. Luo and M. Titirici**, *Queen Mary University of London, UK*

Efficient isolation and deconstruction of lignin in alternative reaction media, **D. S. Zijlstra, H. J. Heeres and P. J. Deuss**, *University of Groningen, Netherlands*

Lignin-derived carbon nanofibres for flexible energy storage, **S. Herou and M. Titirici**, *Queen Mary University of London, UK*

Future biorefineries through integrated chemical and biological technologies, **J. Fan, F. Santomauro, V. Budarin, J. Clark and C. Chuck**, *Green Chemistry Centre of Excellence, University of York, UK*

Catalysed transesterification of triglycerides and methanol by activated bovine bone, **S. M. Smith, S. Chaveanghong, C. Oopathum, V. Weeramongkhonlert,**

P. Ketwong, C. Smith, A. Luengnaruemitchai and S. Boonyuen, *Department of Chemistry, Faculty of Science, Mahidol University, Thailand*

The application and limitations of cyrene as a bio-derived alternative to DMF, **K. Wilson, A. Watson, C. Jamieson and J. Murray**, *University of Strathclyde, UK*

Valorisation of metal-treated waste wood using low-cost ionic liquids, **F. Gschwend, L. Hennequin, J. Hallett, P. Fennell, A. Brandt and G. H. Kelsall**, *Imperial College of London, UK*

Renewable production routes for Nylon-6 precursor: techno-economic analysis for caprolactam, **V. Thaore, I. Graça, D. Chadwick and N. Shah**, *Imperial College London, UK*

Investigation of biomass-based anode electrodes in Na-ion batteries, **H. Alptekin, P. Yilmaz and M. M. Titirici**, *Queen Mary University of London, UK*

Optimisation of enzymatic detoxification of pulp liquors, **L. Sparlinek, V. Leitner and K. Wöss**, *Wood K plus Kompetenzzentrum Holz, Austria*

NEWFERT: nutrient recovery from biobased waste for fertilizer production, **A. Martínez Gracia and A. Garrido-Escudero**, *Drage & Mate International, S.L.*

Intelligent approach to solvent substitution: identification of a new class of levoglucosenone derivatives, **A. Alves Costa Pacheco, J. Sherwood, A. Zhenova, C. R. McElroy, A. J. Hunt, H. L. Parker, T. J. Farmer, A. Constantinou, M. De bruyn, A. C. Whitwood, W. Raverty and J. H. Clark**, *University of York, UK*

Development of bio-based hydrogel carrier for controlled release water and fertilizer, **Y. C. Chen and Y.-H. Chen**, *National Chung Hsing University, Chinese Taipei*

On the effect of acid additives on the catalytic upstream biorefining of lignocelluloses, **C. Chesi, C. Rezende, T. Upcraft and R. Rinaldi**, *Imperial College London, UK*

Confocal microscopy as a new tool for the semi-quantitative analysis of lignocellulosic biomass, **Y. Luo, J. Fan, V. L. Budarin, J. Marrison, C. Hu and J. H. Clark**, *Sichuan University, China*

The Faraday Discussions Poster Prize for the best poster was jointly awarded to Mr Servann Herou of the *Queen Mary University of London, UK*, for his poster on "Lignin-derived carbon nanofibres for flexible energy storage" and Ms Anna Zhenova of the *University of York, UK*, for her poster on "Intelligent approach to solvent substitution: identification of a new class of levoglucosenone derivatives".

List of participants

Dr Jakob Albert, *FAU Erlangen-Nürnberg, Germany*
Dr Leigh Aldous, *King's College London,United Kingdom*
Miss Hande Alptekin, *Queen Mary University of London,United Kingdom*
Prof. Dr Lise Appels, *KU Leuven, Belgium*
Prof. Dimitris Argyropoulos, *North Carolina State University, U.S.A.*
Prof. Dr Harry Bitter, *Biobased Chemistry and Technology, Wageningen University, Netherlands*
Prof. José-Vitor Bomtempo, *UFRJ - Federal University of Rio de Janeiro, Brazil*
Dr Agnieszka Brandt-Talbot, *Imperial College London,United Kingdom*
Dr Vitaliy Budarin, *Green Chemistry Centre of Excellence, University of York, United Kingdom*
Mr Carlos I. Cabrera-Rodríguez, *Delft University of Technology, Netherlands*
Prof. Dr Ramon Canela-Garayoa, *Lleida University, Spain*
Prof. Yi Chun Chen, *National Chung Hsing University, Chinese Taipei*
Mr Claudio Chesi, *Imperial College London,United Kingdom*
Miss Ming Chung, *Johnson Matthey PLC, United Kingdom*
Prof. James Clark, *Green Chemistry Centre of Excellence, University of York, United Kingdom*
Dr Marta Coma, *University of Bath, United Kingdom*
Mr Chris Comerford, *Celignis, Ireland*
Dr David Constable, *American Chemical Society, U.S.A.*
Dr Terence Cooper, *ARGO Group International, U.S.A.*
Mr Hugh Cowley, *Royal Society of Chemistry, United Kingdom*
Prof. Claudia Crestini, *Depatment of Chemical Sciences and Technologies, University of Rome Tor Vergata, Italy*
Prof. Bruce Dale, *Michigan State University, U.S.A.*
Dr Nathan De Geyter, *Ghent University, Belgium*
Prof. Dr Raf Dewil, *KU Leuven, Belgium*
Dr Jiajun Fan, *Green Chemistry Centre of Excellence, University of York, United Kingdom*
Dr Thomas Farmer, *Green Chemistry Centre of Excellence, University of York, United Kingdom*
Mr Amalio Garrido, *Drage & Mate International, S.L., Spain*
Miss Kathryn Gempf, *Royal Society of Chemistry, United Kingdom*
Dr Ines Graca, *Imperial College London, United Kingdom*
Miss Florence Gschwend, *Imperial College London, United Kingdom*
Dr Jason Hallett, *Imperial College London, United Kingdom*
Dr Daniel Hayes, *Celignis, Ireland*
Mr Louis Hennequin, *Imperial College of London,United Kingdom*
Mr Servann Herou, *Queen Mary University of London, United Kingdom*
Ms Anne Horan, *Royal Society of Chemistry, United Kingdom*
Dr Suzanne Howson, *Royal Society of Chemistry, United Kingdom*
Prof. Changwei Hu, *Sichuan University, China*

Dr Xiaoming Huang, *Eindhoven University of Technology, Netherlands*
Prof. George Huber, *University of Wisconsin Madison, U.S.A.*
Mr Trevor Hughes, *Schlumberger Gould Research, United Kingdom*
Dr Andrew Hunt, *University of York, United Kingdom*
Mr Kodai Iwasaki, *Tokyo University of Science, Japan*
Prof. Eero Kontturi, *Aalto University, Finland*
Prof. Alexei Lapkin, *University of Cambridge, United Kingdom*
Mr Daniel Lukamto, *University of Cambridge, United Kingdom*
Dr Gang Luo, *Fudan University, China*
Ms Yiping Luo, *Sichuan University/University of York, China/United Kingdom*
Miss Hui Luo, *Queen Mary University of London, United Kingdom*
Dr Gary Lye, *University College London, United Kingdom*
Dr Duncan Macquarrie, *University of York, United Kingdom*
Prof. Dr Assunta Marrocchi, *Department of Chemistry, Biology and Biotechnology, University of Perugia, Italy*
Miss Alexandra Martínez Gracia, *Drage & Mate International, S.L., Spain*
Prof. Mark Mascal, *University of California Davis, U.S.A.*
Dr Avtar Matharu, *Green Chemistry Centre of Excellence, University of York, United Kingdom*
Prof. Valentina Matveeva, *Tver State Technical University, Russia*
Dr Karl McCarthy, *Biocel Ltd, Ireland*
Prof. Andrew Mount, *The University of Edinburgh, School of Chemistry, United Kingdom*
Prof. Dr Xindong Mu, *Qingdao Institute of Bioenergy and Bioprocess Technology, Chinese Academy of Sciences, China*
Dr Michaela Muehlberg, *Royal Society of Chemistry, United Kingdom*
Prof. Dr Kazuya Nakata, *Tokyo University of Science, Japan*
Mr Xianhong Ouyang, *Eindhoven University of Technology, Netherlands*
Dr Deepak Pant, *VITO, Belgium*
Mr Nikolaos Papaioannou, *Materials Research Institute, Queen Mary University of London, United Kingdom*
Dr Emmanouil Papaioannou, *Lancaster University, United Kingdom*
Miss Sushma Pariyar, *Literary Academy For Dalit of Nepal (LAD-Nepal), Nepal*
Miss Olivia Petrvalska, *Charles University, Czech Republic*
Dr Jon Pillow, *CDT Ltd, United Kingdom*
Ms Kathrin Preuss, *Queen Mary, University of London, United Kingdom*
Dr Roberto Rinaldi, *Imperial College London, United Kingdom*
Miss Gloria Rosetto, *Imperial College London, United Kingdom*
Prof. Dr Gadi Rothenberg, *University of Amsterdam, Netherlands*
Prof. Joseph Samec, *Stockholm University, Sweden*
Prof. Simo Sarkanen, *University of Minnesota, U.S.A.*
Ms Rosalind Searle, *Royal Society of Chemistry, United Kingdom*
Mr Christoph-Maximilian Seidel, *ETH Zurich Institute of Process Engineering, Switzerland*
Mr Takahiro Semachi, *Tokyo University of Science, Japan*
Dr Angela Serpe, *University of Cagliari, Italy*
Dr Brajendra Sharma, *University of Illinois, Urbana-Champaign, U.S.A.*
Dr Christopher Smith, *Mahidol University, Thailand*
Dr Siwaporn Meejoo Smith, *Department of Chemistry, Faculty of Science, Mahidol University, Thailand*
Dr Donna Smith, *Royal Society of Chemistry, United Kingdom*
Ms Laura Sparlinek, *Wood K plus Kompetenzzentrum Holz, Austria*
Ms Katie Springham, *Royal Society of Chemistry, United Kingdom*

Prof. Dr Andrzej Stankiewicz, *Delft University of Technology, Netherlands*
Prof. Dr Christian Stevens, *Ghent University, Belgium*
Dr Cara Sutton, *Royal Society of Chemistry, United Kingdom*
Mr Anthony Szego, *Cardiff University, United Kingdom*
Mr Masaki Tano, *Tokyo University of Science, Japan*
Dr Vaishali Thaore, *Imperial College London, United Kingdom*
Mr William Thiam, *Givaudan (Flavours and Fragrance), Switzerland*
Mr Olivier Thinon, *IFPEN, France*
Prof. Magdalena Titirici, *Queen Mary University of London, United Kingdom*
Dr Keith Waldron, *Quadram Institute Bioscience, United Kingdom*
Prof. Karen Wilson, *EBRI, Aston University, United Kingdom*
Miss Kirsty Wilson, *University of Strathclyde, United Kingdom*
Dr Robert Woodward, *Imperial College London, United Kingdom*
Mr Fei Xie, *Queen Mary University of London, United Kingdom*
Prof. Fang Zhang, *Department of Chemistry, Shanghai Normal University, China*
Ms Anna Zhenova, *University of York, United Kingdom*
Mr Long Zhou, *Green Chemistry Centre of Excellence, University of York, United Kingdom*
Mr Douwe Sjirk Zijlstra, *University of Groningen, Netherlands*
Prof. Dr Vânia Zuin, *DQ-UFSCar, Brazil*